Essentials

of Human Anatomy and Physiology

For my students,
who taught me how to teach

Contents

Expanded Contents

Unit 1
Levels of Organization

Unit 2

Support and
Movement

Unit 3

Integration and Coordination

Unit 5
Reproduction

Preface

Essentials of Human Anatomy and Physiology, Second edition, is designed to provide accurate information about the structure and function of the human body in an interesting and readable manner. It is especially planned for students in one-semester courses in anatomy and physiology who are pursuing careers in allied health fields and who have minimal backgrounds in physical and biological sciences.

Organization

The text is organized in units or groups of related chapters. These chapters are arranged traditionally beginning with a discussion of the physical basis of life and proceeding through levels of increasing complexity.

Unit 1

is concerned with the structure and function of cells and tissues; it introduces membranes as organs and the integumentary system as an organ system.

Unit 2

deals with the skeletal and muscular systems that support and protect body parts and make movements possible.

Unit 3

concerns the nervous and endocrine systems that integrate and coordinate body functions.

Unit 4

discusses the digestive, respiratory, circulatory, lymphatic, and urinary systems. These systems obtain nutrients and oxygen from the external environment; transport these substances internally; utilize them as energy sources, structural materials, and essential components in metabolic reactions; and excrete the resulting wastes.

Unit 5

describes the male and female reproductive systems, their functions in producing offspring, and the early growth and development of this offspring.

Biological Themes

In addition to being organized according to levels of increasing complexity, the narrative emphasizes the following: the complementary nature of structure and function, homeostasis and homeostatic regulating mechanisms, the interaction between humans and their environments, and metabolic processes.

Changes in the Second Edition

The second edition of *Essentials of Human Anatomy and Physiology* has been updated throughout and a special effort has been made to increase the coverage of physiology. Among the topics that have been added or expanded significantly are facilitated diffusion, duplication of DNA, skeletal muscle contraction, recruitment of motor units, processing of nerve impulses, hormone actions, control of hormone secretion, liver functions, immune responses, and kidney functions. Other topics and terms new to the second edition include endocytosis, microvilli, perichondrium, neuronal pools, facilitation, convergence of impulses, divergence of impulses, adenylate cyclase, peridontal ligament, Kupffer cells, hematocrit, and clones of lymphocytes.

The illustrations have been reviewed carefully, and many of the figures have been improved or have had color added to increase their usefulness. In addition, the second edition contains many new illustrations, including a set of color micrographs to accompany the tissue diagram in chapter 5, and a set of full color drawings of the human torso that reveal the structure and location of major internal organs in chapter 7.

Several new charts have been included to summarize sections of the narrative, and a number of new boxed asides have been added to help students relate topics discussed to clinical or practical situations.

In two instances, closely related chapters have been combined. The discussions of digestion and nutrition are presented together in chapter 13; and the topics of reproduction, pregnancy, and early development have been combined in chapter 20.

Pedagogical Devices

The text includes an unusually large number of pedagogical devices intended to involve students in the learning process, to stimulate their interests in the subject matter, and to help them relate their classroom knowledge to their future vocational experiences. These include the following:

Unit introductions

Each unit opens with a brief description of the general content of the unit and a list of chapters included within the unit. (See p. 3 for an example.) This introduction provides an overview of chapters that make up a unit and tells how the unit relates to the other aspects of human anatomy and physiology.

Chapter introductions

Each chapter introduction previews the chapter's contents and relates that chapter to the others within the unit. (See p. 5.)

Chapter outlines

The chapter outline at the beginning of each chapter includes all the major topic headings and subheadings within the body of the chapter. (See p. 6.) It provides an overview of the chapter's contents and helps the reader locate sections dealing with particular topics.

Chapter objectives

The chapter objectives at the beginning of each chapter indicate what the reader should be able to do after mastering the information within the narrative. (See p. 6.) The review activities at the end of each chapter (see p. 15) are phrased like detailed objectives, and it is helpful for the reader to review them before beginning a study of the chapter. Both sets of objectives are guides that indicate important sections of the narrative.

Key terms

A list of terms and their phonetic pronunciations are given at the beginning of each chapter, to help the reader build a scientific vocabulary. The words included in each list are used within the chapter and are likely to be found in subsequent chapters as well. (See p. 7.) There is an explanation of phonetic pronunciation on page G1 of the Glossary.

Aids to understanding words

Aids to understanding words, at the beginning of each chapter, also help to build the reader's vocabulary. This section includes a list of word roots, stems, prefixes, and suffixes to help the reader discover word meanings. Each root and an example word, which uses that root, are defined. (See p. 7.)

Knowing the roots from these lists will help the reader discover and remember scientific word meanings.

Review questions within the narrative

Review questions occur at the ends of major sections within each chapter to test the reader's understanding of previous discussions. (See p. 8.) If the reader has difficulty answering a set of these questions, the previous section should be reread.

Illustrations and tables

Numerous illustrations and tables occur in each chapter and are placed near their related textual discussion. They are designed to help the reader visualize structures and processes, to clarify complex ideas, to summarize sections of the narrative, or to present pertinent data.

Sometimes the figure legends contain questions that will help the reader apply newly acquired knowledge to the object or process the figure illustrates. (See fig. 3.5.) The ability to apply information to new situations is of prime importance, and such questions concerning the illustrations will provide practice in this skill.

Boxed information

Short paragraphs set off in boxes of colored type also occur throughout each chapter. (See p. 10.) These asides often contain information that will help the reader apply the ideas presented in the narrative to clinical situations. Some of the boxes contain information about changes that occur in body structure and function as a result of disease processes.

Clinical terms

At the ends of certain chapters are lists of related terms and phonetic pronunciations that are sometimes used in clinical situations. (See p. 14.) Although these lists and the word definitions are often brief, they will be a useful addition to the reader's understanding of medical terminology.

Chapter summaries

A summary in outline form occurs at the end of each chapter to help the reader review the major ideas presented in the narrative. (See p. 14.)

Application of knowledge

These questions at the end of each chapter (See p. 32) help the reader gain experience in applying information to clinical situations. They may also serve as topics for discussion with other students or with an instructor.

Review activities

The review activities at the end of each chapter (See p. 15) will serve as a check on the reader's understanding of the major ideas presented in the narrative.

Appendixes

The appendixes following chapter 20 contain a variety of useful information. They include the following:

1. Lists of various units of measurements and their equivalents together with a description of how to convert one to another. (See p. A1.)

2. Lists of clinical laboratory tests commonly performed on human blood and urine. These lists include test names, normal adult values, and brief descriptions of their clinical significance. (See p. A3.)

3. Suggestions for additional reading. These lists will help interested readers locate library materials that can extend understanding of topics discussed within the chapters. (See p. R1.)

Readability

Readability is an important asset of this text. The writing style is intentionally informal and easy to read. Technical vocabulary has been minimized, and illustrations, summary tables, and flow diagrams are carefully positioned near the discussions they complement.

Other features provided to increase readability and to aid student understanding include:

Bold and italic type

identifying words and ideas of particular importance.

Glossary of terms

with phonetic pronunciations, supplying easy access to word definitions.

Index

which is complete and comprehensive.

Supplementary Materials

Supplementary materials designed to help the instructor plan class work and presentations and to aid students in their learning activities are also available. They include:

Instructor's resource manual and test item file

by John W. Hole, Jr., which contains chapter overviews, instructional techniques, a suggested schedule, discussions of chapter elements, lists of related films, and directories of suppliers of audiovisual and laboratory materials. It also contains approximately forty test items for each chapter of the text, designed to evaluate student understanding.

A student study guide to accompany Essentials of Human Anatomy and Physiology

by Nancy Corbett that contains overviews, chapter objectives, focus questions, mastery tests, study activities, and answer keys corresponding to the chapters of the text.

Transparencies

a set of 50 acetate transparencies designed to complement lectures or to be used for short quizzes.

Laboratory manual

by John W. Hole, Jr. is designed specifically to accompany *Essentials of Human Anatomy and Physiology,* second edition. It contains 44 exercises planned to stimulate interest in the subject matter, to involve students in the learning process, and to guide them through a variety of laboratory activities.

Acknowledgments

I would like to express my gratitude to the following persons who carefully reviewed this edition of *Essentials of Human Anatomy and Physiology* and made many valuable suggestions for its improvement. They are Professors Bruce E. Beekman, California State University, Long Beach; Wendy Beltz-Wilson, Albuquerque Technical Vocational Institute; Mary K. Banco, Wheeling Park High School; James D. Houston, Westark Community College; Michela M. Gilarde-Green, Greenwich Academy; R. L. Vandergrift, Cypress College; Les Fenderson, Prescott High School.

I also wish to thank Professor Karen Koos, Rio Hondo College, for her contributions to this second edition. She read the entire manuscript critically and provided helpful assistance as it was being put into final form.

A special thanks is extended to the members of the editorial and production staffs of Wm. C. Brown Publishers, who attended to the multitude of complex details that had to be controlled in this project, to Lynne Meyers, Karen Doland, and Terri Webb—three very talented people—who worked closely with me in the planning and production phases, and to Ed Jaffe who supplied continued support and encouragement, for which I am especially grateful.

Essentials
of Human Anatomy and Physiology

Levels of Organization

The chapters of unit 1 introduce the study of human anatomy and physiology. They are concerned with the ways the human is similar to other living organisms, the ways the structure and function of the human body parts are interdependent, and the ways survival of the body is dependent upon interactions between its parts— chemical substances, cells, tissues, organs, and organ systems.

This unit includes

An Introduction
to Human Anatomy
and Physiology

1

Since a human is a particular kind of living organism (*Homo sapiens*), it has a variety of traits that are shared by other organisms. For example, like all organisms, it carries on life processes and thus demonstrates the characteristics of life. It also has needs that must be met if it is to survive, and its life depends upon *homeostasis,* or the maintenance of a stable internal environment.

A discussion of traits that humans have in common with other organisms and the ways their complex bodies are organized can provide a beginning for the study of anatomy and physiology. This is the purpose of chapter 1.

Chapter Outline

Chapter Objectives

After you have studied this chapter, you should be able to

1. Define *anatomy* and *physiology* and explain how they are related.

2. List and describe the major characteristics of life.

3. List and describe the major needs of organisms.

4. Define *homeostasis* and explain its importance to survival.

5. Describe a homeostatic mechanism.

6. Explain what is meant by levels of organization.

7. Complete the review activities at the end of this chapter. Note that the items are worded in the form of specific learning objectives. You may want to refer to them before reading the chapter.

absorption   (ab-sorp'shun)

anatomy   (ah-nat'o-me)

assimilation   (ah-sim''ĭ-la'shun)

atom   (at'om)

cell   (sel)

circulation   (ser-ku-la'shun)

digestion   (di-jes'chun)

excretion   (ek-skre'shun)

growth   (grōth)

homeostasis   (ho''me-o-sta'sis)

metabolism   (mĕ-tab'o-lizm)

molecule   (mol'ĕ-kūl)

organ   (or'gan)

organelle   (or''gan-el')

organism   (or'gah-nizm)

physiology   (fiz''e-ol'o-je)

reproduction   (re''pro-duk'shun)

respiration   (res''pĭ-ra'shun)

system   (sis'tem)

tissue   (tish'u)

ana-, apart: *ana*tomy—the study of structure, which often involves cutting or removing body parts.

cret-, to separate: exe*cret*ion—the separation or removal of wastes from the body.

homeo-, the same: *homeo*stasis—the maintenance of a stable internal environment.

-logy, the study of: physio*logy*—the study of body functions.

meta-, a change: *meta*bolism—the chemical changes that occur within the body.

-stasis, standing still: homeo*stasis*—the maintenance of a stable internal environment.

-tomy, cutting: ana*tomy*—the study of structure, which often involves cutting or removing body parts.

The accent marks used in the pronunciation guides are derived from a simplified system of phonetics standard in medical usage. The single accent (') denotes the major stress. Emphasis is placed on the most heavily pronounced syllable in the word. The double accent ('') indicates secondary stress. A syllable marked with a double accent receives less emphasis than the syllable that carries the main stress, but more emphasis than neighboring unstressed syllables.

The study of the human body probably stretches back to our earliest ancestors, for they must have been as curious about their body parts and functions as we are today. At first their interests most likely concerned injuries and illnesses, because healthy bodies demand little attention from their owners. Even primitive people injured themselves, suffered from aches and pains, bled, broke bones, and developed diseases. When they were sick or felt pain, these people probably sought relief by visiting priests or medicine men. The treatment they received, however, must have been less than satisfactory, because primitive doctors relied heavily on superstitions and notions about magic and religion to help their patients. Still, these early medical workers began to discover useful ways of examining and treating the sick. They observed the effects of injuries, noticed how wounds healed, and attempted to determine causes of death by examining dead bodies. They also began to learn how certain herbs and potions affected body functions and how some could be used to treat coughs, headaches, and other common problems.

In ancient times however, it was generally believed that most natural processes were controlled by spirits and supernatural forces that humans could not understand. About 2,500 years ago attitudes began to change, and the belief that natural processes were caused by forces that humans could understand grew in popularity.

This new idea stimulated people to look more closely at the world around them. They began to ask questions and seek answers. In this way the stage was set for the development of modern science. As techniques were developed for making accurate observations and performing careful experiments, knowledge of the human body expanded rapidly. (See fig. 1.1.)

There is a list of some modern medical and applied sciences on page 14.

---

1. What factors probably stimulated early interest in the human body?
2. What idea sparked the beginning of modern science?
3. What kinds of activities helped promote the development of modern science?

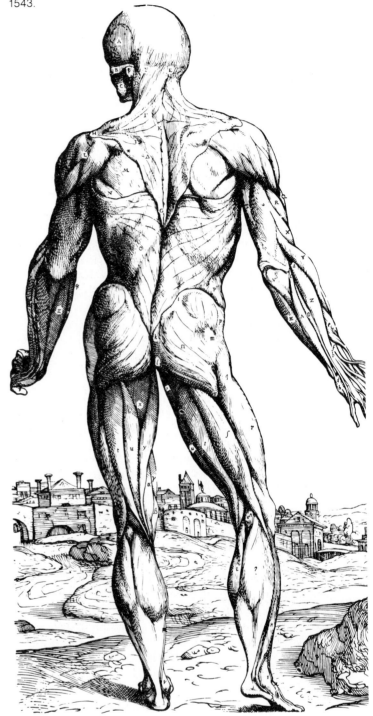

Fig. 1.1 The study of the human body has a long history, as indicated by this illustration from the second book of *De Humani Corporis Fabrica* by Andreas Vesalius, issued in 1543.

Anatomy and Physiology

Anatomy is the branch of science that deals with the structure of body parts—their forms and arrangements. **Physiology,** on the other hand, is concerned with the functions of body parts—what they do and how they do it.

Actually, it is difficult to separate the topics of anatomy and physiology because the structures of body parts are so closely associated with their functions. These parts are arranged to form a well-organized unit—the human organism—and each part plays a role in the operation of the unit. This role, which is the part's function, depends upon the way the part is constructed—that is, the way its subparts are organized. For example, the arrangement of parts in the human hand with its long, jointed fingers is related to the function of grasping objects. The tubular blood vessels of the heart are designed to transport blood to and from its hollow chambers; the shape of the mouth is related to the function of receiving food; and the teeth are designed to break solids into smaller pieces.

While reading about the human body, keep the connection between form and function of body parts in mind. Although the relationship between the anatomy and the physiology of a particular part is not always obvious, such a relationship is sure to exist.

---

1. Why is it difficult to separate the topics of anatomy and physiology?
2. List several examples to illustrate the idea that the structure of a body part is closely related to its function.

The Characteristics of Life

Before beginning a more detailed study of anatomy and physiology, it is helpful to consider some of the traits humans share with other organisms. These traits, which are called *characteristics of life,* include the following:

1. **Movement.** Movement usually refers to a change in an organism's position or to its traveling from one place to another. However, the term also applies to the motion of internal parts, such as the heart.

2. **Responsiveness.** Responsiveness is the ability of an organism to sense changes taking place inside or outside its body and to react to these changes. Pulling away from a hot stove demonstrates responsiveness; drinking water to quench thirst is a response to loss of water from the body tissues.

3. **Growth.** Growth is an increase in body size, usually without any important change in shape. It occurs whenever an organism produces new body materials faster than the old ones are worn out or used up.

4. **Reproduction.** Reproduction is the process of making a new individual, as when parents produce an offspring. It also indicates the process by which microscopic cells produce others like themselves, as they do when body parts are repaired or replaced following an injury.

5. **Respiration.** Respiration is the process of obtaining oxygen, using oxygen in releasing energy from food substances, and removing the resulting gaseous wastes.

6. **Digestion.** Digestion is the process by which various food substances are chemically changed into simpler forms that can be absorbed and used by body parts.

7. **Absorption.** Absorption is the passage of the digestive products through the membranes that line the digestive organs and into the body fluids.

8. **Circulation.** Circulation is the movement of substances from place to place within the body by means of the body fluids.

9. **Assimilation.** Assimilation is the changing of absorbed substances into forms that are chemically different from those that entered the body fluids.

10. **Excretion.** Excretion is the removal of wastes that are produced by body parts as a result of their activities.

Each of these characteristics of life—in fact, everything an organism does—depends upon physical and chemical changes that occur within body parts. Taken together, these changes are referred to as **metabolism.**

Vital signs are among the more common observations made by physicians and nurses working with patients. They include measuring body temperature and blood pressure, and observing rates and types of pulse and breathing movements. There is a close relationship between these signs and the characteristics of life, since vital signs are the results of *metabolic activities*. In fact, death is recognized by the absence of such signs. More specifically, a person who has died will lack spontaneous muscular movements (including those of breathing muscles), responses to stimuli (even the most painful that can be ethically applied), reflexes (such as the knee-jerk reflex and pupillary reflexes of the eye), and brain waves (which reflects a lack of metabolic activity in the brain).

1. What are the characteristics of life?
2. How are the characteristics of life related to metabolism?

The Maintenance of Life

With the exception of an organism's reproductive structures, which function to ensure that its particular form of life will continue into the future, the structures and functions of all body parts are directed toward achieving one goal—the maintenance of the life of the organism.

Needs of Organisms

Life is fragile, and it depends upon the presence of certain environmental factors for its existence. These factors include the following:

1. **Water.** Water is the most abundant chemical substance within the body. It is required for a variety of metabolic processes, and provides the environment in which most chemical reactions take place. Water also transports substances from place to place within the organism and is important in the regulation of body temperature.

2. **Food.** Food is a general term for substances that provide the body with necessary chemicals in addition to water. Some of these chemicals are used as energy sources, others as raw materials for building new living matter, and still others help regulate vital chemical reactions.

3. **Oxygen.** Oxygen is a gas that makes up about one-fifth of ordinary air. It is used in the process of releasing energy from food substances. The energy, in turn, is needed to drive metabolic processes.

4. **Heat.** Heat is a form of energy. It is a by-product of metabolic reactions, and the rate at which these reactions occur is partly governed by the amount of heat present. Generally, the greater the amount of heat, the more rapidly the reactions take place. (The measurement of heat intensity is called *temperature*.)

5. **Pressure.** Pressure is a force that creates a pressing (or compressing) action. For example, the force acting on the outside of the body due to the weight of air is called *atmospheric pressure*. In humans, this pressure plays an important part in breathing. Similarly, organisms living under water are subjected to *hydrostatic pressure* due to the weight of water. In complex organisms, such as humans, heart action creates hydrostatic pressure in blood (blood pressure), which is needed to force blood through the blood vessels.

Although organisms need water, food, oxygen, heat, and pressure, the presence of these factors alone is not enough to ensure their survival. The quantities and the qualities of such factors are important, too. For example, the amount of water entering and leaving an organism must be regulated, as must the concentration of oxygen in the body fluids. Similarly, survival depends on the quality as well as the quantity of food available—that is, food must supply the correct chemicals in adequate amounts.

Homeostasis

As an organism moves from place to place or as the climate changes, the factors in its external environment also change. If the organism is to survive, however, the conditions within the fluids surrounding its body cells must remain relatively stable. In other words, body parts function efficiently only when the concentrations of water, food substances, and oxygen,

Fig. 1.2 A thermostat that can signal an air conditioner and a furnace to turn on or off helps to maintain a relatively stable room temperature.

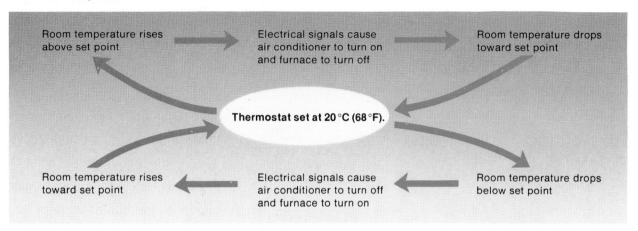

and the conditions of heat and pressure remain within certain limits. Thus, it is not surprising that the metabolic activities of an organism are directed toward maintaining such steady conditions. This tendency to maintain a stable internal environment is called **homeostasis.**

To better understand the idea of maintaining a stable environment, imagine a room equipped with a furnace and an air conditioner. Suppose the room temperature is to remain near 20°C (68°F), so the thermostat is adjusted to a set point of 20°C. Since a thermostat is sensitive to temperature changes, once it is adjusted it will signal the furnace to start whenever the room temperature drops below the set point. If the temperature rises above the set point, the thermostat will cause the furnace to stop and the air conditioner to start. As a result, a relatively constant temperature will be maintained in the room. (See fig. 1.2.)

A *homeostatic mechanism* in the human body achieves similar results in regulating body temperature. The thermostat is a temperature-sensitive region of the brain. In healthy persons the set point of the brain's thermostat is at or near 37°C (98.6°F).

If a person is exposed to a cold environment and the body temperature begins to drop, the brain can sense this change and triggers heat-generating and heat-conserving activities. For example, small groups of muscles may be stimulated to contract involuntarily, causing an action called shivering. Such muscular contractions produce heat as a by-product, which

helps to warm the body. At the same time, blood vessels in the skin may be signaled to constrict so that less warm blood reaches the skin, and heat that might otherwise be lost is held in deeper tissues.

If a person is becoming overheated, the brain may trigger a series of changes that leads to increased loss of body heat. It may stimulate the sweat glands of the skin to secrete watery perspiration; as the water evaporates from the surface, some heat is carried away and the skin is cooled. Also, blood vessels in the skin may be signaled to dilate so blood from deeper tissues reaches the surface in greater volume and blood loses some of its heat to the outside. (See chapter 6.)

Another homeostatic mechanism functions to regulate blood pressure in vessels leading away from the heart. In this instance, pressure-sensitive sensors in the walls of these vessels are stimulated when the blood pressure increases above normal. As this happens, the sensors signal the brain, and the brain signals the heart, causing the heart chambers to contract more slowly and with less force. Because of this decreased heart action, less blood enters the blood vessels, and the pressure inside the vessels decreases. If the blood pressure is dropping below normal, the brain signals the heart to contract more rapidly and with greater force so that the pressure in the vessels increases. (See fig. 1.3.)

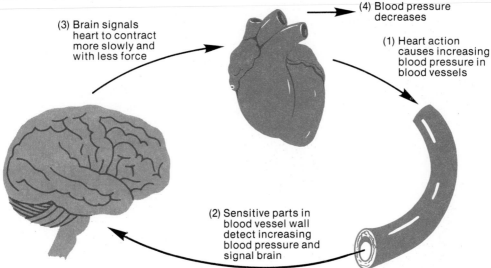

Fig. 1.3 A homeostatic mechanism helps to lower blood pressure if it increases above normal. A similar mechanism helps to increase blood pressure if it decreases below normal.

(3) Brain signals heart to contract more slowly and with less force

(4) Blood pressure decreases

(1) Heart action causes increasing blood pressure in blood vessels

(2) Sensitive parts in blood vessel wall detect increasing blood pressure and signal brain

---

1. *What needs of organisms are provided from the external environment?*
2. *Why is homeostasis important to survival?*
3. *Describe two homeostatic mechanisms.*

Levels of Organization

Early investigators focused their attention on the larger body structures, for they were limited in their ability to observe small parts. Studies of small parts had to wait for the invention of magnifying lenses and microscopes, which came into use about 400 years ago. Once these tools were available, it was discovered that larger body structures were made up of smaller parts, which, in turn, were composed of still smaller parts.

Today scientists recognize that all materials, including those that comprise the human body, are composed of chemical substances. These substances are made up of tiny, invisible particles called **atoms,** which are commonly bound together to form larger particles called **molecules;** small molecules may be combined in complex ways to form larger molecules (macromolecules).

Within the human organism the basic unit of structure and function is a microscopic part called a **cell.** Although individual cells vary in size, shape, and specialized functions, all have certain traits in common. For instance, all cells contain tiny parts called **organelles** that carry on specific activities. These organelles are composed of aggregates of large molecules, including those of such substances as proteins, carbohydrates, lipids, and nucleic acids.

Cells are organized into layers or masses that have common functions. Such a group of cells constitutes a **tissue.** Groups of different tissues form **organs**—complex structures with special functions—while groups of organs are arranged into **organ systems.** Organ systems make up an **organism.**

Thus, various body parts can be thought of as occupying different *levels of organization.* Furthermore, these parts vary in *complexity* from one level to the next. That is, atoms are less complex than molecules, molecules are less complex than organelles, and so forth. (See fig. 1.4.)

Chapters 2 through 7 discuss these levels of organization in more detail. Chapter 2, for example, describes the atomic and molecular levels. Chapter 3 deals with organelles and cellular structures and functions, while chapter 5 describes tissues. Chapter 6 presents membranes as examples of organs, and the skin and its accessory organs as an example of an organ system. Finally, chapter 7 introduces the organ systems that comprise the human organism and at the same time presents some terminology used to describe various body cavities, membranes, and regions.

Beginning with chapter 8, the structures and functions of each of the organ systems are described in detail.

Fig. 1.4 A human body is composed of parts within parts, which vary in complexity.

LEVELS OF ORGANIZATION

LEVEL

EXAMPLES

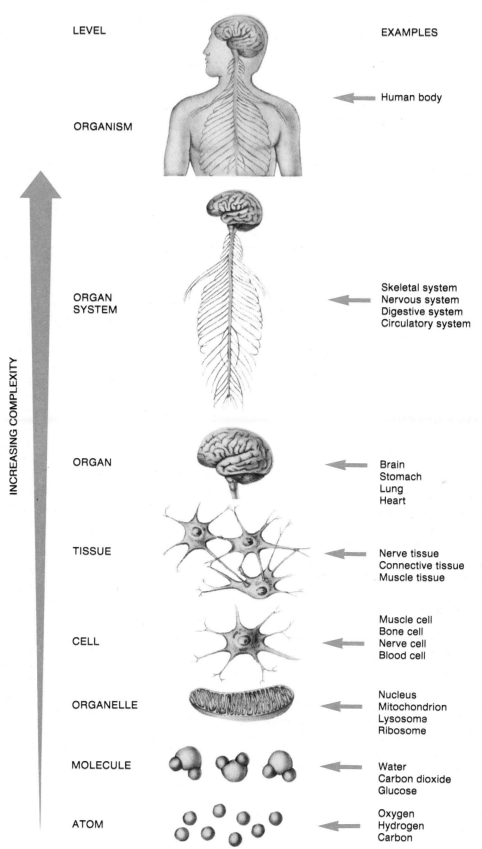

ORGANISM ← Human body

ORGAN SYSTEM ← Skeletal system
Nervous system
Digestive system
Circulatory system

ORGAN ← Brain
Stomach
Lung
Heart

TISSUE ← Nerve tissue
Connective tissue
Muscle tissue

CELL ← Muscle cell
Bone cell
Nerve cell
Blood cell

ORGANELLE ← Nucleus
Mitochondrion
Lysosome
Ribosome

MOLECULE ← Water
Carbon dioxide
Glucose

ATOM ← Oxygen
Hydrogen
Carbon

INCREASING COMPLEXITY

1. How can the human body be used to illustrate the idea of levels of organization?
2. What is an organism?
3. How do body parts that occupy different levels of organization vary in complexity?

Some of the Medical and Applied Sciences

cardiology (kar''de-ol'o-je)—branch of medical science dealing with the heart and heart diseases.

cytology (si-tol'o-je)—study of the structure and function of cells and the diseases involving them.

dermatology (der''mah-tol'o-je)—study of skin and its diseases.

endocrinology (en''do-kri-nol'o-je)—study of hormone-secreting glands and the diseases involving them.

epidemiology (ep''ĭ-de''me-ol'o-je)—study of infectious diseases, their distribution and control.

gastroenterology (gas''tro-en''ter-ol'o-je)—study of the stomach and intestines and diseases involving these organs.

geriatrics (jer''e-at'riks)—branch of medicine dealing with aged persons and their medical problems.

gerontology (jer''on-tol'o-je)—study of the process of aging and the problems of elderly persons.

gynecology (gi''nĕ-kol'o-je)—study of the female reproductive system and its diseases.

hematology (hem''ah-tol'o-je)—study of blood and blood diseases.

histology (his-tol'o-je)—study of the structure and function of tissues and the diseases involving them.

immunology (im''u-nol'o-je)—study of body resistance to disease.

neurology (nu-rol'o-je)—study of the nervous system in health and disease.

obstetrics (ob-stet'riks)—branch of medicine dealing with pregnancy and childbirth.

oncology (ong-kol'o-je)—study of tumors.

ophthalmology (of''thal-mol'o-je)—study of the eye and eye diseases.

orthopedics (or''tho-pe'diks)—branch of medicine dealing with the muscular and skeletal systems and problems of these systems.

otolaryngology (o''to-lar''in-gol'o-je)—study of the ear, throat, larynx, and diseases of these parts.

pathology (pah-thol'o-je)—study of body changes produced by diseases.

pediatrics (pe''de-at'riks)—branch of medicine dealing with children and their diseases.

pharmacology (fahr''mah-kol'o-je)—study of drugs and their uses in the treatment of diseases.

psychiatry (si-ki'ah-tre)—branch of medicine dealing with the mind and its disorders.

radiology (ra''de-ol'o-je)—study of X rays and radioactive substances and their uses in the diagnosis and treatment of diseases.

toxicology (tok''si-kol'o-je)—study of poisonous substances and their effects upon body parts.

urology (u-rol'o-je)—branch of medicine dealing with the kidneys and urinary system and their diseases.

Chapter Summary

Introduction

1. Early interest in the human body probably concerned injuries and illnesses.

2. Primitive doctors began to learn how certain herbs and potions affected body functions.

3. In ancient times it was believed that natural processes were caused by spirits and supernatural forces.

4. The belief that humans could understand forces that caused natural events led to the development of modern science.

Anatomy and Physiology

1. Anatomy deals with the form and arrangement of body parts.

2. Physiology deals with the functions of these parts.

3. The function of a part depends upon the way it is constructed.

The Characteristics of Life

Traits shared by all living things are called characteristics of life.

1. Characteristics of life include:
   a. Movement—changes in body position or motion of internal parts.
   b. Responsiveness—sensing and reacting to internal or external changes.
   c. Growth—increase in size without change in shape.
   d. Reproduction—production of offspring.
   e. Respiration—obtaining oxygen, using oxygen to release energy from foods, and removing gaseous wastes.

f. Digestion—changing food substances into forms that can be absorbed and used by cells.

g. Absorption—passage of digestive products into body fluids.

h. Circulation—movement of substances in body fluids.

i. Assimilation—changing of substances into chemically different forms.

j. Excretion—removal of metabolic wastes.

2. Taken together, the chemical and physical changes that occur in body parts constitute metabolism.

The Maintenance of Life

The structures and functions of body parts are directed toward maintaining the life of the organism.

1. Needs of organisms include:
   a. Water—used in a variety of metabolic processes; provides the environment for metabolic reactions, and transports substances.
   b. Food—needed to supply energy, to provide raw materials for building new living matter, and to supply chemicals necessary in vital reactions.
   c. Oxygen—used in releasing energy from food materials; this energy drives metabolic reactions.
   d. Heat—a by-product of metabolic reactions; helps govern the rates of these reactions.
   e. Pressure—a force; in humans, atmospheric and hydrostatic pressures help breathing and blood movement.
   f. Survival of an organism depends upon the quantities and qualities of these factors.

2. Homeostasis
   a. If an organism is to survive, the conditions within its body fluids must remain stable.
   b. Metabolic activities are largely directed toward maintaining stable internal conditions.
   c. The tendency to maintain a stable internal environment is called homeostasis.
   d. Homeostatic mechanisms help regulate body temperature and blood pressure.

Levels of Organization

The body is composed of parts that occupy different levels of organization.

1. Material substances are composed of atoms.

2. Atoms bond together to form molecules.

3. Organelles contain groups of large molecules.

4. Cells, which are composed of organelles, are the basic units of structure and function within the body.

5. Cells are organized into tissues.

6. Tissues are organized into organs.

7. Organs are arranged into organ systems.

8. Organ systems constitute the organism.

9. These parts vary in complexity from one level to the next.

Review Activities

1. Briefly describe the early development of knowledge about the human body.

2. Distinguish between anatomy and physiology.

3. Explain the relationship between the form and function of body parts.

4. List and describe ten characteristics of life.

5. Define metabolism.

6. List and describe five needs of organisms.

7. Describe two types of pressure that may act on the outside of an organism.

8. Define homeostasis and explain its importance.

9. Explain how body temperature is controlled.

10. Describe a homeostatic mechanism that helps to regulate blood pressure.

11. List the levels of organization within the human body.

The Chemical Basis of Life

A s chapter 1 explained, an organism is composed of parts that vary in levels of organization. These include organ systems, organs, tissues, cells, and cellular organelles.

Organelles are in turn composed of chemical substances made up of atoms and groups of atoms called molecules. Similarly, atoms are composed of still smaller units called protons, neutrons, and electrons.

A study of living material at its lower levels of organization must be concerned with the chemical substances that form the structural basis of all matter and that interact in the metabolic processes carried on within organisms. This is the theme of chapter 2.

2

After you have studied this chapter, you should be able to

1. Explain how the study of living material is dependent on the study of chemistry.

2. Describe the relationships between matter, atoms, and molecules.

3. Discuss how atomic structure is related to the ways in which atoms interact.

4. Explain how molecular and structural formulas are used to symbolize the composition of compounds.

5. Describe two types of chemical reactions.

6. Discuss the concept of pH.

7. List the major groups of inorganic substances that are common in cells.

8. Describe the general roles played in cells by various types of organic substances.

9. Complete the review activities at the end of this chapter. Note that the items are worded in the form of specific learning objectives. You may want to refer to them before reading the chapter.

atom   (at′om)

carbohydrate   (kar″bo-hi′drāt)

decomposition   (de″kom-po-zish′un)

electrolyte   (e-lek′tro-līt)

formula   (for′mu-lah)

inorganic   (in″or-gan′ik)

ion   (i′on)

lipid   (lip′id)

molecule   (mol′ĕ-kūl)

nucleic acid   (nu-kle′ik as′id)

organic   (or-gan′ik)

protein   (pro′te-in)

synthesis   (sin′thĕ-sis)

di-, two: *di*saccharide—a compound whose molecules are composed of two saccharide units bound together.

glyc-, sweet: *glyc*ogen—a complex carbohydrate composed of sugar molecules bound together.

lip-, fat: *lip*ids—group of organic compounds that includes fats.

-lyt, dissolvable: electro*lyte*—substance that dissolves in water and releases ions.

mono-, one: *mono*saccharide—a compound whose molecules consist of a single saccharide unit.

poly-, many: *poly*unsaturated—molecule that has many double bonds between its carbon atoms.

sacchar-, sugar: mono*sacchar*ide—sugar molecule composed of a single saccharide unit.

syn-, together: *syn*thesis—process by which substances are united to form new types of substances.

Chemistry is the branch of science dealing with the composition of substances and changes that take place in their composition. Although it is possible to study anatomy without much reference to this subject, knowledge of chemistry is essential for understanding physiology, because body functions involve chemical changes that occur within cells.

Structure of Matter

Matter is anything that has weight and takes up space. This includes all the solids, liquids, and gases in our surroundings as well as in our bodies.

Elements and Atoms

Studies of matter have revealed that all things are composed of basic substances called **elements.** At present, 106 such elements are known, although most naturally occurring matter includes only about 90 of them. Among these elements are such common materials as iron, copper, silver, gold, aluminum, carbon, hydrogen, and oxygen. Although some elements exist in a pure form, they occur more frequently in mixtures or in chemical combinations of 2 or more elements known as **compounds.**

About 20 elements are needed by living things, and of these oxygen, carbon, hydrogen, and nitrogen make up more than 95% of the human body. A list of the more abundant elements in the body is shown in chart 2.1. Notice that each element is represented by a one- or two-letter *symbol*.

Elements are composed of tiny, invisible particles called **atoms.** Although the atoms that make up each element are very similar to each other, they differ from the atoms that make up other elements. Atoms vary in size, weight, and the ways they interact with each other. Some, for instance, are capable of combining with atoms like themselves or with other kinds of atoms, while others lack this ability.

Atomic Structure

An atom consists of a central core called the **nucleus** and one or more **electrons** that are in constant motion around the nucleus. The nucleus itself contains some relatively large particles called **protons** and **neutrons,** whose weights are about equal. (See fig. 2.1.)

Electrons, which are extremely small and have almost no weight, carry a single, negative electrical charge (e^-), while protons each carry a single, positive electrical charge (p^+). Neutrons are uncharged and thus are electrically neutral (n^0).

Chart 2.1   Most abundant elements in the human body

Major Elements	Symbol	Approximate Percentage of the Human Body
Oxygen	O	65.0
Carbon	C	18.5
Hydrogen	H	9.5
Nitrogen	N	3.2
Calcium	Ca	1.5
Phosphorus	P	1.0
Potassium	K	0.4
Sulfur	S	0.3
Chlorine	Cl	0.2
Sodium	Na	0.2
Magnesium	Mg	0.1
Trace Elements		
Cobalt	Co	
Copper	Cu	
Fluorine	F	
Iodine	I	Less than 0.1
Iron	Fe	
Manganese	Mn	
Zinc	Zn	

Fig. 2.1   An atom of lithium includes 3 electrons in motion around a nucleus, which contains 3 protons and 4 neutrons.

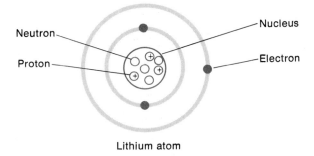

Lithium atom

Since the atomic nucleus contains the protons, this part of an atom is always positively charged. However, the number of electrons outside the nucleus is equal to the number of protons, so a complete atom is electrically uncharged, or neutral.

The atoms of different elements contain different numbers of protons. The number of protons in the atoms of a particular element is called its *atomic number*. Hydrogen, for example, whose atoms contain one proton, has the atomic number 1; and carbon, whose atoms have six protons, has the atomic number 6.

The weight of an atom of an element is primarily due to the protons and neutrons in its nucleus because the electrons have so little weight. For this reason, an atom of carbon with six protons and six neutrons weighs about 12 times as much as an atom of hydrogen, which has only one proton and no neutrons.

Chart 2.2 Atomic structure of elements 1 through 12

Element	Symbol	Atomic Number	Atomic Weight	Protons	Neutrons	Electrons in Shells		
						First	Second	Third
Hydrogen	H	1	1	1	0	1		
Helium	He	2	4	2	2	2	(inert)	
Lithium	Li	3	7	3	4	2	1	
Beryllium	Be	4	9	4	5	2	2	
Boron	B	5	11	5	6	2	3	
Carbon	C	6	12	6	6	2	4	
Nitrogen	N	7	14	7	7	2	5	
Oxygen	O	8	16	8	8	2	6	
Fluorine	F	9	19	9	10	2	7	
Neon	Ne	10	20	10	10	2	8	(inert)
Sodium	Na	11	23	11	12	2	8	1
Magnesium	Mg	12	24	12	12	2	8	2

The number of protons plus the number of neutrons in each of its atoms is approximately equal to the *atomic weight* of an element. Thus, the atomic weight of hydrogen is 1, and the atomic weight of carbon is 12. (See chart 2.2.)

1. What is the relationship between matter and elements?
2. What elements are most common in the human body?
3. How are electrons, protons, and neutrons positioned within an atom?

Bonding of Atoms

When atoms combine with other atoms, they either gain or lose electrons, or share electrons with other atoms.

The electrons of an atom are arranged in one or more *shells* around the nucleus. The maximum number of electrons that each of the first three shells can hold is as follows:

First shell (closest to the nucleus)	2 electrons
Second shell	8 electrons
Third shell	8 electrons

(The third shell can hold 8 electrons for elements up to atomic number 20. More complex atoms may have as many as 18 electrons in the third shell.)

The arrangements of electrons within the shells of atoms can be represented by simplified diagrams such as those in figure 2.2. Notice that the single electron of a hydrogen atom is located in the first shell; the 2 electrons of a helium atom fill its first shell; the 3 electrons of a lithium atom are arranged with 2 in the first shell and 1 in the second shell. (See chart 2.2.)

Fig. 2.2 The single electron of a hydrogen atom is located in its first shell; the 2 electrons of a helium atom fill its first shell; and the 3 electrons of a lithium atom are arranged with 2 in the first shell and 1 in the second shell.

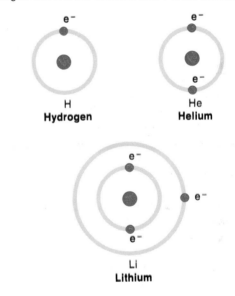

Atoms such as helium, whose outermost electron shells are filled, have stable structures and are chemically inactive (inert). Atoms with incompletely filled outer shells, such as those of hydrogen or lithium, tend to gain, lose, or share electrons in ways that empty or fill their outer shells. In this way they achieve stable structures.

An atom of sodium, for example, has 11 electrons, arranged as shown in figure 2.3. This atom tends to lose the single electron in its outer shell, which leaves the second shell filled and the form stable.

A chlorine atom has 17 electrons arranged with 2 in the first shell, 8 in the second shell, and 7 in the third shell. An atom of this type will tend to accept a single electron, thus filling its outer shell and achieving a stable form.

Fig. 2.3 A diagram of a sodium atom.

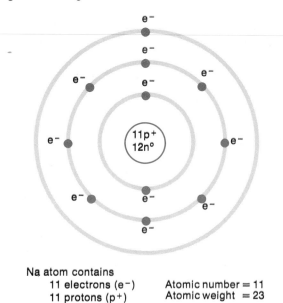

Na atom contains
    11 electrons (e^-)
    11 protons (p^+)
    12 neutrons (n^o)

Atomic number $= 11$
Atomic weight $= 23$

Since each sodium atom tends to lose a single electron and each chlorine atom tends to accept a single electron, sodium and chlorine atoms will react together. During this reaction, a sodium atom loses an electron and is left with 11 protons ($11+$) in its nucleus and only 10 electrons ($10-$). As a result, the atom develops a net electrical charge of $1+$ and is symbolized Na^+. At the same time, a chlorine atom gains an electron, which leaves it with 17 protons ($17+$) in its nucleus and 18 electrons ($18-$). Thus it develops a net electrical charge of $1-$. Such an atom is symbolized Cl^-.

Atoms that have become electrically charged by gaining or losing electrons are called **ions,** and two ions with opposite charges are attracted to one another electrically. When this happens, a chemical bond called an *electrovalent bond* (ionic bond) is created between them, as shown in figure 2.4. When sodium ions (Na^+) and chlorine ions (Cl^-) unite in this manner, the compound sodium chloride ($NaCl$, common salt) is formed.

Fig. 2.4 If a sodium atom loses an electron to a chlorine atom, the sodium atom becomes a sodium ion and the chlorine atom becomes a chloride ion. These oppositely charged particles are attracted electrically to one another and become united by an electrovalent bond.

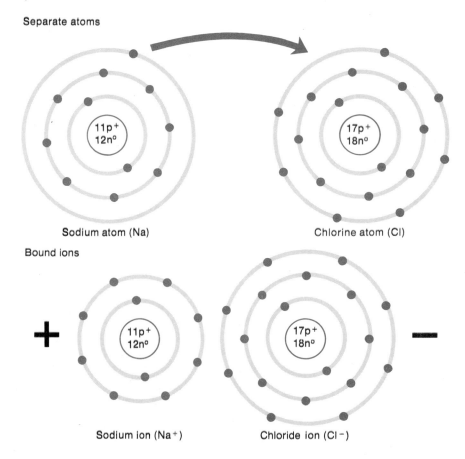

Separate atoms

Sodium atom (Na)

Chlorine atom (Cl)

Bound ions

Sodium ion (Na^+)

Chloride ion (Cl^-)

Atoms can also bond together by sharing electrons rather than by exchanging them. A hydrogen atom, for example, has 1 electron in its first shell, but needs 2 to achieve a stable structure. It may fill this shell by combining with another hydrogen atom in such a way that the 2 atoms share a pair of electrons. As figure 2.5 shows, the 2 electrons then encircle the nuclei of both atoms so that each achieves a stable form. In this case, the atoms are held together by a *covalent bond.*

Carbon atoms, with 2 electrons in their first shells and 4 electrons in their second shells, always form covalent bonds when they unite with other atoms. In fact, carbon atoms may bond to other carbon atoms in such a way that 2 atoms share 1 or more pairs of electrons. If 1 pair of electrons is shared, the resulting bond is called a *single covalent bond;* if 2 pairs of electrons are shared, the bond is called a *double covalent bond.*

Molecules and Compounds

When 2 or more atoms bond together, they form a new kind of particle called a **molecule.** When atoms of the same element combine, they produce molecules of that element. The gases of hydrogen (H_2), oxygen (O_2), and nitrogen (N_2) contain such molecules. (See fig. 2.5.)

When atoms of different elements combine, molecules of substances called **compounds** form. Two atoms of hydrogen, for example, can combine with one atom of oxygen to produce a molecule of the compound water (H_2O), as shown in figure 2.6. Sugar, table salt, natural gas, alcohol, and most drugs are examples of compounds.

A molecule of a compound always contains definite kinds and numbers of atoms. A molecule of water (H_2O), for instance, always contains 2 hydrogen atoms and 1 oxygen atom. If 2 hydrogen atoms combine with 2 oxygen atoms, the compound formed is not water, but hydrogen peroxide (H_2O_2).

Fig. 2.5 Two hydrogen atoms may bond together by sharing a pair of electrons.

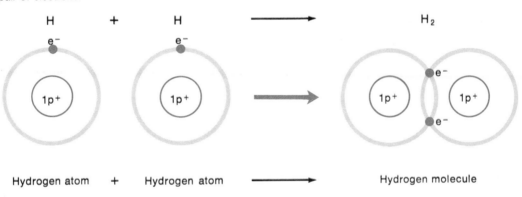

Fig. 2.6 Hydrogen molecules (H_2) combine with oxygen molecules (O_2) to form water molecules (H_2O).

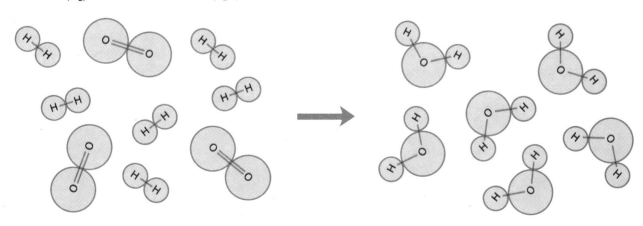

Chart 2.3 Some particles of matter

Name	Characteristic
Atom	Smallest particle of an element that has the properties of that element.
Electron (e^-)	Extremely small particle with almost no weight; carries a negative electrical charge and is in constant motion around an atomic nucleus.
Proton (p^+)	Relatively large atomic particle; carries a positive electrical charge and is found within the nucleus of an atom.
Neutron (n^0)	Particle with about the same weight as a proton; uncharged and thus electrically neutral; found within the nucleus of an atom.
Ion	Atom that is electrically charged because it has gained or lost 1 or more electrons.
Molecule	Particle formed by the chemical union of 2 or more atoms.

From Hole, John W., Jr., *Human Anatomy and Physiology* 3d ed. © 1978, 1981, 1984 Wm. C. Brown Publishers, Dubuque, Iowa. All Rights Reserved. Reprinted by permission.

Chart 2.3 lists some particles of matter and their characteristics.

---

1. *What is an ion?*
2. *Describe two ways in which atoms may combine with other atoms.*
3. *Distinguish between a molecule and a compound.*

Formulas. The numbers and kinds of atoms in a molecule can be represented by a *molecular formula.* Such a formula consists of the symbols of the elements in the molecule together with numbers to indicate how many atoms of each element are present. For example, the molecular formula for water is H_2O, which means there are 2 atoms of hydrogen and 1 atom of oxygen in each molecule. The molecular formula for the sugar called glucose is $C_6H_{12}O_6$, which means there are 6 atoms of carbon, 12 atoms of hydrogen, and 6 atoms of oxygen in a molecule.

Usually the atoms of each element will form a specific number of bonds—hydrogen atoms form single bonds, oxygen atoms form 2 bonds, nitrogen atoms form 3 bonds, and carbon atoms form 4 bonds. The bonding capacity of these atoms can be represented by using symbols and lines as follows:

$$—H \quad —O— \quad \overset{\diagup}{\underset{\diagup}{N}} \quad —\overset{|}{\underset{|}{C}}—$$

These representations can be used to show how atoms are joined and arranged in various molecules. Illustrations of this type are called *structural formulas.* (See fig. 2.7.)

Fig. 2.7 Structural formulas of molecules of hydrogen (H_2), oxygen (O_2), water (H_2O), and carbon dioxide (CO_2).

H—H	O=O	H—O—H	O=C=O
H_2	O_2	H_2O	CO_2

Chemical Reactions

When atoms or molecules react together, bonds between atoms are formed or broken. As a result, new combinations of atoms are created.

Two kinds of chemical reactions are **synthesis** and **decomposition.** For example, when 2 or more atoms bond together to form a more complex structure, as when atoms of hydrogen and oxygen bond together to form molecules of water, the reaction is called *synthesis.* Such a reaction is symbolized in this way:

$$A + B \longrightarrow AB$$

In a *decomposition* reaction the bonds within a molecule break so that simpler molecules, atoms, or ions form. Thus, molecules of water can decompose to yield hydrogen and oxygen. Decomposition is symbolized as follows:

$$AB \longrightarrow A + B$$

Synthetic reactions are particularly important in the growth of body parts and the repair of worn or damaged tissues, which involve the buildup of larger molecules from smaller ones. When food substances are digested or energy is released from food substances through respiration, the processes are largely decomposition reactions.

Acids and Bases

Some compounds release ions (ionize) when they dissolve in water or react with water molecules. Sodium chloride (NaCl), for example, releases sodium ions (Na^+) and chlorine ions (Cl^-) when it dissolves, as is represented by the following:

$$NaCl \longrightarrow Na^+ + Cl^-$$

Since the resulting solution contains electrically charged particles (ions), it will conduct an electric current. Substances that ionize in water are, therefore, known as **electrolytes.** Electrolytes that release hydrogen ions (H^+) in water are called **acids.** For example, the compound carbonic acid (H_2CO_3) ionizes in water to release hydrogen ions and bicarbonate ions (HCO_3^-).[1]

$$H_2CO_3 \longrightarrow H^+ + HCO_3^-$$

---

1. Some ions, such as HCO_3^- ions, contain two or more atoms. However, such a group behaves like a single atom and usually remains unchanged in a chemical reaction.

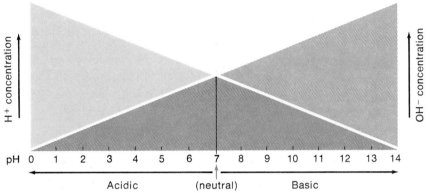

Fig. 2.8 As the concentration of hydrogen ions (H⁺) increases, a solution becomes more acidic, and the pH value decreases. As the concentration of hydroxyl ions (OH⁻) increases, a solution becomes more basic, and the pH value increases.

Electrolytes that release ions that combine with hydrogen ions are called **bases.** The compound sodium hydroxide (NaOH), for example, ionizes in water to release hydroxyl ions (OH⁻).

$$NaOH \longrightarrow Na^+ + OH^-$$

The hydroxyl ions, in turn, can combine with hydrogen ions to form water; thus, sodium hydroxide is a base.

The chemical reactions involved with life processes often are affected by the presence of hydrogen and hydroxyl ions; therefore the concentrations of these ions in body fluids are important. Such concentrations are measured in units of **pH.**

The pH scale includes values from 0 to 14. A solution with a pH of 7.0, the midpoint of the scale, contains equal numbers of hydrogen and hydroxyl ions and is said to be *neutral.* (See fig. 2.8.) A solution that contains more hydrogen than hydroxyl ions has a pH less than 7.0 and is *acidic,* while one with fewer hydrogen than hydroxyl ions has a pH above 7.0 and is *basic* (alkaline). Figure 2.9 indicates the pH values of some common substances.

There is a tenfold difference in the hydrogen ion concentration between each whole number on the pH scale. For example, a solution of pH 4.0 contains 0.0001 grams of hydrogen ions per liter, while a solution of pH 3.0 contains 0.001 grams of hydrogen ions per liter.

Note in figure 2.9 that the pH of human blood is normally 7.4. If this pH value drops below 7.4, the person is said to have *acidosis;* if it rises above 7.4, the condition is called *alkalosis.* Usually a person cannot survive if the pH drops to 7.0 or rises to 7.8 for more than a few minutes.

Fig. 2.9 Approximate pH values of some common substances.

Approximate pH values of some common substances

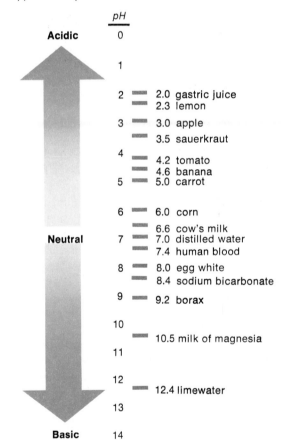

pH	
Acidic 0	
1	
2	2.0 gastric juice
	2.3 lemon
3	3.0 apple
	3.5 sauerkraut
4	4.2 tomato
	4.6 banana
5	5.0 carrot
6	6.0 corn
	6.6 cow's milk
Neutral 7	7.0 distilled water
	7.4 human blood
8	8.0 egg white
	8.4 sodium bicarbonate
9	9.2 borax
10	
	10.5 milk of magnesia
11	
12	12.4 limewater
13	
Basic 14	

1. What is meant by a formula?
2. Describe two kinds of chemical reactions.
3. Compare the characteristics of an acid with those of a base.
4. What is meant by pH?

Chemical Constituents of Cells

Chemicals that enter into metabolic reactions or are produced by them can be divided into two large groups. Except for a few simple molecules, those that contain carbon atoms are called **organic,** and those that lack carbon are called **inorganic.**

Generally, inorganic substances will dissolve in water or react with water to produce ions; thus they are electrolytes. Some organic compounds will dissolve in water also, but they are more likely to dissolve in organic liquids like ether or alcohol. Those that will dissolve in water usually do not release ions, and are therefore called *nonelectrolytes.*

Inorganic Substances

Among the inorganic substances common in cells are water, oxygen, carbon dioxide, and a group called salts.

Water. Water (H_2O) is the most abundant compound in living material and is responsible for about two-thirds of the weight of an adult human. It is the major ingredient of blood and other body fluids, including those within cells.

When substances dissolve in water, relatively large pieces of the material break into smaller ones, and eventually molecular-sized particles or ions result. These tiny particles are much more likely to react with one another than were the original large pieces. Consequently, most metabolic reactions occur in water.

Water also plays an important role in the transportation of chemicals within the body. The aqueous portion of blood, for example, carries many vital substances, such as oxygen, sugars, salts, and vitamins, from the organs of digestion and respiration to the body cells.

In addition, water can absorb and transport heat. For instance, heat released from muscle cells during exercise can be carried from deeper parts to the surface by the blood.

Oxygen. Molecules of oxygen (O_2) enter the body through the respiratory organs and are transported to cells by the blood. The oxygen is used by cellular organelles in the process of releasing energy from glucose and certain other molecules. The energy is needed to drive the cell's metabolic activities.

Carbon Dioxide. Carbon dioxide (CO_2) is one of the simple carbon-containing compounds of the inorganic group. It is produced as a waste product when energy is released during respiration and is given off into the air in the lungs.

Inorganic Salts. Various kinds of inorganic salts are common in body parts and fluids. They are the sources of many necessary ions, including ions of sodium (Na^+), chlorine (Cl^-), potassium (K^+), calcium (Ca^{+2}), magnesium (Mg^{+2}), phosphate (PO_4^{-3}), carbonate (CO_3^{-2}), bicarbonate (HCO_3^-), and sulfate (SO_4^{-2}). These ions play important roles in metabolic processes.

1. What is the difference between an organic and an inorganic molecule?
2. What is the difference between an electrolyte and a nonelectrolyte?
3. Name the inorganic substances common in body fluids.

Organic Substances

Important groups of organic substances found in cells include carbohydrates, lipids, proteins, and nucleic acids.

Carbohydrates. These compounds provide much of the energy that cells need. They also supply materials to build certain cell structures and often are stored as reserve energy supplies.

Carbohydrate molecules contain atoms of carbon, hydrogen, and oxygen. These molecules usually have twice as many hydrogen as oxygen atoms, the same ratio of hydrogen to oxygen that occurs in water molecules (H_2O). This ratio is easy to see in the molecular formulas of glucose ($C_6H_{12}O_6$) and sucrose or table sugar ($C_{12}H_{22}O_{11}$).

The carbon atoms of carbohydrate molecules are joined in chains whose lengths vary with the kind of carbohydrate. For example, those with shorter chains are called **sugars.**

Sugars with 6 carbon atoms are known as *simple sugars* or *monosaccharides,* and they are the building blocks of more complex carbohydrate molecules. The simple sugars include glucose, fructose, and galactose. Figure 2.10 illustrates the molecular structure of glucose.

In complex carbohydrates, the simple sugars are bound to form molecules of varying sizes, as shown in figure 2.11. Some, like sucrose, maltose, and lactose, are *double sugars* or *disaccharides,* whose molecules each contain two building blocks. Others are made up of many simple sugar units joined together to form *polysaccharides,* such as plant starch.

Animals, including humans, synthesize a polysaccharide similar to starch called *glycogen.*

Lipids. **Lipids** include a number of substances, such as phospholipids and cholesterol, that have vital functions in cells. The most common lipids, however, are *fats.* Fats are used to build cell parts and to supply energy for cellular activities. In fact, fat molecules can supply more energy, gram for gram, than carbohydrate molecules.

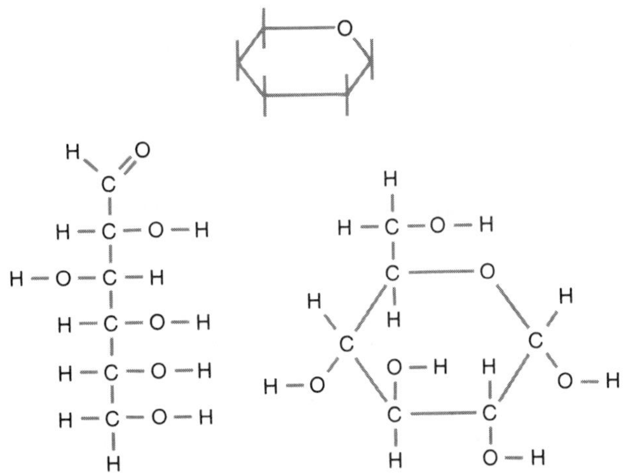

Like carbohydrates, fat molecules are composed of carbon, hydrogen, and oxygen atoms. They contain, however, a much smaller proportion of oxygen than do carbohydrates. This fact can be illustrated by the formula for the fat, tristearin, $C_{57}H_{110}O_6$.

The building blocks of fat molecules are **fatty acids** and **glycerol**. These smaller molecules are united so that each glycerol molecule is combined with 3 fatty acid molecules. The result is a single fat or *triglyceride* molecule, as shown in figure 2.12.

Although the glycerol portion of every fat molecule is the same, there are many kinds of fatty acids and, therefore, many kinds of fats. Fatty acid molecules differ in the lengths of their carbon atom chains and in the way the carbon atoms are combined. In some cases, the carbon atoms are all joined by single carbon-carbon bonds. A fat that contains this type of fatty acid is *saturated;* that is, each carbon atom is bound to as many hydrogen atoms as possible and is thus saturated with hydrogen atoms. Other fatty acids have one or more double bonds between carbon atoms, and the fats that contain them are *unsaturated.* Fat molecules with many double-bonded carbon atoms are called *polyunsaturated.*

Chart 2.4 lists three important groups of lipids and their characteristics.

Proteins. **Proteins** serve as structural materials, are often used as energy sources, and play vital roles in all metabolic processes by acting as **enzymes.** Enzymes are proteins that have special properties and can speed up specific chemical reactions without being changed or used up themselves. (Enzymes are discussed in more detail in chapter 4.)

Fig. 2.10 *Left,* some glucose molecules ($C_6H_{12}O_6$) have a straight chain of carbon atoms; *right,* more commonly, glucose molecules form a ring structure, which can be symbolized with this shape:

Fig. 2.11 (a) A simple sugar molecule, such as glucose, consists of one 6-carbon atom building block; (b) a double sugar molecule, such as sucrose (table sugar), consists of two building blocks; (c) a complex carbohydrate molecule, such as starch, consists of many building blocks.

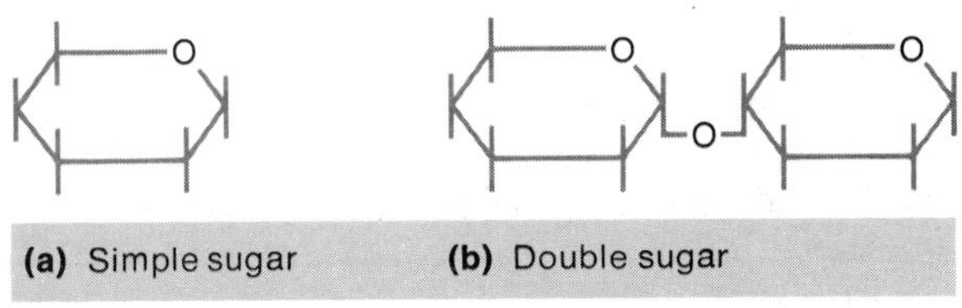

(a) Simple sugar

(b) Double sugar

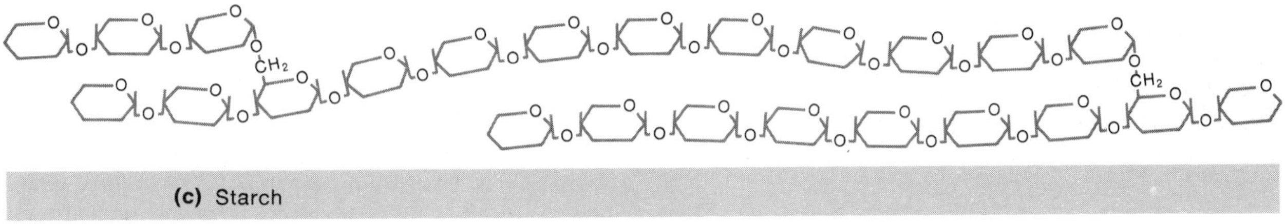

(c) Starch

Fig. 2.12 A fat molecule consists of a glycerol portion and three fatty acid portions.

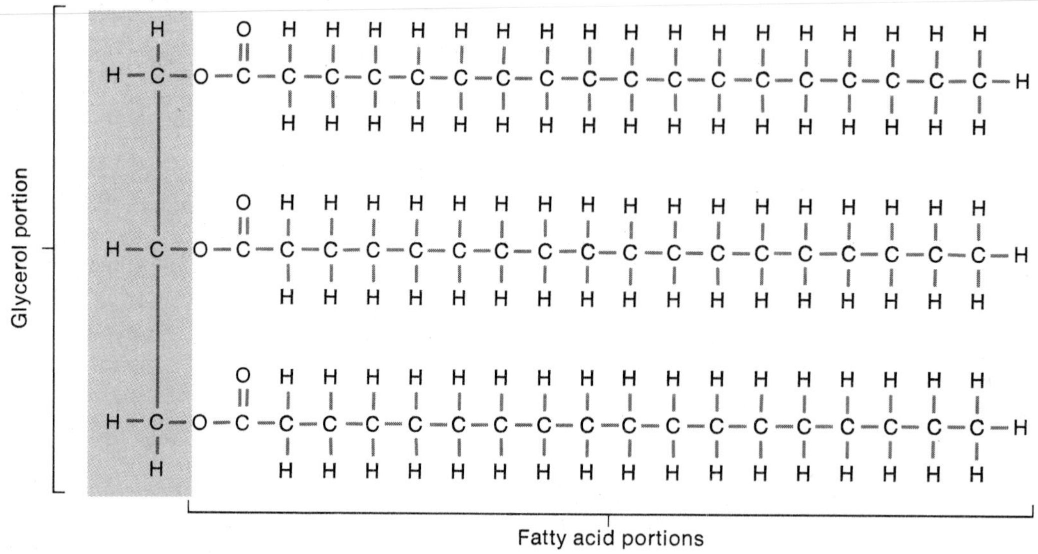

Glycerol portion

Fatty acid portions

Fig. 2.13 Some representative amino acids and their structural formulas. Each amino acid molecule has a particular shape due to the arrangement of its parts.

Amino acid	Structural formulas	Amino acid	Structural formulas
Alanine		Phenylalanine	
Valine		Tyrosine	
Cysteine		Histidine	

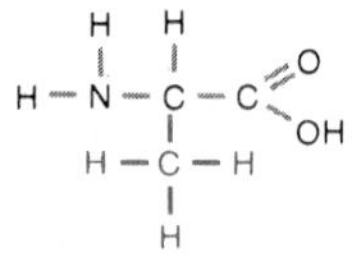

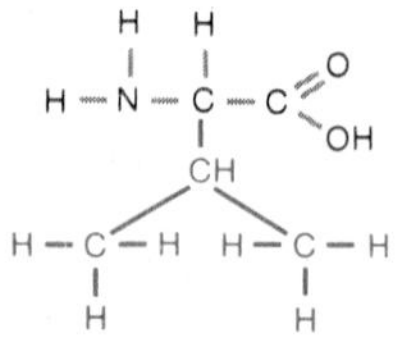

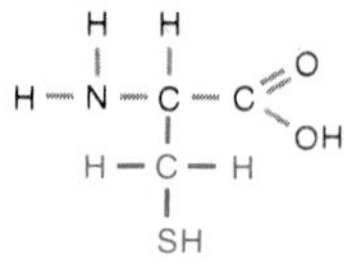

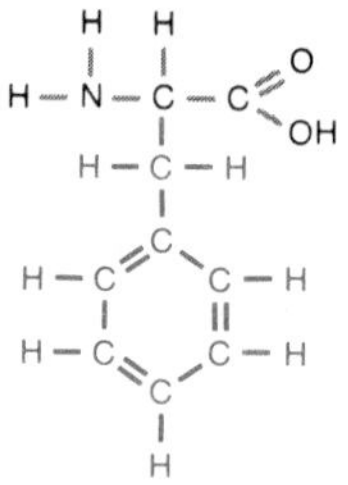

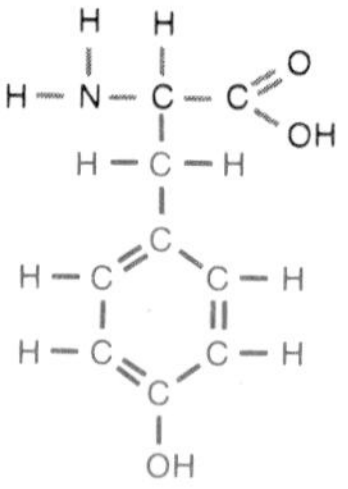

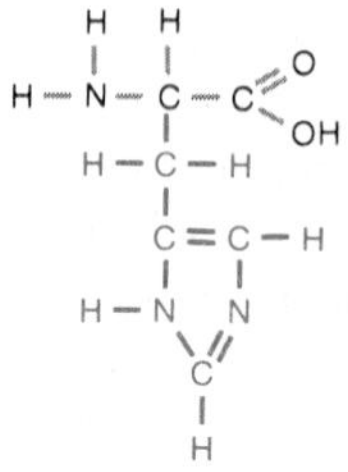

Chart 2.4 Some important groups of lipids

Group	Basic Molecular Structure	Characteristics
Triglycerides	Three fatty acid molecules bound to a glycerol molecule	Most common lipids in the human body; stored in fat tissue as an energy supply. Also provide heat insulation beneath the skin
Phospholipids	Two fatty acid molecules and a phosphate group bound to a glycerol molecule. (May also include a nitrogen-containing molecule attached to the phosphate group)	Used as structural components in cell membranes. Large amounts occur in the liver and parts of the nervous system
Steroids	Four interconnected rings of carbon atoms	Widely distributed in the body with a variety of functions. Include cholesterol, hormones of the adrenal cortex, sex hormones, bile acids, and vitamin D

Like carbohydrates and lipids, proteins contain atoms of carbon, hydrogen, and oxygen. In addition, they always contain nitrogen atoms, and sometimes sulfur atoms as well. The building blocks of proteins are smaller molecules called **amino acids.** (See fig. 2.13.) About 20 different amino acids occur commonly in proteins. They are joined together in chains, varying in length from less than 100 to more than 50,000 amino acids.

Each type of protein contains specific numbers and kinds of amino acids arranged in particular sequences. Consequently, different kinds of protein molecules have different shapes.

The special shapes of protein molecules can be altered by exposure to excessive amounts of heat, radiation, electricity, or various chemicals. When this occurs, the molecules lose their unique shapes, become disorganized, and are said to be *denatured*. At the same time, they lose their unique properties and behave differently. For example, the protein in egg white is denatured when it is heated. This treatment causes the egg white to change from a liquid to a solid—a change that cannot be reversed. Similarly, if cellular proteins are denatured, they are permanently changed and become nonfunctional. Such actions may threaten the life of cells.

Nucleic Acids. Of all the compounds present in cells, the most fundamental are the **nucleic acids.** These organic substances control cellular activities in very special ways.

Nucleic acid molecules are generally very large and complex. They contain atoms of carbon, hydrogen, oxygen, nitrogen, and phosphorus, which are bound into building blocks called **nucleotides.** Each nucleotide consists of a 5-carbon *sugar* (ribose or deoxyribose), a *phosphate group,* and one of several *organic bases* (fig. 2.14). A nucleic acid molecule consists of many nucleotides united in a chain, as shown in figure 2.15.

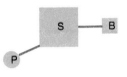

Fig. 2.14 A nucleotide consists of a 5-carbon sugar (S), a phosphate group (P), and an organic base (B).

Fig. 2.15 A nucleic acid molecule consists of nucleotides joined in a chain.

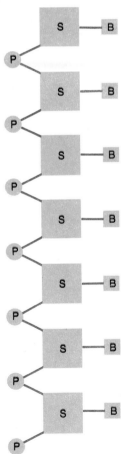

Chart 2.5 Organic compounds in cells

Compound	Elements Present	Building Blocks	Functions	Examples
Carbohydrates	C,H,O	Simple sugar	Energy, cell structure	Glucose, starch
Fats	C,H,O	Glycerol, fatty acids	Energy, cell structure	Tristearin, tripalmitin
Proteins	C,H,O,N (often S)	Amino acids	Cell structure, enzymes, energy	Albumins, hemoglobin
Nucleic acids	C,H,O,N,P	Nucleotides	Stores information, controls cell activities	RNA, DNA

From Hole, John W., Jr., *Human Anatomy and Physiology 3d ed.* © 1978, 1981, 1984 Wm. C. Brown Publishers, Dubuque, Iowa. All Rights Reserved. Reprinted by permission.

There are two major types of nucleic acids. One type, composed of molecules whose nucleotides contain ribose sugar, is called **RNA** (ribonucleic acid). The nucleotides of the second type contain deoxyribose sugar, and this type is called **DNA** (deoxyribonucleic acid).

DNA molecules store information in a kind of molecular code. This information tells cell parts how to construct specific protein molecules, and these proteins, acting as enzymes, are responsible for all of the metabolic reactions that occur in cells. RNA molecules help to synthesize these proteins. (Nucleic acids are discussed in more detail in chapter 4.)

Chart 2.5 summarizes these 4 groups of organic compounds.

---

1. *Compare the chemical composition of carbohydrates, lipids, proteins, and nucleic acids.*
2. *How does an enzyme affect a chemical reaction?*
3. *What is likely to happen to a protein molecule that is exposed to excessive heat or radiation?*
4. *What is the function of the nucleic acids?*

Chapter Summary

Introduction

Chemistry deals with the composition of substances and various changes in their composition.

Structure of Matter

1. Elements and atoms
   a. Naturally occurring matter is composed of about 90 elements.
   b. Some elements occur in pure form, but more frequently they are found in mixtures or are chemically combined in compounds.
   c. Elements are composed of atoms.
   d. Atoms of different elements vary in size, weight, and ways of interacting.

2. Atomic structure
   a. An atom consists of electrons surrounding an atomic nucleus, which contains protons and neutrons.
   b. Electrons are negatively charged, protons positively charged, and neutrons uncharged.
   c. A complete atom is electrically neutral.
   d. The atomic number of an element is equal to the number of protons in each atom; the atomic weight is equal to the number of protons plus the number of neutrons.

3. Bonding of atoms
   a. When atoms combine they gain, lose, or share electrons.
   b. Electrons are arranged in shells around atomic nuclei.
   c. Atoms with filled outer shells are inactive, while atoms with incompletely filled outer shells tend to gain, lose, or share electrons and thus achieve a stable structure.
   d. Atoms that lose electrons become positively charged; atoms that gain electrons become negatively charged.
   e. Ions with opposite charges are attracted to one another and become bound by electrovalent bonds; atoms that share electrons become bound by covalent bonds.

4. Molecules and compounds
   a. When 2 or more atoms of the same element are united, a molecule of that element is formed; when atoms of different elements are united, a molecule of a compound is formed.
   b. Molecules contain definite kinds and numbers of atoms.
   c. The numbers and kinds of atoms in a molecule can be represented by a molecular formula.
   d. The arrangement of atoms within a molecule can be represented by a structural formula.

5. Chemical reactions
   a. When a chemical reaction occurs, bonds between atoms are broken or created.
   b. Two kinds of chemical reaction are synthesis, in which larger molecules are formed from smaller particles; and decomposition, in which smaller particles are formed from larger molecules.

6. Acids and bases
   a. Compounds that ionize when they dissolve in water are electrolytes.
   b. Electrolytes that release hydrogen ions are acids, and those that release hydroxyl or other ions that react with hydrogen ions are bases.
   c. The concentration of hydrogen (H^+) and hydroxyl (OH^-) ions in a solution can be represented by pH.
   d. A solution with equal numbers of H^+ and OH^- is neutral, and has a pH of 7.0; a solution with more H^+ than OH^- is acidic; a solution with fewer H^+ than OH^- is basic.
   e. There is a tenfold difference in the hydrogen ion concentration between each whole number on the pH scale.

Chemical Constituents of Cells

Molecules containing carbon atoms are organic and are usually nonelectrolytes; those lacking carbon atoms are inorganic and are usually electrolytes.

1. Inorganic substances
   a. Water is the most abundant compound in cells and serves as a substance in which chemical reactions occur; it also transports chemicals and heat.
   b. Oxygen is used in releasing energy from glucose and other molecules.
   c. Carbon dioxide is produced when energy is released during respiration.
   d. Inorganic salts provide ions needed in metabolic processes.

2. Organic substances
   a. Carbohydrates provide much of the energy needed by cells; their basic building blocks are simple sugar molecules.
   b. Lipids supply energy also; their basic building blocks are molecules of glycerol and fatty acids.
   c. Proteins serve as structural materials, energy sources, and enzymes.
      (1) Enzymes speed up chemical reactions.
      (2) The building blocks of proteins are amino acids.
      (3) Different kinds of proteins vary in the numbers and kinds of aminio acids they contain and in the sequence in which these amino acids are arranged.
      (4) Protein molecules can be denatured.
   d. Nucleic acids are the most fundamental chemicals in cells because they control cell activities.
      (1) The 2 major kinds are RNA and DNA.
      (2) They are composed of nucleotides.
      (3) DNA molecules store information that tells a cell how to construct protein molecules, including enzymes; RNA molecules help to synthesize proteins.

Application of Knowledge

1. What acidic and alkaline substances do you encounter in your living activities? What acidic foods do you eat regularly? What alkaline foods do you eat?
2. Read the nutritional information on the packages of several food products. Which of these foods contain relatively high concentrations of carbohydrates? Of lipids? Of proteins?

Review Activities

1. Define *chemistry*.
2. Define *matter*.
3. Explain the relationship between elements and atoms.
4. List the 4 most abundant elements in the human body.
5. Describe the major parts of an atom.
6. Explain why a complete atom is electrically neutral.
7. Distinguish between atomic number and atomic weight.
8. Explain how electrons are arranged within an atom.
9. Distinguish between an electrovalent bond and a covalent bond.
10. Explain the relationship between molecules and compounds.
11. Distinguish between a molecular formula and a structural formula.
12. Describe two major types of chemical reactions.
13. Define *acid*, *base*, and *electrolyte*.
14. Explain what is meant by pH, and describe the pH scale.
15. Distinguish between organic and inorganic substances.
16. Describe the roles played by water and oxygen in the human body.
17. List several ions found in body fluids.
18. Describe the general characteristics of carbohydrates.
19. Distinguish between simple and complex carbohydrates.
20. Describe the general characteristics of lipids.
21. Distinguish between saturated and unsaturated fats.
22. Describe the general characteristics of proteins.
23. Define *enzyme*.
24. Explain how protein molecules may become denatured.
25. Describe the general characteristics of nucleic acids.
26. Explain the general functions of nucleic acids.

The Cell

The human body is composed entirely of cells, the products of cells, and various fluids. These cells represent the structural units of the body in that they are the building blocks from which all larger parts are formed. They are also the functional units, since whatever a body part can do is the result of activities within its cells.

Cells account for the shape, organization, and construction of the body and carry on its life processes. In addition, they can reproduce, and thus provide new cells needed for growth and development and for replacement of worn and injured tissues.

3

Chapter Outline

Chapter Objectives

After you have studied this chapter, you should be able to

1. Explain how cells vary from one another.

2. Describe the general characteristics of a composite cell.

3. Explain how the structure of a cell membrane is related to its function.

4. Describe each kind of cytoplasmic organelle and explain its function.

5. Describe the cell nucleus and its parts.

6. Explain how substances move through cell membranes by physical and physiological processes.

7. Describe the life cycle of a cell.

8. Explain how a cell reproduces.

9. Complete the review activities at the end of this chapter. Note that the items are worded in the form of specific learning objectives. You may want to refer to them before reading the chapter.

active transport   (ak′tiv trans′port)

centrosome   (sen′tro-sōm)

chromosome   (kro′mo-sōm)

cytoplasm   (si′to-plazm)

differentiation   (dif″er-en″she-a′shun)

diffusion   (dĭ-fu′zhun)

endocytosis   (en″do-si-to′sis)

endoplasmic reticulum
   (en′do-plaz′mik rĕ-tik′u-lum)

equilibrium   (e″kwĭ-lib′re-um)

facilitated diffusion   (fah-sil″′ĭ-tat′ed dĭ-fu′zhun)

filtration   (fil-tra′shun)

Golgi apparatus   (gol′je ap″ah-ra′tus)

lysosome   (li′so-sōm)

mitochondrion   (mi″′to-kon′dre-un); plural,
   mitochondria   (mi-″to-kon′dre-ah)

mitosis   (mi-to′sis)

nucleolus   (nu-kle′o-lus)

nucleus   (nu′kle-us)

osmosis   (oz-mo′sis)

permeable   (per′me-ah-bl)

phagocytosis   (fag″o-si-to′sis)

pinocytosis   (pi″no-si-to′sis)

ribosome   (ri′bo-sōm)

cyt-, cell: *cyt*oplasm—fluid that occupies the space between the cell membrane and the nuclear membrane.

endo-, within: *endo*plasmic reticulum—network of membranes within the cytoplasm.

hyper-, above: *hyper*tonic—a solution that has a greater concentration of dissolved particles than another solution.

hypo-, below: *hypo*tonic—a solution that has a lesser concentration of dissolved particles than another solution.

inter-, between: *inter*phase—stage that occurs between mitotic divisions of a cell.

iso-, equal: *iso*tonic—a solution that has a concentration of dissolved particles equal to that of another solution.

mit-, thread: *mit*osis—process during which threadlike chromosomes appear within a cell.

phag-, to eat: *phag*ocytosis—process by which a cell takes in solid particles.

pino-, to drink: *pino*cytosis—process by which a cell takes in tiny droplets of liquid.

-som, body: ribo*som*e—tiny, spherical organelle.

The estimated 75 trillion cells that make up an adult human body have much in common, yet those in different tissues vary in a number of ways. For example, they vary considerably in size.

Cells also vary in shape, and typically their shapes are closely related to their functions. For instance, nerve cells often have long, threadlike extensions that transmit nerve impulses from one part of the body to another. The epithelial cells that line the inside of the mouth serve to shield underlying cells. These protective cells are thin, flattened, and tightly packed, somewhat like the tiles of a floor. Muscle cells, which function to pull parts closer together, are slender and rodlike, with their ends attached to the parts they move. (See fig. 3.15.)

A Composite Cell

Since cells vary so greatly, it is not possible to describe a typical cell. For purposes of discussion, however, it is convenient to imagine that one exists. Thus, the cell shown in figure 3.1 and described in the following sections is not real. Instead, it is a composite cell—one that includes many known cell structures.

Commonly a cell consists of two major parts, one within the other and each surrounded by a thin membrane. The inner portion is called the *cell nucleus,* and it is enclosed by a *nuclear membrane.* A mass of fluid called *cytoplasm* surrounds the nucleus and is encircled by a cytoplasmic or *cell membrane.*

Within the cytoplasm are networks of membranes and membranes that mark off tiny, distinct parts called *cytoplasmic organelles.* These organelles perform specific metabolic functions necessary for cell survival. The nucleus, on the other hand, directs the overall activities of the cell.

---

1. *Give two examples to illustrate the idea that the shape of a cell is related to its function.*
2. *Name the two major parts of a cell.*
3. *What are the general functions of these two parts?*

Cell Membrane

The **cell membrane** is the outermost limit of the cell, but it is more than a simple envelope surrounding the cellular contents; it is an actively functioning part of the living material and many important metabolic reactions take place on its surface. The membrane is extremely thin—visible only with the aid of an electron microscope—but is flexible and somewhat elastic. It typically has a complex surface with many outpouchings and infoldings that provide extra surface area.

In addition to maintaining the wholeness of the cell, the membrane serves as a "gateway" through which chemicals enter and leave. This gate acts in a special manner; it allows some substances to pass and excludes others. When a membrane functions in this way, it is called *selectively permeable.* A *permeable* membrane, on the other hand, is one that allows all materials to pass through.

The mechanism by which the membrane accomplishes its selective function is not well understood. It is known, however, that the mechanism involves the chemical nature of the membrane—which is largely protein and lipid (usually phospholipid and cholesterol), with a small amount of carbohydrate.

The way molecules are positioned within the cell membrane is not well understood either. Evidence indicates, however, that the structural qualities of the membrane are due primarily to the molecules of phospholipids, which are arranged in two layers.

Although there appears to be only a few types of lipid molecules in the membrane, there are many kinds of proteins. Some of these protein molecules are quite large and seem to extend across the lipid layer; other kinds are located on the outer surface of the lipid layer and project outward from it; still other kinds reside on the cytoplasmic side of the membrane. (See fig. 3.2.)

The carbohydrates of the membrane all seem to be associated with the outer surface, where some of them are combined with lipids and others are combined with proteins.

Since different kinds of molecules are located on the inner and outer surfaces of the membrane, it is not surprising that the two surfaces have functional differences. For instance, various molecules on the outer surface function as receptor sites that can combine with specific chemicals, such as hormones. Many of the proteins of the inner surface function as enzymes that accelerate chemical reactions, such as those that help vital substances pass through the membrane.

Since the membrane is largely phospholipid, molecules that are soluble in lipids, such as oxygen and carbon dioxide, can pass through it easily. On the other hand, molecules of substances like water, which do not dissolve in lipids, cannot penetrate the phospholipid layers. Water and certain other small molecules, however, can pass through some regions where large protein molecules span the thickness of the membrane and create minute passageways or "pores."

Fig. 3.1 A composite cell and its major organelles.

Lysosome

Cell membrane

Nuclear membrane

Endoplasmic reticulum

Chromatin

Nucleolus

Ribosomes

Nucleus

Mitochondrion

Small vacuoles

Centrosome

Cytoplasm

Centrioles

Microtubules

Golgi apparatus

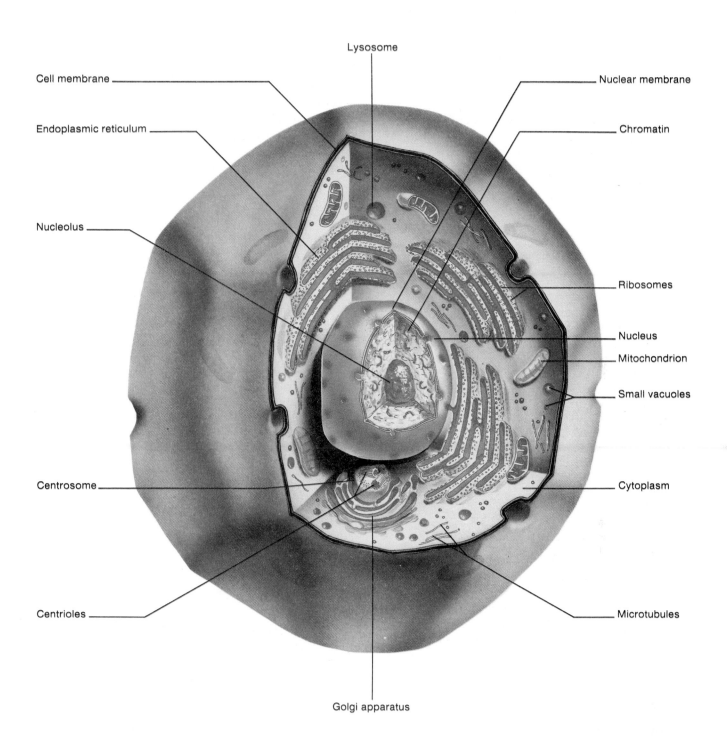

Fig. 3.2 The cell membrane is composed primarily of phospholipids with proteins scattered throughout the lipid layer and associated with its surfaces.

Fig. 3.3 Endoplasmic reticulum is a complex network of membranes and often has ribosomes attached to its surface.

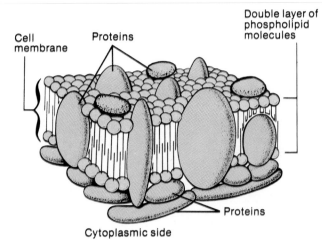

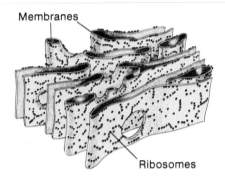

Other pores serve as selective channels which allow only particular substances to pass through. For example, some channels control the movements of sodium and potassium ions and thus play important roles in the functions of muscle and nerve cells.

Cytoplasm

When viewed through an ordinary laboratory microscope, **cytoplasm** usually appears as a clear liquid with specks scattered throughout. An electron microscope, which produces much greater magnification and resolution, however, reveals it to be highly structured and filled with networks of membranes and other organelles. (See fig. 3.1.)

The activities of a cell occur largely in its cytoplasm. It is there that food molecules are received, processed, and used. In other words, cytoplasm is a site of metabolic reactions in which the following **cytoplasmic organelles** play specific roles:

1. **Endoplasmic reticulum.** The endoplasmic reticulum is a complex network of interconnected membranes, which form flattened sacs and elongated canals. This membranous network is connected to the cell membrane, the nuclear membrane, and certain cytoplasmic organelles. It seems to function as a tubular communication system through which molecules can be transported from one cell part to another.

The endoplasmic reticulum also functions in the synthesis of certain types of molecules. Often a large proportion of its outer membranous surface has numerous tiny, spherical organelles called *ribosomes* attached to it, and for this reason it is termed *rough* endoplasmic reticulum. Endoplasmic reticulum that lacks such ribosomes is called *smooth* endoplasmic reticulum. (See fig. 3.3.)

Ribosomes synthesize proteins, and the resulting molecules may be transported through the tubules of the endoplasmic reticulum to the Golgi apparatus for secretion to the outside of the cell. Smooth endoplasmic reticulum contains enzymes involved in the manufacture of various lipid molecules and certain hormones (steroids).

2. **Ribosomes.** Although many ribosomes are attached to membranes of the endoplasmic reticulum, others occur as free particles scattered throughout the cytoplasm. In both cases, these tiny, spherical particles are composed of protein and RNA molecules and function in the synthesis of protein molecules. (This function is described in more detail in chapter 4.)

3. **Golgi apparatus.** The Golgi apparatus is usually located near the nucleus. It consists of a stack of flattened, membranous sacs whose membranes are continuous with those of the endoplasmic reticulum.

The Golgi apparatus is involved in the refining and "packaging" of proteins for secretion to the outside of the cell. For example, the Golgi apparatus seems to manufacture certain complex carbohydrate molecules that are combined with protein molecules synthesized by ribosomes. The resulting *glycoproteins* are then packaged in bits of the membrane which bud off from the Golgi apparatus and are moved to the cell membrane, where the contents are released to the outside. (See fig. 3.4.)

1. What is meant by a selectively permeable membrane?
2. Describe the chemical structure of a cell membrane.
3. What are the functions of the endoplasmic reticulum?
4. Describe the functions of the Golgi apparatus.

Fig. 3.4 The Golgi apparatus functions in packaging proteins, which then may be moved to the cell membrane and released as a secretion.

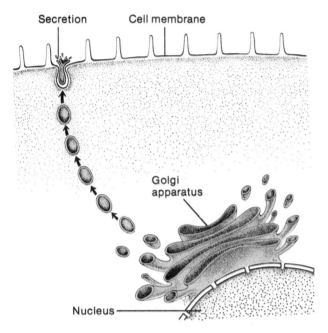

Fig. 3.5 What is the function of the enzymes within a mitochondrion?

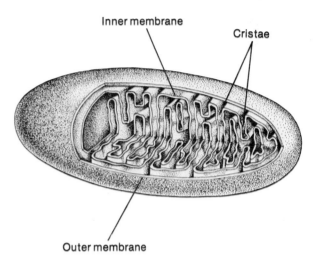

4. **Mitochondria.** Mitochondria are relatively large, fluid-filled sacs, which vary in size and shape. They often move about slowly in the cytoplasm and can produce others like themselves by dividing.

The membrane surrounding a mitochondrion has an outer and an inner layer. The inner layer is folded to form partitions called *cristae*. Connected to the cristae are enzymes that control some of the chemical reactions by which energy is released from glucose and other molecules. (See figs. 3.1 and 3.5.) The mitochondria also function in transforming this energy into a form that is usable by cell parts. (This energy-releasing function is described in more detail in chapter 4.)

5. **Lysosomes.** Lysosomes appear as tiny, membranous sacs. (See fig. 3.1.) They contain powerful enzymes that are capable of breaking down molecules of protein, carbohydrates, and nucleic acids. The enzymes function to digest various particles that enter cells. Certain white blood cells, for example, can engulf bacteria, which are then digested by the lysosomal enzymes. Consequently, white blood cells aid in preventing bacterial infections. Lysosomes also function in the destruction of worn cell parts. In fact, sometimes they destroy entire, injured cells that have been engulfed by scavenger cells.

Lysosomal digestive activity seems to be responsible for decreasing the size of body tissues at certain times. Such regression in size occurs in the uterus following birth, in the breasts after the weaning of an infant, and in skeletal muscles during periods of prolonged inactivity.

6. **Centrosome.** The centrosome is located in the cytoplasm near the Golgi apparatus and nucleus. It is nonmembranous and consists of two hollow cylinders called *centrioles* (fig. 3.1). The centrioles lie at right angles to each other and function in cell reproduction. During this process, they aid in the distribution of parts in the nucleus (chromosomes) to the newly forming daughter cells.

7. **Vacuoles.** Vacuoles are membranous sacs formed by an action of the cell membrane in which a portion of the membrane folds inward and pinches off. As a result, a small, bubblelike vacuole, containing some liquid or solid material that was outside the cell a moment before, appears in the cytoplasm. (See fig. 3.1.)

8. **Microfilaments** and **microtubules.** Two types of thin, threadlike processes likely to be found in the cytoplasm are microfilaments and microtubules. Microfilaments are tiny rods of protein, arranged in meshworks or bundles. They seem to function in causing various kinds of cellular movement. They are most highly developed in muscle cells as *myofibrils*, which help these cells to shorten or contract. (See chapter 9.)

Chart 3.1 Structure and function of cellular organelles

Organelle	Structure	Function
Cell membrane	Membrane composed of protein and lipid molecules	Maintains wholeness of cell and controls passage of materials in and out of cell
Endoplasmic reticulum	Network of interconnected membrane forming sacs and canals	Transports materials within the cell, provides attachment for ribosomes, and synthesizes lipids
Ribosomes	Particles composed of protein and RNA molecules	Synthesize proteins
Golgi apparatus	Group of flattened, membranous sacs	Synthesizes carbohydrates and packages protein molecules for secretion
Mitochondria	Membranous sacs with inner partitions	Release energy from food molecules and transform energy into usable form
Lysosomes	Membranous sacs	Digest substances that enter cells
Centrosome	Nonmembranous structure composed of two rodlike centrioles	Helps distribute chromosomes to daughter cells during cell reproduction
Vacuoles	Membranous sacs	Contain various substances that recently entered the cell
Microfilaments and microtubules	Thin rods and tubules	Provide support to cytoplasm and help move objects within the cytoplasm
Nuclear membrane	Porous membrane that separates nuclear contents from cytoplasm	Maintains wholeness of the nucleus and controls passage of materials between nucleus and cytoplasm
Nucleolus	Dense, nonmembranous body composed of protein and RNA molecules	Forms ribosomes
Chromatin	Fibers composed of protein and DNA molecules	Contains cellular information for carrying on life processes

From Hole, John W., Jr., *Human Anatomy and Physiology 3d ed.* © 1978, 1981, 1984 Wm. C. Brown Publishers, Dubuque, Iowa. All Rights Reserved. Reprinted by permission.

Microtubules are long, slender tubes composed of globular proteins. They are usually stiff and form an "internal skeleton" that helps to maintain the shape of a cell or its parts. Microtubules also aid in moving organelles. For instance, they appear during cellular reproduction and are involved in the distribution of chromosomes to the newly forming daughter cells. (Cell reproduction is described in more detail in a later section of this chapter.)

1. Describe a mitochondrion.
2. What is the function of a lysosome?
3. Distinguish between a centrosome and a centriole.
4. How do microfilaments and microtubules differ?

Cell Nucleus

A **cell nucleus** (fig. 3.1) usually is located near the center of the cytoplasm. It is a relatively large, spherical structure enclosed in a **nuclear membrane.** The membrane is porous and allows substances to move between the nucleus and the cytoplasm. The nucleus contains a fluid in which other structures float. These structures include the following:

1. **Nucleolus.** A nucleolus ("little nucleus") is a small, dense body composed largely of RNA and protein. It has no surrounding membrane and is formed in specialized regions of certain chromosomes. It functions in the production of ribosomes. Once the ribosomes are formed, they migrate through the pores in the nuclear membrane and enter the cytoplasm.

2. **Chromatin.** Chromatin consists of loosely coiled fibers. When the cell begins to undergo the reproductive process, these fibers become more tightly coiled and are transformed into tiny, rodlike *chromosomes*. Chromatin fibers are composed of protein and DNA molecules, which in turn contain the information that directs the cell in carrying out its life processes.

Chart 3.1 summarizes the structures and functions of the cellular organelles.

1. How are the nuclear contents separated from the cytoplasm of a cell?
2. What is the function of the nucleolus?
3. What is chromatin?

Movements through Cell Membranes

The cell membrane provides a surface through which various substances enter and leave the cell. These movements involve diffusion, facilitated diffusion, osmosis, filtration, active transport, and endocytosis.

Diffusion

Diffusion is the process by which molecules or ions become scattered or are spread from regions where they are in higher concentrations toward regions where they are in lower concentrations.

These molecules and ions are constantly moving at high speeds. Each particle travels in a separate path along a straight line until it collides and bounces off some other particle. Then it moves in another direction, only to collide again and change direction once more. Such motion is haphazard, but it accounts for the mixing of molecules that commonly occurs when different kinds of substances are put together.

For example, when some sugar (a solute) is put into a glass of water (a solvent), as illustrated in figure 3.6, the sugar will seem to remain at the bottom for a while. Then it slowly disappears into solution. As this happens, the moving water and sugar molecules are colliding with one another, and in time the sugar and water molecules will be evenly mixed. This mixing occurs by diffusion—the sugar molecules spread from where they are in higher concentration toward the regions where they are less concentrated. Eventually the sugar becomes uniformly distributed in the water. This condition is called *equilibrium*. Although the molecules continue to move after equilibrium is achieved, their concentrations no longer change.

To better understand how diffusion accounts for the movement of various molecules through a cell membrane, imagine a container of water that is separated into two compartments by a permeable membrane (fig. 3.7). This membrane has numerous pores that are large enough for water and sugar molecules to pass through. Sugar molecules are placed in one compartment (*A*) but not in the other (*B*). Although the sugar molecules will move in all directions, more will diffuse from compartment *A* (where they are in greater concentration) through the pores in the membrane and into compartment *B* (where they are in lesser concentration) than will move in the other direction. At the same time, water molecules will tend to diffuse from compartment *B* (where they are in

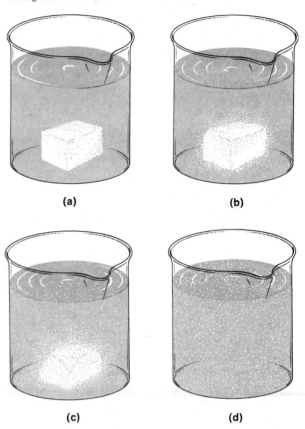

Fig. 3.6 (a, b, c) If a sugar cube is placed in water, it slowly disappears. As this happens, the sugar molecules diffuse from regions where they are more concentrated toward regions where they are less concentrated. (d) Eventually, the sugar molecules are distributed evenly throughout the water.

(a) **(b)**

(c) **(d)**

greater concentration) through the pores into compartment *A* (where they are in lesser concentration). Eventually, equilibrium will be achieved when there are equal numbers of water and sugar molecules in each compartment.

Similarly, oxygen molecules diffuse through cell membranes and enter cells if these molecules are more highly concentrated on the outside than on the inside. Carbon dioxide molecules, too, diffuse through cell membranes and leave cells if they are more concentrated on the inside than on the outside. Thus, it is by diffusion that oxygen and carbon dioxide molecules are exchanged between the air and the blood in the lungs and between the blood and the cells of various tissues.

Fig. 3.7 (1) The container is separated into two compartments by a permeable membrane. Compartment A contains water and sugar molecules, while compartment B contains only water molecules. (2) As a result of molecular motions, sugar molecules will tend to diffuse from compartment A into compartment B. Water molecules will tend to diffuse from compartment B into compartment A. (3) Eventually, equilibrium will be achieved.

• Water molecule
o Sugar molecule

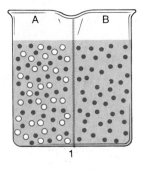

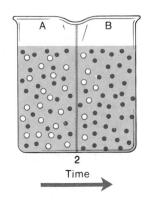

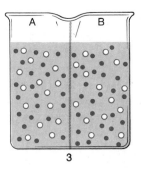

1 2 3

Time

Diffusion can be used to separate smaller molecules from larger ones in a liquid. This process, called *dialysis*, is employed in the artificial kidney, which is often used with patients suffering from kidney damage or failure. When this device is operating, blood from the patient is passed through a long, coiled tubing composed of porous cellophane. The size of the pores allows smaller molecules in blood, such as urea, to pass out through the cellophane, while larger molecules, like those of blood proteins, remain inside. The tubing is submerged in a tank of dialyzing fluid (wash solution) which contains varying concentrations of chemicals. The solution is low, for instance, in concentrations of substances that should leave the blood and higher in concentrations of those that should remain in the blood.

Since it is desirable for an artificial kidney to remove blood urea, the dialyzing fluid must have a lower concentration of urea than the blood does; it is also desirable to maintain the blood glucose concentration, so the concentration of glucose in the wash solution must be kept at least equal to that of the blood. Thus, by altering the concentrations of molecules in the dialyzing fluid, it is possible to control those that will diffuse out of the blood and those that will remain in it.

Facilitated Diffusion

Since a cell membrane is composed primarily of lipids, molecules that are soluble in lipids such as oxygen and carbon dioxide can readily diffuse through it. Other substances that are not soluble in lipids, such as water and certain ions, cannot easily diffuse through the membranes but may pass through membrane pores.

On the other hand, most sugars are insoluble in lipids and have molecular sizes that prevent them from passing through membrane pores. Some of these, including glucose, can still enter through the lipid portion of the membrane by a process called **facilitated diffusion.** In this process, the glucose combines with a special carrier molecule at the surface of the membrane. This union of glucose and carrier molecule creates a compound that is soluble in lipid and it can diffuse to the other side. There, the glucose portion is released, and the carrier molecule then can return to the opposite side of the membrane and pick up another glucose molecule.

Facilitated diffusion is similar to simple diffusion in that it only can cause movement of molecules from regions of higher concentration toward regions of lower concentration. The rate at which facilitated diffusion can occur, however, is limited by the number of carrier molecules in the cell membrane. (See fig. 3.8.)

Fig. 3.8 Some substances are moved through the cell membrane by facilitated diffusion in which carrier molecules transport the substance from a region of higher concentration to one of lower concentration.

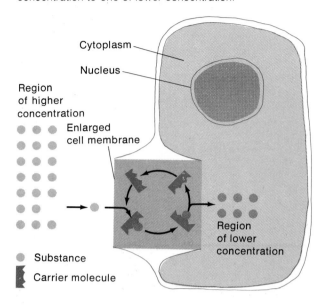

Cytoplasm

Nucleus

Region of higher concentration

Enlarged cell membrane

Region of lower concentration

● Substance

▌ Carrier molecule

Fig. 3.9 (1) The container is separated into two compartments by a selectively permeable membrane. At first, compartment A contains water and sugar molecules, while compartment B contains only water molecules. As a result of molecular motions, water molecules will tend to diffuse by osmosis from compartment B into compartment A. The sugar molecules will remain in compartment A because they are too large to pass through the pores of the membrane. (2) Since more water is entering compartment A than is leaving it, water will accumulate in this compartment, and the level of the liquid will rise on this side.

● Water molecule
○ Sugar molecule

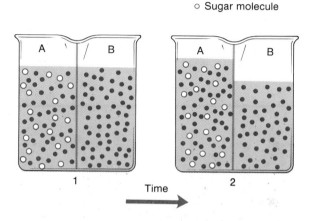

1   Time   2

Osmosis

Osmosis is a special case of diffusion. It occurs whenever *water* molecules diffuse from a region of higher water concentration (where the solute concentration is lower) to a region of lower water concentration (where the solute concentration is higher) through a selectively permeable membrane.

Ordinarily, the concentrations of water molecules are equal on either side of a cell membrane. Sometimes, however, the water on one side has more solute dissolved in it than the water on the other side. For example, if there is a greater concentration of glucose in the water outside a cell, there must be a lesser concentration of water there because the glucose molecules occupy space that would otherwise contain water molecules. Under such conditions, water molecules diffuse from inside the cell (where they are in higher concentration) toward the outside (where they are in lesser concentration).

This process, shown in figure 3.9, is similar to the one in figure 3.7, which illustrates the diffusion of water and glucose molecules through a permeable membrane. In the case of osmosis, however, the membrane is selectively permeable—water molecules pass through readily, but glucose molecules do not.

Note in figure 3.9 that as osmosis occurs, the volume of water on side *A* increases. This increase in volume would be resisted if pressure were applied to the surface of the liquid on side *A*. The amount of pressure needed to stop osmosis in such a case is called *osmotic pressure.* Thus, the osmotic pressure of a solution is potential pressure and is due to the presence of nondiffusable solute particles in that solution. Furthermore, the greater the number of solute particles in the solution, the greater the osmotic pressure of that solution.

When osmosis occurs, water tends to move toward the region of greater osmotic pressure. Since the amount of osmotic pressure depends upon the difference in concentration of solute particles on opposite sides of the membrane, the greater the difference, the greater the tendency for water to move toward the region of higher solute concentration.

If some cells were put into a water solution that had a greater concentration of solute particles (higher osmotic pressure) than did the cells, the cells would begin to shrink and are said to become *crenated.* A solution of this type, in which more water leaves a cell than enters it because the concentration of solute particles is greater outside the cell, is called **hypertonic.**

Conversely, if there is a greater concentration of solute particles inside a cell than in the water around it, water will diffuse into the cell. As this happens, water accumulates within the cell, and it begins to swell. A solution in which more water enters a cell than leaves it because of a lesser concentration of solute particles outside the cell is called **hypotonic.**

A solution that contains the same concentration of solute particles as a cell is said to be **isotonic** to that cell. In such a solution, water enters and leaves a cell in equal amounts, so the cell size remains unchanged.

It is important to control the concentration of solute in solutions that are infused into body tissues or blood. Otherwise, osmosis may cause cells to swell or shrink, and they may be damaged. For instance, if red blood cells are exposed to distilled water (which is hypotonic to them), water will diffuse into the cells and they will burst (hemolyze). On the other hand, if red blood cells are exposed to 0.9% NaCl solution (normal saline), the cells will remain unchanged because this solution is isotonic to human cells. Similarly, a 5% solution of glucose is isotonic to human cells.

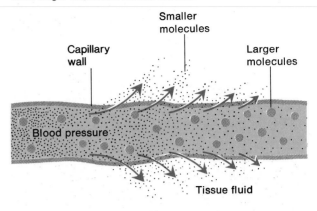

Fig. 3.10 In this example of filtration, smaller molecules are forced through the wall of a capillary by blood pressure, while larger molecules remain behind.

1. *What kinds of substances move most readily through a cell membrane by diffusion?*
2. *Explain the differences between diffusion and osmosis.*
3. *Distinguish between hypertonic, isotonic, and hypotonic solutions.*
4. *Explain how filtration occurs within the body.*

Filtration

The passage of substances through membranes by diffusion or osmosis is due to movements of the molecules of those substances. In other instances, molecules are forced through membranes by hydrostatic pressure or blood pressure that is greater on one side of the membrane than on the other. This process is called **filtration.**

In the laboratory, filtration is commonly used to separate solids from a liquid. One method is to pour a mixture onto some filter paper in a funnel. The paper serves as a porous membrane through which the liquid can pass, leaving the solids behind. Hydrostatic pressure, which is created by the weight of water on the paper, forces the water through to the other side.

Similarly, in the body, tissue fluid is formed when water and dissolved substances are forced out through the thin walls of blood capillaries, while larger particles such as blood protein molecules are left inside (fig. 3.10). The force for this movement comes from blood pressure, created largely by heart action, which is greater within the vessel than outside.

Active Transport

When molecules or ions pass through cell membranes by diffusion, facilitated diffusion, or osmosis, their net movements are from regions of higher concentration to regions of lower concentration. Sometimes, however, the net movement of particles passing through membranes is in the opposite direction; that is, they move from a region of lower concentration to one of higher concentration.

It is known, for example, that sodium ions can diffuse through cell membranes. Yet the concentration of these ions typically remains many times greater on the outside of cells than on the inside. Furthermore, sodium ions are continually moved from the regions of lower concentration (inside) to regions of higher concentration (outside). Movement of this type is called **active transport.** It depends on life processes within cells and requires energy. In fact, it is estimated that up to 40% of a cell's energy supply may be used for active transport of particles through its membranes.

Although the details are poorly understood, the mechanism of active transport involves specific carrier molecules, probably protein or lipoprotein molecules, found within cell membranes. (A lipoprotein

Fig. 3.11 What process is illustrated in this diagram?

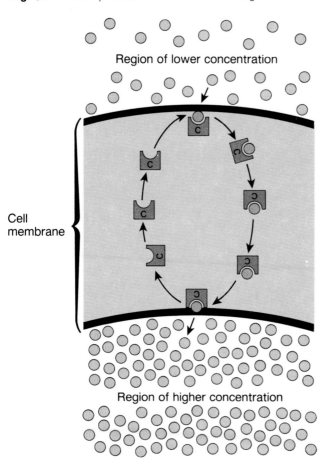

Region of lower concentration

Cell membrane

Region of higher concentration

The word **pinocytosis,** which means *cell drinking,* refers to the process by which cells take in tiny droplets of liquid from their surroundings. When this happens, a small portion of a cell membrane becomes indented. The open end of the tubelike part thus formed soon seals off, and the result is a small vacuole that becomes detached from the surface and moves into the cytoplasm.

Eventually the vacuolar membrane breaks down, and the liquid inside becomes part of the cytoplasm. In this way, a cell is able to take in water and dissolved particles, such as proteins, that otherwise might be unable to pass through the cell membrane because of large molecular size.

Phagocytosis (*cell eating*) describes a process that is essentially the same as pinocytosis. In phagocytosis, however, the material taken into the cell is solid rather than liquid.

Certain kinds of white blood cells are called *phagocytes* because they can take in tiny solid particles, such as bacterial cells, by phagocytosis. When such a phagocyte first encounters a particle, the object becomes attached to its cell membrane. Portions of the phagocytic cell project outward, surround the object, and slowly draw it inside. The part of the cell membrane surrounding the solid detaches from the surface, and a vacuole containing the particle is formed (fig. 3.12).

Commonly, a lysosome then combines with such a vacuole, and its digestive enzymes cause the enclosed solid to be decomposed. The products of this decomposition may then diffuse out of the lysosome and into the cell's cytoplasm. Any remaining residue usually is expelled from the cell.

Chart 3.2 summarizes the various types of movement through cell membranes.

---

1. *What type of mechanism is responsible for maintaining unequal concentrations of ions on opposite sides of a cell membrane?*
2. *What factors are necessary for the action of this mechanism?*
3. *What is the difference between pinocytosis and phagocytosis?*

---

consists of a lipid and a protein combined into one molecule.) As figure 3.11 shows, these carrier molecules combine at the membrane's surface with the particles being transported. Then they move through the membrane with their "passengers" attached. Once on the other side, the transported molecules are released, and the carriers return for another load.

Particles that are carried across cell membranes by active transport include various sugars and amino acids as well as a variety of ions such as those of sodium, potassium, calcium, and hydrogen.

Endocytosis

Another process by which substances may move through cell membranes is called **endocytosis.** In this case, molecules or other particles that are too large to enter a cell by diffusion or active transport may be conveyed within a tiny vacuole (vesicle) formed by a section of the cell membrane. Two forms of endocytosis are called pinocytosis and phagocytosis.

Fig. 3.12 A cell may take in a solid particle from its surroundings by phagocytosis.

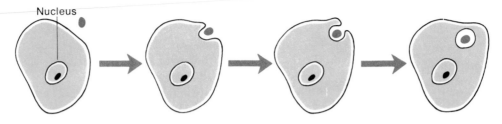

Nucleus

Chart 3.2 Movements through cell membranes

Process	Characteristics	Source of Energy	Example
Diffusion	Molecules or ions move from regions of higher concentration toward regions of lower concentration	Molecular motion	Exchange of gases in lungs
Facilitated diffusion	Molecules are moved through a membrane from a region of higher concentration to one of lower concentration by carrier molecules	Molecular motion	Movement of glucose through a cell membrane
Osmosis	Water molecules move from regions of higher concentration to regions of lower concentration through a selectively permeable membrane	Molecular motion	Water enters a cell placed in distilled water
Filtration	Molecules are forced from regions of higher pressure to regions of lower pressure	Blood pressure	Molecules leave blood capillaries
Active transport	Molecules or ions are carried through membranes by other molecules	Cellular energy	Movement of various ions and amino acids through membranes
Pinocytosis	Membrane acts to engulf minute droplets of liquid from surroundings	Cellular energy	Membrane forms vacuole containing liquid and dissolved particles
Phagocytosis	Membrane acts to engulf solid particles from surroundings	Cellular energy	White blood cell membrane engulfs bacterial cell

From Hole, John W., Jr., *Human Anatomy and Physiology 3d ed.* © 1978, 1981, 1984 Wm. C. Brown Publishers, Dubuque, Iowa. All Rights Reserved. Reprinted by permission.

Life Cycle of a Cell

The series of changes that a cell undergoes from the time it is formed until it reproduces is called its *life cycle*. Superficially, this cycle seems rather simple—a newly formed cell grows for a time and then splits in half to form two daughter cells, which in turn may grow and reproduce. Yet the details of the cycle are quite complex, involving mitosis, cytoplasmic division, interphase, and differentiation.

Mitosis

Cell reproduction involves the dividing of a cell into two portions and includes two separate processes: (1) division of the nuclear parts, which is called **mitosis;** and (2) division of the cytoplasm.

Division of the nuclear parts by mitosis is, of necessity, very precise, since the nucleus contains information, in the form of DNA molecules, that "tells" cell parts how to function. Each daughter cell must have a copy of this information in order to survive. Thus, the DNA molecules of the parent cell must be duplicated, and the duplicate sets must be distributed equally to the daughter cells. Once this has been accomplished, the cytoplasm and its parts can be divided.

Although mitosis is often described in terms of stages, the process is really a continuous one without marked changes between one step and the next. (See fig. 3.13.) The idea of stages is useful, however, to indicate the sequence in which major events occur. These stages include the following:

1. **Prophase.** One of the first indications that a cell is going to reproduce is the appearance of *chromosomes*. These structures form from chromatin in the nucleus as chromatin threads condense into tightly coiled, rodlike parts. The resulting chromosomes contain DNA and protein molecules. Sometime earlier (during interphase, which is described on page 48) the DNA molecules became duplicated, and each chromosome is

Fig. 3.13 Mitosis is a continuous process during which the nuclear parts of a cell are divided into two equal portions. After reading about mitosis, identify the phases of the process in this diagram.

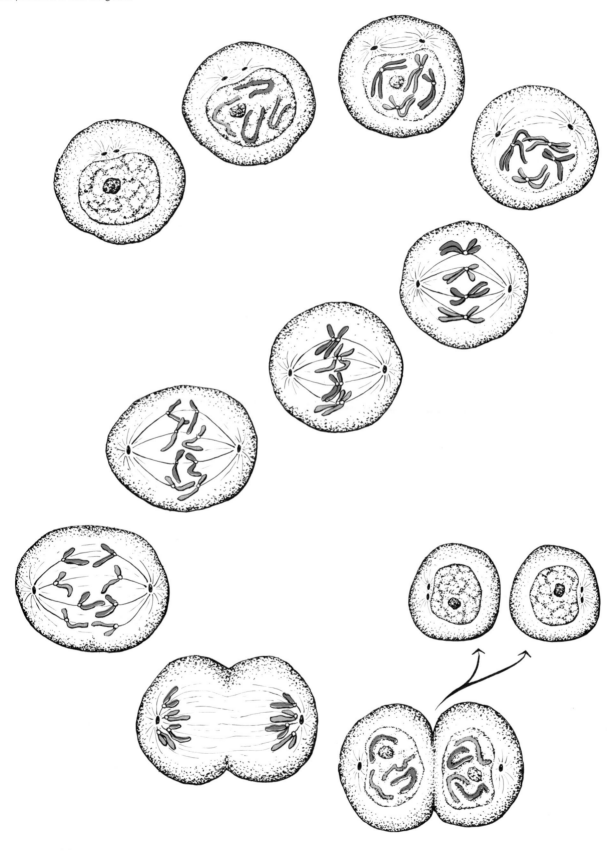

consequently composed of two identical portions (chromatids). These parts are temporarily fastened together by a portion of the chromosome called the *centromere.*

The centrioles of the centrosome move apart and travel to opposite sides of the cytoplasm. Soon the nuclear membrane and the nucleolus disappear. Microtubules (spindle fibers) appear in the cytoplasm, and they become associated with the centrioles and chromosomes.

2. **Metaphase.** The chromosomes line up in an orderly fashion about midway between the centrioles, apparently as a result of microtubule activity. Microtubules also are attached to the centromeres so that a microtubule accompanying one centriole is attached to one side of a centromere, while a microtubule accompanying the other centriole is attached to the other side. Soon the centromeres divide, and the identical chromosome parts, which have been fastened together, become separated as individual chromosomes.

3. **Anaphase.** The separated chromosomes now move in opposite directions, and once again, the movement seems to result from microtubule activity. In this case, the microtubules shorten and pull their attached chromosomes toward the centrioles at opposite sides of the cell.

4. **Telophase.** The final stage of mitosis is said to begin when the chromosomes complete their migration toward the centrioles. It is much like prophase, but in reverse. As the chromosomes approach the centrioles, they begin to elongate and change from rodlike into threadlike structures. A nuclear membrane forms around each chromosome set, and nucleoli appear within the newly formed nuclei. Finally, the microtubules disappear and the centrioles become duplicated. Telophase is accompanied by cytoplasmic division in which the cytoplasm is split into two portions. (See fig. 3.14.)

Cytoplasmic Division

The mechanism responsible for cytoplasmic division is not well understood, but the cell membrane begins to constrict during telophase and then pinches inward between the daughter nuclei. As a result, about half of the cytoplasmic organelles are distributed to each new cell.

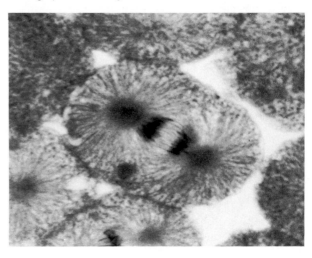

Fig. 3.14 Note the two groups of darkly stained chromosomes within the cell near the center of this micrograph. What stage of mitosis does this cell illustrate?

Although the newly formed cells may differ slightly in size and number of cytoplasmic parts, they have identical chromosomes and thus contain identical DNA information. Except for size, they are copies of the parent cell.

Interphase

Once formed, the daughter cells usually begin growing. This requires that the young cells obtain nutrients, utilize them in the manufacture of new living materials, and synthesize many vital compounds. At the same time, various cell parts become duplicated. In the nucleus, the chromosomes may be doubling (at least in cells that will soon reproduce); and in the cytoplasm, new ribosomes, lysosomes, mitochondria, and membranes are appearing. This stage in the life cycle is called **interphase,** and it lasts until the cell begins to undergo mitosis.

Many kinds of cells within the body are constantly growing and reproducing. Such activity is responsible for the growth and development of an embryo into a child, and a child into an adult. It is also responsible for the replacement of cells that are worn out or lost due to injury or disease.

Cell Differentiation

Since all body cells are formed by mitosis and contain the same DNA information, it might be expected that they look and act alike; obviously, they do not.

Fig. 3.15 During development, the numerous body cells are produced from a single fertilized egg cell by mitosis. As these cells undergo differentiation, they become different kinds of cells that carry on specialized functions.

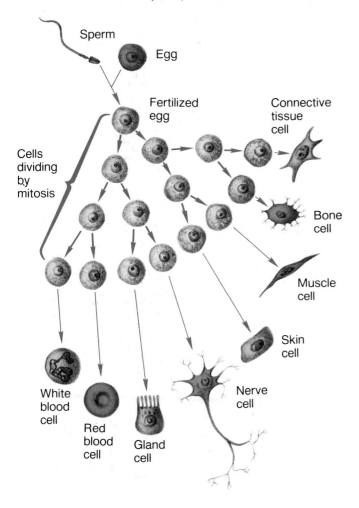

A human begins life as a single cell—a fertilized egg cell. This cell reproduces to form 2 daughter cells; they in turn divide into 4 cells, the 4 become 8, and so forth. Then, sometime during development, the cells begin to *specialize*. That is, they develop special structures or begin to function in different ways. Some become skin cells, others become bone cells, and still others become nerve cells. (See fig. 3.15.)

The process by which cells develop different characteristics in structure and function is called **differentiation**. The mechanism responsible for this phenomenon is not well understood, but it seems to involve the repression of some of the DNA information. Thus, the DNA information needed for general cell activities may be "switched on" in both nerve and bone cells.

The information related to specific bone cell functions, however, may be repressed or "switched off" in the nerve cells. Similarly, the information related to specific nerve cell functions may be repressed in bone cells.

Although the mechanism of differentiation is obscure, the results are obvious. Cells of many kinds are produced, and each kind carries on specialized functions: skin cells protect underlying tissues, red blood cells carry oxygen, nerve cells transmit impulses. Each type of cell somehow helps the others, and aids in the survival of the organism.

---

1. *Why is it important that the division of nuclear materials during mitosis be so precise?*
2. *Describe the events that occur during mitosis.*
3. *Name the process by which some cells become muscle cells and others become nerve cells.*

Sometimes cells give rise to cancers. Although cancers represent a group of closely related diseases rather than a single disease, these conditions have certain characteristics in common. They include the following:

1. **Hyperplasia.** Hyperplasia is the uncontrolled reproduction of cells. Although the rate of reproduction among cancer cells is usually unchanged, they are not responsive to normal controls on cell numbers. As a result, cancer cells eventually give rise to large cell populations.

2. **Anaplasia.** The word anaplasia is used to describe the appearance of abnormalities in cellular structure. Typically, cancer cells resemble undifferentiated or primitive cells. That is, they fail to develop the specialized structure of the kind of cell they represent, and they fail to carry on the usual specialized function.

3. **Metastasis.** Metastasis is a tendency to spread. Normal cells usually stick together in groups of similar kinds. Cancer cells often become detached from their cellular mass and may then be carried away from their place of origin. Consequently, new cancerous growths may be established in other parts of the body.

Chapter Summary

Introduction

The shapes of cells are closely related to their functions.

A Composite Cell

A cell includes a nucleus, cytoplasm, and a cell membrane.

Cytoplasmic organelles perform specific vital functions, while the nucleus controls the overall activities of the cell.

1. Cell membrane
   a. The cell membrane forms the outermost limit of the living material.
   b. It acts as a selectively permeable passageway.
   c. It is composed of protein, lipid, and carbohydrate molecules.
   d. Some substances pass through the lipid portion; others pass through protein "pores."

2. Cytoplasm
   a. Endoplasmic reticulum provides a tubular communication system in the cytoplasm and functions in the synthesis of lipids and proteins.
   b. Ribosomes function in protein synthesis.
   c. Golgi apparatus manufactures carbohydrates and packages glycoproteins for secretion.
   d. Mitochondria contain enzymes involved with releasing energy from food molecules.
   e. Lysosomes contain digestive enzymes that can destroy substances that enter cells.
   f. The centrosome aids in the distribution of chromosomes during cell reproduction.
   g. Vacuoles contain substances that recently entered the cell.
   h. Microfilaments and microtubules aid cellular movements and provide support and stability to cytoplasm.

3. Cell nucleus
   a. The nucleus is enclosed in a nuclear membrane.
   b. It contains a nucleolus that functions in the production of ribosomes.
   c. It contains chromatin that is composed of loosely coiled fibers of protein and DNA and become chromosomes during cellular reproduction.

Movements through Cell Membranes

The cell membrane provides a surface through which substances enter and leave a cell.

1. Diffusion
   a. Diffusion is a scattering of molecules or ions from regions of higher toward regions of lower concentration.
   b. It is responsible for exchanges of oxygen and carbon dioxide within the body.

2. Facilitated diffusion
   a. Facilitated diffusion involves the use of carrier molecules in the cell membrane.
   b. This process can only move substances from regions of higher toward regions of lower concentration.

3. Osmosis
   a. Osmosis is a case of diffusion in which water molecules diffuse from regions of higher water concentration to regions of lower water concentration through a selectively permeable membrane.
   b. Osmotic pressure increases as the number of particles dissolved in a solution increases.
   c. Cells lose water when placed in hypertonic solutions and gain water when placed in hypotonic solutions.
   d. Solutions that contain the same concentration of solute particles as a cell are called isotonic.

4. Filtration
   a. Filtration involves movement of molecules from regions of higher toward regions of lower hydrostatic or blood pressure.
   b. Blood pressure causes filtration through capillary walls.

5. Active transport
   a. Active transport is responsible for movement of molecules or ions from regions of lower to regions of higher concentration.
   b. It requires cellular energy and involves action of carrier molecules.

6. Endocytosis
   a. Endocytosis is a process by which relatively large particles may be conveyed through a cell membrane.
   b. A cell membrane may engulf tiny droplets of liquid by pinocytosis.
   c. Solid particles may be engulfed by phagocytosis.

Life Cycle of a Cell

The life cycle of a cell includes mitosis, cytoplasmic division, interphase, and differentiation.

1. Mitosis
   a. Mitosis involves the division and distribution of nuclear parts to daughter cells during cellular reproduction.
   b. The stages of mitosis include prophase, metaphase, anaphase, and telophase.

2. Cytoplasmic division is a process by which cytoplasm is divided into two portions following mitosis.

3. Interphase
   a. Interphase is the stage in the life cycle when a cell grows and forms new organelles.
   b. It terminates when the cell begins to undergo mitosis.

4. Cell differentiation involves the development of specialized structures and functions.

Application of Knowledge

1. Which process—diffusion, osmosis, or filtration—is most closely related to each of the following situations?
   a. A person is given an injection of isotonic glucose solution.
   b. The concentration of urea in the wash solution of an artificial kidney is decreased.

2. What characteristic of cell membranes may account for the observation that fat-soluble substances like chloroform and ether cause rapid effects upon cells?

3. A person who has been exposed to excessive amounts of X ray may develop a decrease in white blood cell number and an increase in susceptibility to infections. In what way are these effects related?

Review Activities

1. Describe how the shapes of nerve and muscle cells are related to their functions.

2. Name the major portions of a cell and describe their relationship to one another.

3. Discuss the structure and functions of a cell membrane.

4. Define selectively permeable.

5. Describe the structure and functions of each of the following:
   a. endoplasmic reticulum
   b. ribosome
   c. Golgi apparatus
   d. mitochondrion
   e. lysosome
   f. centrosome
   g. vacuole
   h. microfilament and microtubule

6. Describe the structure of the nucleus and the functions of its parts.

7. Explain how diffusion aids in the exchange of gases within the body.

8. Distinguish between diffusion and facilitated diffusion.

9. Define osmosis.

10. Define osmotic pressure.

11. Distinguish between solutions that are hypertonic, hypotonic, and isotonic.

12. Define filtration.

13. Explain how filtration aids the movement of substances through capillary walls.

14. Explain the function of carrier molecules in active transport.

15. Distinguish between pinocytosis and phagocytosis.

16. List the phases in the life cycle of a cell.

17. Name the two processes included in cell reproduction.

18. Describe the major events of mitosis.

19. Explain what happens during interphase.

20. Define differentiation.

Cellular Metabolism

Cellular metabolism includes all the chemical reactions that occur within cells. These reactions, for the most part, involve the utilization of food substances and are of two major types. In one type of reaction, molecules of nutrients are converted into simpler forms by changes accompanied by the release of energy. In the other type, the nutrient molecules are used in constructive processes that produce cell structural and functional materials or molecules that store energy.

In either case, metabolic reactions are controlled by proteins called *enzymes,* which are synthesized in cells. The production of these enzymes is, in turn, controlled by genetic information held within molecules of DNA, which instructs a cell how to synthesize specific kinds of enzyme molecules.

4

Chapter Objectives

After you have studied this chapter, you should be able to

1. Define *anabolic* and *catabolic metabolism.*

2. Explain how enzymes control metabolic processes.

3. Explain how chemical energy is released by respiratory processes.

4. Describe how energy is made available for cellular activities.

5. Describe the general metabolic pathways of carbohydrates, lipids, and proteins.

6. Explain how genetic information is stored within nucleic acid molecules.

7. Explain how genetic information is used in the control of cellular processes.

8. Describe how DNA molecules are duplicated.

9. Complete the review activities at the end of this chapter. Note that the items are worded in the form of specific learning objectives. You may want to refer to them before reading the chapter.

aerobic respiration
(a″-er-o′bik res″pĭ-ra′shun)

anabolic metabolism
(an″ah-bol′ik mĕ-tab′o-lizm)

anaerobic respiration
(an″a-er-o′bik res″pĭ-ra′shun)

catabolic metabolism
(kat″ah-bol′ik mĕ-tab′o-lizm)

dehydration synthesis
(de″hi-dra′shun sin′thĕ-sis)

enzyme   (en′zĭm)

gene   (jēn)

hydrolysis   (hi-drol′ĭ-sis)

oxidation   (ok″sĭ-da′shun)

substrate   (sub′strāt)

an-, without: *an*aerobic respiration—respiratory process that proceeds without oxygen.

ana-, up: *ana*bolic metabolism—cellular processes in which smaller molecules are used to build larger ones.

cata-, down: *cata*bolic metabolism—cellular processes in which larger molecules are broken down into smaller ones.

de-, undoing: *de*amination—process by which nitrogen-containing portions of amino acid molecules are removed.

mut-, change: *mut*ation—change in the genetic information of a cell.

zym-, causing to ferment: en*zym*e—protein that initiates or speeds up a chemical reaction without being changed itself.

Although a living cell may appear to be idle, it is actually very active because the numerous metabolic processes needed to maintain its life are occurring all the time.

Metabolic Processes

Metabolic processes can be divided into two major types. One type, called *anabolic,* involves the buildup of larger molecules from smaller ones and utilizes energy. The other, called *catabolic,* involves the breakdown of larger molecules into smaller ones and releases energy.

Anabolic Metabolism

Anabolic metabolism includes all the constructive processes used to manufacture substances needed for cellular growth and repair. For example, cells often join simple sugar molecules (monosaccharides) to form larger molecules of glycogen by an anabolic process called *dehydration synthesis.* In this process, the larger carbohydrate molecule is built by bonding monosaccharide molecules together into a complex chain. As the adjacent monosaccharide units are joined, an –OH (hydroxyl group) from one monosaccharide molecule and an –H (hydrogen atom) from another are removed. These particles react to produce a water molecule, and the monosaccharides are united by a shared oxygen atom, as shown in figure 4.1. As this process is repeated, the molecular chain becomes longer.

Similarly, glycerol and fatty acid molecules are joined by dehydration synthesis in fat tissue cells to form fat molecules. In this case, 3 hydrogen atoms are removed from a glycerol molecule, and an –OH group is removed from each of 3 fatty acid molecules, as shown in figure 4.2. The result is 3 water molecules and a single fat molecule, whose glycerol and fatty acid portions are bound by shared oxygen atoms.

Cells also build protein molecules by joining amino acids by means of dehydration synthesis. When 2 amino acid molecules are united, an –OH is removed from one and a hydrogen atom from the other. A water molecule is formed, and the amino acid molecules are joined by a bond between a carbon atom

and a nitrogen atom (fig. 4.3). This type of bond, which is called a *peptide bond,* holds the amino acids together. Two amino acids bound together form a *dipeptide,* while many joined in a chain form a *polypeptide.* Generally, a polypeptide consisting of 100 or more amino acid molecules is called a *protein.*

Catabolic Metabolism

Physiological processes in which larger molecules are broken down into smaller ones constitute **catabolic metabolism.** An example of such a process is *hydrolysis,* which can bring about the decomposition of carbohydrates, lipids, and proteins.

When a molecule of one of these substances is hydrolyzed, a water molecule is used, and the original molecule is split into two simpler parts. The hydrolysis of a disaccharide such as sucrose, for instance, results in molecules of 2 monosaccharides, glucose and fructose.

$$C_{12}H_{22}O_{11} + H_2O \rightarrow C_6H_{12}O_6 + C_6H_{12}O_6$$
(sucrose) (water) (glucose) (fructose)

In this case, the bond between the simple sugars within the sucrose molecule is broken, and the water molecule supplies a hydrogen atom to one sugar molecule and a hydroxyl group to the other. Thus, hydrolysis is the reverse of dehydration synthesis, illustrated in figures 4.1, 4.2, and 4.3.

Hydrolysis

Disaccharide + Water $\rightleftharpoons$ Monosaccharide + Monosaccharide

Dehydration

synthesis

Hydrolysis, more commonly called *digestion,* occurs in various regions of the digestive tract and is responsible for the breakdown of carbohydrates into monosaccharides, fats into glycerol and fatty acids, and proteins into amino acids. (Digestion is discussed in more detail in chapter 13.)

---

1. *What general functions does anabolic metabolism serve? Catabolic metabolism?*
2. *Name a substance formed by the anabolic metabolism of monosaccharides. Of amino acids. Of glycerol and fatty acids.*
3. *Distinguish between dehydration synthesis and hydrolysis.*

Fig. 4.1 What type of reaction is represented in this diagram?

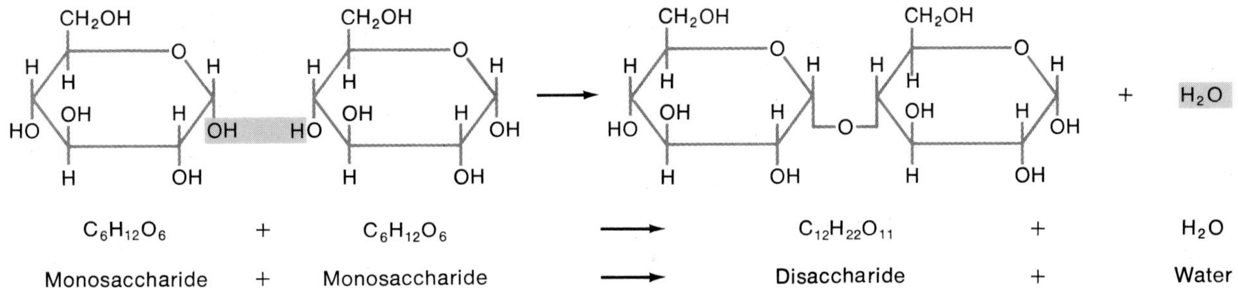
$C_6H_{12}O_6$ + $C_6H_{12}O_6$ → $C_{12}H_{22}O_{11}$ + H_2O
Monosaccharide + Monosaccharide → Disaccharide + Water

Fig. 4.2 A glycerol molecule and three fatty acid molecules may be joined by dehydration synthesis to form a fat molecule.

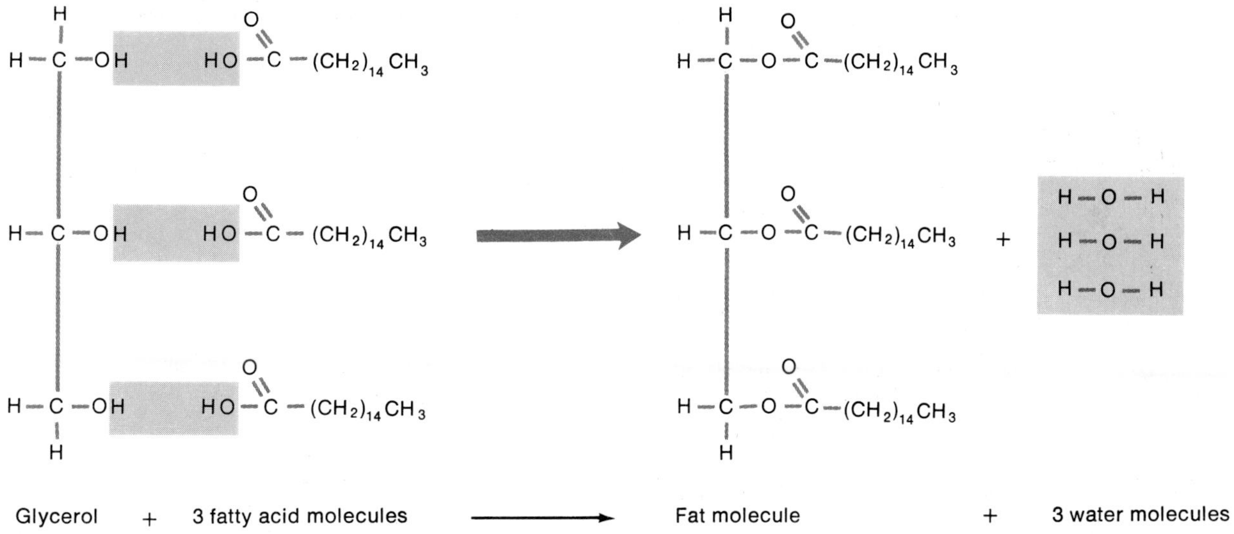
Glycerol + 3 fatty acid molecules → Fat molecule + 3 water molecules

Fig. 4.3 When two amino acid molecules are united by dehydration synthesis, a peptide bond is formed between a carbon atom and a nitrogen atom.

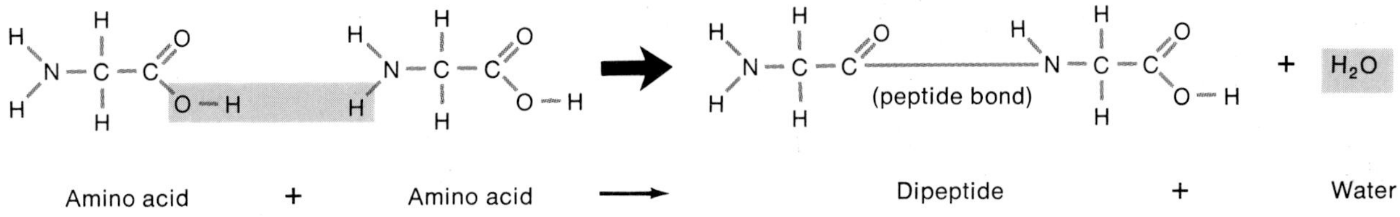
(peptide bond)
Amino acid + Amino acid → Dipeptide + Water

Control of Metabolic Reactions

Although different kinds of cells may conduct specialized metabolic processes, all cells perform certain basic reactions, such as the buildup and breakdown of carbohydrates, lipids, proteins, and nucleic acids. These reactions actually include hundreds of very specific chemical changes which must occur in orderly fashion.

Enzymes and Their Actions

Like other chemical reactions, metabolic reactions generally require a certain amount of energy, such as heat, before they occur. The temperature conditions that exist in cells, however, are usually too mild to promote the reactions needed to support life; but this is not a problem because cells contain enzymes.

Enzymes are proteins that promote chemical reactions within cells by somehow lowering the amount of energy (activation energy) needed to initiate these reactions. Thus, in the presence of enzymes, metabolic reactions are speeded up. Enzymes are needed in very small quantities because as they function, they are not changed or used up, and can, therefore, function again and again.

The reaction promoted by a particular enzyme is very specific. Each enzyme acts only on a particular substance, which is called its **substrate.** For example, the substrate of one enzyme called *catalase* is hydrogen peroxide, a harmful substance sometimes produced in cells. This enzyme's only function is to cause the decomposition of hydrogen peroxide into water and oxygen. In this way it helps to prevent an accumulation of hydrogen peroxide that might damage cells.

Cellular metabolism includes hundreds of different chemical reactions, and each of these reactions is controlled by a specific kind of enzyme. Thus, there must be hundreds of different kinds of enzymes present in every cell. How each enzyme "recognizes" its specific substrate is not well understood, but this ability seems to involve the shapes of molecules. That is, each enzyme's polypeptide chain is thought to be twisted and coiled into a unique three-dimensional form that fits the special shape of a substrate molecule. In short, an enzyme molecule is thought to fit a molecule of its substrate much as a key fits a particular lock.

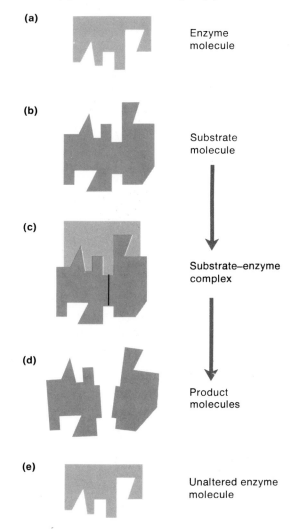

Fig. 4.4 The shape of an enzyme molecule (a) fits the shape of the substrate molecule (b). When the substrate molecule becomes temporarily combined with the enzyme (c) a chemical reaction occurs. The result is some product molecules (d) and an unaltered enzyme (e).

(a) Enzyme molecule

(b) Substrate molecule

(c) Substrate–enzyme complex

(d) Product molecules

(e) Unaltered enzyme molecule

During an enzyme-controlled reaction, particular regions of the enzyme molecule called *active sites* temporarily combine with portions of the substrate, creating a substrate-enzyme complex. At the same time, the interaction between the molecules seems to distort or strain chemical bonds within the substrate, increasing the likelihood of a change in the substrate molecule. When the substrate is changed, the product of the reaction appears, and the enzyme is released unaltered. (See fig. 4.4.)

Often the protein portion of an enzyme molecule is inactive until it is combined with an additional substance. This nonprotein part is needed to complete the proper shape of the active site of the enzyme molecule or help bind the enzyme to its substrate. Such a substance is called a **cofactor**. A cofactor may be an ion of an element such as copper, iron, or zinc, or it may be a relatively small organic molecule, in which case it is called a **coenzyme**. Most coenzymes are obtained from vitamins.

Vitamins are essential substances that cannot be synthesized (or cannot be synthesized in sufficient quantities) by human cells and therefore must be obtained in the diet. Since vitamins provide coenzymes that can, like enzymes, be used again and again, they are needed by cells in very small quantities. (Vitamins are discussed in more detail in chapter 13.)

Factors That Alter Enzymes

Enzymes are proteins, and like other proteins, they can be denatured by exposure to excessive heat, radiation, electricity, or certain chemicals. For example, many enzymes become inactive at 45°C, and nearly all of them are denatured at 55°C. Some poisons are chemical substances that cause enzymes to be denatured. Cyanides, for instance, interfere with respiratory enzymes and damage cells by halting their energy-releasing processes.

1. What is an enzyme?
2. How does an enzyme recognize its substrate?
3. What factors are likely to denature enzymes?

Energy for Metabolic Reactions

Energy is the capacity to produce changes in matter or the ability to move something, that is, to do work. Therefore, energy is recognized by what it can do; it is utilized whenever changes take place. Common forms of energy include heat, light, sound, electrical energy, mechanical energy, and chemical energy.

Release of Chemical Energy

Most metabolic processes use chemical energy. This form of energy is held in the bonds between the atoms of molecules and is released when these bonds are broken. For example, the chemical energy of many substances can be released by burning. Such a reaction must be started, and this is usually accomplished by applying heat to activate the burning process. As the substance burns, molecular bonds are broken, and energy escapes as heat and light.

Similarly, glucose molecules are "burned" in cells, although the process is more correctly called **oxidation.** The energy released by the oxidation of glucose is used to promote cellular metabolism. There are, however, some important differences between the oxidation of substances inside cells and the burning of substances outside them.

Burning usually requires a relatively large amount of energy to activate the process, and most of the energy released escapes as heat or light. In cells, the oxidation process is initiated by enzymes that reduce the amount of energy (activation energy) needed. Also, by transferring energy to special energy-carrying molecules, cells are able to capture about half of the energy released. The rest escapes as heat, which helps maintain body temperature.

1. What is energy?
2. How does cellular oxidation differ from burning?

Anaerobic Respiration

When a 6-carbon glucose molecule is decomposed in cellular respiration, enzymes control a series of reactions that break it into two 3-carbon pyruvic acid molecules. This phase of respiration occurs in the cytoplasm, and because it takes place in the absence of oxygen, it is called **anaerobic respiration.**

Although some energy is needed to activate the reactions of anaerobic respiration, more energy is released than is used. A portion of the excess is transferred to molecules of an energy-carrying substance called **ATP** (adenosine triphosphate). (See fig. 4.5.)

Following the anaerobic phase of respiration, oxygen must be available for further molecular breakdown to occur. For this reason, the second phase is called **aerobic respiration.** It takes place within the mitochondria, and as a result, considerably more energy is transferred to ATP molecules.

When the decomposition of a glucose molecule is complete, carbon dioxide molecules and hydrogen atoms remain. The carbon dioxide diffuses out of the cell as a waste, and the hydrogen atoms combine with oxygen to form water molecules. Thus, the final products of glucose oxidation are carbon dioxide, water, and energy. (See fig. 4.5.)

For each glucose molecule that is decomposed, about 38 molecules of ATP can be produced. Two of these ATP molecules are the result of anaerobic respiration, while the rest are formed during the aerobic phase.

Each ATP molecule contains 3 phosphates in a chain. (See fig. 4.6.) As energy is released during cellular respiration, some of it is captured in the bond of the end phosphate. When energy is needed for some metabolic process, the terminal phosphate bond of an ATP molecule is broken, and the energy stored in this bond is released. Such energy is used for a variety of cellular functions including muscle contractions, active transport mechanisms, and manufacture of various compounds.

An ATP molecule that has lost its terminal phosphate becomes an **ADP** (adenosine diphosphate) molecule. The ADP molecule can be converted back into an ATP, however, by capturing some energy and a phosphate. Thus, as figure 4.6 shows, ATP and ADP molecules shuttle back and forth between the energy-releasing reactions of cellular respiration and the energy-utilizing reactions of the cell.

Fig. 4.5 Anaerobic respiration occurs in the cytoplasm in the absence of oxygen, while aerobic respiration occurs in the mitochondria in the presence of oxygen.

1. *What is meant by anaerobic respiration? Aerobic respiration?*
2. *What are the final products of these reactions?*
3. *What is the function of ATP molecules?*

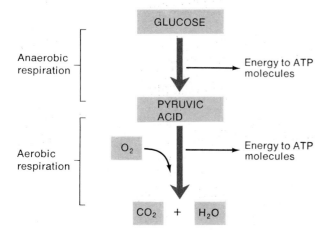

Fig. 4.6 What is the significance of this cyclic process?

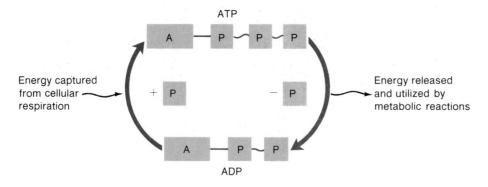

Metabolic Pathways

The anabolic and catabolic reactions that occur in cells usually involve a number of different steps that must occur in a particular sequence. For example, the anaerobic phase of cellular respiration, by which glucose is converted to pyruvic acid, involves 10 separate reactions. Since each reaction is controlled by a specific kind of enzyme, these enzymes must act in proper order. Such precision of activity suggests that the enzymes are arranged in precise ways. It is believed that those responsible for aerobic respiration are located in tiny stalked particles on the membranes (cristae) within the mitochondria and that they are positioned in the exact sequence as the reactions they control.

Such a sequence of enzyme-controlled reactions that leads to the production of particular products is called a **metabolic pathway.** (See fig. 4.7.)

Carbohydrate Pathways

The average human diet consists largely of carbohydrates, which are changed by digestion to monosaccharides such as glucose. These substances are used primarily as cellular energy sources, which means they usually enter the catabolic pathways of cellular respiration. As was discussed previously, the first phase of this process occurs in the cytoplasm and is anaerobic. It involves the conversion of glucose into molecules of *pyruvic acid*. During the second phase of the process, the pyruvic acid is transformed into a substance called *acetyl coenzyme A*. It, in turn, is transported into a mitochondrion and is changed into a number of intermediate products by a complex series of chemical reactions known as the **citric acid cycle** (Kreb's cycle). (See fig. 4.8.) As these changes occur, energy is released, and some of it is transferred to molecules of ATP, while the rest is lost as heat. Any excess glucose may enter anabolic pathways and be converted into storage forms such as glycogen or fat.

Lipid Pathways

Although foods may contain lipids in the form of phospholipids or cholesterol, the most common dietary lipids are fats. As was explained in chapter 2, the molecule of a fat (triglyceride) consists of a glycerol portion and 3 fatty acids.

Lipids provide for a variety of physiological functions; however, they are used mainly to supply energy. Gram for gram, fats contain more than twice as much chemical energy as carbohydrates or proteins.

Before energy can be released from a fat molecule it must undergo hydrolysis. As shown in figure 4.9, some of the resulting fatty acid portions can then be converted into molecules of acetyl coenzyme A by a series of reactions called *beta oxidation*. Others can be converted into compounds called *ketone bodies*. Later these may be changed to acetyl coenzyme A, which in turn is oxidized by means of the citric acid cycle. The glycerol portions of the triglyceride molecules also can enter metabolic pathways leading to the citric acid cycle, or they can be used to synthesize glucose.

Another possibility for glycerol and fatty acid molecules resulting from the hydrolysis of fats, is to be changed back into fat molecules by anabolic processes and stored in fat tissue.

Fig. 4.7 A metabolic pathway consists of a series of enzyme-controlled reactions leading to a product.

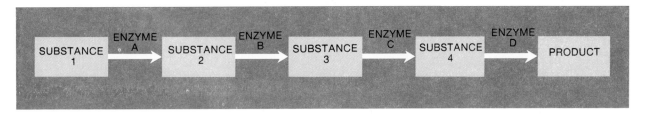

Fig. 4.8 Aerobic respiration, the second phase of cellular respiration, includes a complex series of chemical reactions called the citric acid cycle.

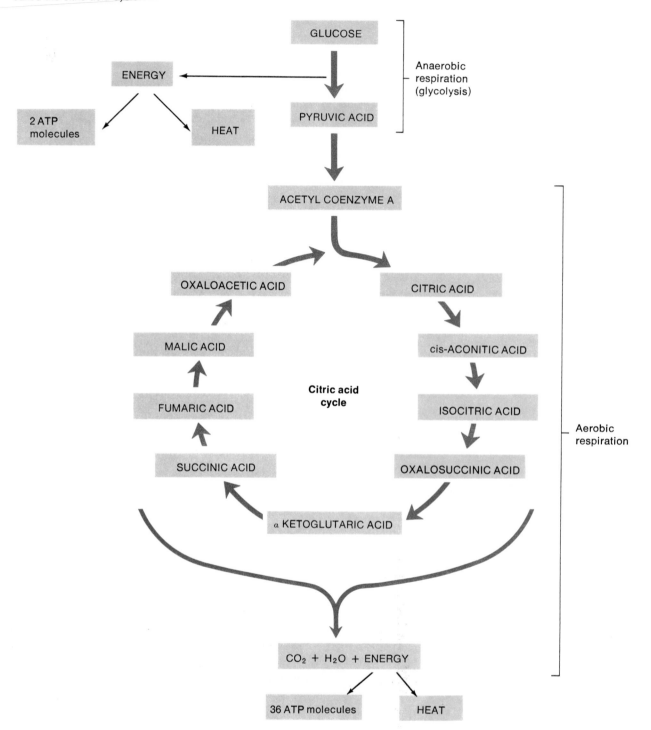

Fig. 4.9 Fats from foods are digested into glycerol and fatty acids. These molecules may enter catabolic pathways and be used as energy sources.

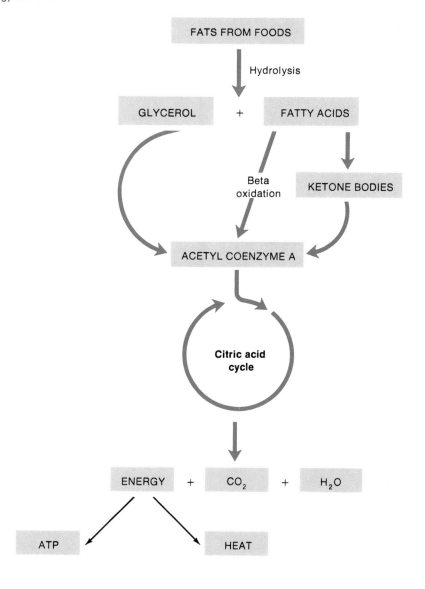

When ketone bodies are formed faster than they can be decomposed, some of them are eliminated through the lungs and kidneys. Consequently, the breath and urine may develop a fruity odor due to the presence of a ketone called *acetone.* This sometimes happens when a person fasts or diets to lose weight, thus forcing cells to metabolize body fat. Persons suffering from *diabetes mellitus* also are likely to metabolize excessive amounts of fats, and they too may have acetone in the breath and urine. At the same time, they may develop a serious imbalance in pH called *acidosis* due to an accumulation of acidic ketone bodies.

Protein Pathways

When dietary proteins are digested, the resulting amino acids are absorbed and transported by the blood to various body cells. Many of these amino acids are joined to form protein molecules again, which may be incorporated into cell parts or used as enzymes. Others may be decomposed and used to supply energy.

When protein molecules are used as energy sources, they first must be broken down into amino acids. The amino acids then undergo *deamination,* a process that occurs in the liver and involves removing the nitrogen-containing portions ($-NH_2$ groups) from the amino acids. (See fig. 2.13.) These $-NH_2$ groups are converted later into a waste substance called *urea,* which is excreted in the urine.

Fig. 4.10 Proteins from foods are digested into amino acids, but before these smaller molecules can be used as energy sources, they must be deaminated.

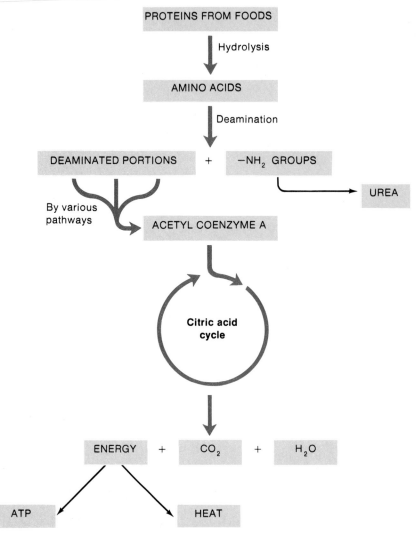

Depending upon the amino acids involved, the remaining deaminated portions of the amino acid molecules are decomposed by one of several pathways—all of which lead to the formation of acetyl coenzyme A. As before, the acetyl coenzyme A can enter the citric acid cycle, and as energy is released, some of it is captured in molecules of ATP. (See fig. 4.10.) If energy is not needed immediately, the deaminated portions of the amino acids may be changed into glucose or fat molecules by still other metabolic pathways.

---

1. What is meant by a metabolic pathway?
2. How are carbohydrates used within cells?
3. What must happen to fat molecules before they can be used as energy sources?
4. How do cells use proteins?

Nucleic Acids and Protein Synthesis

Since enzymes control the metabolic processes that enable cells to survive, the cells must possess information for producing these specialized proteins. Such information is held in DNA molecules in the form of a genetic code. This code "instructs" cells how to synthesize specific protein molecules.

Genetic Information

Children resemble their parents because of inherited traits; but what is actually passed from parents to a child is *genetic information*. This information is received in the form of DNA molecules from the parents' sex cells, and as an offspring develops, the

Fig. 4.11 The nucleotides of a DNA strand are joined to form a sugar-phosphate backbone (P-S). The organic bases of the nucleotides (B) extend from this backbone and are weakly bound to the bases of the second strand.

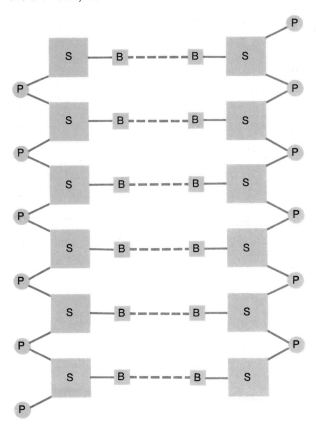

Fig. 4.12 The ladder of a double-stranded DNA molecule is twisted into the form of a double helix.

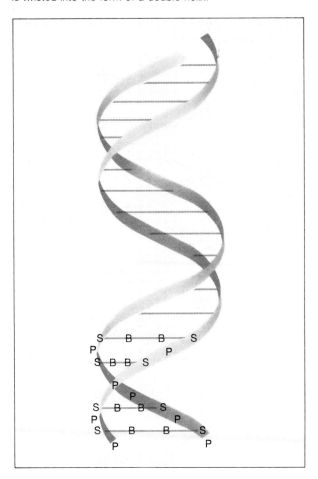

information is passed from cell to cell by mitosis. Genetic information "tells" the cells of the developing body how to construct specific protein molecules, which in turn function as structural materials, enzymes, or other vital substances.

The portion of a DNA molecule that contains the genetic information for making one kind of protein is called a **gene.** Thus, inherited traits are determined by the genes contained in the parents' sex cells that fused to form the first cell of the offspring's body. These genes instructed cells to synthesize the particular enzymes needed to control metabolic pathways.

DNA Molecules

As described in chapter 2, the building blocks of nucleic acids (nucleotides) are joined together so that the sugar and phosphate portions form a long chain, or "backbone." (See fig. 2.15.)

In a DNA molecule, the organic bases project out from this backbone and are bound weakly to those of the second strand. (See fig. 4.11.) The resulting structure is something like a ladder in which the uprights represent the sugar and phosphate backbones of the 2 strands, and the crossbars represent the organic bases. In addition, the molecular ladder is twisted to form a *double helix* (fig. 4.12).

The organic base of a DNA nucleotide can be one of four kinds: *adenine* (A), *thymine* (T), *cytosine* (C), or *guanine* (G). Therefore, there are only four kinds of DNA nucleotides: adenine nucleotide, thymine nucleotide, cytosine nucleotide, and guanine nucleotide.

A strand of DNA consists of nucleotides arranged in a particular sequence (fig. 4.13). Moreover, the nucleotides of one strand are paired in a special way with those of the other strand. Only certain organic bases have the molecular shapes needed to fit together so that their nucleotides can bond with one

Fig. 4.13 A strand of DNA consists of a chain of nucleotides arranged in a particular sequence. Within the chain there are four kinds of nucleotides: adenine nucleotide (A); thymine nucleotide (T); cytosine nucleotide (C); and guanine nucleotide (G).

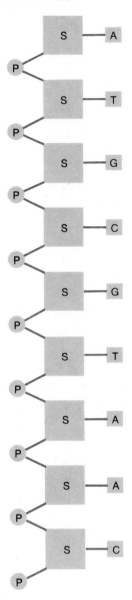

Chart 4.1 Genetic code for certain amino acids

Amino Acid	DNA Code	Messenger RNA Code
Alanine	CGA	GCU
Arginine	GCA	CGU
Asparagine	TTA	AAU
Glutamine	GTT	CAA
Histidine	GTA	CAU
Phenylalanine	AAA	UUU

The Genetic Code

Genetic information contains instructions for synthesizing proteins, and since proteins consist of 20 different amino acids joined in particular sequences, the genetic information must tell how to position the amino acids correctly in a polypeptide chain.

Each of the 20 different amino acids is represented in a DNA molecule by a particular sequence of 3-nucleotides. That is, the sequence C, G, A in a DNA strand represents one kind of amino acid; the sequence G, C, A represents another kind, and T, T, A still another kind. (See chart 4.1.) Other nucleotide sequences represent instructions for beginning or ending the synthesis of a protein molecule. Thus, the sequence in which the nucleotide groups are arranged within a DNA molecule can denote the arrangement of amino acids within a protein molecule as well as indicate where to start or stop the synthesis of a molecule. This method of storing information for the synthesis of protein molecules is termed the **genetic code.**

Although DNA molecules are located in the chromatin within a cell's nucleus, protein synthesis occurs in the cytoplasm. Therefore the genetic information must somehow be transferred from the nucleus into the cytoplasm. This transfer of information is the function of certain RNA molecules.

RNA Molecules

RNA (ribonucleic acid) molecules differ from DNA molecules in several ways. For example, RNA molecules are usually single-stranded, and their nucleotides contain ribose rather than deoxyribose sugar. Like DNA, RNA nucleotides each contain one of four organic bases; but while adenine, cytosine, and guanine nucleotides occur in both DNA and RNA, thymine nucleotides are found only in DNA. In its place, RNA molecules contain *uracil* (U) nucleotides.

another. Specifically, an adenine will bond only to a thymine, and a cytosine will bond only to a guanine. As a consequence of such base pairing, a DNA strand possessing the base sequence T, C, A, G would have to be joined to a second strand with the complementary base sequence A, G, T, C (fig. 4.14). It is the particular sequence of base pairs that encodes the genetic information held in a DNA molecule.

Fig. 4.14 The nucleotides of a double-stranded molecule of DNA are paired so that an adenine nucleotide of one strand is joined to a thymine nucleotide of the other strand; and a guanine nucleotide of one is joined to a cytosine nucleotide of the other. The dotted lines represent weak bonds between the paired bases of nucleotides.

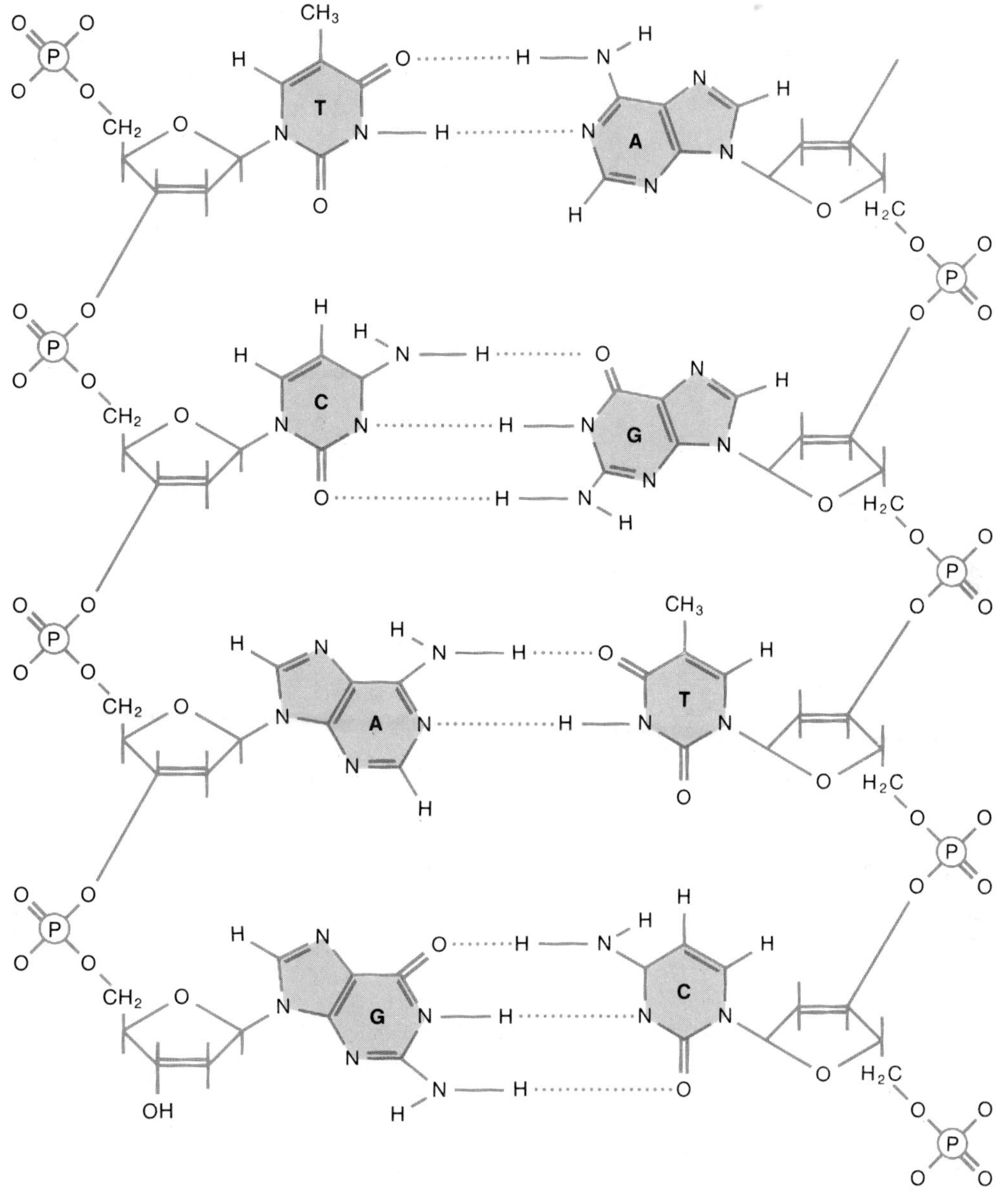

One step in the transfer of information from the nucleus to the cytoplasm involves the synthesis of a type of RNA called **messenger RNA** (mRNA). As it is produced, a double-stranded section of a DNA molecule unwinds and pulls apart. At the same time, the weak bonds between the base pairs of this section are broken. A molecule of messenger RNA is then formed of nucleotides that are complementary to those arranged along one of the exposed strands of DNA. (The other strand is not used in this process; however,

it is important in the duplication of the DNA molecule.) For example, if the sequence of DNA bases is A, T, G, C, G, T, A, A, C, then the complementary bases in the developing RNA molecule would be U, A, C, G, C, A, U, U, G, as shown in figure 4.15. In this way, an RNA molecule is synthesized that contains the information for arranging the amino acids of a protein molecule in the sequence dictated by the DNA "master information."

Fig. 4.15 When an RNA molecule is synthesized beside a strand of DNA, complementary nucleotides bond as in a double-stranded molecule of DNA with one exception: RNA contains uracil nucleotides (U) in place of thymine nucleotides (T).

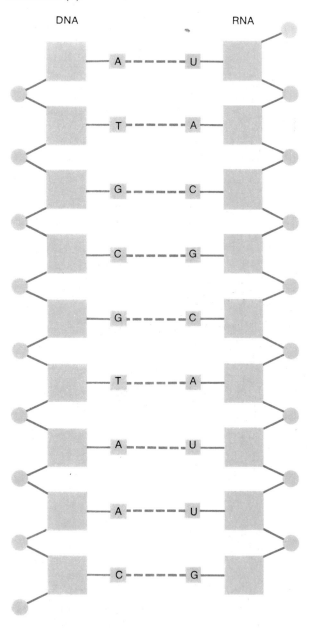

DNA   RNA

Fig. 4.16 After copying a section of DNA information, a messenger RNA molecule (mRNA) moves out of the nucleus and enters the cytoplasm. There it becomes associated with a ribosome.

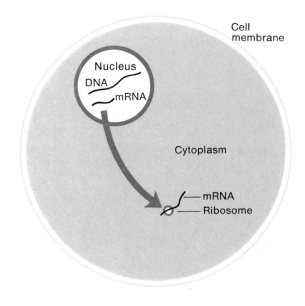

Furthermore, these amino acids must be positioned in the proper locations along a strand of messenger RNA. The positioning of amino acid molecules is the function of a second kind of RNA molecule called **transfer RNA** (tRNA).

Since 20 different kinds of amino acids are involved in protein synthesis, there must be at least 20 different kinds of transfer RNA molecules to serve as guides.

Each type of transfer RNA, which is a relatively small molecule, contains 3 nucleotides in a particular sequence. These nucleotides can only bond to the complementary set of 3 nucleotides of a messenger RNA molecule. In this way, the transfer RNA carries its amino acid to a correct position on a messenger RNA strand. This action occurs within a ribosome.

The ribosome binds to a molecule of messenger RNA near a set of 3 nucleotides, allowing a transfer RNA molecule with the complementary set of 3 nucleotides to recognize its correct location on the messenger RNA. The transfer RNA molecule brings the amino acid it carries into position and becomes temporarily joined to the ribosome. Then the transfer RNA molecule releases its amino acid and returns to the cytoplasm. (See fig. 4.17.)

This process is repeated again and again as the messenger RNA moves through the ribosome. The amino acids, which are released by the transfer RNA molecules, are added one at a time to a developing protein molecule.

Once they are formed, messenger RNA molecules (each of which consists of hundreds or even thousands of nucleotides) can move out of the nucleus through the tiny pores in the nuclear membrane and enter the cytoplasm (fig. 4.16). There they become associated with ribosomes and act as patterns or templates for the synthesis of protein molecules.

Protein Synthesis. Before a protein molecule can be synthesized, the correct amino acids must be present in the cytoplasm to serve as building blocks.

Fig. 4.17 Molecules of transfer RNA bring specific amino acids and place them in the sequence determined by the nucleotides in the messenger RNA molecule. These amino acids bond and form the polypeptide chain of a protein molecule.

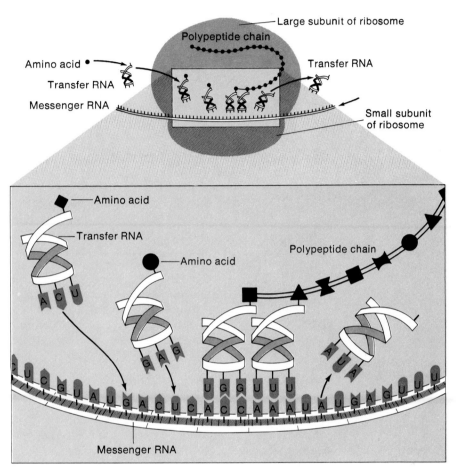

As the protein molecule forms, it folds into its unique shape, and when it is completed it is released to become a separate functional molecule. The transfer RNA molecules can pick up other amino acids from the cytoplasm, and the messenger RNA molecules can function again and again.

---

1. *What is the function of DNA?*
2. *How is information carried from the nucleus to the cytoplasm?*
3. *How are protein molecules synthesized?*

Duplication of DNA

When a cell reproduces, each daughter cell needs a copy of the parent cell's genetic information so that it will be able to synthesize the proteins necessary to build cellular parts and carry on metabolism.

DNA molecules can be duplicated (replicated) and this duplication takes place during the interphase of the cell's life cycle.

As the duplication process begins, bonds are broken between complementary base pairs of the double strands in each DNA molecule. Then the double-stranded structure pulls apart and unwinds, exposing the organic bases of its nucleotides. New nucleotides of the four types found in DNA become paired with the exposed bases, and enzymes cause the complementary bases to join together. In this way a new strand of complementary nucleotides is constructed along each of the old strands. As a result, two complete DNA molecules are produced, each with one old strand of the original molecule and one new strand.

These two DNA molecules become incorporated into chromosomes and are separated during mitosis so that one passes to each of the newly forming daughter cells. (See fig. 4.18.)

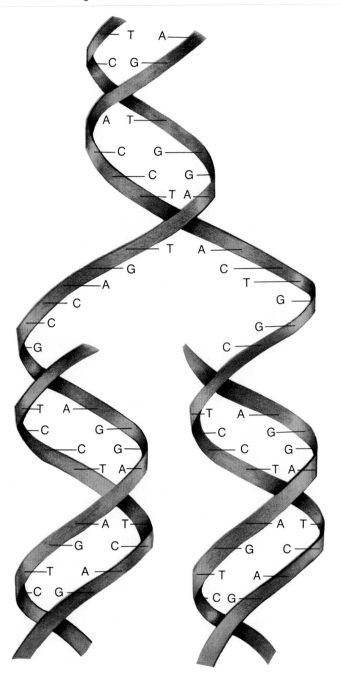

The amount of genetic information contained within a set of human chromosomes is very large, and since each of the trillions of cells in an adult body results from mitosis, this large amount of information must be duplicated many times during development. Although the duplication process usually is very precise, occasionally a mistake occurs and the information is altered. Such a change in genetic information is called a *mutation*.

The effects of mutations vary greatly. At one extreme there is little or no effect; at the other extreme the mutation causes cell death. The mutations that cause the most concern are those between the two extremes—they result in a decrease in cell efficiency, and although the cell may live and reproduce, it has difficulty functioning normally. In time, an increasing number of such mutated cells may produce abnormal effects in the structure and functions of tissues and organs.

1. *Why must DNA molecules be duplicated?*
2. *How is this duplication accomplished?*

Chapter Summary

Introduction

A cell continuously carries on metabolic processes.

Metabolic Processes

1. Anabolic metabolism
   a. Anabolic metabolism consists of constructive processes in which smaller molecules are used to build larger ones.
   b. In dehydration synthesis, water is formed, and smaller molecules become bound by shared atoms.
   c. Complex carbohydrates are synthesized from monosaccharides, fats are synthesized from glycerol and fatty acids, and proteins are synthesized from amino acids.

2. Catabolic metabolism
   a. Catabolic metabolism consists of decomposition processes in which larger molecules are broken down into smaller ones.
   b. In hydrolysis, a water molecule is used, and the bond between these two portions of the molecule is broken.
   c. Complex carbohydrates are decomposed into monosaccharides, fats are decomposed into glycerol and fatty acids, and proteins are decomposed into amino acids.

Control of Metabolic Reactions

Metabolic reactions usually involve many specific chemical changes.

1. Enzymes and their actions
   a. Enzymes are proteins that initiate or speed up metabolic reactions without being changed themselves.
   b. An enzyme acts upon a specific substrate.
   c. The shape of an enzyme molecule seems to fit the shape of its substrate molecule.
   d. When an enzyme combines with its substrate, the substrate is changed, resulting in a product, while the enzyme is unaltered.

2. Factors that alter enzymes
   a. Enzymes are proteins and can be denatured.
   b. Factors that may denature enzymes include excessive heat, radiation, electricity, and certain chemicals.

Energy for Metabolic Reactions

Energy is a capacity to produce change or to do work.

Common forms of energy include heat, light, sound, electrical energy, mechanical energy, and chemical energy.

1. Release of chemical energy
   a. Most metabolic processes utilize chemical energy that is released when molecular bonds are broken.
   b. The energy released from glucose during respiration is used to promote cellular metabolism.

2. Anaerobic respiration
   a. The first phase of glucose decomposition is anaerobic.
   b. Some of the energy released is transferred to molecules of ATP.

3. Aerobic respiration
   a. The second phase of glucose decomposition is aerobic.
   b. Considerably more energy is transferred to ATP molecules during this phase than during the anaerobic phase.
   c. The final products of glucose decomposition are carbon dioxide, water, and energy.

4. ATP molecules
   a. Energy is captured in the bond of the terminal phosphate of each ATP molecule.
   b. When energy is needed by a cellular process, the terminal phosphate bond of an ATP molecule is broken and the stored energy is released.
   c. An ATP molecule that loses its terminal phosphate becomes an ADP molecule.
   d. An ADP can be converted to an ATP by capturing some energy and a phosphate.

Metabolic Pathways

Metabolic processes usually involve a number of steps that must occur in the correct sequence.

A sequence of enzyme-controlled reactions is called a metabolic pathway.

1. Carbohydrate pathways
   a. Carbohydrates may enter catabolic pathways and be used as energy sources.
   b. Carbohydrates may enter anabolic pathways and be converted into glycogen or fat.

2. Lipid pathways
   a. Before fats can be used as an energy source they must be converted into glycerol and fatty acids.
   b. Fatty acids can be changed to acetyl coenzyme A, which in turn can be oxidized by the citric acid cycle.

3. Protein pathways
   a. Proteins are used as building materials for cellular parts, as enzymes, and as energy sources.
   b. Before proteins can be used as energy sources, they must be decomposed into amino acids, and the amino acids must be deaminated.
   c. The deaminated portions of amino acids can be broken down into carbon dioxide and water, or can be converted into glucose or fat.

Nucleic Acids and Protein Synthesis

DNA molecules contain information that instructs a cell how to synthesize proteins.

1. Genetic information
   a. Inherited traits result from DNA information that is passed from parents to child.
   b. A gene is a portion of a DNA molecule that contains the genetic information for making one kind of protein.

2. DNA molecules
   a. A DNA molecule consists of two strands of nucleotides twisted into a double helix.
   b. The nucleotides of a DNA strand are arranged in a particular sequence.
   c. The nucleotides of each strand are paired with those of the other strand in a complementary fashion.

3. The genetic code
   a. The sequence of nucleotides in a DNA molecule represents the sequence of amino acids in a protein molecule.
   b. Genetic information is transferred from the nucleus to the cytoplasm by RNA molecules.

4. RNA molecules
   a. RNA molecules are usually single strands, contain ribose instead of deoxyribose, and contain uracil nucleotides in place of thymine nucleotides.
   b. Messenger RNA molecules contain a nucleotide sequence that is complementary to that of an exposed strand of DNA.
   c. Messenger RNA molecules become associated with ribosomes, and act as patterns for the synthesis of protein molecules.
   d. Protein synthesis
      (1) Molecules of transfer RNA serve to position amino acids along a strand of messenger RNA.
      (2) A ribosome binds to a messenger RNA molecule and allows a transfer RNA molecule to recognize its correct position on the messenger RNA.
      (3) Amino acids released from the transfer RNA molecules become joined together and form a protein molecule with a unique shape.
5. Duplication of DNA
   a. Each daughter cell needs a copy of the parent cell's genetic information.
   b. DNA molecules are duplicated during interphase of the cell's life cycle.
   c. Each new DNA molecule contains one old strand and one new strand.

Application of Knowledge

1. Since enzymes are proteins, they may be denatured. Relate this to the fact that a very high fever accompanying an illness may be life-threatening.

2. Some reducing diets drastically limit the dieter's intake of carbohydrates but allow liberal use of fat and protein foods. What changes would such a diet cause in the cellular metabolism of the dieter? What changes might be noted in the urine of such a person?

Review Activities

1. Distinguish between anabolic and catabolic metabolism.
2. Distinguish between dehydration synthesis and hydrolysis.
3. Define *enzyme*.
4. Describe how an enzyme is thought to interact with its substrate.
5. Explain how an enzyme may be denatured.
6. Explain how oxidation of molecules inside cells differs from burning of substances.
7. Distinguish between anaerobic and aerobic respiration.
8. Explain the importance of ATP to cellular processes.
9. Describe the relationship between ATP and ADP.
10. Explain what is meant by a metabolic pathway.
11. Describe what happens to carbohydrates that enter catabolic pathways.
12. Explain how fats may serve as energy sources.
13. Define *deamination* and explain its importance.
14. Explain what is meant by genetic information.
15. Describe the relationship between a DNA molecule and a gene.
16. Explain how genetic information is stored in a DNA molecule.
17. Distinguish between messenger RNA and transfer RNA.
18. Explain the function of a ribosome in protein synthesis.
19. Explain why a daughter cell needs a copy of the parent cell's genetic information.
20. Describe the process by which DNA molecules are duplicated.

Tissues

5

Cells, the basic units of structure and function within the human organism, are organized into groups and layers called *tissues*.

Each type of tissue is composed of similar cells that are specialized to carry on particular functions. For example, epithelial tissues form protective coverings and function in secretion and absorption. Connective tissues provide support for softer body parts and bind structures together. Muscle tissues are responsible for producing body movements, and nerve tissue is specialized to conduct impulses that help to control and coordinate body activities.

After you have studied this chapter, you should be able to

1. Describe the general characteristics and functions of epithelial tissue.

2. Name the major types of epithelium and identify an organ in which each is found.

3. Describe the general characteristics of connective tissue.

4. List the major types of connective tissues that occur within the body.

5. Describe the major functions of each type of connective tissue.

6. Distinguish between the three types of muscle tissue.

7. Describe the general characteristics and functions of nerve tissue.

8. Complete the review activities at the end of this chapter. Note that the items are worded in the form of specific learning objectives. You may want to refer to them before reading the chapter.

adipose tissue   (ad′ĭ-pōs tish′u)

cartilage   (kar′tĭ-lij)

connective tissue   (kŏ-nek′tiv tish′u)

epithelial tissue   (ep′′ĭ-the′le-al tish′u)

fibroblast   (fi′bro-blast)

fibrous tissue   (fi′brus tish′u)

macrophage   (mak′ro-fāj)

muscle tissue   (mus′el tish′u)

nerve tissue   (nerv tish′u)

neuroglia   (nu-rog′le-ah)

neuron   (nu′ron)

osteocyte   (os′′te-o-sīt′′)

reticuloendothelial tissue
    (rĕ-tik′′u-lo-en′′do-the′le-al tish′u)

adip-, fat: *adip*ose tissue—tissue that stores fat.

-cyt, cell: osteo*cyt*e—a bone cell.

epi-, upon: *epi*thelial tissue—tissue that covers all free body surfaces.

-glia, glue: neuro*glia*—cells that bind nerve tissue together.

inter-, between: *inter*calated disc—band located between the ends of adjacent cardiac muscle cells.

macro-, large: *macro*phage—a large phagocytic cell.

pseudo-, false: *pseudo*stratified epithelium—tissue whose cells appear to be arranged in layers, but are not.

squam-, scale: *squam*ous epithelium—tissue whose cells appear flattened or scalelike.

stratum, layer: *strat*ified epithelium—tissue whose cells occur in layers.

Within the body, groups of cells perform specialized structural and functional roles. Such a group constitutes a **tissue.** Although the cells of different tissues vary in size, shape, arrangement, and function, those within each kind of tissue are quite similar.

The tissues of the human body include four major types: *epithelial tissues, connective tissues, muscle tissues,* and *nerve tissue.*

---

1. What is meant by a tissue?
2. List the four major types of tissues.

Epithelial Tissues

General Characteristics

Epithelial tissues are widespread throughout the body. They cover all body surfaces—inside and out. They are also the major tissues of glands.

Since epithelium covers organs, forms the inner lining of body cavities, and lines hollow organs, it always has a free surface—one that is exposed to the outside or to an open space internally. The underside of the tissue always is anchored to connective tissue by a thin, nonliving layer called the *basement membrane.*

As a rule, epithelial tissues lack blood vessels, however, epithelial cells are nourished by substances that diffuse from underlying connective tissues which are well supplied with blood vessels.

Although the cells of some tissues have limited abilities to reproduce, those of epithelium reproduce readily. Injuries to epithelium are likely to heal rapidly as new cells replace lost or damaged ones. Skin cells and the cells that line the stomach and intestines, for example, are continually being damaged and replaced.

Epithelial cells are tightly packed, and there is little intercellular material between them. Consequently, these cells can provide effective protective barriers, such as the outer layer of the skin and the lining of the mouth. Other epithelial functions include secretion, absorption, excretion, and sensory reception.

Simple Squamous Epithelium

Simple squamous epithelium consists of a single layer of thin, flattened cells. These cells fit tightly together, somewhat like floor tiles, and their nuclei are usually broad and thin. (See fig. 5.1.)

As a rule, substances pass rather easily through this type of tissue, and it often occurs where diffusion and filtration are taking place. For instance, simple squamous epithelium lines the air sacs of the lungs where oxygen and carbon dioxide are exchanged. It also forms the walls of capillaries, lines the insides of blood vessels, and covers the membranes that line body cavities.

Simple Cuboidal Epithelium

Simple cuboidal epithelium consists of a single layer of cube-shaped cells. These cells usually have centrally located spherical nuclei. (See fig. 5.2.)

This tissue covers the ovaries, lines the kidney tubules, and lines the ducts of glands, such as the salivary glands, thyroid gland, pancreas, and liver. In the kidneys, it functions in secretion and absorption, while in the glands it is concerned with the secretion of the glandular products.

Simple Columnar Epithelium

The cells of **simple columnar epithelium** are elongated; that is, they are longer than they are wide. This tissue is composed of a single layer of cells whose nuclei are usually located at about the same level near the basement membrane.

Simple columnar epithelium occurs in the linings of the uterus and various organs of the digestive tract, including the stomach and intestines. Since its cells are elongated, the tissue is relatively thick, providing protection for underlying tissues. It also functions in the secretion of digestive fluids and in the absorption of nutrient molecules resulting from the digestion of foods.

Columnar cells, whose principal function is absorption, often have numerous tiny, cylindrical processes extending outward from their cell surfaces. These processes, called *microvilli,* function to increase the surface of the cell membrane where it is exposed to the substances being absorbed.

Typically, there are specialized, flask-shaped glandular cells scattered along the columnar cells of this tissue. These cells, called *goblet cells,* secrete a thick, protective fluid called *mucus* onto the free surface of the tissue. (See fig. 5.3.)

Fig. 5.1 (a) Simple squamous epithelium consists of a single layer of tightly packed flattened cells. (b) What features of the tissue can you identify in this micrograph?

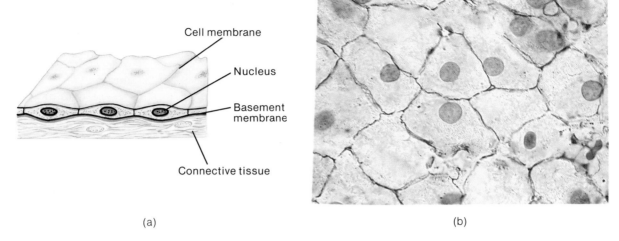

Cell membrane

Nucleus

Basement membrane

Connective tissue

(a)

(b)

Fig. 5.2 (a) Simple cuboidal epithelium consists of a single layer of tightly packed cube-shaped cells. (b) Note the layer of simple cuboidal cells indicated by the arrow in this micrograph.

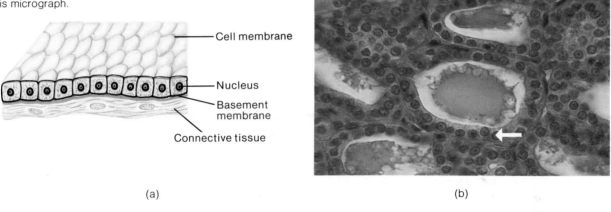

Cell membrane

Nucleus

Basement membrane

Connective tissue

(a)

(b)

Fig. 5.3 (a) Simple columnar epithelium consists of a single layer of elongated cells. (b) What is the function of the goblet cell indicated by the arrow in this micrograph?

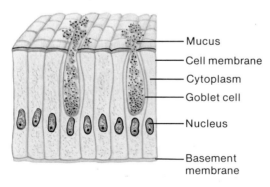

Mucus

Cell membrane

Cytoplasm

Goblet cell

Nucleus

Basement membrane

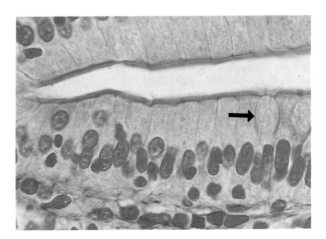

(a)

(b)

Fig. 5.4 (a) Pseudostratified columnar epithelium appears stratified because the nuclei of various cells are located at different levels. (b) What features of the tissue can you identify in this micrograph?

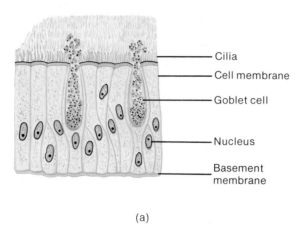

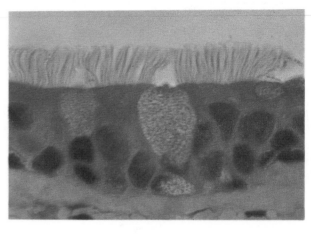

Cilia

Cell membrane

Goblet cell

Nucleus

Basement membrane

(a)

(b)

Pseudostratified Columnar Epithelium

The cells of **pseudostratified columnar epithelium** appear to be stratified or layered, but they are not. The layered effect occurs because nuclei are located at two or more levels within the cells of the tissue.

These cells commonly possess microscopic hair-like projections called *cilia*. The cilia extend from the free surfaces of the cells and move constantly. Goblet cells are also scattered throughout this tissue, and the mucus they secrete is swept along by the activity of the cilia. (See fig. 5.4.)

Pseudostratified columnar epithelium is found lining the passages of the respiratory system and in various tubes of the reproductive systems. In the respiratory passages, the mucus-covered linings are sticky and tend to trap particles of dust and microorganisms that enter with the air. The cilia move the mucus and its captured particles upward and out of the airways. In the reproductive tubes, the cilia aid in moving sex cells from one region to another.

Stratified Squamous Epithelium

Stratified squamous epithelium consists of many layers of cells, making this tissue relatively thick. Cellular reproduction occurs in the deeper layers, and as the newer cells grow, older ones are pushed further and further outward. (See fig. 5.5.)

This tissue forms the outer layer of the skin (epidermis). As the skin cells age, they accumulate a protein called *keratin* and become hardened and die. This action produces a covering of dry, tough, protective material that prevents the escape of water from underlying tissues and the entrance of various microorganisms.

Stratified squamous epithelium also lines the mouth cavity, throat, vagina, and anal canal. In these parts, the tissue is not keratinized; it remains moist, and the cells on the free surfaces remain alive.

Transitional Epithelium

Transitional epithelium is specialized to undergo changes in tension. It forms the inner lining of the urinary bladder and the passageways of the urinary system. When the wall of one of these organs is contracted, the tissue consists of several layers of cuboidal cells; however, when the organ is distended, the tissue is stretched and may have only a few layers. (See fig. 5.6.)

In addition to providing a distensible lining, transitional epithelium is thought to form a barrier that helps to prevent the contents of the urinary tract from diffusing out of its passageways.

Chart 5.1 summarizes the characteristics of epithelial tissues.

Fig. 5.5 (a) Stratified squamous epithelium consists of many layers of cells. (b) In this micrograph, note how the cells near the surface of the tissue have become flattened.

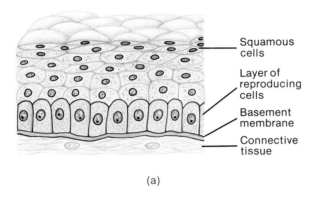

Squamous cells

Layer of reproducing cells

Basement membrane

Connective tissue

(a)

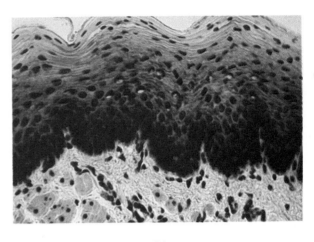

(b)

Fig. 5.6 (a) Micrograph of transitional epithelium; (b) this tissue consists of many layers when the wall of an organ is contracted; (c) the tissue is thin when the wall is stretched.

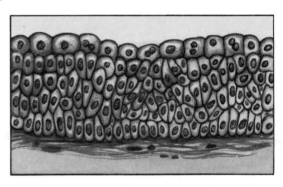

(b)

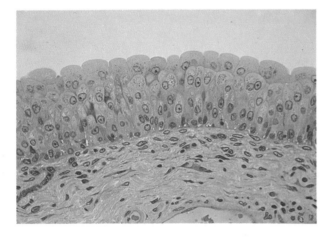

(a)

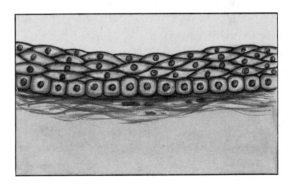

(c)

Chart 5.1 Epithelial tissues

Type	Function	Location
Simple squamous epithelium	Filtration, diffusion,osmosis	Air sacs of lungs, walls of capillaries, linings of blood vessels
Simple cuboidal epithelium	Secretion, absorption	Surface of ovaries, linings of kidney tubules, and linings of ducts of various glands
Simple columnar epithelium	Protection, secretion, absorption	Linings of uterus and tubes of the digestive tract
Pseudostratified columnar epithelium	Protection, secretion, movement of mucus and sex cells	Linings of respiratory passages and various tubes of the reproductive systems
Stratified squamous epithelium	Protection	Outer layers of skin, linings of mouth cavity, throat, vagina, and anal canal
Transitional epithelium	Distensibility, protection	Inner lining of urinary bladder and passageways of urinary tract

From Hole, John W., Jr., *Human Anatomy and Physiology 3d ed.* © 1978, 1981, 1984 Wm. C. Brown Publishers, Dubuque, Iowa. All Rights Reserved. Reprinted by permission.

A cancer arising from epithelium is called a *carcinoma*. It is estimated that up to 90% of all human cancers are of this type. Most of these cancers begin on surfaces which contact the external environment, such as skin, linings of the airways in the respiratory tract, or linings of the stomach or intestines in the digestive tract. This observation suggests that the more common cancer-causing agents may not penetrate tissues very deeply.

1. List the general characteristics of epithelial tissue.
2. Describe the structure of each type of epithelium.
3. Describe the special functions of each type of epithelium.

Connective Tissues

General Characteristics

Connective tissues occur in all parts of the body. They bind structures together, provide support, serve as frameworks, fill spaces, store fat, produce blood cells, provide protection against infections, and help to repair tissue damage. Connective tissue cells are usually further apart than epithelial cells, and they have an abundance of intercellular material called *matrix* between them. This material consists of fibers and a ground substance whose consistency varies from fluid, to semisolid or solid.

Connective tissue cells are able to reproduce. In most cases, they have good blood supplies and are well nourished. Although some connective tissues, such as bone and cartilage, are quite rigid, loose connective tissue, adipose connective tissue, and fibrous connective tissue are more flexible.

Major Cell Types. Connective tissues contain a variety of cell types. Some of them are called resident cells because they are usually present in relatively stable numbers. These include **fibroblasts** and **macrophages.** Another group known as wandering cells appear temporarily in the tissues, usually in response to an injury or infection. The wandering cells include several types of white blood cells.

The fibroblast is the most common kind of cell in connective tissue. It is relatively large and usually star-shaped. Fibroblasts function to produce white and yellow fibers.

Macrophages, which are almost as numerous as fibroblasts in some connective tissues, are specialized to carry on phagocytosis. They can move about and function as scavenger cells that clear foreign particles from tissues.

Types of Fibers. The white fibers produced by fibroblasts are composed of the protein *collagen,* which is the major structural protein of the body. Collagenous fibers occur in bundles and are flexible, but only slightly elastic. More importantly, they have great tensile strength—that is, they are strong in resisting pulling forces.

Yellow fibers are composed of the protein *elastin,* which has less strength than collagen but is highly elastic. They commonly branch to form dense networks and are abundant in tissues that normally must stretch.

Fig. 5.7 (a) Loose connective tissue contains numerous fibroblasts that produce white and yellow fibers. (b) Micrograph of loose connective tissue.

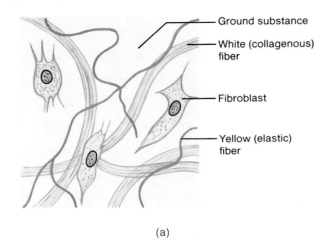

Ground substance
White (collagenous) fiber
Fibroblast
Yellow (elastic) fiber

(a)

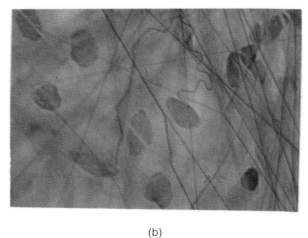

(b)

Fig. 5.8 (a) Adipose tissue cells contain large fat droplets that cause the nuclei to be pushed close to the cell membranes. (b) Note the nucleus indicated by the arrow in this micrograph.

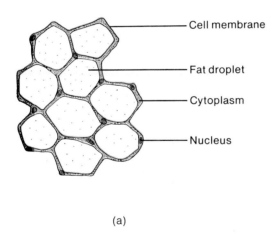

Cell membrane
Fat droplet
Cytoplasm
Nucleus

(a)

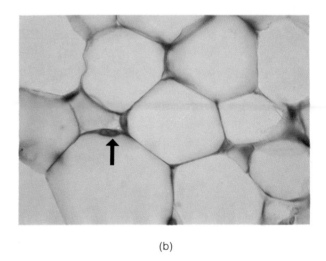

(b)

Loose Connective Tissue

Loose connective tissue forms delicate, thin membranes throughout the body. The cells of this tissue are mainly fibroblasts. They are located some distance apart and are separated by a gellike ground substance that contains many white and yellow fibers (fig. 5.7).

This tissue binds the skin to the underlying organs and fills spaces between muscles. It lies beneath most layers of epithelium, where its numerous blood vessels provide nourishment for the epithelial cells.

Adipose Tissue

Adipose tissue, commonly called fat, is really a specialized form of loose connective tissue. Certain cells within loose connective tissue store fat in droplets within their cytoplasm and become swollen. (See fig. 5.8.) When they occur in such large numbers that other cell types are crowded out, they form adipose tissue.

Adipose tissue is found beneath the skin and in the spaces between the muscles. It also occurs around the kidneys, behind the eyeballs, in certain abdominal membranes, on the surface of the heart, and around various joints.

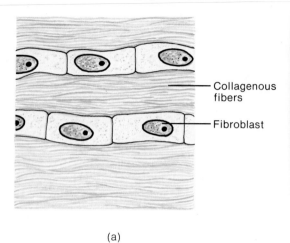

Collagenous fibers

Fibroblast

(a)

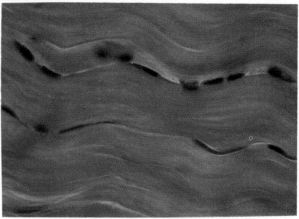

(b)

Adipose tissue serves as a protective cushion for joints and some organs, such as the kidneys. It also functions as a heat insulator beneath the skin and stores energy in fat molecules.

The amount of adipose tissue present in the body is usually related to a person's diet, for excess food substances are likely to be converted to fat and stored. During a period of fasting, adipose cells may lose their fat droplets, shrink in size, and become more like fibroblasts again.

Fibrous Connective Tissue

Fibrous connective tissue contains many, closely packed, thick, collagenous fibers and a fine network of elastic fibers. It has relatively few cells, almost all of which are fibroblasts (fig. 5.9).

Since white fibers are very strong, this type of tissue can withstand pulling forces, and it often functions to bind body parts together. **Tendons,** which connect muscles to bones, and **ligaments,** which connect bones to bones at joints, are composed of fibrous connective tissue. This type of tissue also occurs in the protective white layer of the eyeball and in the deeper portions of the skin.

The blood supply to fibrous connective tissue is relatively poor, so tissue repair occurs slowly.

1. What are the general characteristics of connective tissue?
2. What are the characteristics of collagen and elastin?
3. How is loose connective tissue related to adipose tissue?
4. Distinguish between a tendon and a ligament.

Cartilage

Cartilage is one of the rigid connective tissues. It supports parts, provides frameworks and attachments, protects underlying tissues, and forms structural models for many developing bones.

The intercellular material of cartilage is abundant and is composed largely of fibers embedded in a gellike ground substance. The cartilage cells occupy small chambers called *lacunae,* which are completely surrounded by matrix.

As a rule, a cartilaginous structure is enclosed in a covering of fibrous connective tissue called the *perichondrium.* Although cartilage tissue lacks a direct blood supply, there are blood vessels in the perichondrium that surrounds it. The cartilage cells obtain nutrients from these vessels by diffusion. This lack of a direct blood supply is related to the slow rate of cellular reproduction and repair that is characteristic of cartilage.

There are three kinds of cartilage, and each contains a different type of intercellular material. Hyaline cartilage has very fine white fibers in its matrix, elastic cartilage contains a dense network of yellow fibers, and fibrocartilage contains many large white fibers.

Fig. 5.10 (a) Hyaline cartilage cells are located in lacunae surrounded by intercellular material. (b) Micrograph of hyaline cartilage.

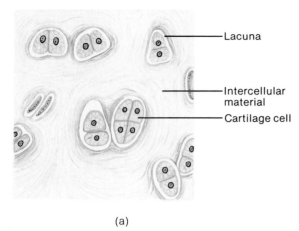

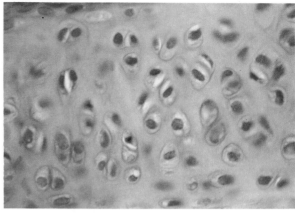

(a)

(b)

Fig. 5.11 (a) Elastic cartilage contains fine, white fibers and many yellow elastic fibers in its intercellular material. (b) Micrograph of elastic cartilage.

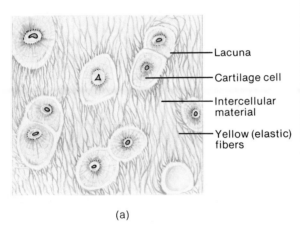

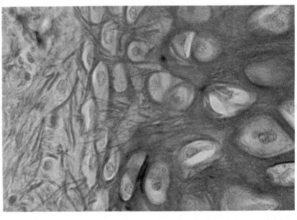

(a)

(b)

Hyaline cartilage (fig. 5.10), the most common type of cartilage, looks somewhat like milk glass. It occurs on the ends of bones in many joints, in the soft part of the nose, and in the supporting rings of the respiratory passages.

Elastic cartilage (fig. 5.11), is more flexible than hyaline cartilage because it contains numerous yellow fibers in its matrix. It provides the framework for the external ears and parts of the larynx.

Fibrocartilage (fig. 5.12), a very tough tissue, contains many white fibers. It often serves as a shock absorber for structures that are subjected to pressure. For example, fibrocartilage forms pads (intervertebral discs) between the individual parts of the backbone. It also forms protective cushions between certain bones in the knees and the pelvic girdle.

Bone

Bone is the most rigid of the connective tissues. Its hardness is due largely to the presence of mineral salts in the intercellular material. This matrix also contains a considerable amount of collagen.

Bone provides an internal support for body structures. It protects vital parts in various cavities and serves as an attachment for muscles. It also functions in forming blood cells and in storing certain inorganic salts.

Bone matrix is deposited in thin layers called *lamellae,* which most commonly are arranged in concentric patterns around tiny longitudinal tubes called

Fig. 5.12 (a) Fibrocartilage contains many large, white fibers in its intercellular material. (b) Micrograph of fibrocartilage.

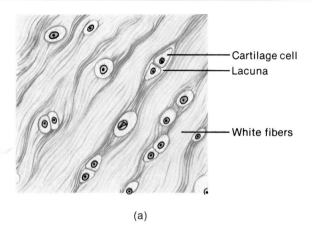

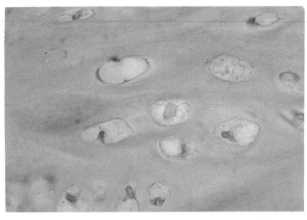

(a)   (b)

Fig. 5.13 (a) Bone matrix is deposited in concentric layers around haversian canals. (b) What features of the tissue can you identify in this micrograph?

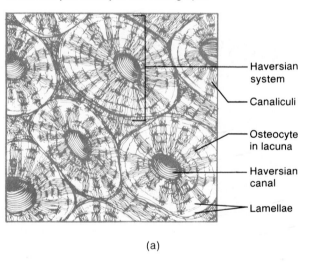

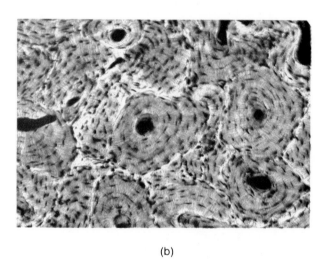

(a)   (b)

haversian canals. Bone cells called **osteocytes** are located in lacunae, rather evenly spaced between the lamellae. Consequently, they too are arranged in patterns of concentric circles.

In a bone, the osteocytes and layers of intercellular material clustered concentrically around a haversian canal form a cylinder-shaped unit called a *haversian system.* Many of these units cemented together form the substance of the bone. (See fig. 5.13.)

Each haversian canal contains a blood vessel, so that every bone cell is fairly close to a nutrient supply. In addition, the bone cells have numerous cytoplasmic processes that extend outward and pass through minute tubes called *canaliculi* in the matrix.

These cellular processes are attached to the membranes of nearby cells or connect to haversian canals. As a result, materials can move rapidly between blood vessels and bone cells. Thus, in spite of its inert appearance, bone is a very active tissue that heals much more rapidly than cartilage when it is damaged.

Chart 5.2 lists the characteristics of the major types of connective tissue.

Other Connective Tissues

Other important connective tissues include blood (vascular tissue) and reticuloendothelial tissue.

Blood is composed of cells that are suspended in fluid intercellular matrix called *blood plasma.*

Chart 5.2 Connective tissues

Type	Function	Location
Loose connective tissue	Binds organs together, holds tissue fluids	Beneath the skin, between muscles, beneath most epithelial layers
Adipose tissue	Protection, insulation, and storage of fat	Beneath the skin, around the kidneys, behind the eyeballs, on the surface of the heart
Fibrous connective tissue	Binds organs together	Tendons, ligaments, skin
Hyaline cartilage	Support, protection, provides framework	Ends of bones, nose, rings in walls of respiratory passages
Elastic cartilage	Support, protection, provides framework	Framework of external ear and part of the larynx
Fibrocartilage	Support, protection	Between bony parts of backbone and knee
Bone	Support, protection, provides framework	Bones of skeleton

From Hole, John W., Jr., *Human Anatomy and Physiology 3d ed.* © 1978, 1981, 1984 Wm. C. Brown Publishers, Dubuque, Iowa. All Rights Reserved. Reprinted by permission.

Fig. 5.14 (a) Blood tissue consists of an intercellular fluid in which red blood cells, white blood cells, and platelets are suspended. (b) Micrograph of human blood cells.

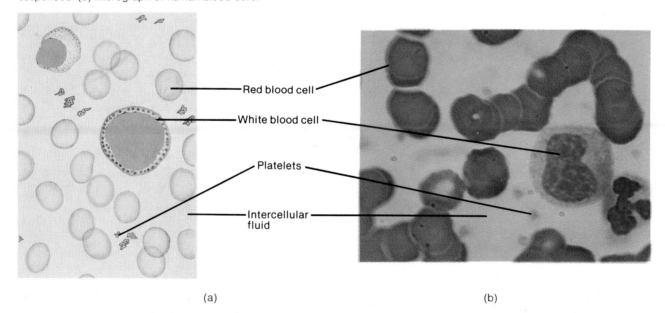

Red blood cell
White blood cell
Platelets
Intercellular fluid

(a)   (b)

These cells include red blood cells, white blood cells, and some cellular fragments called platelets. Most of the cells are formed by special tissues found in the hollow parts of certain bones. (See fig. 5.14.) Blood is described in more detail in chapter 15.

Reticuloendothelial tissue is composed of a variety of specialized cells that are widely scattered throughout the body. As a group, these cells are phagocytic, and they function to ingest and destroy foreign particles, such as microorganisms. Thus, they are particularly important in defending the body against infection.

The reticuloendothelial cells include types found in the blood, lungs, brain, bone marrow, and lymph glands. The most common ones, however, are *macrophages*.

Reticuloendothelial tissue is discussed in more detail in chapter 17.

1. Describe the general characteristics of cartilage.
2. Explain why injured bone heals more rapidly than cartilage.
3. Name the kinds of cells found in blood.
4. What do the cells of reticuloendothelial tissue have in common?

Fig. 5.15 (a) Skeletal muscle tissue is composed of striated muscle fibers that contain many nuclei. (b) Micrograph of skeletal muscle tissue.

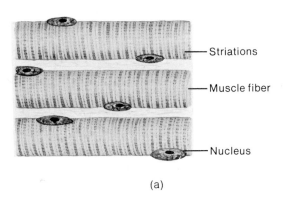

— Striations

— Muscle fiber

— Nucleus

(a)

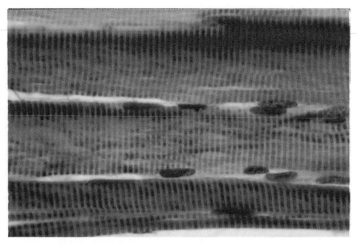

(b)

Muscle Tissues

General Characteristics.

Muscle tissues are contractile—their elongated cells or *muscle fibers* can change shape by becoming shorter and thicker. As they contract, the fibers pull at their attached ends and cause body parts to move. The three types of muscle tissue are skeletal muscle, smooth muscle, and cardiac muscle.

Skeletal Muscle Tissue

Skeletal muscle (fig. 5.15) is found in muscles that usually are attached to bones and can be controlled by conscious effort. For this reason, it is sometimes called *voluntary* muscle. The cells or muscle fibers are long and threadlike, with alternating light and dark cross-markings called *striations*. Each fiber has many nuclei, located just beneath its cell membrane. When the muscle fiber is stimulated by an action of a nerve fiber, it contracts and then relaxes.

Skeletal muscles are responsible for moving the head, trunk, and limbs. They also produce the movements involved with facial expressions, writing, talking, and singing, and with such actions as chewing, swallowing, and breathing.

Smooth Muscle Tissue

Smooth muscle (fig. 5.16) is called smooth because it lacks striations. This tissue is found in the walls of hollow internal organs such as the stomach, intestines, urinary bladder, uterus, and blood vessels. Unlike skeletal muscle, smooth muscle usually cannot be stimulated to contract by conscious effort; therefore it is a type of *involuntary* muscle. Smooth muscle cells are shorter than those of skeletal muscle, and they each have a single, centrally located nucleus. Smooth muscle tissue is responsible for the movements that force food through the digestive tube, constrict blood vessels, and empty the urinary bladder.

Cardiac Muscle Tissue

Cardiac muscle tissue (fig. 5.17) occurs only in the heart. Its cells, which are striated, are joined end to end. The resulting fibers are branched and interconnected in complex networks. Each cell has a single nucleus. At its end, where it touches another cell, there is a specialized intercellular connection called an *intercalated disc*.

Cardiac muscle, like smooth muscle, is controlled *involuntarily*. This tissue makes up the bulk of the heart and is responsible for pumping blood through the heart chambers and into the blood vessels.

---

1. List the general characteristics of muscle tissue.
2. Distinguish between skeletal, smooth, and cardiac muscle tissue.

Fig. 5.16 (a) Smooth muscle tissue is formed of spindle-shaped cells, each containing a single nucleus. (b) Note the nucleus indicated by the arrow in this micrograph.

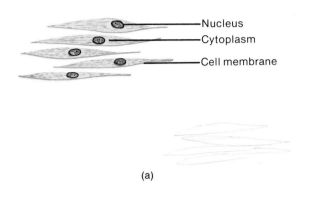

Nucleus
Cytoplasm
Cell membrane

(a)

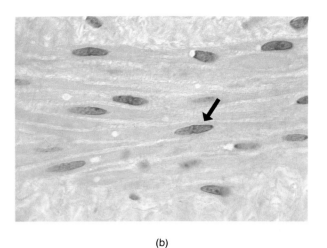

(b)

Fig. 5.17 (a) How is this cardiac muscle tissue similar to skeletal muscle? How is it similar to smooth muscle? (b) Note the intercalated disc indicated by the arrow in this micrograph.

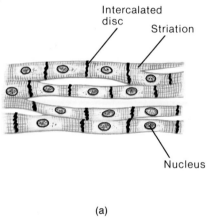

Intercalated disc

Striation

Nucleus

(a)

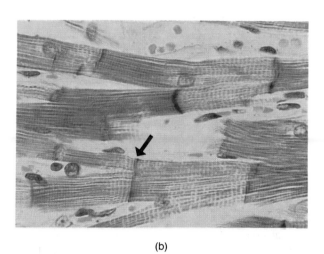

(b)

Nerve Tissue

Nerve tissue is found in the brain, spinal cord, and associated nerves. The basic cells of this tissue are called nerve cells, or *neurons*. Neurons are sensitive to certain types of changes in their surroundings, and they respond by transmitting nerve impulses along cytoplasmic extensions (nerve fibers) to other neurons or to muscles or glands (fig. 5.18).

Because neurons are connected with each other and various body parts by extremely complex patterns, they are able to coordinate and regulate many body functions.

In addition to neurons, nerve tissue contains **neuroglial cells.** (See fig. 10.4.) These cells function to support and bind the nerve tissue together, carry on phagocytosis, and aid in supplying nutrients to neurons by connecting them to blood vessels. Nerve tissue is discussed in more detail in chapter 10.

Chart 5.3 summarizes the general characteristics of muscle and nerve tissues.

1. *Describe the general characteristics of nerve tissue.*
2. *Distinguish between neurons and neuroglial cells.*

Fig. 5.18 (a) Neurons function to transmit impulses to other neurons or to muscles or glands. (b) What features of a neuron can you identify in this micrograph?

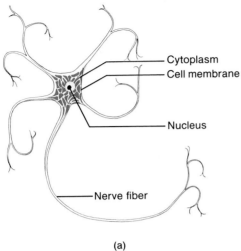

Cytoplasm
Cell membrane
Nucleus
Nerve fiber

(a)

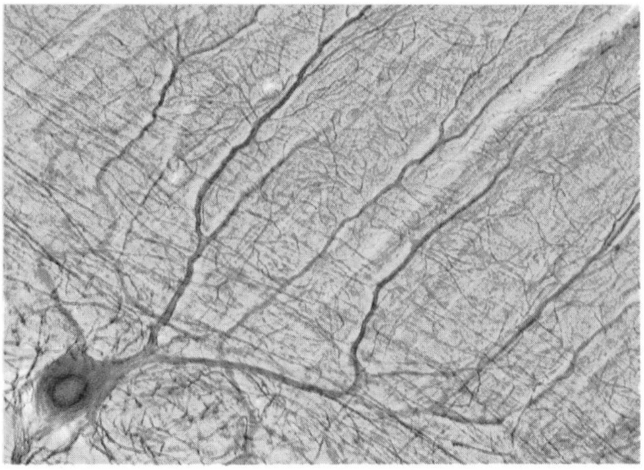

(b)

Chart 5.3   Muscle and nerve tissues

Type	Function	Location
Skeletal muscle tissue	Voluntary movements of skeletal parts	Muscles attached to bones
Smooth muscle tissue	Involuntary movements of internal organs	Walls of hollow internal organs
Cardiac muscle tissue	Heart movements	Heart muscle
Nerve tissue	Sensitivity and conduction of nerve impulses	Brain, spinal cord, and nerves

Chapter Summary

Introduction

Cells are arranged in tissues that perform specialized structural and functional roles.

The four major types of human tissue are epithelial tissue, connective tissue, muscle tissue, and nerve tissue.

Epithelial Tissues

1. General characteristics
   a. Epithelial tissue covers all free body surfaces and is the major tissue of glands.
   b. It is anchored to connective tissue by a basement membrane, lacks blood vessels, contains little intercellular material, and is replaced continuously.
   c. It functions in protection, secretion, absorption, excretion, and sensory reception.

2. Simple squamous epithelium
   a. This tissue consists of a single layer of flattened cells.
   b. It functions in the exchange of gases in the lungs, and lines blood vessels and various membranes.

3. Simple cuboidal epithelium
   a. This tissue consists of a single layer of cube-shaped cells.
   b. It carries on secretion and absorption in the kidneys and various glands.

4. Simple columnar epithelium
   a. This tissue is composed of elongated cells whose nuclei are located near the basement membrane.
   b. It lines the uterus and digestive tract.
   c. Absorbing cells often possess microvilli.
   d. This tissue contains goblet cells that secrete mucus.

5. Pseudostratified columnar epithelium
   a. This tissue appears stratified because the nuclei are located at two or more levels within the cells.
   b. Its cells may have cilia that function to move mucus or sex cells.
   c. It lines various tubes of the respiratory and reproductive systems.

6. Stratified squamous epithelium
   a. This tissue is composed of many cell layers.
   b. It protects underlying cells.
   c. It covers the skin and lines the mouth, throat, vagina, and anal canal.

7. Transitional epithelium
   a. This tissue is specialized to undergo changes in tension.
   b. It occurs in the walls of various organs of the urinary system.

Connective Tissues

1. General characteristics
   a. Connective tissue connects, supports, protects, provides frameworks, fills spaces, stores fat, and produces blood cells.
   b. It consists of cells, usually some distance apart, and a considerable amount of intercellular material.
   c. Major cell types
      (1) Connective tissue cells include resident and wandering cells.
      (2) Fibroblasts produce collagenous and elastic fibers.
      (3) Macrophages function as phagocytes.

2. Loose connective tissue
   a. This tissue forms thin membranes between organs.
   b. It is found beneath the skin and between muscles.

3. Adipose tissue
   a. Adipose tissue is a specialized form of loose connective tissue that stores fat.
   b. It is found beneath the skin, in certain abdominal membranes, and around the kidneys, heart, and various joints.

4. Fibrous connective tissue
   a. This tissue is composed largely of strong, collagenous fibers.
   b. It is found in tendons, ligaments, eyes, and skin.

5. Cartilage
   a. Cartilage provides support and framework for various parts.
   b. Its intercellular material is largely composed of fibers and a gellike ground substance.
   c. Cartilaginous structures are usually enclosed in a perichondrium.
   d. Cartilage lacks a direct blood supply and is slow to heal following an injury.
   e. Major types are hyaline cartilage, elastic cartilage, and fibrocartilage.

6. Bone
   a. The intercellular matrix of bone contains mineral salts and collagen.
   b. Its cells are arranged in concentric circles around haversian canals and are interconnected by canaliculi.
   c. It is an active tissue that heals rapidly following an injury.

7. Other connective tissues
   a. Blood
      (1) Blood is composed of red cells, white cells, and platelets suspended in plasma.
      (2) The cells are formed by special tissue in the hollow parts of bones.
   b. Reticuloendothelial tissue
      (1) This tissue is composed of phagocytic cells that are widely distributed in body organs.
      (2) It defends the body against invasion by microorganisms.

Muscle Tissues

1. General characteristics
   a. Muscle tissue is contractile tissue that moves parts that are attached to it.
   b. Three types are skeletal, smooth, and cardiac muscle tissues.

2. Skeletal muscle tissue
   a. This tissue is attached to bones and is controlled by conscious effort.
   b. Cells or muscle fibers are long and threadlike.
   c. Muscle fibers contract when stimulated by nerve action and relax immediately.

3. Smooth muscle tissue
   a. This tissue is found in walls of hollow internal organs.
   b. It is usually controlled by involuntary activity.

4. Cardiac muscle tissue
   a. This tissue is found only in the heart.
   b. Its cells are joined by intercalated discs and arranged in branched, interconnecting networks.

Nerve Tissue

1. Nerve tissue is found in the brain, spinal cord, and nerves.
2. Neurons
   a. These cells are sensitive to changes and respond by transmitting nerve impulses to other neurons or body parts.
   b. They function in coordinating and regulating body activities.
3. Neuroglial cells
   a. Some forms bind and support nerve tissue.
   b. Others carry on phagocytosis.
   c. Still others connect neurons to blood vessels.

Application of Knowledge

1. The sweeping action of the ciliated epithelium that lines the respiratory passages is inhibited by excessive exposure to tobacco smoke or airborne pollutants. What special problems is this likely to create?
2. Joints, such as the elbow, shoulder, and knee, contain considerable amounts of cartilage and fibrous connective tissue. How is this related to the fact that joint injuries are often very slow to heal?
3. There is a group of disorders called collagenous diseases that are characterized by deterioration of connective tissues. Why would you expect such diseases to produce widely varying symptoms?

Review Activities

1. Define *tissue*.
2. Name the four major types of tissue found in the human body.
3. Describe the general characteristics of epithelial tissue.
4. Explain how the structure of simple squamous epithelium is related to its function.
5. Name an organ in which each of the following tissues is found, and give the function of the tissue in each case:
   a. Simple squamous epithelium
   b. Simple cuboidal epithelium
   c. Simple columnar epithelium
   d. Pseudostratified columnar epithelium
   e. Stratified squamous epithelium
   f. Transitional epithelium
6. Describe the general characteristics of connective tissue.
7. Distinguish between resident cells and wandering cells.
8. Distinguish between collagen and elastin.
9. Explain the relationship between loose connective tissue and adipose tissue.
10. Explain why injured fibrous connective tissue and cartilage are usually slow to heal.
11. Name the major types of cartilage, and describe their differences and similarities.
12. Describe how bone cells are arranged in bone tissue.
13. Describe the composition of blood.
14. Define *reticuloendothelial tissue*.
15. Describe the general characteristics of muscle tissue.
16. Distinguish between skeletal, smooth, and cardiac muscle tissues.
17. Describe the general characteristics of nerve tissue.
18. Distinguish between neurons and neuroglial cells.

The Skin and the Integumentary System

The previous chapters dealt with the lower levels of organization within the human organism—tissues, cells, cellular organelles, and the chemical substances that form these parts.

Chapter 6 explains how tissues are grouped to form organs and how organs comprise organ systems. In this explanation various membranes, including the skin, are used as examples of organs. Since the skin acts with hair follicles, sebaceous glands, and sweat glands to provide a variety of functions, these organs together constitute the integumentary organ system.

Chapter Outline

Chapter Objectives

After you have studied this chapter, you should be able to

1. Describe the four major types of membranes.

2. Describe the structure of the various layers of the skin.

3. List the general functions of each of these layers of skin.

4. Describe the accessory organs associated with the skin.

5. Explain how the skin functions in regulating body temperature.

6. Complete the review activities at the end of this chapter. Note that the items are worded in the form of specific learning objectives. You may want to refer to them before reading the chapter.

cutaneous membrane   (ku-ta′ne-us mem′brān)

dermis   (der′mis)

epidermis   (ep′′ĭ-der′mis)

hair follicle   (hār fol′ĭ-kl)

integumentary   (in-teg-u-men′tar-e)

keratinization   (ker′′ah-tĭn′′ĭ-za′shun)

melanin   (mel′ah-nin)

mucous membrane   (mu′kus mem′brān)

sebaceous gland   (se-ba′shus gland)

serous membrane   (se′rus mem′brān)

subcutaneous   (sub′′ku-ta′ne-us)

sweat gland   (swet gland)

cut-, skin: sub*cut*aneous—beneath the skin.

derm-, skin: *derm*is—inner layer of the skin.

epi-, upon: *epi*dermis—outer layer of the skin.

follic-, small bag: hair *follic*le—tubelike depression in which a hair develops.

kerat-, horn: *kerat*in—protein produced as epidermal cells die and harden.

melan-, black: *melan*in—dark pigment produced by certain epidermal cells.

seb-, grease: *seb*aceous gland—gland that secretes an oily substance.

Two or more kinds of tissues grouped together and performing specialized functions constitute an organ. Thus, the thin, sheetlike structures called *membranes* that cover body surfaces and line body cavities are organs.

One of these membranes, the skin, together with various accessory organs, makes up the **integumentary system.**

Types of Membranes

There are four major types of membranes: *serous, mucous, cutaneous,* and *synovial.*

Serous membranes line the body cavities that lack openings to the outside. They form the inner linings of the thorax and abdomen, and they cover the organs within these cavities. Such a membrane consists of a layer of simple squamous epithelium covering a thin layer of loose connective tissue. It secretes a watery *serous fluid,* which helps to lubricate the surfaces of the membrane.

Mucous membranes line the cavities and tubes that open to the outside of the body. These include the oral and nasal cavities and the tubes of the digestive, respiratory, urinary, and reproductive systems. Mucous membrane consists of epithelium overlying a layer of loose connective tissue. Specialized cells within a mucous membrane secrete the thick fluid called *mucus.*

The **cutaneous membrane** is an organ of the integumentary system and is more commonly called *skin.* It is described in detail in the next section of this chapter.

Synovial membranes form the inner linings of joint cavities between the ends of bones at freely movable joints. These membranes usually include fibrous connective tissue overlying loose connective tissue and adipose tissue. They secrete a thick, colorless *synovial fluid* into the joint cavity, which lubricates the ends of the bones within the joint.

The Skin and Its Tissues

The skin, which is the largest and one of the more versatile organs of the body, plays vital roles in the maintenance of homeostasis. It functions as a protective covering, aids in the regulation of body temperature, retards water loss from deeper tissues, houses sensory receptors, synthesizes various chemical substances, and excretes small quantities of waste substances.

The skin includes two distinct layers of tissues. The outer layer, called the **epidermis,** is composed of stratified squamous epithelium. The inner layer, or **dermis,** is thicker than the epidermis and includes a variety of tissues, such as epithelial tissue, fibrous connective tissue, smooth muscle tissue, nerve tissue, and blood. (See fig. 6.1.)

Beneath the dermis are masses of loose connective and adipose tissues that bind the skin to underlying organs. They form the **subcutaneous layer.**

Subcutaneous injections are administered into the layer beneath the skin. Intradermal injections, on the other hand, are injected between layers of tissues within the skin. Subcutaneous injections and intramuscular injections, which are administered into muscles, are sometimes called hypodermic injections.

1. Name the four types of membranes and explain how they differ.
2. List the general functions of the skin.
3. Name the tissue (or tissues) found in the outer layer of the skin; the inner layer.

The Epidermis

Since the epidermis is composed of epithelium, it lacks blood vessels; however the deepest cells of the epidermis, which are next to the dermis are nourished by dermal blood vessels and are capable of reproducing. They form a layer called the *stratum germinativum.* As the cells of this layer divide, older epidermal cells are pushed slowly away from the dermis toward the surface of the skin. The further the cells travel, the poorer their nutrient supply becomes, and in time they die.

Meanwhile, the maturing cells undergo a hardening process called **keratinization,** during which the cytoplasm develops strands of tough, fibrous, waterproof protein called *keratin.* As a result, many layers of tough, dead cells accumulate in the outer portion of the epidermis. This outermost layer is called the *stratum corneum,* and the dead cells that compose it are easily rubbed away. (See fig. 6.1.)

Fig. 6.1 A section of skin. Why is the skin considered to be an organ?

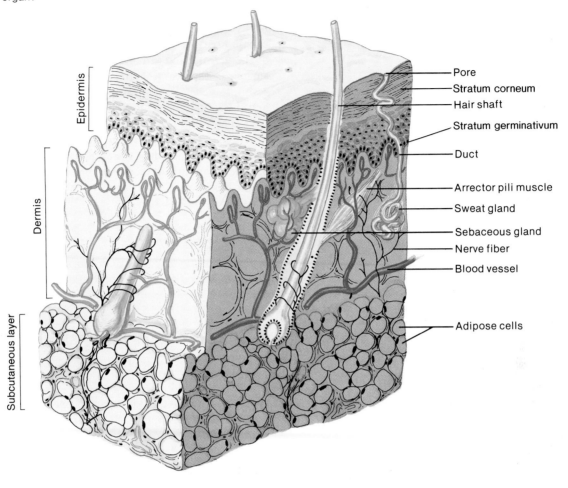

Epidermis

Dermis

Subcutaneous layer

Pore
Stratum corneum
Hair shaft
Stratum germinativum
Duct
Arrector pili muscle
Sweat gland
Sebaceous gland
Nerve fiber
Blood vessel
Adipose cells

In healthy skin, the production of epidermal cells is closely balanced with the loss of stratum corneum, so that skin seldom wears away completely. In fact, the rate of cellular reproduction tends to increase in regions where the skin is being rubbed or pressed regularly. This response causes thickened regions called *calluses* to develop on the palms and soles.

The epidermis has important protective functions. It shields the moist underlying tissues against excess water loss, mechanical injury, and the effects of harmful chemicals. When it is unbroken, the epidermis also prevents the entrance of many disease-causing microorganisms. **Melanin** is a pigment that occurs in deeper layers of the epidermis and is produced by specialized cells known as *melanocytes*. It absorbs light energy, and in this way, helps to protect deeper cells from the damaging effects of ultraviolet rays of sunlight.

1. *Explain how the epidermis is formed.*
2. *Distinguish between stratum germinativum and stratum corneum.*
3. *What is the function of melanin?*

The Dermis

The dermis binds the epidermis to the underlying tissues. It is composed largely of fibrous connective tissue, which includes tough white (collagenous) fibers and yellow (elastin) fibers surrounded by a gel-like substance. Networks of these fibers give the dermis its strength and elasticity.

Blood vessels in the dermis supply nutrients to all skin cells, including those of the epidermis. These dermal blood vessels also aid in regulating body temperature, as is explained in a subsequent section of this chapter.

Fig. 6.2 Various types of sensory receptors, such as touch receptors, are located in the dermis.

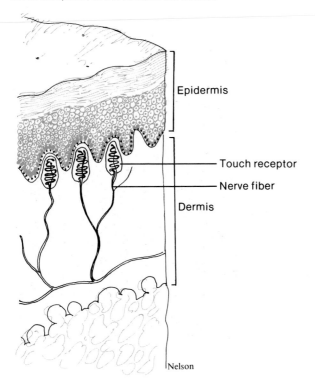

Epidermis

Touch receptor

Nerve fiber

Dermis

Nelson

Since blood vessels in the dermis supply nutrients to the epidermis, any interference with blood flow is likely to result in the death of epidermal cells. For example, when a person lies in one position for a prolonged time, the weight of the body pressing against the bed interferes with the skin's blood supply. If cells die, the tissues begin to break down and a *pressure ulcer* (also called decubitus ulcer or bedsore) may appear.

Pressure ulcers usually occur in the skin overlying bony projections, such as on the hip, heel, elbow, or shoulder. These ulcers often can be prevented by changing the body position frequently or by massaging the skin to stimulate blood flow in regions associated with bony prominences.

There are numerous nerve fibers scattered throughout the dermis. Some of them (motor fibers) carry impulses to dermal muscles and glands, causing these structures to react. Others (sensory fibers) carry impulses away from specialized sensory receptors located within the dermis (fig. 6.2).

The Subcutaneous Layer

As was mentioned, the subcutaneous layer lies beneath the dermis and consists largely of loose connective and adipose tissues. The collagenous and elastic fibers in this layer are continuous with those of the dermis, and although most run parallel to the surface of the skin, they travel in all directions. As a result, there is no sharp boundary between the dermis and the subcutaneous layer.

The adipose tissue of the subcutaneous layer functions as a heat insulator—helping to conserve body heat and impeding the entrance of heat from the outside. The subcutaneous layer also contains the major blood vessels that supply the skin and the underlying adipose tissue.

1. *Describe the dermis.*
2. *What are the functions of this layer of skin?*
3. *What are the functions of the subcutaneous layer?*

Accessory Organs of the Skin

Hair Follicles

Hair is present on all skin surfaces except the palms, soles, lips, nipples, and various parts of the external reproductive organs.

Each hair develops from a group of epidermal cells at the base of a tubelike depression called a **hair follicle.** This follicle extends from the surface into the dermis and may pass into the subcutaneous layer. The epidermal cells at its base receive nourishment from dermal blood vessels, which occur in a projection of connective tissue at the base of the follicle. As these cells divide and grow, older cells are pushed toward the surface. The cells that move upward and away from their nutrient supply become keratinized and die. Their remains constitute the shaft of a developing hair. In other words, a hair is composed of dead epidermal cells. (See figs. 6.3 and 6.4.)

A bundle of smooth muscle cells, forming the *arrector pili muscle,* is attached to each hair follicle. This muscle is positioned so that the hair within the follicle tends to stand on end when the muscle contracts. If a person is emotionally upset or very cold, nerve impulses may stimulate the arrector pili muscles to contract, causing gooseflesh or goose bumps.

Fig. 6.3 A hair grows from the base of a hair follicle when epidermal cells undergo cell division and older cells move outward and become keratinized.

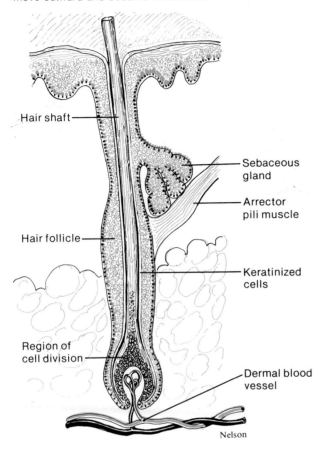

Hair shaft

Sebaceous gland

Arrector pili muscle

Hair follicle

Keratinized cells

Region of cell division

Dermal blood vessel

Nelson

Fig. 6.4 Note how the hairs penetrate the surface layer in this scanning electron micrograph of mammalian skin (260 ×). What other features of the skin do you recognize?
From *Tissues and Organs: A Text-Atlas of Scanning Electron Microscopy* by Richard G. Kessel and Randy H. Kardon. W. H. Freeman and Company. Copyright © 1979.

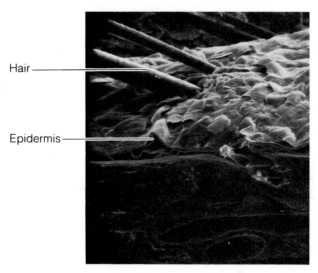

Hair

Epidermis

Sebaceous Glands

Sebaceous glands (fig. 6.3) contain groups of specialized epithelial cells and are usually associated with hair follicles. They produce an oily secretion called *sebum,* which is a mixture of fatty material and cellular debris. Sebum is secreted through small ducts into the hair follicles and helps to keep the hair and skin soft, pliable, and relatively waterproof.

Sebaceous glands are responsible for acne, a common adolescent disorder. In this condition, the sebaceous glands in certain body regions become overactive in response to hormonal changes, and their excessive secretions cause the openings of hair follicles to become dilated and plugged. The result is blackheads; if pus-forming bacteria are present, pimples or boils may develop in the follicles.

Nails

Nails are protective coverings on the ends of fingers and toes. Each nail is produced by layers of specialized epithelial cells that are continuous with the epithelium of the skin. The whitish, half-moonshaped region at the base of a nail is its most active growing portion. The epithelial cells formed in this region undergo keratinization and become part of the nail. The keratin they contain, however, is harder than that formed in the epidermis and hair follicles.

Sweat Glands

Sweat glands occur in nearly all regions of the skin, but are most numerous in the palms and soles. Each gland consists of a tiny tube that originates as a ball-shaped coil in the dermis or subcutaneous layer. The coiled portion of the gland is closed at its end and is lined with sweat-secreting epithelial cells.

Some sweat glands, the *apocrine glands,* respond to emotional stress and become active when a person is frightened or suffering pain. They are most numerous in the armpits and groin and are usually connected with hair follicles. The development of these glands is stimulated by sex hormones, so they begin to function at puberty.

Other sweat glands, the *eccrine glands,* are not associated with hair follicles. They function throughout life and respond primarily to elevated body temperatures associated with environmental heat or physical exercise. These glands are common on the forehead, neck, and back, where they produce profuse sweating on hot days and during physical exertion. (See fig. 6.1.)

The fluid secreted by these glands is carried away by a tubular part that opens at the surface as a pore. Although it is mostly water, the fluid contains small quantities of salts and certain wastes, such as urea and uric acid. Thus, the secretion of sweat is, to a limited degree, an excretory function.

---

1. Explain how a hair forms.
2. What is the function of the sebaceous glands?
3. Distinguish between apocrine and eccrine glands.

Regulation of Body Temperature

The regulation of body temperature is of vital importance because metabolic processes involve chemical reactions that are speeded or slowed by changing heat conditions. As a result, even slight shifts in body temperature can disrupt the normal rates of such reactions and produce metabolic disorders.

Normally, the temperature of deeper body parts remains close to 37°C (98.6°F), and the maintenance of a stable temperature requires that the amount of heat lost by the body be balanced by the amount it produces. The skin plays a key role in the homeostatic mechanism associated with the regulation of body temperature.

Heat Production and Loss

Heat is a by-product of cellular respiration. Thus, the more active cells of the body are the major heat producers. They include muscle cells and the cells of certain glands such as the liver.

When heat production is excessive, nerve impulses stimulate structures in the skin and other organs to react in ways that promote heat loss. For example, during physical exercise, active muscles release heat, and the blood carries away the excess. A part of the brain senses the increasing blood temperature and signals muscles in the walls of dermal blood vessels to relax. As the blood vessels dilate, more heat-carrying blood enters the dermis, and some of the heat is likely to escape to the outside of the body.

At the same time, the nervous system stimulates eccrine sweat glands to become active and to release fluid onto the surface of the skin. As this fluid evaporates (changes from a liquid to a gas), it carries heat away from the surface, cooling the skin.

If heat is lost excessively, as may occur in a very cold environment, the brain triggers different responses in the skin structures. For example, muscles in the walls of dermal blood vessels are stimulated to contract; this decreases the flow of heat-carrying blood through the skin and helps to reduce heat loss. Also, the sweat glands remain inactive, decreasing heat loss by evaporation. If the body temperature continues to drop, the nervous system may stimulate muscle fibers in the skeletal muscles throughout the body to contract slightly. This action requires an increase in the rate of cellular respiration and produces heat as a by-product. If this response does not raise the body temperature to normal, small groups of muscles may be stimulated to contract rhythmically with still greater force, and the person begins to shiver, generating more heat.

In addition to the skin, the circulatory and respiratory systems function in body temperature regulation. For example, if the body temperature is rising above normal, the heart is stimulated to beat faster. This causes more blood to be moved from the deeper, heat-generating tissues into the skin. At the same time, the breathing rate is increased, and more heat is lost from the lungs as an increased volume of air is moved in and out.

Figure 6.5 summarizes some of the actions involved in the body temperature-regulating mechanism.

Unfortunately, the body's temperature-regulating mechanism does not always operate satisfactorily, and the consequences may be unpleasant and dangerous. For example, since air can hold only a limited amount of water vapor, on a hot, humid day the air may become saturated with it. At such times, the sweat glands may be activated, but the sweat will not evaporate. The skin becomes wet, but the person remains hot and uncomfortable. In addition, if the air temperature is high, heat is lost less effectively. In fact, if the air temperature is higher than the body temperature, the person may gain heat from the surroundings, and the body temperature may rise even higher.

---

1. Why is the regulation of body temperature so important?
2. How does the body lose heat?
3. What actions help the body to conserve heat?

Fig. 6.5 Some actions involved in body temperature regulation.

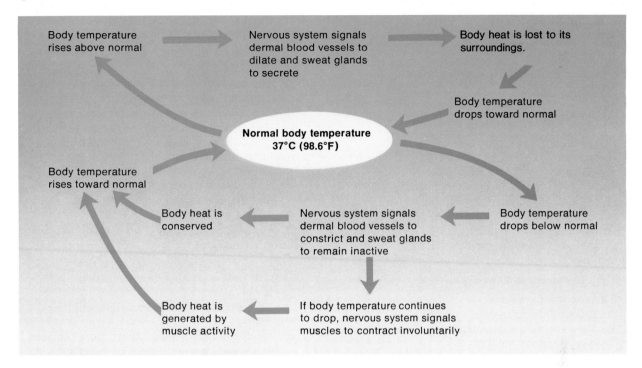

Common Skin Disorders

acne (ak′ne)—a disease of the sebaceous glands accompanied by blackheads and pimples.

alopecia (al″o-pe′she-ah)—loss of hair.

athlete's foot (ath′-lētz foot) (tinea pedis)—a fungus infection usually involving the skin of the toes and soles.

birthmark (berth′mark)—vascular tumor involving the skin and subcutaneous tissues.

boil (boil) (furuncle)—a bacterial infection involving a hair follicle and/or sebaceous glands.

carbuncle (kar′bung-kl)—a bacterial infection similar to a boil, spreading into the subcutaneous tissues.

cyst (sist)—a liquid-filled sac or capsule.

dermatitis (der″mah-ti′tis)—an inflammation of the skin.

eczema (ek′zĕ-mah)—a noncontagious skin rash, often accompanied by itching, blistering, and scaling.

erythema (er″ĭ-the′mah)—reddening of the skin, due to dilation of dermal blood vessels, in response to injury or inflammation.

herpes (her′pēz)—an infectious disease of the skin usually caused by the virus *Herpes simplex* and characterized by recurring formations of small clusters of vesicles.

impetigo (im″pĕ-ti′go)—a contagious disease of bacterial origin characterized by pustules that rupture and become covered with loosely held crusts.

keloids (ke′loidz)—fibrous tumors usually initiated by an injury.

mole (mōl) (nevi)—skin tumor that usually is pigmented; colors range from yellow to brown or black.

pediculosis (pĕ-dik″u-lo′sis)—a disease produced by infestation of lice.

pruritus (proo-ri′tus)—itching of the skin without eruptions.

psoriasis (so-ri′ah-sis)—a chronic skin disease characterized by red patches covered with silvery scales.

pustule (pus′tūl)—elevated, pus-filled area.

scabies (ska′bēz)—a disease resulting from an infestation of mites.

seborrhea (seb″o-re′ah)—a disease characterized by hyperactivity of the sebaceous glands and accompanied by greasy skin and dandruff.

shingles (shing′gelz) (herpes zoster)—a disease caused by a viral infection of certain spinal nerves; characterized by severe localized pain and blisters in the skin areas supplied by the affected nerves.

ulcer (ul′ser)—an open sore.

wart (wort)—a flesh-colored, raised area caused by a virus infection.

Chapter Summary

Introduction

Organs, such as membranes, are composed of two or more kinds of tissues.

The skin and its accessory organs constitute the integumentary system.

Types of Membranes

1. Serous membranes
   a. Serous membranes line body cavities that lack openings to the outside.
   b. They secrete watery serous fluid that lubricates membrane surfaces.

2. Mucous membranes
   a. Mucous membranes line cavities and tubes that open to the outside.
   b. They secrete mucus.

3. The cutaneous membrane is the skin.

4. Synovial membranes
   a. Synovial membranes line joint cavities.
   b. They secrete synovial fluid that lubricates the ends of bones at joints.

The Skin and Its Tissues

Skin functions as a protective covering, aids in regulating body temperature, retards water loss, houses sensory receptors, synthesizes various chemicals, and excretes wastes.

It is composed of an epidermis, and a dermis.

1. The epidermis
   a. The deepest layer of the epidermis contains cells undergoing mitosis.
   b. Epidermal cells undergo keratinization as they mature and are pushed toward the surface.
   c. The outermost layer is composed of dead epidermal cells.
   d. Epidermis functions to protect underlying tissues against water loss, mechanical injury, and the effects of harmful chemicals.

2. The dermis
   a. The dermis is a layer that binds the epidermis to underlying tissues.
   b. Dermal blood vessels supply nutrients to all skin cells and aid in regulating body temperature.
   c. Nerve tissue is scattered throughout the dermis.
      (1) Some dermal nerve fibers carry impulses to muscles and glands of the skin.
      (2) Other dermal nerve fibers are associated with various sensory receptors in the skin.

3. The subcutaneous layer
   a. The subcutaneous layer is composed of loose connective tissue and adipose tissue.
   b. Adipose tissue helps to conserve body heat.
   c. This layer contains blood vessels that supply the skin.

Accessory Organs of the Skin

1. Hair follicles
   a. Each hair develops from epidermal cells at the base of a hair follicle.
   b. As newly formed cells develop and grow, older cells are pushed toward the surface and undergo keratinization.
   c. A bundle of smooth muscle cells is attached to each hair follicle.

2. Sebaceous glands
   a. Sebaceous glands usually are associated with hair follicles.
   b. They secrete sebum, which helps keep skin and hair soft and waterproof.
3. Nails
   Nails are produced by epidermal cells that undergo keratinization.
4. Sweat glands
   a. Each sweat gland consists of a coiled tube.
   b. Apocrine glands respond to emotional stress, while eccrine glands respond to elevated body temperature.
   c. Sweat is primarily water, but also contains salts and waste products.

Regulation of Body Temperature

Regulation of body temperature is vital because heat affects the rates of metabolic reactions.
Normal temperature of deeper body parts is about 37°C (98.6°F).

1. Heat production and loss
   a. When the body temperature rises above normal, dermal blood vessels dilate and sweat glands secrete sweat.
   b. If the body temperature drops below normal, dermal blood vessels constrict and sweat glands become inactive.
   c. During excessive heat loss, skeletal muscles are stimulated to contract involuntarily.

Application of Knowledge

1. What special problems would result from a loss of 50% of a person's functional skin surface? How might this person's environment be modified to compensate partially for such a loss?
2. A premature infant typically lacks subcutaneous adipose tissue. Also, the surface area of an infant's small body is relatively large compared to its volume. How do you think these factors influence such an infant's ability to regulate its body temperature?

Review Activities

1. Explain why a membrane is an organ.
2. Define *integumentary system*.
3. Distinguish between serous and mucous membranes.
4. List six functions of the skin.
5. Distinguish between the epidermis and the dermis.
6. Describe the subcutaneous layer and its function.
7. Explain what happens to epidermal cells as they undergo keratinization.
8. Describe the function of melanocytes.
9. Review the functions of dermal nerve tissue.
10. Distinguish between a hair and a hair follicle.
11. Explain the function of sebaceous glands.
12. Describe how nails are formed.
13. Distinguish between apocrine and eccrine glands.
14. Explain how body heat is produced.
15. Explain how sweat glands function in regulating body temperature.
16. Describe the body's responses to decreasing body temperature.

Body Organization

Chapters 1 through 6 discussed different levels of organization—atoms, molecules, organelles, cells, tissues, organs, and an organ system. This chapter deals with the structure of the whole human organism and the special set of terms used to describe its major parts. It also presents brief descriptions of the organ systems that comprise the human organism. The chapters that follow will discuss the structure and function of these organ systems in more detail.

7

Chapter Outline

Chapter Objectives

After you have studied this chapter, you should be able to

1. Describe the location of the major body cavities.

2. List the organs located in each of the body cavities.

3. Name the membranes associated with the thoracic and abdominopelvic cavities.

4. Name the major organ systems of the body and list the organs associated with each.

5. Describe the general functions of each organ system.

6. Properly use the terms that describe relative positions, body sections, and body regions.

7. Complete the review activities at the end of this chapter. Note that the items are worded in the form of specific learning objectives. You may want to refer to them before reading the chapter.

abdominal (ab-do′mĭn-al)

appendicular (ap″en-dik′u-lar)

axial (ak′se-al)

cranial (kra′ne-al)

parietal (pah-ri′ĕ-tal)

pelvic (pel′vik)

pericardial (per″ĭ-kar′de-al)

peritoneal (per″ĭ-to-ne-′al)

pleural (ploo′ral)

spinal (spi′nal)

thoracic (tho-ras′ik)

visceral (vis′er-al)

append-, to hang something: *append*icular—pertaining to the arms and legs.

cran-, helmet: *cran*ial—pertaining to the portion of the skull that surrounds the brain.

dors-, back: *dors*al—a position toward the back of the body.

mediastin-, middle: *mediastin*um—region that separates the thorax into two compartments.

pariet-, wall: *pariet*al membrane—membrane that lines the wall of a cavity.

pelv-, basin: *pelv*ic cavity—basin-shaped cavity enclosed by the pelvic bones.

peri-, around: *peri*cardial membrane—membrane that surrounds the heart.

pleur-, rib: *pleur*al membrane—membrane that encloses the lungs within the rib cage.

thorac-, chest: *thorac*ic—pertaining to the chest.

ventr-, belly: *ventr*al—a position toward the front of the body.

Fig. 7.1 Major body cavities.

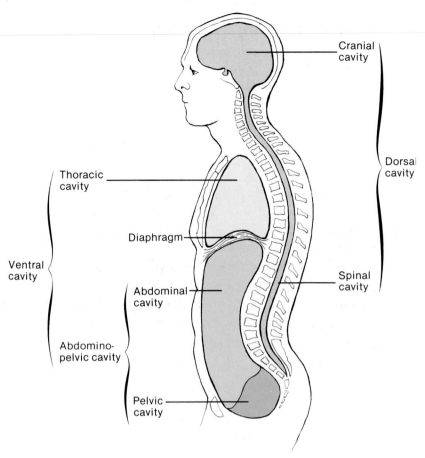

The human organism is a complex structure composed of many parts. In order to communicate with each other more effectively, investigators over the ages have developed a set of terms to name body features—cavities, membranes, organs, and organ systems. An additional set of terms has been devised to describe positions of body parts with respect to each other, imaginary planes passing through these structures, and various body regions.

Body Cavities and Membranes

Body Cavities

The human organism can be divided into an **axial portion,** which includes the head, neck, and trunk, and an **appendicular portion,** which includes the arms and legs. Within the axial portion of the body there are two major cavities—a **dorsal cavity** and a larger **ventral cavity.** These cavities are filled with organs, which as a group are called *visceral organs.* The dorsal cavity can be subdivided into two parts—the **cranial cavity**

within the skull, which houses the brain; and the **spinal cavity,** which contains the spinal cord and is surrounded by sections of the backbone (vertebrae). The ventral cavity consists of a **thoracic cavity** and an **abdominopelvic cavity.** These major body cavities are shown in figure 7.1.

The thoracic cavity is separated from the lower abdominopelvic cavity by a broad, thin muscle called the **diaphragm.** The wall of the thoracic cavity is composed of skin, skeletal muscles, and various bones. The viscera within it include the lungs and a region called the *mediastinum.* The mediastinum separates the thorax into two compartments that contain the right and left lungs. The rest of the thoracic viscera—heart, esophagus, trachea, and thymus gland—are located within the mediastinum.

The abdominopelvic cavity, which includes an upper abdominal portion and a lower pelvic portion, extends from the diaphragm to the floor of the pelvis. Its wall consists primarily of skin, skeletal muscles, and bones. The visceral organs within the abdominal cavity include the stomach, liver, spleen, gallbladder, and most of the small and large intestines.

Fig. 7.2 A transverse section through the thorax reveals the serous membranes associated with the heart and lungs.

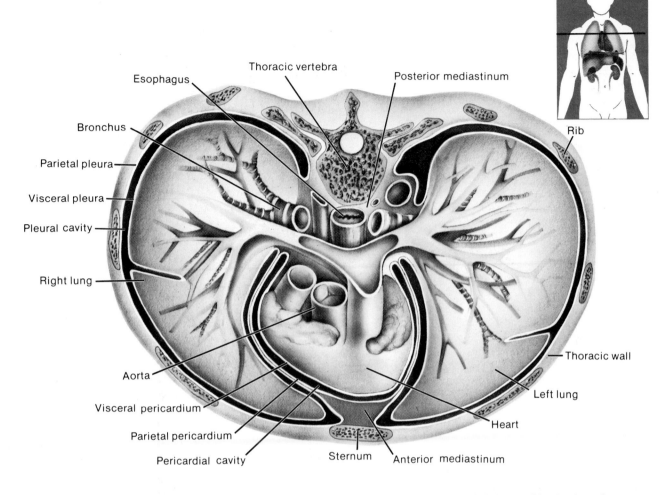

The **pelvic cavity** is the portion of the abdominopelvic cavity enclosed by the pelvic bones. It contains the terminal end of the large intestine, the rectum, the urinary bladder, and the internal reproductive organs.

Thoracic and Abdominopelvic Membranes

The walls of the right and left thoracic compartments, which contain the lungs, are lined with a serous membrane called the *parietal pleura*. The lungs themselves are covered by a similar membrane called the *visceral pleura.*

The parietal and visceral **pleural membranes** are separated by a thin film of watery serous fluid that they secrete, and although there is normally no actual space between these membranes, a possible or potential space called the *pleural cavity* exists between them.

The heart, which is located in the broadest portion of the mediastinum, is surrounded by serous membranes called **pericardial membranes.** A thin *visceral pericardium* covers the heart's surface and is separated from the much thicker, fibrous *parietal pericardium* by a small amount of serous fluid. The potential space between these membranes is called the *pericardial cavity.* Figure 7.2 shows the membranes associated with the heart and lungs.

In the abdominopelvic cavity, the serous membranes are called **peritoneal membranes.** A *parietal peritoneum* lines the wall, and a *visceral peritoneum* covers each organ in the abdominal cavity. The potential space between these membranes is called the *peritoneal cavity* (fig. 7.3).

1. What is meant by visceral organ?
2. What organs occupy the dorsal cavity? The ventral cavity?
3. Describe the membranes associated with the thoracic and abdominopelvic cavities.

Fig. 7.3 A transverse section through the abdomen at the level of the stomach. How would you distinguish between the parietal and the visceral peritoneum?

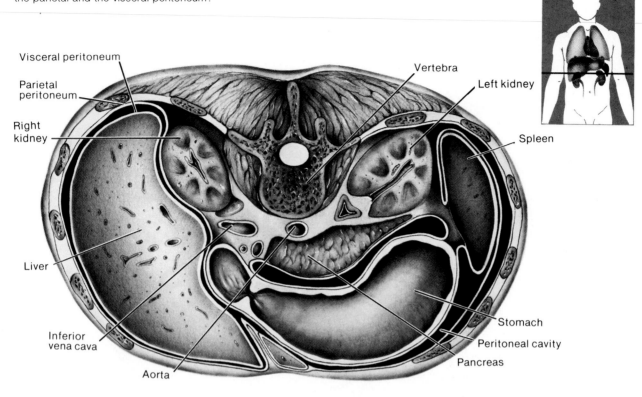

Visceral peritoneum
Parietal peritoneum
Right kidney
Liver
Inferior vena cava
Aorta
Vertebra
Left kidney
Spleen
Stomach
Peritoneal cavity
Pancreas

Organ Systems

The human organism is composed of several organ systems. Each of these systems includes a set of interrelated organs that work together to provide specialized functions, vital to the survival of the organism. As you read about each system, you may want to consult the illustrations of the human torso, and locate some of the organs listed in the description. (See figs. 7.4–7.10.)

Body Covering

The organs of the **integumentary system** (discussed in chapter 6) include the skin and accessory organs such as the hair, nails, sweat glands, and sebaceous glands. These parts protect underlying tissues, help regulate body temperature, house a variety of sensory receptors, and synthesize certain cellular products.

Support and Movement

The organs of the skeletal and muscular systems (discussed in chapters 8 and 9) function to support and move body parts.

The **skeletal system** consists of the bones as well as the ligaments and cartilages that bind the bones together. These parts provide frameworks and protective shields for softer tissues, serve as attachments for muscles, and act together with muscles when body parts move. Tissues within bones also produce blood cells and store inorganic salts.

The muscles are the organs of the **muscular system.** By contracting and pulling their ends closer together, they furnish the forces that cause body movements. They also function in maintaining posture and are largely responsible for the production of body heat.

Fig. 7.4 Human torso, anterior surface, with superficial muscle layer exposed.

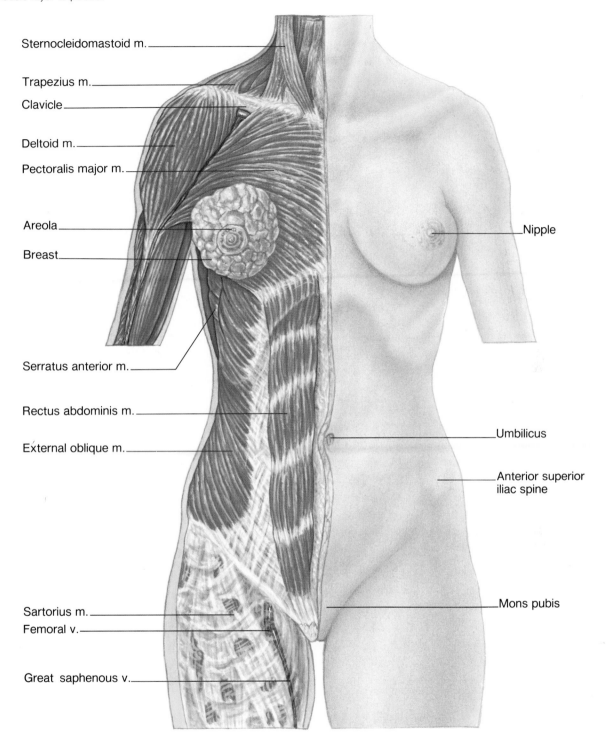

Sternocleidomastoid m.

Trapezius m.

Clavicle

Deltoid m.

Pectoralis major m.

Areola

Breast

Serratus anterior m.

Rectus abdominis m.

External oblique m.

Sartorius m.

Femoral v.

Great saphenous v.

Nipple

Umbilicus

Anterior superior iliac spine

Mons pubis

Fig. 7.5 Human torso with deeper muscle layers exposed.

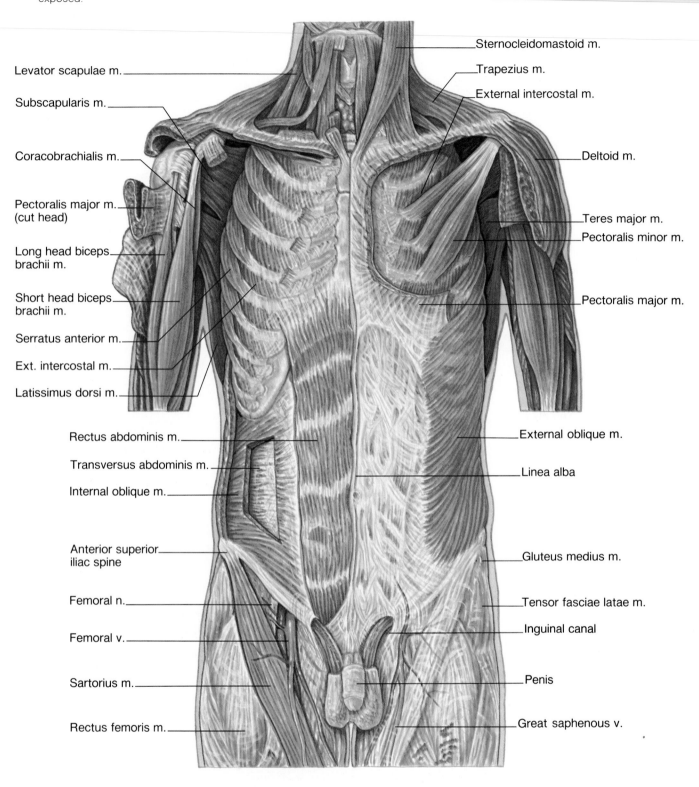

Levator scapulae m.

Subscapularis m.

Coracobrachialis m.

Pectoralis major m. (cut head)

Long head biceps brachii m.

Short head biceps brachii m.

Serratus anterior m.

Ext. intercostal m.

Latissimus dorsi m.

Rectus abdominis m.

Transversus abdominis m.

Internal oblique m.

Anterior superior iliac spine

Femoral n.

Femoral v.

Sartorius m.

Rectus femoris m.

Sternocleidomastoid m.

Trapezius m.

External intercostal m.

Deltoid m.

Teres major m.

Pectoralis minor m.

Pectoralis major m.

External oblique m.

Linea alba

Gluteus medius m.

Tensor fasciae latae m.

Inguinal canal

Penis

Great saphenous v.

Fig. 7.6 Human torso with abdominal viscera exposed.

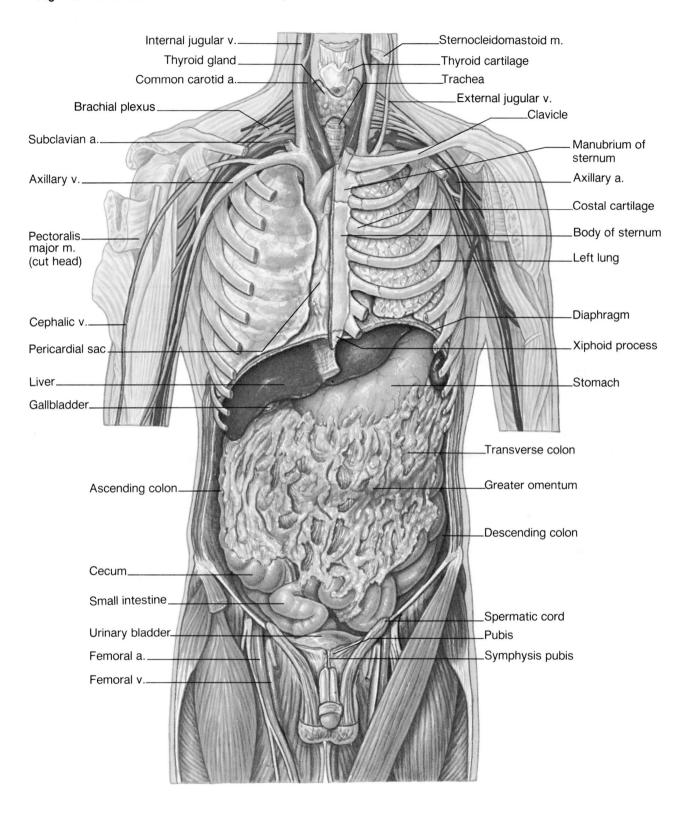

Internal jugular v.

Thyroid gland

Common carotid a.

Brachial plexus

Subclavian a.

Axillary v.

Pectoralis major m. (cut head)

Cephalic v.

Pericardial sac

Liver

Gallbladder

Ascending colon

Cecum

Small intestine

Urinary bladder

Femoral a.

Femoral v.

Sternocleidomastoid m.

Thyroid cartilage

Trachea

External jugular v.

Clavicle

Manubrium of sternum

Axillary a.

Costal cartilage

Body of sternum

Left lung

Diaphragm

Xiphoid process

Stomach

Transverse colon

Greater omentum

Descending colon

Spermatic cord

Pubis

Symphysis pubis

Fig. 7.7 Human torso with thoracic viscera exposed.

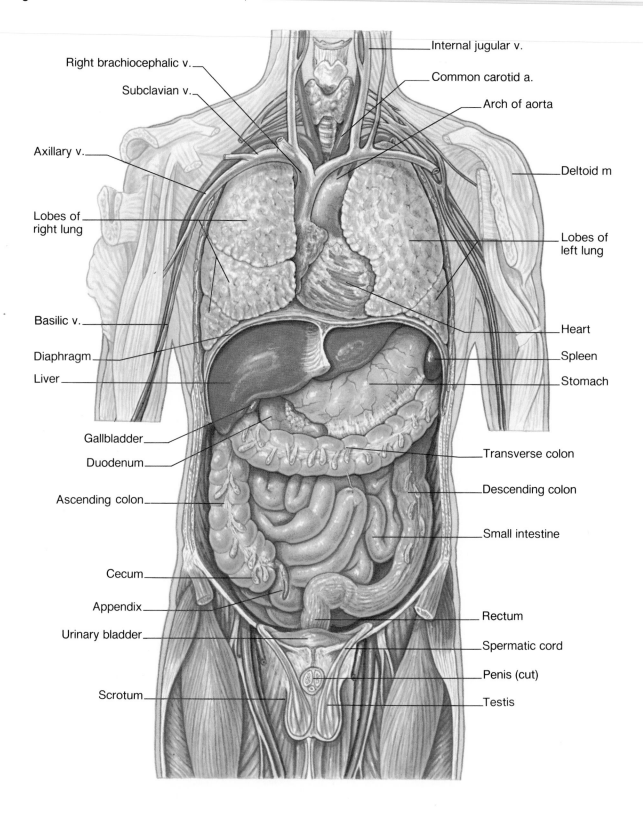

Right brachiocephalic v.

Subclavian v.

Axillary v.

Lobes of right lung

Basilic v.

Diaphragm

Liver

Gallbladder

Duodenum

Ascending colon

Cecum

Appendix

Urinary bladder

Scrotum

Internal jugular v.

Common carotid a.

Arch of aorta

Deltoid m

Lobes of left lung

Heart

Spleen

Stomach

Transverse colon

Descending colon

Small intestine

Rectum

Spermatic cord

Penis (cut)

Testis

Fig. 7.8 Human torso with lungs sectioned, heart and small intestine removed.

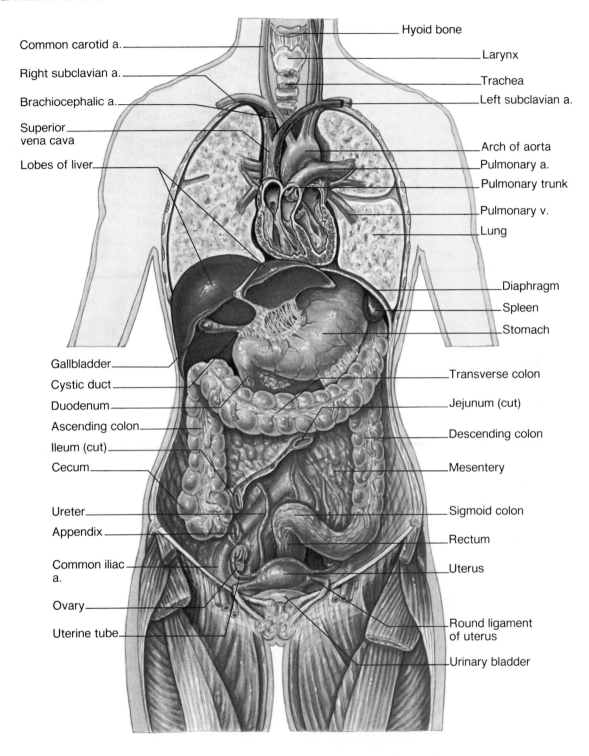

Common carotid a.

Right subclavian a.

Brachiocephalic a.

Superior vena cava

Lobes of liver

Gallbladder

Cystic duct

Duodenum

Ascending colon

Ileum (cut)

Cecum

Ureter

Appendix

Common iliac a.

Ovary

Uterine tube

Hyoid bone

Larynx

Trachea

Left subclavian a.

Arch of aorta

Pulmonary a.

Pulmonary trunk

Pulmonary v.

Lung

Diaphragm

Spleen

Stomach

Transverse colon

Jejunum (cut)

Descending colon

Mesentery

Sigmoid colon

Rectum

Uterus

Round ligament of uterus

Urinary bladder

Fig. 7.9 Human torso with stomach and parts of intestines and lungs removed.

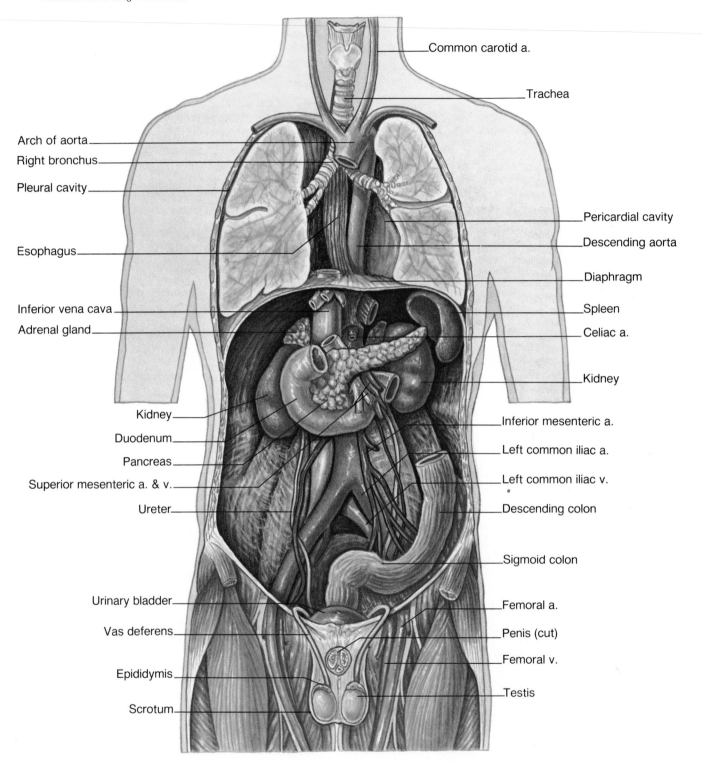

Common carotid a.

Trachea

Arch of aorta

Right bronchus

Pleural cavity

Pericardial cavity

Descending aorta

Esophagus

Diaphragm

Inferior vena cava

Spleen

Adrenal gland

Celiac a.

Kidney

Kidney

Inferior mesenteric a.

Duodenum

Left common iliac a.

Pancreas

Left common iliac v.

Superior mesenteric a. & v.

Descending colon

Ureter

Sigmoid colon

Urinary bladder

Femoral a.

Vas deferens

Penis (cut)

Femoral v.

Epididymis

Testis

Scrotum

Fig. 7.10 Human torso with posterior walls of thorax and abdominal and pelvic cavities exposed.

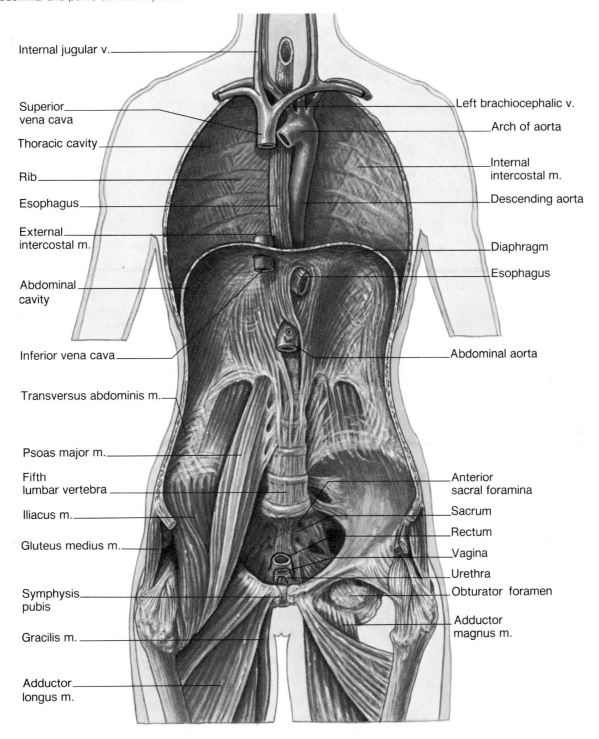

Internal jugular v.

Superior vena cava

Thoracic cavity

Rib

Esophagus

External intercostal m.

Abdominal cavity

Inferior vena cava

Transversus abdominis m.

Psoas major m.

Fifth lumbar vertebra

Iliacus m.

Gluteus medius m.

Symphysis pubis

Gracilis m.

Adductor longus m.

Left brachiocephalic v.

Arch of aorta

Internal intercostal m.

Descending aorta

Diaphragm

Esophagus

Abdominal aorta

Anterior sacral foramina

Sacrum

Rectum

Vagina

Urethra

Obturator foramen

Adductor magnus m.

In order to act as a unit, body parts must be integrated and coordinated. That is, their activities must be controlled and adjusted so that homeostasis is maintained. This is the general function of the nervous and endocrine systems.

The **nervous system** (discussed in chapter 10) consists of the brain, spinal cord, and associated nerves. Nerve cells within these organs use electrochemical signals called nerve impulses to communicate with one another and with muscles and glands. Some nerve cells are specialized to detect changes that occur inside and outside the body. Others receive the impulses transmitted from these sensory units and interpret and act on this information. Still others carry impulses from the brain or spinal cord to muscles or glands and stimulate these parts to contract or to secrete various products.

The **endocrine system** (discussed in chapter 12) includes all the glands that secrete **hormones.** The hormones, in turn, travel away from the gland in some body fluid such as blood. Usually a particular hormone only affects a particular group of cells, which is called its target tissue. The effect of a hormone is to alter the function of the target tissue in some way and thus aid in controlling metabolism.

The organs of the endocrine system include the pituitary, thyroid, parathyroid, and adrenal glands, as well as the pancreas, ovaries, testes, pineal gland, and thymus gland.

Processing and Transporting

The organs of several systems are involved with processing and transporting nutrients, oxygen, and various wastes. The organs of the **digestive system** (discussed in chapter 13) receive foods from the outside. Then, they convert some food molecules into forms that can pass through cell membranes and be absorbed. Materials that fail to be absorbed are eliminated by being transported to the outside of the body. These organs also produce a variety of hormones.

The digestive system includes the mouth, teeth, salivary glands, pharynx, esophagus, stomach, liver, gallbladder, pancreas, small intestine, and large intestine.

The organs of the **respiratory system** (discussed in chapter 14) provide for the intake and output of air and for the exchange of gases between the blood and the air. The nasal cavity, pharynx, larynx, trachea, bronchi, and lungs are parts of this system.

The **circulatory system** (discussed in chapters 15 and 16) includes the heart, arteries, veins, capillaries, and blood. The heart functions as a muscular pump that forces blood through the blood vessels. The blood serves as a fluid for transporting oxygen, nutrients, hormones, and wastes. It carries oxygen from the lungs and nutrients from the digestive organs to all body cells, where these substances are used in metabolic processes. It also transports hormones from various endocrine glands to their target tissues and wastes from body cells to the excretory organs, where they are removed from the blood and released to the outside.

The **lymphatic system** (discussed in chapter 17) is sometimes considered to be part of the circulatory system. It is composed of the lymphatic vessels, lymph fluid, lymph nodes, thymus gland, and spleen. This system transports fluid from the tissues back to the bloodstream and carries certain lipids away from digestive organs. It also aids in defending the body against infections by removing particles such as microorganisms from the tissue fluid. In addition, it supports the activities of the cells (lymphocytes) responsible for immune reactions against specific disease-causing agents.

The **urinary system** (discussed in chapter 18) consists of the kidneys, ureters, urinary bladder, and urethra. The kidneys remove various wastes from the blood, and assist in maintaining water and electrolyte balance. The product of these activities is urine. Other portions of the urinary system are concerned with storing urine and transporting it to the outside of the body.

Reproduction

Reproduction is the process of producing offspring. Cells reproduce when they undergo mitosis and give rise to daughter cells. The **reproductive system** of an organism, however, is concerned with the production of new organisms. (See chapter 20.)

The male reproductive system includes the scrotum, testes, epididymides, vasa deferentia, seminal vesicles, prostate gland, bulbourethral glands, penis, and urethra. These parts are concerned with the production and maintenance of the male sex cells or sperm. They also function to transfer these cells into the female reproductive tract.

The female reproductive system consists of the ovaries, fallopian tubes, uterus, vagina, and vulva. These parts produce and maintain the female sex cells or eggs, receive the male cells, and transport these cells within the female system. The female parts also provide for the support and development of embryos and function in the birth process.

1. Name each of the major organ systems and list the organs of each system.
2. Describe the general functions of each organ system.

Descriptive Terms

In describing body parts and the relationships between them, it is convenient to use terms with precise meanings. Some of these terms concern the relative positions of parts, others relate to imaginary planes along which cuts may be made, and still others are used to describe various regions of the body.

When such terms are used, it is assumed that the body is in the **anatomical position,** that is, it is standing erect, the face is forward, and the arms are at the sides, with the palms forward.

Relative Position

Terms of relative position are used to describe the location of one part with respect to another. They include the following:

1. **Superior** means a part is above, or closer to, the head than another part. (The thoracic cavity is superior to the abdominopelvic cavity.)

2. **Inferior** means situated below something, or toward the feet. (The neck is inferior to the head.)

3. **Anterior** (or *ventral*) means toward the front. (The eyes are anterior to the brain.)

4. **Posterior** (or *dorsal*) is the opposite of anterior; it means toward the back. (The pharynx is posterior to the oral cavity.)

5. **Medial** relates to an imaginary midline dividing the body into equal right and left halves. A part is medial if it is closer to this line than another part. (The nose is medial to the eyes.)

6. **Lateral** means toward the side with respect to the imaginary midline. (The ears are lateral to the eyes.)

7. **Proximal** is used to describe a part that is closer to a point of attachment or to the trunk of the body than something else. (The elbow is proximal to the wrist.)

8. **Distal** is the opposite of proximal. It means something is further from the point of attachment or the trunk than something else. (The fingers are distal to the wrist.)

Fig. 7.11 Some descriptive terms used to indicate the relative positions of body parts.

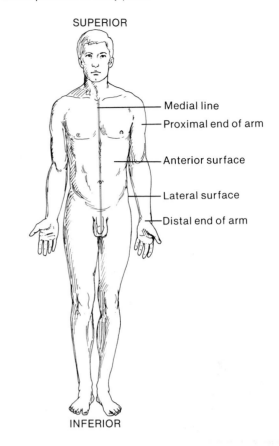

SUPERIOR

— Medial line
— Proximal end of arm

— Anterior surface

— Lateral surface

— Distal end of arm

INFERIOR

9. **Superficial** means situated near the surface. (The epidermis is the superficial layer of the skin.) **Peripheral** also means outward or near the surface. It is used to describe the location of certain blood vessels and nerves. (Nerves that branch from the brain and spinal cord are peripheral nerves.)

10. **Deep** is used to describe more internal parts. (The dermis is the deep layer of the skin.)

Figure 7.11 illustrates some of these terms.

Body Sections

To observe the relative locations and arrangements of internal parts, it is necessary to cut or section the body along various planes (fig. 7.12). The following terms are used to describe such planes and sections.

1. **Sagittal.** This refers to a lengthwise cut that divides the body into right and left portions. If the section passes along the midline and divides the body into equal halves, it is called *mid-sagittal*.

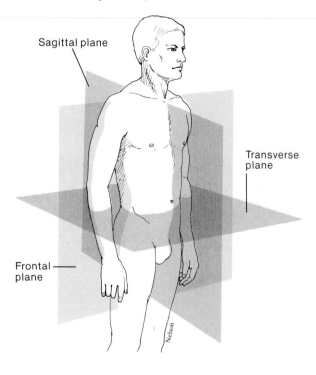

Fig. 7.14 CAT scans of (a) the head and (b) the abdomen. What organs can you identify?

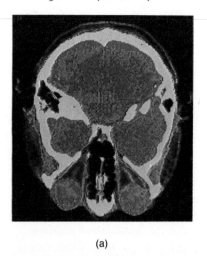

(a)

Fig. 7.13 Tubular parts, such as blood vessels, may be cut in (a) cross section, (b) oblique section, or (c) longitudinal section.

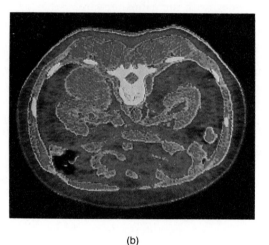

(b)

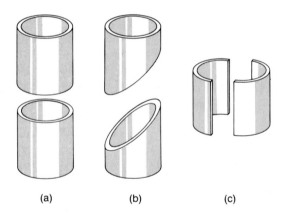

(a)   (b)   (c)

Sections of internal organs can be visualized by using a procedure called *computerized axial tomography* or CAT scanning. In this procedure, an X ray-emitting part is moved around the body region being examined. At the same time, an X ray detector is moved in the opposite direction on the other side of the body. As the parts move, an X ray beam is passed through the body from hundreds of different angles. Since tissues and organs of varying densities within the body absorb X rays differently, the amount of X ray reaching the detector varies from position to position.

The measurements made by the detector are recorded in the memory of a computer, which later combines them mathematically and creates a sectional image of the internal body parts on a viewing screen. (See fig. 7.14.)

2. **Transverse.** A transverse cut is one that divides the body into superior and inferior portions.

3. **Frontal** (or *coronal*). A frontal section divides the body into anterior and posterior portions.

Sometimes a tubular organ such as a blood vessel is sectioned. In this case, a cut across the structure is called a *cross section,* an angular cut is called an *oblique section,* and a lengthwise cut is called a *longitudinal section* (fig. 7.13).

Fig. 7.15 The abdominal area is subdivided into nine regions. How do the names of these regions describe their locations?

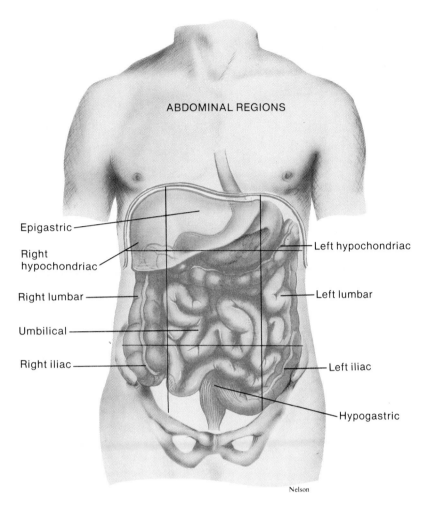

ABDOMINAL REGIONS

Epigastric

Right hypochondriac

Right lumbar

Umbilical

Right iliac

Left hypochondriac

Left lumbar

Left iliac

Hypogastric

Nelson

Body Regions

A number of terms designate body regions or refer to specific parts. The abdomen, for example, is subdivided as shown in figure 7.15. The upper middle portion is called the **epigastric region,** the central portion is the **umbilical region,** and the lower middle portion is the **hypogastric region.** On either side of the epigastric region is a **hypochondriac region;** on each side of the umbilical region is a **lumbar region;** on each side of the hypogastric region is an **iliac region.**

Some other terms commonly used to indicate body regions include the following (fig. 7.16):

antebrachium (an''te-bra'ke-um)—pertaining to the forearm.

antecubital (an''te-ku'bĭ-tal)—pertaining to the space in front of the elbow.

axillary (ak'sĭ-ler''e)—pertaining to the armpit.

brachial (bra'ke-al)—pertaining to the arm.

buccal (buk'al)—pertaining to the mouth or inner surfaces of the cheeks.

celiac (se'le-ak)—pertaining to the abdomen.

cephalic (sĕ-fal'ik)—pertaining to the head.

cervical (ser'vĭ-kal)—pertaining to the neck.

costal (kos'tal)—pertaining to the ribs.

cubital (ku'bĭ-tal)—pertaining to the elbow.

cutaneous (ku-ta'ne-us)—pertaining to the skin.

dorsum (dor'sum)—pertaining to the back.

frontal (frun'tal)—pertaining to the forehead.

gluteal (gloo'te-al)—pertaining to the buttocks.

groin (groin)—pertaining to the depressed area of the abdominal wall near the thigh.

inguinal ing'gwĭ-nal)—pertaining to the groin.

lumbar (lum'bar)—pertaining to the loins, the region of the lower back between the ribs and the pelvis.

mammary (mam'er-e)—pertaining to the breasts.

occipital (ok-sip'ĭ-tal)—pertaining to the back of the head.

Fig. 7.16 Some terms used to describe various body regions.

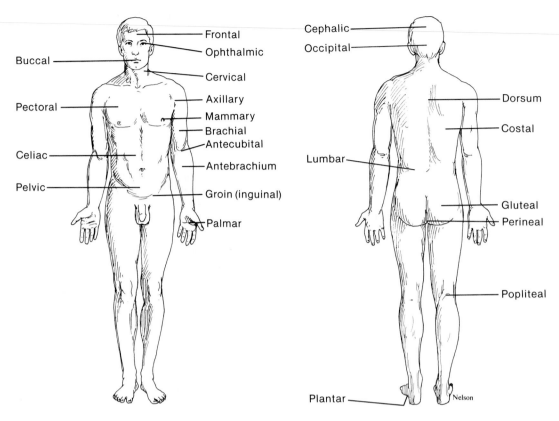

ophthalmic (of-thal'mik)—pertaining to the eyes.
palmar (pahl'mar)—pertaining to the palm.
pectoral (pek'tor-al)—pertaining to the chest or breast.
pelvic (pel'vik)—pertaining to the pelvis.
perineal (per''i-ne'al)—pertaining to the perineum, the region between the anus and the external reproductive organs.
plantar (plan'tar)—pertaining to the sole.
popliteal (pop''li-te'al)—pertaining to the area behind the knee.

---

1. Describe the anatomical position.
2. Using the appropriate terms, describe the relative positions of several body parts.
3. Describe three types of body sections.

Chapter Summary

Introduction

Precise terms have been devised to describe body features, positions of parts, imaginary planes, and body regions.

Body Cavities and Membranes

1. Body cavities
   a. The axial portion of the body contains the dorsal and ventral cavities.
      (1) Dorsal cavity includes the cranial and spinal cavities.
      (2) Ventral cavity includes the thoracic and abdominopelvic cavities.
   b. Body cavities are filled with visceral organs.

2. Thoracic and abdominopelvic membranes
   a. Thoracic membranes
      (1) Pleural membranes line the thoracic cavity and cover the lungs.
      (2) Pericardial membranes surround the heart and cover its surface.
      (3) The pleural and pericardial cavities are potential spaces between these membranes.

b. Abdominopelvic membranes
 (1) Peritoneal membranes line the abdominopelvic cavity and cover the organs inside.
 (2) The peritoneal cavity is a potential space between these membranes.

Organ Systems

The human organism is composed of several organ systems. Each system consists of a set of interrelated organs that carry on vital functions.

1. Body covering
 a. The integumentary system includes the skin, hair, nails, sweat glands, and sebaceous glands.
 b. It functions to protect underlying tissues, regulate body temperature, house sensory receptors, and synthesize various cellular products.

2. Support and movement
 a. Skeletal system
 (1) The skeletal system is composed of bones and parts that bind bones together.
 (2) It provides framework, protective shields, and attachments for muscles; also produces blood cells and stores inorganic salts.
 b. Muscular system
 (1) The muscular system includes the muscles of the body.
 (2) It is responsible for body movements, maintenance of posture, and production of body heat.

3. Integration and coordination
 a. Nervous system
 (1) The nervous system consists of the brain, spinal cord, and nerves.
 (2) It functions to receive impulses from sensory parts, interpret these impulses, and act on them by causing muscles or glands to respond.
 b. Endocrine system
 (1) The endocrine system consists of glands that secrete hormones.
 (2) Hormones help regulate metabolism.
 (3) It includes the pituitary, thyroid, parathyroid, and adrenal glands; the pancreas, ovaries, testes, pineal gland, and thymus gland.

4. Processing and transporting
 a. Digestive system
 (1) The digestive system receives foods, converts their molecules into forms that can pass through cell membranes, and eliminates materials that are not absorbed.
 (2) It includes the mouth, teeth, salivary glands, pharynx, esophagus, stomach, liver, gallbladder, pancreas, small intestine, and large intestine.

b. Respiratory system
 (1) The respiratory system provides for the intake and output of air and for the exchange of gases.
 (2) It includes the nasal cavity, pharynx, larynx, trachea, bronchi, and lungs.
c. Circulatory system
 (1) The circulatory system includes the heart, which pumps blood, and the blood vessels, which carry blood to and from body parts.
 (2) Blood transports oxygen, nutrients, hormones, and wastes.
d. Lymphatic system
 (1) The lymphatic system is composed of lymphatic vessels, lymph nodes, thymus, and spleen.
 (2) It transports lymph from tissues to the bloodstream, carries lipids away from digestive organs, and aids in defending the body against infections.
e. Urinary system
 (1) The urinary system includes the kidneys, ureters, urinary bladder, and urethra.
 (2) It filters wastes from the blood and helps maintain water and electrolyte balance.

5. Reproduction
 a. The reproductive systems are concerned with the production of new organisms.
 b. Male reproductive system includes the scrotum, testes, epididymides, vasa deferentia, seminal vesicles, prostate gland, bulbourethral glands, penis, and urethra, which produce, maintain, and transport male sex cells.
 c. Female reproductive system includes the ovaries, fallopian tubes, uterus, vagina, and vulva, which produce, maintain, and transport female sex cells.

Descriptive Terms

1. Relative position
 These terms are used to describe the location of one part with respect to another part.

2. Body sections
 Body sections are planes along which the body may be cut to observe the relative locations and arrangements of internal parts.

3. Body regions
 Various body regions are designated by special terms.

Application of Knowledge

1. When a woman is pregnant, what changes would occur in the relative sizes of the abdominal and pelvic portions of the abdominopelvic cavity?

2. Suppose two individuals are afflicted with benign (noncancerous) tumors that produce symptoms because they occupy space and crowd adjacent organs. If one of these persons has the tumor in the ventral cavity and the other has it in the dorsal cavity, which would be likely to develop symptoms first? Why?

Review Activities

Part A

1. Distinguish between the axial and the appendicular portions of the body.

2. Distinguish between the dorsal and ventral body cavities and name the smaller cavities that occur within each.
3. Explain what is meant by a visceral organ.
4. Describe the mediastinum and its contents.
5. Distinguish between a parietal and a visceral membrane.
6. Name the major organ systems and describe the general functions of each.
7. List the major organs that comprise each organ system.

Part B

1. Name the body cavity in which each of the following organs is located:
   a. Stomach
   b. Heart
   c. Brain
   d. Liver
   e. Trachea
   f. Rectum
   g. Spinal cord
   h. Esophagus
   i. Spleen
   j. Urinary bladder

2. Write complete sentences using each of the following terms correctly:
   a. Superior
   b. Inferior
   c. Anterior
   d. Posterior
   e. Medial
   f. Lateral
   g. Proximal
   h. Distal
   i. Superficial
   j. Peripheral
   k. Deep

3. Prepare a sketch of a human body and use lines to indicate each of the following sections:
   a. Midsagittal
   b. Transverse
   c. Frontal

4. Prepare a sketch of the abdominal area and indicate the location of each of the following regions:
   a. Epigastric
   b. Umbilical
   c. Hypogastric
   d. Hypochondriac
   e. Lumbar
   f. Iliac

5. Name the region or body part to which each of the following terms pertains:
   a. Antebrachium
   b. Antecubital
   c. Axillary
   d. Brachial
   e. Buccal
   f. Celiac
   g. Cephalic
   h. Cervical
   i. Costal
   j. Cubital
   k. Cutaneous
   l. Dorsum
   m. Frontal
   n. Gluteal
   o. Groin
   p. Inguinal
   q. Lumbar
   r. Mammary
   s. Occipital
   t. Ophthalmic
   u. Palmar
   v. Pectoral
   w. Pelvic
   x. Perineal
   y. Plantar
   z. Popliteal

Support and Movement

The chapters of unit 2 deal with the structures and functions of the skeletal and muscular systems. They describe how the organs of the skeletal system support and protect other body parts and how they function with organs of the muscular system to enable body parts to move. They also describe how skeletal structures participate in the formation of blood and in the storage of inorganic salts, and how the muscular tissues act to produce body heat and to move body fluids.

This unit includes

Chapter 8 The Skeletal System
Chapter 9 The Muscular System

The Skeletal System

The bones of the skeleton are composed of several kinds of tissues and thus are the organs of the *skeletal system*. Since bones are rigid structures, they provide support and protection for softer tissues, and they act together with skeletal muscles to make body movements possible. They also manufacture blood cells and store inorganic salts.

The shapes of individual bones are closely related to their functions. Projections provide places for the attachments of muscles, tendons, and ligaments, openings serve as passageways for blood vessels and nerves, and the ends of bones are modified to form joints with other bones.

8

Chapter Outline

Chapter Objectives

After you have studied this chapter, you should be able to

1. Describe the general structure of a bone and list the functions of its parts.

2. Distinguish between intramembranous and endochondral bones and explain how such bones develop and grow.

3. Discuss the major functions of bones.

4. Distinguish between the axial and appendicular skeletons and name the major parts of each.

5. Locate and identify the bones and the major features of the bones that comprise the skull, vertebral column, thoracic cage, pectoral girdle, upper limb, pelvic girdle, and lower limb.

6. List three types of joints, describe their characteristics, and name an example of each.

7. List six types of freely movable joints and describe the actions of each.

8. Explain how skeletal muscles produce movements at joints and identify several types of such movements.

9. Complete the review activities at the end of this chapter. Note that the items are worded in the form of specific learning objectives. You may want to refer to them before reading the chapter.

articular cartilage (ar-tik′u-lar kar′tĭ-lij)

compact bone (kom′pakt bōn)

diaphysis (di-af′ĭ-sis)

endochondral bone (en″do-kon′dral bōn)

epiphyseal disk (ep″ĭ-fiz′e-al disk)

epiphysis (e-pif′ĭ-sis)

hematopoiesis (hem″ah-to-poi-e′sis)

intramembranous bone
  (in″trah-mem′brah-nus bōn)

lever (lev′er)

marrow (mar′o)

medullary cavity (med′u-lar″e kav′ĭ-te)

osteoblast (os′te-o-blast)

osteoclast (os′te-o-klast)

osteocyte (os′te-o-sĭt)

periosteum (per″e-os′te-um)

spongy bone (spun′je bōn)

acetabul-, vinegar cup: *acetabul*um—depression of a coxal bone that articulates with the head of a femur.

ax-, an axis: *ax*ial skeleton—upright portion of the skeleton that supports the head, neck, and trunk.

-blast, budding: osteo*blast*—cell that forms bone tissue.

carp-, wrist: *carp*als—bones of the wrist.

-clast, broken: osteo*clast*—cell that breaks down bone tissue.

condyl-, knob: *condyl*e—a rounded, bony process.

corac-, beaklike: *corac*oid process—beaklike process of the scapula.

cribr-, sievelike: *cribr*iform plate—portion of the ethmoid bone with many small openings.

crist-, ridge: *crist*a galli—a bony ridge that projects upward into the cranial cavity.

fov-, pit: *fov*ea capitis—a pit in the head of a femur.

glen-, joint socket: *glen*oid cavity—a depression in the scapula that articulates with the head of the humerus.

hema-, blood: *hema*toma—a blood clot.

meat-, passage: auditory *meat*us—canal of the temporal bone that leads inward to parts of the ear.

odont-, tooth: *odont*oid process—a toothlike process of the second cervical vertebra.

poie-, making: hemato*poie*sis—process by which blood cells are formed.

An individual bone is composed of a variety of tissues, including bone tissue, cartilage, fibrous connective tissue, blood, and nerve tissue. Because there is so much nonliving material present in the matrix of the bone tissue, the whole organ may appear to be inert. A bone, however, contains very active tissues.

Bone Structure

Although the various bones of the skeletal system vary greatly in size and shape, they are similar in their structure, the way they develop, and the functions they provide.

Parts of a Long Bone

In describing the structure of bone, a long bone, such as one found in an arm or leg, will be used as an example (fig. 8.1). At each end of the bone there is an expanded portion called an **epiphysis,** which articulates (forms a joint) with another bone. On its outer surface, the articulating portion of the epiphysis is coated with a layer of hyaline cartilage called **articular cartilage.** The shaft of the bone, which is located between the epiphyses, is called the **diaphysis.**

Except for the articular cartilage on its ends, the bone is completely enclosed by a tough, vascular covering of fibrous tissue called the **periosteum.** This membrane is firmly attached to the bone, and its fibers are continuous with various ligaments and tendons that are connected to it. The periosteum also functions in the formation and repair of bone tissue.

Each bone has a shape closely related to its functions. Bony projections called *processes,* for example, provide sites for the attachment of ligaments and tendons; grooves and openings serve as passageways for blood vessels and nerves; a depression of one bone might articulate with a process of another.

The wall of the diaphysis is composed of tightly packed tissue called *compact bone.* This type of bone is solid, strong, and resistant to bending. The epiphyses, on the other hand, are composed largely of *spongy* (cancellous) *bone* with thin layers of compact bone on their surfaces. Spongy bone consists of numerous branching bony plates. Irregular interconnected spaces occur between these plates and help reduce the weight of the bone. Spongy bone provides strength, and its bony plates are most highly developed in the regions of the epiphyses that are subjected to forces of compression.

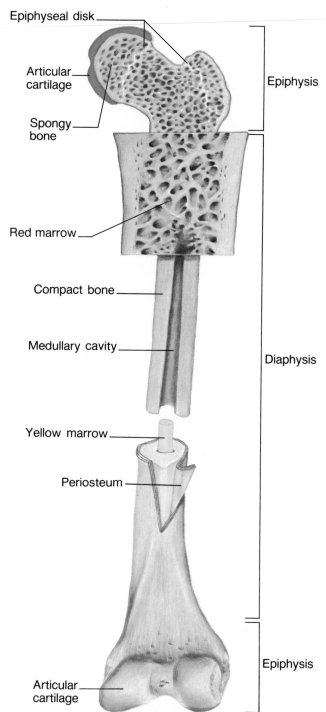

Fig. 8.1 The major parts of a long bone.

Epiphyseal disk

Articular cartilage

Spongy bone

Red marrow

Compact bone

Medullary cavity

Yellow marrow

Periosteum

Articular cartilage

Epiphysis

Diaphysis

Epiphysis

Fig. 8.2 Compact bone is composed of haversian systems cemented together.

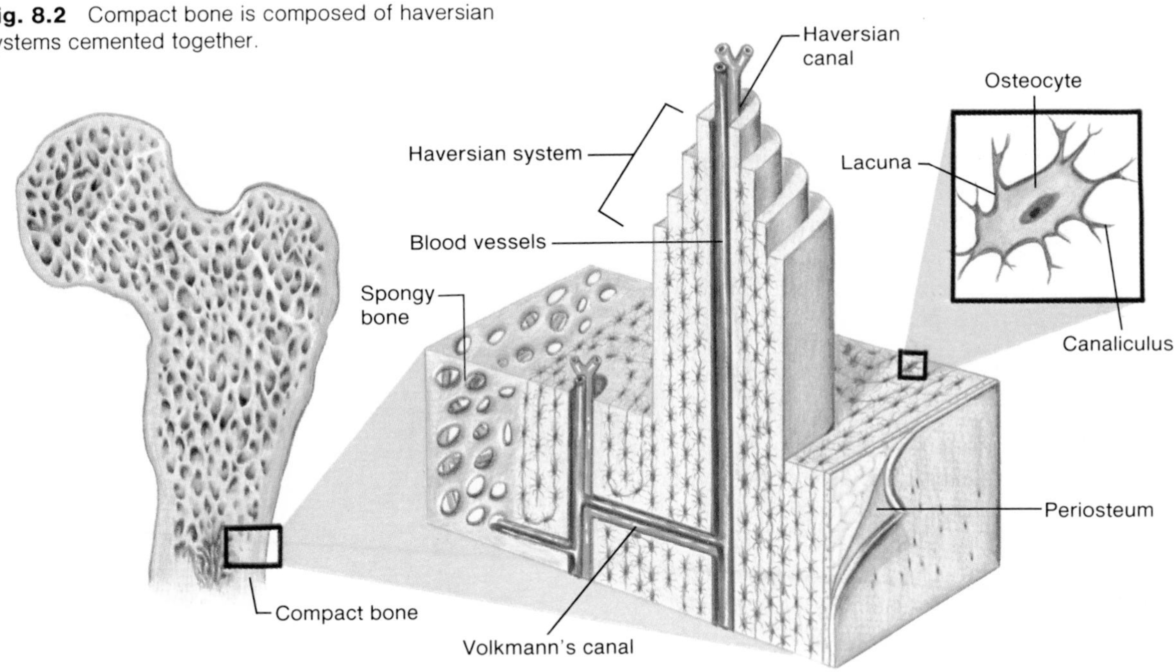

The compact bone in the diaphysis of a long bone forms a rigid tube with a hollow chamber called the **medullary cavity.** This cavity is continuous with the spaces of spongy bone. All of these areas are lined with a thin layer of squamous cells called *endosteum* and are filled with a specialized type of soft connective tissue called *marrow.*

Microscopic Structure

As was discussed in chapter 5, bone cells (osteocytes) are located in minute, bony chambers called *lacunae* which are arranged in concentric circles around haversian canals. These cells communicate with nearby cells by means of cellular processes passing through canaliculi. The intercellular material of bone tissue is largely collagen, which gives it strength and resilience, and inorganic salts, which make it hard and resistant to crushing.

In compact bone, the osteocytes and layers of intercellular material clustered concentrically around a haversian canal form a cylinder-shaped unit called a **haversian system.** Many of these units cemented together form the substance of compact bone (fig. 8.2).

Each haversian canal contains one or two small blood vessels (usually capillaries) surrounded by some loose connective tissue. Blood in these vessels provides nourishment for the cells associated with the haversian canal.

Haversian canals travel longitudinally through bone tissue. They are interconnected by transverse channels called *Volkmann's canals.* These canals contain larger blood vessels by which the vessels in the haversian canals communicate with the surface of the bone and the medullary cavity (fig. 8.2).

Although spongy bone also is composed of osteocytes and intercellular material, its bony plates are usually very thin, and the cells are not arranged around haversian canals. They are nourished by diffusion of substances into canaliculi that lead from the bone cells to the surface of the bony plates.

1. List five major parts of a long bone.
2. How do compact and spongy bone differ in structure?
3. Describe the microscopic structure of compact bone.

Bone Development and Growth

Although parts of the skeletal system begin to form during the first few weeks of life, bony structures continue to develop and grow into adulthood. These bones form by the replacement of existing connective tissues, in one of two ways. Some appear between sheet-like layers of connective tissues, and are called intramembranous bones. Others begin as masses of cartilage that are later replaced by bone tissue. They are called endochondral bones.

Intramembranous Bones

The **intramembranous bones** include certain broad, flat bones of the skull. During their development, sheet-like masses of connective tissues appear in the sites of the future bones. Then some of the primitive connective tissue cells enlarge and differentiate into bone-forming cells called **osteoblasts.** The osteoblasts become active within the membranes and deposit bony intercellular materials around themselves. As a result, spongy bone tissue is produced in all directions within the membranes. Eventually the cells of the membranous tissues that persist outside the developing bone give rise to the periosteum. At the same time osteoblasts on the inside of the periosteum form a layer of compact bone over the surface of the newly built spongy bone. When the osteoblasts are surrounded by matrix, they are called **osteocytes.**

Endochondral Bones

Most of the bones of the skeleton are **endochondral bones.** They develop from masses of hyaline cartilage with shapes similar to the future bony structures. (See fig. 8.3.) These cartilaginous models grow rapidly for a while, and then begin to undergo extensive changes. In a long bone, for example, the changes begin in the center of the diaphysis where the cartilage slowly breaks down and disappears. At about the same time, a periosteum forms from connective tissues on the outside of the developing diaphysis. Blood vessels and osteoblasts from the periosteum invade the disintegrating cartilage, and spongy bone is formed in its place. This region of bone formation is called the *primary ossification center,* and bone tissue develops from it toward the ends of the cartilaginous structure.

Meanwhile, osteoblasts from the periosteum deposit a thin layer of compact bone around the primary ossification center. The epiphyses of the developing bone remain cartilaginous and continue to grow. Later *secondary ossification centers* appear in the epiphyses, and spongy bone forms in all directions

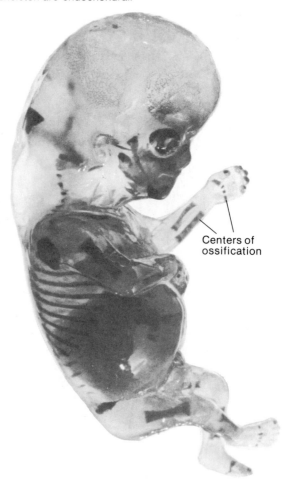

Fig. 8.3 The tissues of this miscarried fetus have been cleared, and the developing bones have been selectively stained. Some of the flat bones of the skull are intramembranous bones, while the other bones of the skeleton are endochondral.

Centers of ossification

from them. As bone is deposited in the diaphysis and in the epiphysis, a band of cartilage, called the **epiphyseal disk,** is left between the two ossification centers.

The cartilaginous cells of the epiphyseal disk include rows of young cells that are undergoing mitosis and producing new daughter cells. As these cells enlarge and intercellular material is formed around them, the cartilaginous disk thickens. Consequently, the length of the entire bone increases. At the same time, calcium salts accumulate in the intercellular substance of the oldest cartilaginous cells, and they begin to die.

Later, this calcified intercellular substance is broken down by the action of giant cells called **osteoclasts.** Soon bone-building osteoblasts invade this region and deposit bone tissue in place of the calcified cartilage.

A long bone will grow in length while the cartilaginous cells of the epiphyseal disk are active.

Fig. 8.4 Major stages in the development of an endochondral bone.

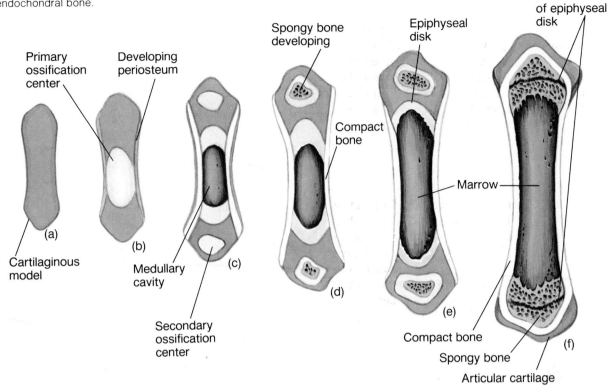

However, once the ossification centers of the diaphysis and epiphysis come together and the epiphyseal disk becomes ossified, growth in length is no longer possible.

If an epiphyseal disk is damaged before it becomes ossified, growth of the long bone may cease prematurely, or if growth continues, it may be uneven. For this reason, injuries to the epiphyses of a young person's bones are of special concern. On the other hand, an epiphysis is sometimes altered surgically in order to equalize growth of bones that are developing at very different rates.

A developing long bone grows in thickness as a result of compact bone tissue being deposited on the outside, just beneath the periosteum. As this is occurring, other bone tissue is being eroded away on the inside by osteoclasts. The space that is created becomes the medullary cavity of the diaphysis, and it later fills with marrow.

The bone in the central regions of the epiphyses remains spongy, and the hyaline cartilage on the ends of the epiphyses persists throughout life as articular cartilage. Figure 8.4 illustrates the stages of endochondral bone development.

1. Describe the development of an intramembranous bone.
2. Explain how endochondral bones develop.

Functions of Bones

Skeletal parts provide shape and support for body structures. They also aid body movements, produce blood cells, and store various inorganic salts.

Support and Protection

Bones give shape to structures such as the head, face, thorax, and limbs. They also provide support and protection. For example, the bones of the feet, legs, pelvis, and backbone support the weight of the body. The bones of the skull protect the eyes, ears, and brain. Those of the rib cage and shoulder girdle protect the heart and lungs, while the bones of the pelvic girdle protect the lower abdominal and internal reproductive organs.

Fig. 8.5 When the arm is (a) bent or (b) straightened at the elbow, a lever is employed

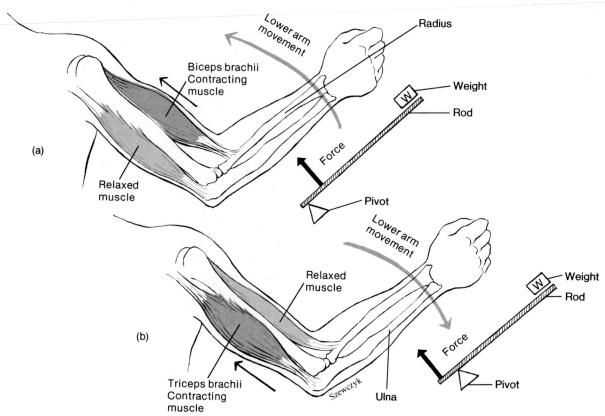

Body Movement

Whenever limbs or other body parts are moved, bones and muscles function together as simple mechanical devices called *levers*. Such a lever has four basic components: (a) a rigid bar or rod; (b) a pivot, or fulcrum, on which the bar turns; (c) an object or weight that is moved; (d) a force that supplies energy for the movement of the bar.

The actions of bending and straightening the arm at the elbow, for example, involve bones and muscles functioning together as levers, as illustrated in figure 8.5. When the arm is bent, the lower arm bones represent the rigid bar, the elbow joint is the pivot, the hand is the weight that is moved, and the force is supplied by muscles on the anterior side of the upper arm. One of these muscles, the *biceps brachii,* is attached by a tendon to a process on the *radius* bone in the lower arm, a short distance below the elbow.

When the arm is straightened at the elbow, the lower arm bones again serve as a rigid bar, and the elbow joint as the pivot. However, this time the force is supplied by the *triceps brachii,* a muscle located on the posterior side of the upper arm. A tendon of this muscle is attached to a process of the *ulna* bone at the point of the elbow.

Many other bone-muscle lever systems function to make body movements possible; however, the parts of the lever systems are sometimes difficult to identify.

Blood Cell Formation

The process of blood cell formation is called **hematopoiesis.** Very early in life it occurs in a structure called a *yolk sac,* which lies outside the body of a human embryo. (See chapter 20.) Later in development, blood cells are manufactured in the liver and spleen, and still later they are formed in the marrow within bones.

Marrow is a soft mass of connective tissue found in the medullary cavities of long bones, in the irregular spaces of spongy bone, and in the larger haversian canals of compact bone.

There are two kinds of marrow—red and yellow. *Red marrow* functions in the formation of red blood cells, white blood cells, and blood platelets. It is red because of the red, oxygen-carrying pigment called **hemoglobin** contained within the red blood cells.

Red marrow occupies the cavities of most bones in an infant. With age however, more and more of it is replaced by *yellow marrow* that functions as fat storage tissue and is inactive in blood cell production.

In an adult, red marrow is found primarily in spongy bone of the skull, ribs, sternum, clavicles, vertebrae, and pelvis. Blood cell formation is described in more detail in chapter 15.

Red marrow may be damaged or destroyed by excessive exposure to X ray, adverse drug reactions, or the presence of cancerous tissues. The treatment of this condition sometimes involves a bone marrow transplant.

In this procedure, normal red marrow cells are removed from the spongy bone of a donor by means of a hollow needle and syringe. These cells are then injected into the recipient's blood with the hope that they will lodge in the bone spaces normally inhabited by red marrow and will, in time, replace the damaged tissue.

Storage of Inorganic Salts

The intercellular matrix of bone tissue contains large quantities of calcium salts. These are mostly in the form of tiny crystals of a type of calcium phosphate.

Calcium is needed for a number of vital metabolic processes. When a low blood calcium ion concentration exists, calcium salts are released from bone matrix into the blood. On the other hand, if the blood calcium ion concentration is excessively high, osteoblasts are stimulated to form bone tissue. As a result, the excessive calcium is stored in the matrix. (The details of this mechanism are described in chapter 12.)

In addition to storing calcium and phosphorus, bone tissue contains lesser amounts of magnesium, sodium, potassium, and carbonate ions. Bones also tend to accumulate certain metallic elements such as lead, radium, or strontium, which are not normally present in the body but are sometimes ingested accidentally.

---

1. Name three functions of bones.
2. Distinguish between red marrow and yellow marrow.
3. List the inorganic salts normally stored in bone.

Organization of the Skeleton

Number of Bones

Although the number of bones in a human skeleton is often reported to be 206, the number actually varies from person to person. Some people may lack certain bones, while others have extra ones. For example, the flat bones of the skull usually grow together and become tightly joined along irregular lines called **sutures.** Occasionally extra bones called *wormian* or *sutural bones* develop in these sutures. Also extra small, round bones called *sesamoid bones* may develop in tendons. (See fig. 8.27.)

Divisions of the Skeleton

For purposes of study, it is convenient to divide the skeleton into two major portions—an axial skeleton and an appendicular skeleton. (See fig. 8.6.)

The **axial skeleton** consists of the bony and cartilaginous parts that support and protect the organs of the head, neck, and trunk. These parts include the following:

1. **Skull.** The skull is composed of the brain case, or *cranium,* and the facial bones.

2. **Hyoid bone.** The hyoid bone is located in the neck between the lower jaw and the larynx. It supports the tongue and serves as an attachment for certain muscles that help to move the tongue and function in swallowing.

3. **Vertebral column.** The vertebral column, or backbone, consists of many vertebrae separated by cartilaginous *intervertebral disks.* Near its distal end, several vertebrae are fused to form the **sacrum,** which is part of the pelvis. A small, rudimentary tailbone called the **coccyx** is attached to the end of the sacrum.

4. **Thoracic cage.** The thoracic cage protects the organs of the thorax and the upper abdomen. It is composed of twelve pairs of ribs that articulate posteriorly with thoracic vertebrae. It also includes the **sternum** (breastbone), to which most of the ribs are attached anteriorly.

The **appendicular skeleton** consists of the bones of the limbs and those that anchor the limbs to the axial skeleton. It includes the following:

1. **Pectoral girdle.** The pectoral girdle is formed by a **scapula** (shoulder blade) and a **clavicle** (collarbone) on both sides of the body. The pectoral girdle connects the bones of the arms to the axial skeleton and aids in arm movements.

Fig. 8.6 The human skeleton. (a) Anterior view; (b) posterior view. Note that the axial skeleton appears in color, while the appendicular skeleton is shown as white.

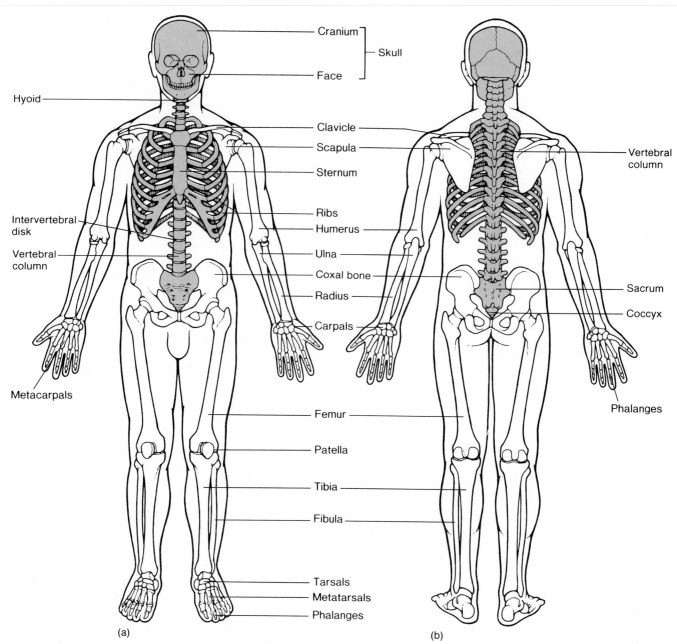

Cranium
Skull
Face

Hyoid

Clavicle
Scapula
Sternum

Ribs
Humerus

Intervertebral disk

Ulna

Vertebral column

Coxal bone
Radius

Carpals

Metacarpals

Femur

Patella

Tibia

Fibula

Tarsals
Metatarsals
Phalanges

(a)

Vertebral column

Sacrum

Coccyx

Phalanges

(b)

Chart 8.1   Terms used to describe skeletal structures

Term	Definition	Example
Condyle (kon′dil)	A rounded process that usually articulates with another bone	Occipital condyle of the occipital bone (fig. 8.10)
Crest (krest)	A narrow, ridgelike projection	Iliac crest of the ilium (fig. 8.29)
Epicondyle (ep″ĭ-kon′dil)	A projection situated above a condyle	Medial epicondyle of the humerus (fig. 8.24)
Facet (fas′et)	A small, nearly flat surface	Costal facet of a thoracic vertebra (fig. 8.15)
Fontanel (fon″tah-nel′)	A soft spot in the skull where membranes cover the space between bones	Anterior fontanel between the frontal and parietal bones (fig. 8.14)
Foramen (fo-ra′men)	An opening through a bone that usually serves as a passageway for blood vessels, nerves, or ligaments	Foramen magnum of the occipital bone (fig. 8.11)
Fossa (fos′ah)	A relatively deep pit or depression	Olecranon fossa of the humerus (fig. 8.24)
Fovea (fo′ve-ah)	A tiny pit or depression	Fovea capitis of the femur (fig. 8.31)
Head (hed)	An enlargement of the end of a bone	Head of the humerus (fig. 8.24)
Meatus (me-a′tus)	A tubelike passageway within a bone	Auditory meatus of the ear (fig. 8.9)
Process (pros′es)	A prominent projection on a bone	Mastoid process of the temporal bone (fig. 8.9)
Sinus (si′nus)	A cavity or hollow space within a bone	Frontal sinus of the frontal bone (fig. 8.12)
Spine (spīn)	A thornlike projection	Spine of the scapula (fig. 8.23)
Suture (soo′cher)	A line of union between bones	Lambdoidal suture between the occipital and parietal bones (fig. 8.9)
Trochanter (tro-kan′ter)	A relatively large process	Greater trochanter of the femur (fig. 8.31)
Tubercle (tu′ber-kl)	A small, knoblike process	Greater tubercle of a humerus (fig. 8.24)
Tuberosity (tu″bĕ-ros′ĭ-te)	A knoblike process usually larger than a tubercle	Radial tuberosity of the radius (fig. 8.25)

From Hole, John W., Jr., *Human Anatomy and Physiology 3d ed.* © 1978, 1981, 1984 Wm. C. Brown Publishers, Dubuque, Iowa. All Rights Reserved. Reprinted by permission.

2. **Upper limbs** (arms). Each upper limb consists of a **humerus,** or upper arm bone, and two lower arm bones—a **radius** and an **ulna.** These three bones articulate at the elbow joint. At the distal end of the radius and ulna, there are eight **carpals** (wrist bones). The bones of the palm are called **metacarpals,** and the finger bones are **phalanges.**

3. **Pelvic girdle.** The pelvic girdle is formed by two **coxal bones** (hip bones), which are attached to each other anteriorly and to the sacrum posteriorly. They connect the bones of the legs to the axial skeleton and, with the sacrum and coccyx, form the **pelvis.**

4. **Lower limbs** (legs). Each lower limb consists of a **femur** (thigh bone) and two lower leg bones—a large **tibia** (shinbone) and a slender **fibula** (calf bone). These three bones articulate at the knee joint, where the **patella** (kneecap) covers the anterior surface. At the distal ends of the tibia and fibula, there are seven **tarsals** (ankle bones). The bones of the foot are called **metatarsals,** while those of the toes (like the fingers) are **phalanges.**

Chart 8.1 includes some terms that are used to describe skeletal structures.

1. Distinguish between the axial and the appendicular skeletons.
2. Name the bones that each includes.

The Skull

A human skull usually consists of twenty-two bones that, except for the lower jaw, are firmly interlocked along sutures. Eight of these immovable bones make up the cranium, and thirteen form the facial skeleton. The **mandible** (lower jawbone) is a movable bone held to the cranium by ligaments. (See fig. 8.7.)

The Cranium

The **cranium** encloses and protects the brain, and its surface provides attachments for various muscles that make chewing and head movements possible. Some of the cranial bones contain air-filled cavities called *sinuses,* which are lined with mucous membranes and are connected by passageways to the nasal cavity. Sinuses reduce the weight of the skull and increase the intensity of the voice by serving as resonant sound chambers. (See fig. 8.8.)

Fig. 8.7 Anterior view of the skull.

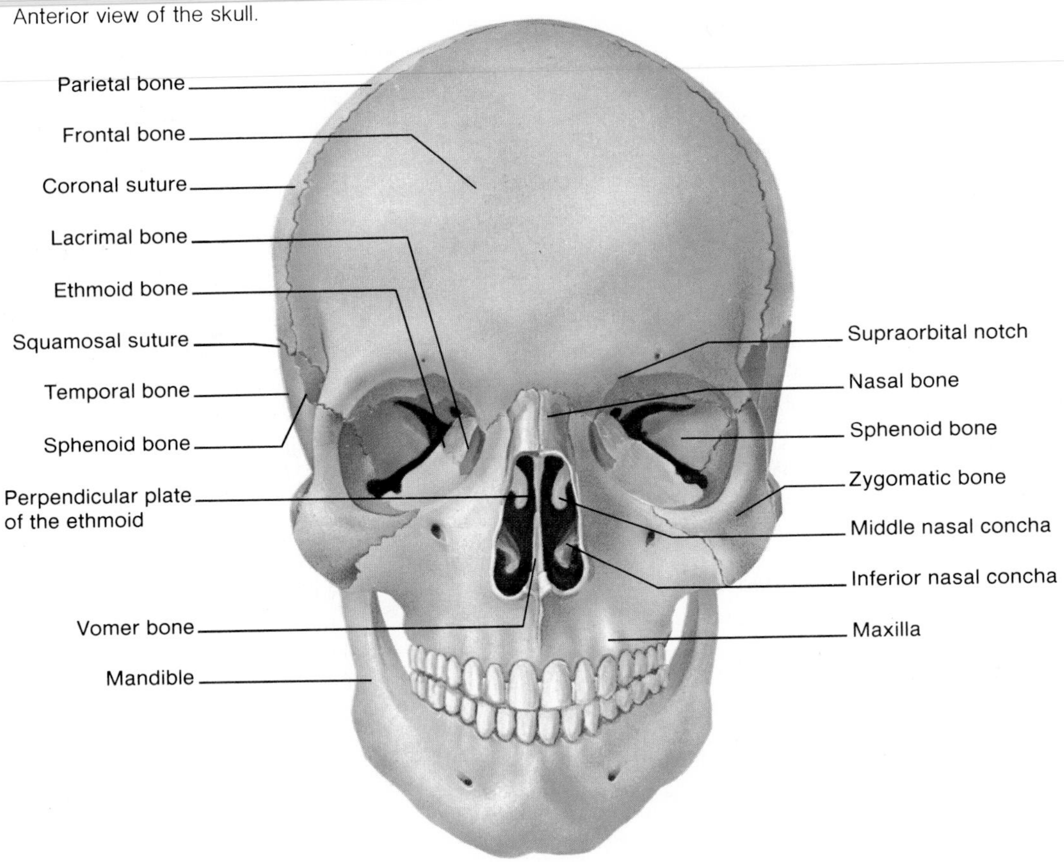

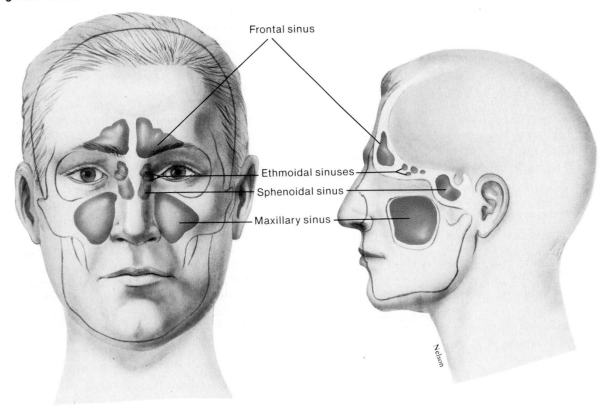

Parietal bone

Frontal bone

Coronal suture

Lacrimal bone

Ethmoid bone

Squamosal suture

Temporal bone

Sphenoid bone

Perpendicular plate
of the ethmoid

Vomer bone

Mandible

Supraorbital notch

Nasal bone

Sphenoid bone

Zygomatic bone

Middle nasal concha

Inferior nasal concha

Maxilla

Fig. 8.8 Locations of the sinuses.

Frontal sinus

Ethmoidal sinuses

Sphenoidal sinus

Maxillary sinus

Nelson

Fig. 8.9 Lateral view of the skull.

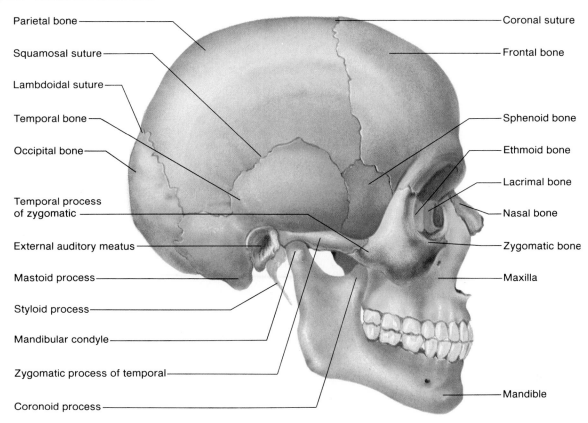

Parietal bone

Squamosal suture

Lambdoidal suture

Temporal bone

Occipital bone

Temporal process of zygomatic

External auditory meatus

Mastoid process

Styloid process

Mandibular condyle

Zygomatic process of temporal

Coronoid process

Coronal suture

Frontal bone

Sphenoid bone

Ethmoid bone

Lacrimal bone

Nasal bone

Zygomatic bone

Maxilla

Mandible

The eight bones of the cranium (figs. 8.7 and 8.9) are as follows:

1. **Frontal bone.** The frontal bone forms the anterior portion of the skull above the eyes. On the upper margin of each orbit (the bony socket of the eye), the frontal bone is marked by a *supraorbital foramen* (or *supraorbital notch* in some skulls) through which blood vessels and nerves pass to the tissues of the forehead. Within the frontal bone are two *frontal sinuses,* one above each eye near the midline (fig. 8.8).

2. **Parietal bones.** One *parietal bone* (fig. 8.9) is located on each side of the skull just posterior to the frontal bone. Together, the parietal bones form the bulging sides and roof of the cranium. They are fused in the midline along the *sagittal suture,* and they meet the frontal bone along the *coronal suture.*

3. **Occipital bone.** The occipital bone (figs. 8.9 and 8.10) joins the parietal bones along the *lambdoidal suture.* It forms the back of the skull and the base of the cranium. There is a large opening on its lower surface called the *foramen magnum,* through which nerve fibers from the brain pass and enter the vertebral canal. Rounded processes called *occipital condyles,* which are located on each side of the foramen magnum, articulate with the first vertebra of the vertebral column.

4. **Temporal bones.** A temporal bone (fig. 8.9) on each side of the skull joins the parietal bone along a *squamosal suture.* The temporal bones form parts of the sides and the base of the cranium. Located near the inferior margin is an opening, the *external auditory meatus,* which leads inward to parts of the ear. The temporal bones have depressions, the *mandibular fossae,* that articulate with processes of the mandible. Below each external auditory meatus, there are two projections—a rounded *mastoid process* and a long, pointed *styloid process.* The mastoid process provides an attachment for certain muscles of the neck, while the styloid process serves as an anchorage for muscles associated with the tongue and pharynx.

A *zygomatic process* projects anteriorly from the temporal bone, joins the *zygomatic bone,* and helps form the prominence of the cheek.

Fig. 8.10 Inferior view of the skull.

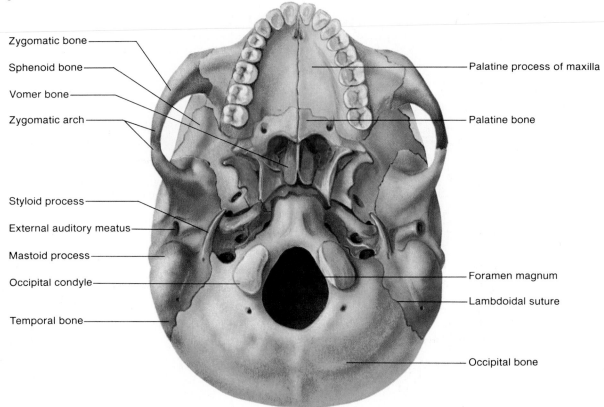

Zygomatic bone

Sphenoid bone

Vomer bone

Zygomatic arch

Styloid process

External auditory meatus

Mastoid process

Occipital condyle

Temporal bone

Palatine process of maxilla

Palatine bone

Foramen magnum

Lambdoidal suture

Occipital bone

5. **Sphenoid bone.** The sphenoid bone (figs. 8.10 and 8.11) is wedged between several other bones in the anterior portion of the cranium. It consists of a central part and two winglike structures that extend laterally toward each side of the skull. This bone helps form the base of the cranium, the sides of the skull, and the floors and sides of the orbits. Along the midline within the cranial cavity, a portion of the sphenoid bone rises up and forms a saddle-shaped mass called *sella turcica* (Turk's saddle). This depression is occupied by the pituitary gland.

The sphenoid bone also contains two *sphenoidal sinuses.* (See fig. 8.8.)

6. **Ethmoid bone.** The ethmoid bone (figs. 8.9 and 8.11) is located in front of the sphenoid bone. It consists of two masses, one on each side of the nasal cavity, which are joined horizontally by thin *cribriform plates.* These plates form part of the roof of the nasal cavity.

Projecting upward into the cranial cavity between the cribriform plates is a triangular process of the ethmoid bone called the *crista galli* (cock's comb). This process serves as an attachment for membranes that enclose the brain.

Portions of the ethmoid bone also form sections of the cranial floor, orbital walls, and nasal cavity walls. A *perpendicular plate* projects downward in the midline from the cribriform plates to form the bulk of the nasal septum.

Delicate scroll-shaped plates called *superior* and *middle nasal conchae* project inward from the lateral portions of the ethmoid bone toward the perpendicular plate. (See fig. 8.7.) The lateral portions of the ethmoid bone contain many small air spaces, the *ethmoidal sinuses.* (See fig. 8.8.)

The Facial Skeleton

The **facial skeleton** consists of thirteen immovable bones and a movable lower jawbone. In addition to forming the basic shape of the face, these bones provide attachments for various muscles that move the jaw and control facial expressions.

The bones of the facial skeleton are as follows:

1. **Maxillary bones.** The maxillary bones (maxillae) (figs. 8.9 and 8.10) form the upper jaw. Portions of these bones comprise the anterior roof of the mouth (*hard palate*), the floors of the orbits, and the sides and floor of the nasal cavity. They also contain the sockets of the upper teeth. Inside the

Fig. 8.11 Floor of the cranial cavity viewed from above.

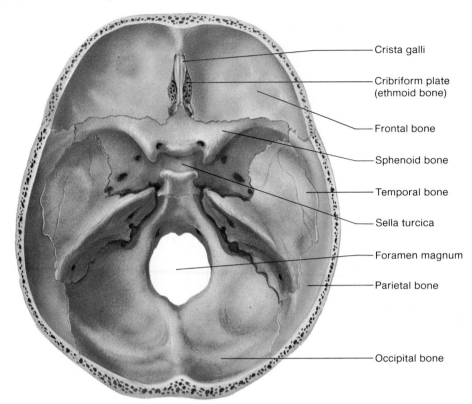

Crista galli

Cribriform plate
(ethmoid bone)

Frontal bone

Sphenoid bone

Temporal bone

Sella turcica

Foramen magnum

Parietal bone

Occipital bone

maxillae, lateral to the nasal cavity, are *maxillary sinuses,* which are the largest of the sinuses. (See fig. 8.8.)

During development, portions of the maxillae called *palatine processes* grow together and fuse along the midline to form the anterior section of the hard palate.

The inferior border of each maxillary bone projects downward forming an *alveolar process.* Together these processes create a horseshoe-shaped *alveolar arch* (dental arch) (fig. 8.12). Cavities in this arch are occupied by the teeth, which are attached to these bony sockets by connective tissues.

Sometimes the fusion of the palatine processes of the maxillae is incomplete at the time of birth; the result is called a *cleft palate.* Infants with this deformity may have trouble sucking because of the opening that remains between the oral and nasal cavities.

2. **Palatine bones.** The palatine bones (figs. 8.10 and 8.12) are located behind the maxillae. Each bone is roughly L-shaped. The horizontal portions serve as both the posterior

section of the hard palate and the floor of the nasal cavity. The perpendicular portions help form the lateral walls of the nasal cavity.

3. **Zygomatic bones.** The zygomatic bones (figs. 8.9 and 8.10) are responsible for the prominences of the cheeks below and to the sides of the eyes. These bones also help form the lateral walls and the floors of the orbits. Each bone has a *temporal process,* which extends posteriorly to join the zygomatic process of a temporal bone. Together these processes form a *zygomatic arch.*

4. **Lacrimal bones.** A lacrimal bone (figs. 8.7 and 8.9) is a thin, scalelike structure located in the medial wall of each orbit between the ethmoid bone and the maxilla.

5. **Nasal bones.** The nasal bones (figs. 8.7 and 8.9) are long, thin, and nearly rectangular. They lie side by side and are fused at the midline, where they form the bridge of the nose.

6. **Vomer bone.** The thin, flat vomer bone (figs. 8.7 and 8.12) is located along the midline within the nasal cavity. Posteriorly it joins the perpendicular plate of the ethmoid bone, and together they form the nasal septum.

Fig. 8.12 Sagittal section of the skull.

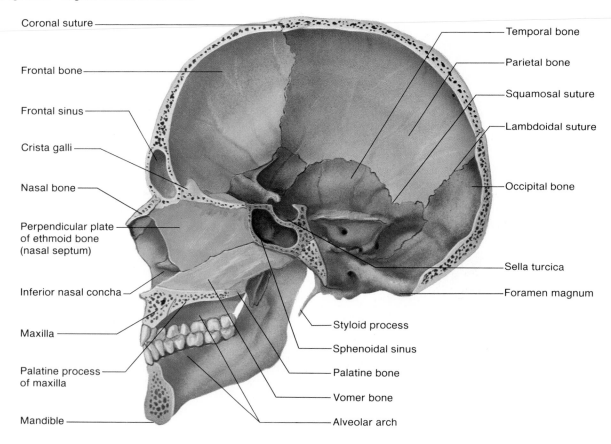

Coronal suture

Frontal bone

Frontal sinus

Crista galli

Nasal bone

Perpendicular plate of ethmoid bone (nasal septum)

Inferior nasal concha

Maxilla

Palatine process of maxilla

Mandible

Temporal bone

Parietal bone

Squamosal suture

Lambdoidal suture

Occipital bone

Sella turcica

Foramen magnum

Styloid process

Sphenoidal sinus

Palatine bone

Vomer bone

Alveolar arch

Fig. 8.13 X-ray film of the skull from the side. What features of the cranium and facial skeleton do you recognize?

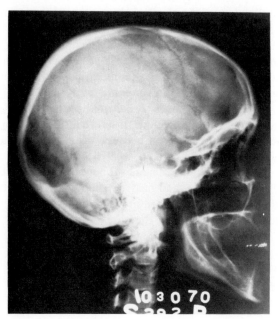

7. **Inferior nasal conchae.** The inferior nasal conchae (figs. 8.7 and 8.12) are fragile, scroll-shaped bones attached to the lateral walls of the nasal cavity. Like the superior and middle conchae, the inferior conchae provide support for mucous membranes within the nasal cavity.

8. **Mandible.** The mandible (figs. 8.7 and 8.12) consists of a horizontal, horseshoe-shaped body with a flat portion projecting upward at each end. This projection is divided into two processes—a posterior *mandibular condyle* and an anterior *coronoid process.* The mandibular condyles articulate with the mandibular fossae of the temporal bones, while the coronoid processes serve as attachments for muscles used in chewing. Other large chewing muscles are inserted on the lateral surface of the mandible. A curved bar of bone on the superior border of the mandible, the *alveolar arch,* contains the sockets of the lower teeth (fig. 8.12).

Various features of the skull can be seen in the X-ray film in figure 8.13.

Fig. 8.14 (a) Lateral view and (b) superior view of the infantile skull.

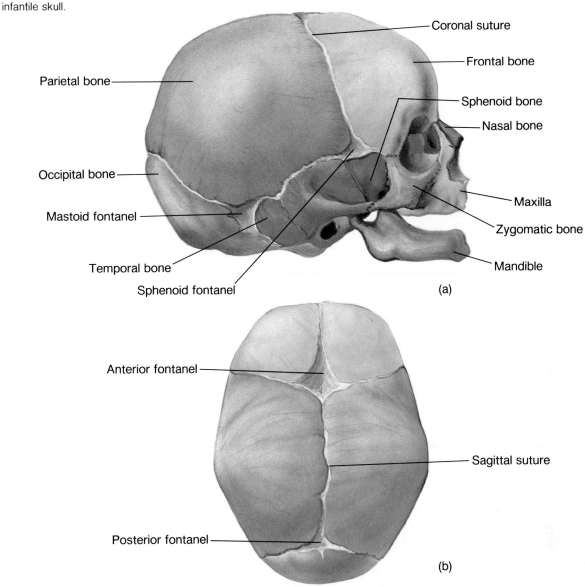

Coronal suture

Frontal bone

Sphenoid bone

Nasal bone

Parietal bone

Occipital bone

Mastoid fontanel

Maxilla

Zygomatic bone

Temporal bone

Mandible

Sphenoid fontanel

(a)

Anterior fontanel

Sagittal suture

Posterior fontanel

(b)

The Infantile Skull

At birth the skull is incompletely developed, and the cranial bones are separated by fibrous membranes. These membranous areas are called **fontanels** or, more commonly, soft spots. (See fig. 8.14.) They permit some movement between the bones, so that the developing skull is partially compressible and can change shape slightly. This action enables an infant's skull to pass more easily through the birth canal. Eventually the fontanels close as the cranial bones grow together.

Other characteristics of an infantile skull include a relatively small face with a prominent forehead and large orbits. The jaw and nasal cavity are small, the sinuses are incompletely formed, and the frontal bone is in two parts. The skull bones are thin, but they are also somewhat flexible and thus are less easily fractured than adult bones.

1. Locate and name each of the bones of the cranium.
2. Locate and name each of the bones of the face.
3. Explain how an adult skull differs from that of an infant.

The Vertebral Column

The **vertebral column** extends from the skull to the pelvis and forms the vertical axis of the skeleton. It is composed of many bony parts called **vertebrae.** These are separated by masses of fibrocartilage called *intervertebral disks* and are connected to one another by ligaments. The vertebral column supports the head and the trunk of the body. It also protects the spinal cord, which passes through a *vertebral canal* formed by openings in the vertebrae.

Normally the vertebral column has four curvatures that give it a degree of resiliency. The names of the curves correspond to the regions in which they occur, as shown in figure 8.15.

A Typical Vertebra

Although the vertebrae in different regions of the vertebral column have special characteristics, they also have features in common. Thus, a typical vertebra (fig. 8.16) has a drum-shaped *body* that forms a thick, anterior portion of the bone. A longitudinal row of these bodies supports the weight of the head and trunk. The intervertebral disks, which separate adjacent vertebral bodies, cushion and soften the forces created by such movements as walking and jumping.

Projecting posteriorly from each vertebral body are two short stalks called *pedicles.* Two plates called *laminae* arise from the pedicles and fuse in the back to become a *spinous process.* The pedicles, laminae, and spinous process together complete a bony *vertebral arch* around a *vertebral foramen,* through which the spinal cord passes.

If the laminae of the vertebrae fail to unite during development, the vertebral arch remains incomplete. This condition is called *spina bifida.* As a result of it, the contents of the vertebral canal may protrude outward. This problem occurs most frequently in the lumbosacral region.

Between the pedicles and laminae of a typical vertebra is a *transverse process* that projects laterally and toward the back. Various ligaments and muscles are attached to the dorsal spinous process and the transverse processes. Projecting upward and downward from each vertebral arch are *superior* and *inferior articulating processes.* These processes bear cartilage-covered facets by which each vertebra is joined to the one above and the one below it.

Fig. 8.15 The curved vertebral column consists of many vertebrae separated by intervertebral disks.

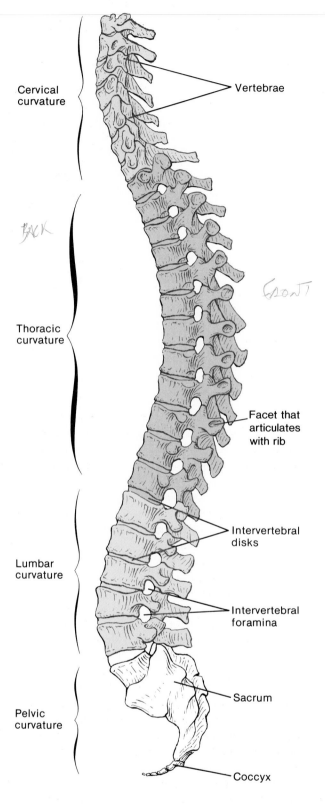

Cervical curvature

Vertebrae

Thoracic curvature

Facet that articulates with rib

Intervertebral disks

Lumbar curvature

Intervertebral foramina

Sacrum

Pelvic curvature

Coccyx

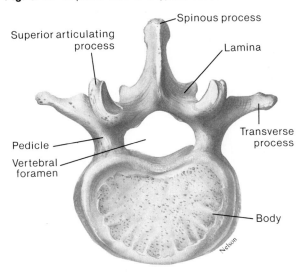

Fig. 8.16 Superior view of a typical vertebra.

- Spinous process
- Superior articulating process
- Lamina
- Transverse process
- Pedicle
- Vertebral foramen
- Body

On the lower surfaces of the vertebral pedicles are notches that align to create openings, called *intervertebral foramina*. These openings provide passageways for spinal nerves that proceed between adjacent vertebrae and connect to the spinal cord. (See fig. 8.16.)

Cervical Vertebrae

Seven **cervical vertebrae** comprise the bony axis of the neck. The transverse processes of these vertebrae are distinctive because they have *transverse foramina*, which serve as passageways for arteries leading to the brain. (See fig. 8.17.)

Two of the cervical vertebrae, shown in figure 8.17, are of special interest. The first vertebra, or **atlas,** supports and balances the head. On its upper surface, it has two kidney-shaped facets that articulate with the occipital condyles of the skull.

The second cervical vertebra, or **axis,** bears a toothlike *odontoid process* on its body. This process projects upward and lies in the ring of the atlas. As the head is turned from side to side, the atlas pivots around the odontoid process.

Thoracic Vertebrae

The twelve **thoracic vertebrae** (fig. 8.15) are larger than those in the cervical region. They have long, pointed spinous processes that slope downward, and facets on the sides of their bodies that articulate with ribs.

Beginning with the third thoracic vertebra and downward, the bodies of these bones increase in size. This reflects the stress placed on them by the increasing amounts of body weight they bear.

Lumbar Vertebrae

There are five **lumbar vertebrae** (fig. 8.15) in the small of the back (loins). Since the lumbars support more weight than the vertebrae above them, these bones have developed larger and stronger bodies.

The Sacrum

The **sacrum** is a triangular structure, composed of five fused vertebrae, that forms the base of the vertebral column. The spinous processes of these fused bones are represented by a ridge of tubercles. To the sides of the tubercles are rows of openings, the *dorsal sacral foramina,* through which nerves and blood vessels pass. (See fig. 8.18.)

The vertebral foramina of the sacral vertebrae form the *sacral canal,* which continues through the sacrum to an opening of variable size at the tip called the *sacral hiatus.*

Although the sacral hiatus is normally covered by fibrous tissue, an anesthetic is sometimes injected through it into the sacral canal in order to reduce pain during childbirth. This procedure is called *caudal anesthesia* or *caudal block.*

The Coccyx

The **coccyx** is the lowest part of the vertebral column and is composed of four fused vertebrae. (See fig. 8.18.) It is attached by ligaments to the margins of the sacral hiatus.

A common vertebral problem involves changes in the intervertebral disks. Each disk is composed of a tough, outer layer of fibrocartilage and an elastic central mass. As a person ages, these disks tend to undergo degenerative changes in which the central masses lose their firmness and the outer layers become thinner and weaker, and develop cracks. Extra pressure, as produced when a person falls or lifts a heavy object, can break the outer layers of the disks and allow the central masses to squeeze out. Such a rupture may cause pressure on the spinal cord or on spinal nerves that branch from it. This condition—a slipped or ruptured disk—may cause back pain and numbness or loss of muscular function in the parts innervated by the affected spinal nerves.

Fig. 8.17 How do the structures of the (a) atlas and (b) axis function together to allow movement of the head?

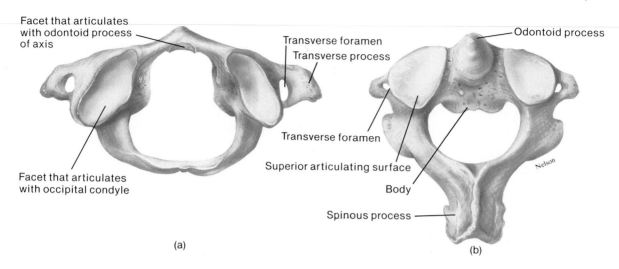

Facet that articulates with odontoid process of axis

Transverse foramen
Transverse process

Odontoid process

Facet that articulates with occipital condyle

Transverse foramen

Superior articulating surface

Body

Spinous process

Nelson

(a)

(b)

Fig. 8.18 Posterior view of the sacrum and coccyx.

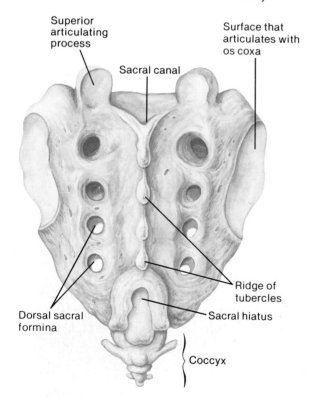

Superior articulating process

Sacral canal

Surface that articulates with os coxa

Dorsal sacral formina

Ridge of tubercles

Sacral hiatus

Coccyx

1. Describe the structure of the vertebral column.
2. Describe a typical vertebra.
3. How do the structures of a cervical, a thoracic, and a lumbar vertebra differ?

The Thoracic Cage

The **thoracic cage** includes the ribs, the thoracic vertebrae, the sternum (breastbone), and the costal cartilages by which the ribs are attached to the sternum. These parts support the shoulder girdle and arms, protect the visceral organs in the thoracic and upper abdominal cavities, and play a role in breathing. (See figs. 8.19 and 8.20.)

The Ribs

Regardless of sex, each person usually has twelve pairs of ribs—one pair attached to each of the twelve thoracic vertebrae.

The first seven rib pairs, the *true ribs,* join the sternum directly by their costal cartilages. The remaining five pairs are called *false ribs* because their cartilages do not reach the sternum directly. The last two rib pairs are sometimes called *floating ribs,* because they have no cartilaginous attachments to the sternum.

Fig. 8.19 The thoracic cage includes the thoracic vertebrae, the sternum, the ribs, and the costal cartilages that attach the ribs to the sternum.

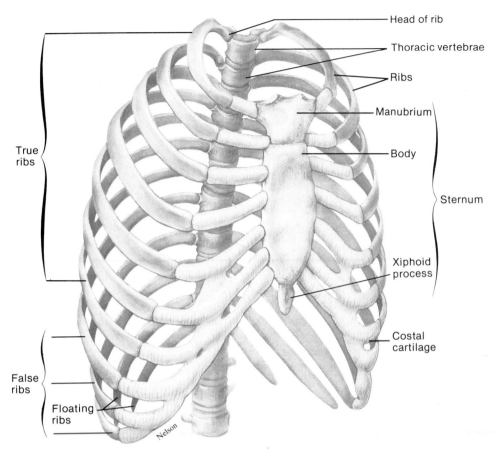

A typical rib has a long, slender shaft that curves around the chest and slopes downward. On the posterior end is an enlarged *head* by which the rib articulates with a facet on the body of its own vertebra and usually with the body of the next higher vertebra. Also near the head are *tubercles* that articulate with the transverse process of the vertebra (fig. 8.19).

The Sternum

The **sternum** is located along the midline in the anterior portion of the thoracic cage. (See fig. 8.19.) It is a flat, elongated bone that develops in three parts—an upper *manubrium,* a middle *body,* and a lower *xiphoid process* that projects downward. The manubrium articulates with the clavicles by facets on its superior border.

Fig. 8.20 X-ray film of the thoracic cage viewed from the front. Note the shadow of the heart behind the sternum and above the diaphragm.

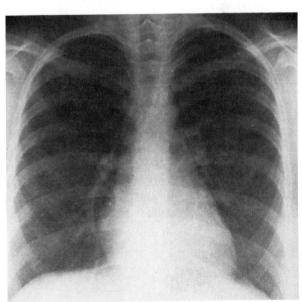

The red marrow within the spongy bone of the sternum functions in blood-cell formation into adulthood. Since the sternum has a thin covering of compact bone and is easy to reach, samples of its blood-cell-forming tissue may be removed for use in diagnosing diseases. This procedure, a *sternal puncture,* involves suctioning (aspirating) some marrow through a hollow needle.

1. What bones make up the thoracic cage?
2. Describe a typical rib.
3. What are the differences between true, false, and floating ribs?

The Pectoral Girdle

The **pectoral girdle** (shoulder girdle) is composed of four parts—two clavicles and two scapulae. Although the word *girdle* suggests a ring-shaped structure, the pectoral girdle is an incomplete ring. It is open in the back between the scapulae, and its bones are separated in the front by the sternum. However, the pectoral girdle supports the arms and serves as an attachment for several muscles that move the arms. (See figs. 8.21 and 8.22.)

The Clavicles

The **clavicles** (fig. 8.21) are slender, rodlike bones with elongated S-shapes. They are located at the base of the neck and run horizontally between the manubrium and scapulae.

The clavicles act as braces for the freely movable scapulae and thus help to hold the shoulders in place. They also provide attachments for muscles of the arms, chest, and back.

The Scapulae

The **scapulae** (figs. 8.21 and 8.23) are broad, somewhat triangular bones located on either side of the upper back. The posterior surface of each scapula is divided into unequal portions by a *spine.* This spine leads to two processes—an *acromion process* that forms the tip of the shoulder and a *coracoid process*

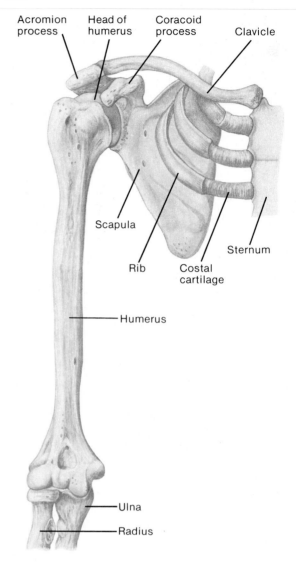

Fig. 8.21 The pectoral girdle, to which the arms are attached, consists of a clavicle and a scapula on either side.

Acromion process Head of humerus Coracoid process Clavicle

Scapula

Rib Costal cartilage Sternum

Humerus

Ulna

Radius

that curves forward and downward below the clavicle. The acromion process articulates with a clavicle and provides attachments for muscles of the arm and chest. The coracoid process also provides attachments for arm and chest muscles.

Between the processes is a depression called the *glenoid cavity.* It articulates with the head of the upper arm bone (humerus).

1. What bones form the pectoral girdle?
2. What is the function of the pectoral girdle?

Fig. 8.22 X-ray film of the right shoulder region viewed from the front.

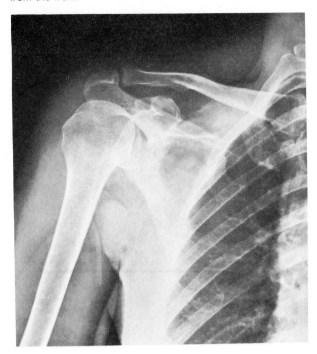

Fig. 8.23 (a) Posterior surface of the scapula; (b) medial view showing the glenoid cavity that articulates with the head of the humerus.

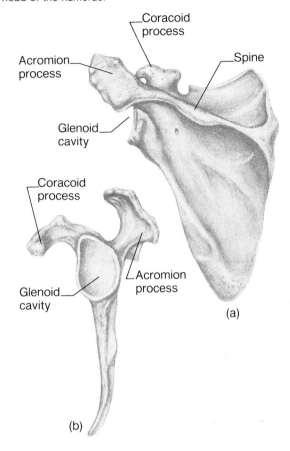

The Upper Limb

The bones of the upper limb form the framework of the arm, wrist, palm, and fingers. They also provide attachments for muscles and function in levers for moving limb parts. These bones include a humerus, a radius, an ulna, and several carpals, metacarpals, and phalanges. (See fig. 8.6.)

The Humerus

The **humerus** (fig. 8.24) is a heavy bone that extends from the scapula to the elbow. At its upper end, it has a smooth, rounded *head* that fits into the glenoid cavity of the scapula. Just below the head, there are two processes—a *greater tubercle* on the lateral side and a *lesser tubercle* anteriorly. These tubercles provide attachments for muscles that move the arm at the shoulder. Between them is a narrow furrow, the *intertubercular groove*.

Just below the head and the tubercles of the humerus is a region called the *surgical neck,* so named because fractures commonly occur there. Near the middle of the bony shaft on the lateral side, there is a rough V-shaped area called the *deltoid tuberosity*. It provides an attachment for the muscle (deltoid) that raises the arm horizontally to the side.

At the lower end of the humerus, there are two smooth *condyles* that articulate with the radius on the lateral side and the ulna on the medial side.

Above the condyles on either side are *epicondyles,* which provide attachments for muscles and ligaments of the elbow. Between the epicondyles anteriorly there is a depression, the *coronoid fossa,* that receives a process of the ulna (coronoid process) when the elbow is bent. Another depression on the posterior surface, the *olecranon fossa,* receives an ulnar process (olecranon process) when the arm is straightened at the elbow.

The Radius

The **radius** (fig. 8.25) extends from the elbow to the wrist and crosses over the ulna when the hand is turned so that the palm faces backward.

A thick, disklike *head* at the upper end of the radius articulates with the humerus and a notch of the ulna (radial notch). This arrangement allows the radius to rotate freely.

On the radial shaft, just below the head, is a process called the *radial tuberosity*. It serves as an attachment for a muscle (biceps brachii). At the lower end of the radius, a lateral *styloid process* provides attachments for ligaments of the wrist.

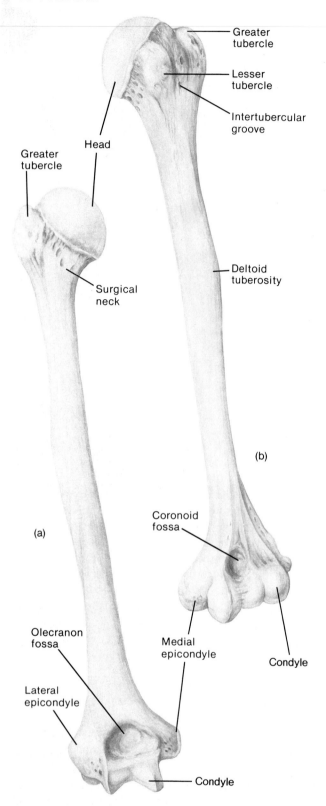

Greater
tubercle

Lesser
tubercle

Intertubercular
groove

Head

Greater
tubercle

Surgical
neck

Deltoid
tuberosity

(b)

Coronoid
fossa

(a)

Olecranon
fossa

Medial
epicondyle

Condyle

Lateral
epicondyle

Condyle

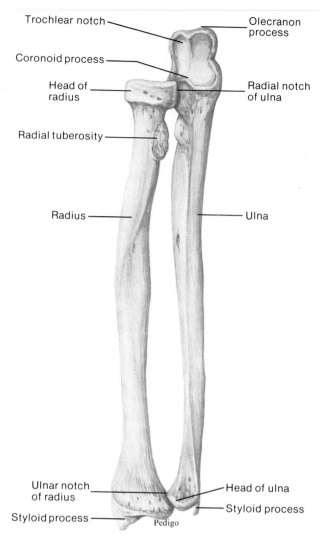

Trochlear notch

Olecranon
process

Coronoid process

Head of
radius

Radial notch
of ulna

Radial tuberosity

Radius

Ulna

Ulnar notch
of radius

Head of ulna

Styloid process

Styloid process

Pedigo

The Ulna

The **ulna** (fig. 8.25) overlaps the end of the humerus posteriorly. At its upper end, the ulna has a wrench-like opening, the *trochlear notch,* that articulates with the humerus. There are two processes, one on either side of this notch, the *olecranon process* and the *coronoid process,* which provide attachments for muscles.

At the lower end, the knoblike *head* of the ulna articulates with a notch of the radius (ulnar notch) laterally and with a disk of fibrocartilage inferiorly. This disk, in turn, joins a wrist bone. A medial *styloid process* at the distal end of the ulna provides attachments for ligaments of the wrist.

Fig. 8.26 The left hand viewed from the back.

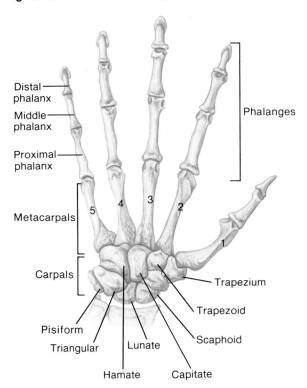

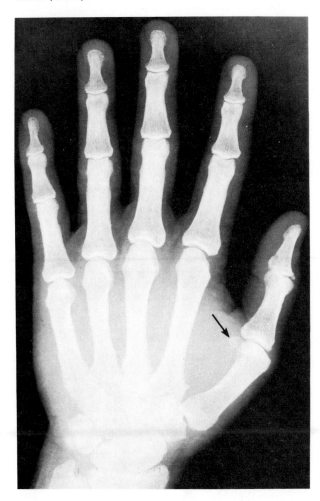

The Hand

The hand is composed of a wrist, a palm, and five fingers. (See figs. 8.26 and 8.27.) The skeleton of the wrist consists of eight small **carpal bones** that are firmly joined in two rows of four bones each. The resulting compact mass is called a *carpus*. The carpus articulates with the radius and with the fibrocartilaginous disk on the ulnar side. Its distal surface articulates with the metacarpal bones. The individual bones of the carpus are named in figure 8.26.

Five **metacarpal bones,** one in line with each finger, form the framework of the palm. These bones are cylindrical, with rounded distal ends that make the knuckles on a clenched fist. They are numbered 1 to 5, beginning with the metacarpal of the thumb. (See fig. 8.26.) The metacarpals articulate proximally with the carpals and distally with the phalanges.

The **phalanges** are the bony elements of the fingers. There are three in each finger—a proximal, a middle, and a distal phalanx—and two in the thumb (it lacks a middle phalanx).

---

1. *Locate and name each of the bones of the upper limb.*
2. *Explain how these bones articulate with one another.*

The Pelvic Girdle

The **pelvic girdle** consists of the two coxal bones that articulate with each other anteriorly and with the sacrum posteriorly. (See fig. 8.28.) With the sacrum and coccyx, the pelvic girdle forms the ringlike pelvis, which provides a stable support for the trunk of the body and attachments for the legs.

The Coxal Bones

Each coxal bone (os coxae) develops from three parts—an ilium, an ischium, and a pubis. These parts fuse in the region of a cup-shaped cavity called the *acetabulum*. This depression is on the lateral surface of the hip bone, and it receives the rounded head of the femur (thigh bone). (See figs. 8.28 and 8.29.)

Fig. 8.28 Anterior view of the pelvic girdle.

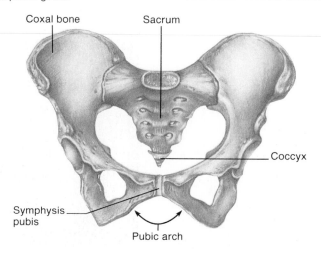

Coxal bone

Sacrum

Coccyx

Symphysis pubis

Pubic arch

Fig. 8.29 Lateral surface of the right coxal bone.

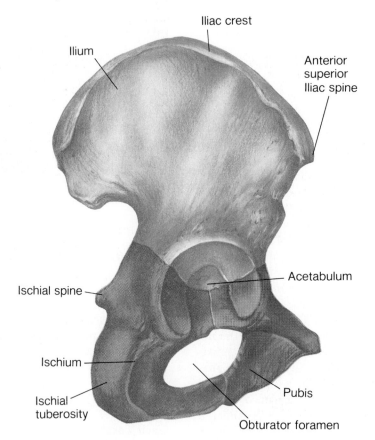

Iliac crest

Ilium

Anterior superior Iliac spine

Ischial spine

Acetabulum

Ischium

Ischial tuberosity

Pubis

Obturator foramen

Chart 8.2 Some sexual differences of
the skeleton

Part	Sexual Differences
Skull	Female skull is relatively smaller and lighter, and its muscular attachments are less conspicuous. The female forehead is longer vertically, the facial area is rounder, the jaw is smaller, and the mastoid process is less prominent than that of a male.
Pelvis	Female pelvic bones are lighter, thinner, and have less obvious muscular attachments. The obturator foramina and the acetabula are smaller and farther apart than those of a male.
Pelvic cavity	Female pelvic cavity is wider in all diameters and is shorter, roomier, and less funnel-shaped. The distances between the ischial spines and between the ischial tuberosities are greater than in the male.
Sacrum	Female sacrum is relatively wider, the first sacral vertebra projects forward to a lesser degree, and the sacral curvature is bent more sharply posteriorly than in a male.
Coccyx	Female coccyx is more movable than that of a male.

From Hole, John W. Jr., *Human Anatomy and Physiology 3d ed.* © 1978, 1981, 1984 Wm. C. Brown Publishers, Dubuque, Iowa. All Rights Reserved. Reprinted by permission.

Fig. 8.30 What differences are apparent in the X-ray films of this female pelvis (a) and this male pelvis (b)?

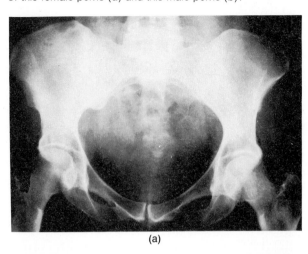

(a)

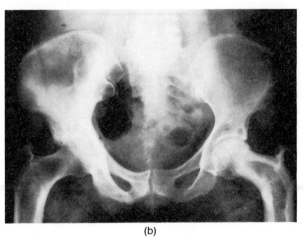

(b)

The **ilium,** which is the largest and uppermost portion of the coxal bone, flares outward to form the prominence of the hip. The margin of this prominence is called the *iliac crest.*

Posteriorly, the ilium joins the sacrum at the *sacroiliac joint.* A projection of the ilium, the *anterior superior iliac spine,* which can be felt lateral to the groin, provides attachments for ligaments and muscles.

The **ischium,** which forms the lowest portion of the coxal bone, is L-shaped with its angle, the *ischial tuberosity,* pointing posteriorly and downward. This tuberosity has a rough surface that provides attachments for ligaments and leg muscles. It also supports the weight of the body when a person is sitting. Above the ischial tuberosity, near the junction of the ilium and ischium, is a sharp projection called the *ischial spine.*

The **pubis** constitutes the anterior portion of the coxal bone. The two pubic bones come together in the midline to create a joint called the *symphysis pubis.* The angle formed by these bones below the symphysis is the *pubic arch.* (See fig. 8.28.)

A portion of each pubis passes posteriorly and downward to join an ischium. Between the bodies of these bones on either side there is a large opening, the *obturator foramen,* which is the largest foramen in the skeleton.

Some sexual differences in the male and female pelves and other skeletal structures are summarized in chart 8.2. (See fig. 8.30.)

1. Locate and name each of the bones of the pelvis.
2. Describe a coxal bone.

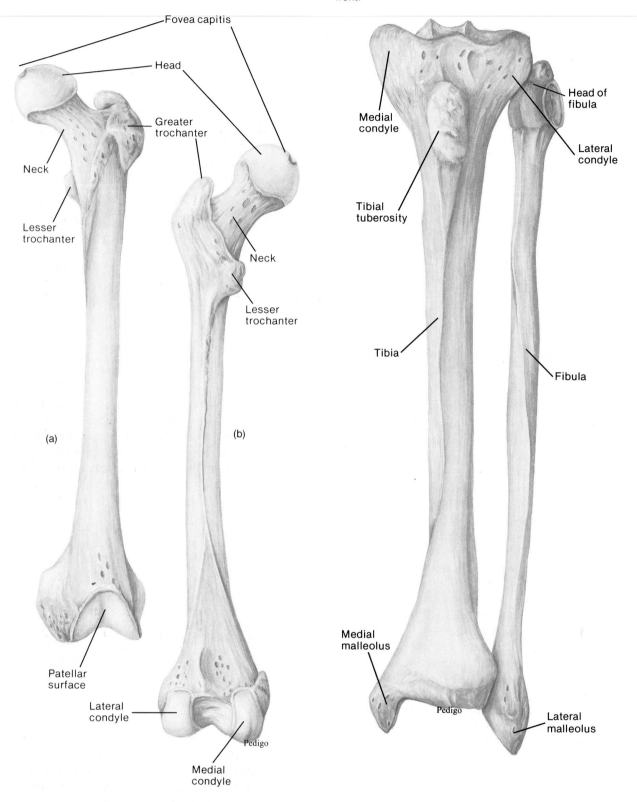

Fig. 8.31 (a) Anterior surface and (b) posterior surface of the left femur.

Fig. 8.32 Bones of the left lower leg viewed from the front.

Fovea capitis

Head

Greater trochanter

Neck

Lesser trochanter

Neck

Lesser trochanter

(a)

(b)

Patellar surface

Lateral condyle

Pedigo

Medial condyle

Medial condyle

Tibial tuberosity

Tibia

Medial malleolus

Head of fibula

Lateral condyle

Fibula

Pedigo

Lateral malleolus

Fig. 8.33 The talus moves freely where it articulates with the tibia and fibula.

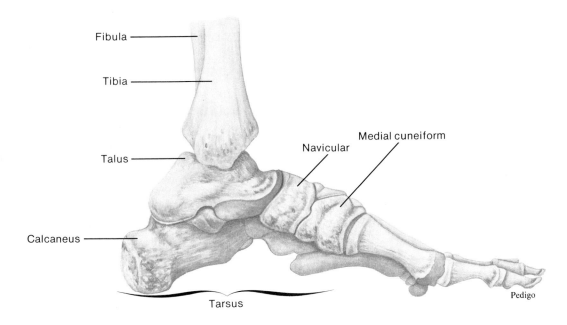

Fibula

Tibia

Talus

Calcaneus

Navicular

Medial cuneiform

Pedigo

Tarsus

The Lower Limb

The bones of the lower limb form the frameworks of the leg, ankle, instep, and toes. They include a femur, a tibia, a fibula, and several tarsals, metatarsals, and phalanges. (See fig. 8.6.)

The Femur

The **femur** is the longest bone in the body and extends from the hip to the knee. (See fig. 8.31.) A large, rounded *head* at its upper end projects medially into the acetabulum of the coxal bone. On the head, a pit called the *fovea capitis* marks the attachment of a ligament. Just below the head, there is a constriction, or *neck,* and two large processes—an upper, lateral *greater trochanter* and a lower, medial *lesser trochanter.* These processes provide attachments for muscles of the legs and buttocks.

At the lower end of the femur, two rounded processes, the *lateral* and *medial condyles,* articulate with the tibia of the lower leg. A **patella** also articulates with the femur on its distal anterior surface. It is a flat sesamoid bone located in a tendon that passes anteriorly over the knee. (See fig. 8.6.)

The Tibia

The **tibia** is the larger of the two lower leg bones and is located on the medial side. (See fig. 8.32.) Its upper end is expanded into *medial* and *lateral condyles,* which have concave surfaces and articulate with the

condyles of the femur. Below the condyles, on the anterior surface, is a process called the *tibial tuberosity,* which provides an attachment for the *patellar ligament*—a continuation of the patella-bearing tendon.

At its lower end, the tibia expands to form a prominence on the inner ankle called the *medial malleolus,* which serves as an attachment for ligaments. On its lateral side is a depression that articulates with the fibula. The inferior surface of its distal end articulates with a large bone (the talus) in the foot.

The Fibula

The **fibula** (fig. 8.32) is a long, slender bone located on the lateral side of the tibia. Its ends are slightly enlarged into an upper *head* and a lower *lateral malleolus.* The head articulates with the tibia just below the lateral condyle; however, it does not enter into the knee joint and does not bear any body weight. The lateral malleolus articulates with the ankle and forms a prominence on the lateral side.

The Foot

The foot consists of an ankle, an instep, and five toes. The ankle is composed of seven **tarsal bones,** forming a group called the *tarsus.* These bones are arranged so that one of them, the **talus,** can move freely where it joins the tibia and fibula. The remaining tarsal bones are bound firmly together, forming a mass on which the talus rests. (See figs. 8.33 and 8.34.)

Fig. 8.34 The left foot viewed from above.

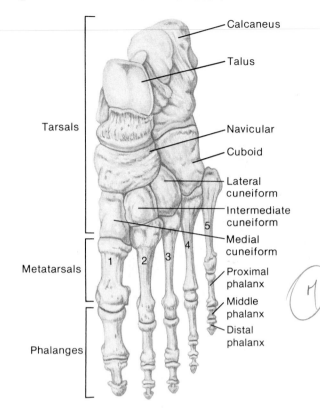

Tarsals

Calcaneus

Talus

Navicular

Cuboid

Lateral cuneiform

Intermediate cuneiform

Medial cuneiform

Metatarsals

Proximal phalanx

Middle phalanx

Distal phalanx

Phalanges

1. Locate and name each of the bones of the lower limb.
2. Explain how these bones articulate with one another.
3. Describe how the foot is adapted to support the body.

The individual bones of the tarsus are named in figure 8.34.

The largest of the ankle bones, the **calcaneus** or heel bone, is located below the talus where it projects backward to form the base of the heel. The calcaneus helps support the weight of the body and provides an attachment for muscles that move the foot.

The instep consists of five, elongated **metatarsal bones** that articulate with the tarsus. They are numbered 1 to 5, beginning on the medial side. (See fig. 8.34.) The heads at the distal ends of these bones form the ball of the foot. The tarsals and metatarsals are arranged and bound by ligaments to form the arches of the foot. A longitudinal arch extends from the heel to the toe, and a transverse arch stretches across the foot. These arches provide a stable, springy base for the body.

The **phalanges** of the toes are similar to those of the fingers. They are in line with the metatarsals and articulate with them. There are three phalanges in each toe—a proximal, a middle, and a distal phalanx—except the great toe which lacks a middle phalanx.

The Joints

Joints (articulations) are functional junctions between bones. Although they vary considerably in structure, they can be classified according to the amount of movement they make possible. On this basis, three general groups can be identified—immovable joints, slightly movable joints, and freely movable joints.

Immovable Joints

Immovable joints occur between bones that come into close contact with one another. The bones at such joints are separated by a thin layer of fibrous tissue or cartilage, as in the case of a *suture* between a pair of flat bones of the cranium (fig. 8.35). No active movement takes place at an immovable joint.

Slightly Movable Joints

The bones of *slightly movable joints* are connected by disks of fibrocartilage or by ligaments. The vertebrae of the vertebral column, for instance, are separated by joints of this type. The articulating surfaces of the vertebrae are covered by thin layers of hyaline cartilage. This cartilage, in turn, is attached to the intervertebral disk that separates the adjacent vertebral bodies.

Each intervertebral disk is composed of a band of particularly fibrous fibrocartilage surrounding a pulpy or gelatinous core (nucleus pulposus). The disk acts as a shock absorber and helps to equalize pressures between adjacent vertebral bodies during body movements (fig. 8.15).

Due to the slight flexibility of the disks, these joints allow a limited amount of movement, as when the back is bent forward or to the side, or is twisted. Other examples of slightly movable joints include the symphysis pubis, sacroiliac joint, and the joint in the lower leg between the distal ends of the tibia and fibula.

Fig. 8.35 (a) The joints between the bones of the cranium are immovable and are called sutures; (b) the bones at a suture are separated by a thin layer of connective tissue.

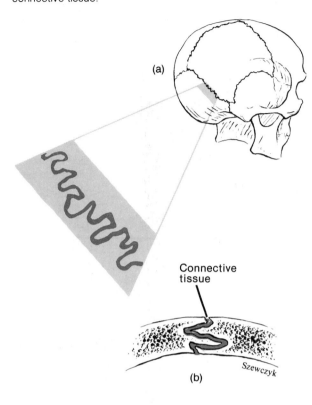

Connective tissue

Szewczyk

(b)

Fig. 8.36 The generalized structure of a freely movable joint. What is the function of the synovial fluid within this type of joint?

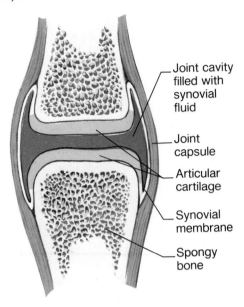

Joint cavity filled with synovial fluid

Joint capsule

Articular cartilage

Synovial membrane

Spongy bone

Freely Movable Joints

Most joints within the skeletal system are *freely movable* and have more complex structures than immovable or slightly movable joints.

The ends of the bones at a freely movable joint are covered with hyaline cartilage (articular cartilage) and are held together by a surrounding, tubelike capsule of dense fibrous tissue. This *joint capsule* is composed of an outer layer of ligaments and an inner lining of synovial membrane, which secretes synovial fluid. For this reason, freely movable joints are often called *synovial joints*. The synovial fluid has a consistency somewhat like eggwhite, and it acts as a joint lubricant. (See fig. 8.36.)

Some freely movable joints have flattened, shock-absorbing pads of fibrocartilage between the articulating surfaces of the bones. The knee joint, for example, contains pads called *semilunar cartilages* (menisci). (See fig. 8.38.) Such joints may also have closed, fluid-filled sacs called **bursae** associated with them. Bursae are lined with synovial membrane, which may be continuous with the synovial membranes of nearby joint cavities.

Bursae are commonly located between the skin and underlying bony prominences, as in the case of the patella of the knee or the olecranon process of the elbow. They aid in the movement of tendons that pass over these bony parts or over other tendons. Figures 8.37 and 8.38 shows some of the bursae associated with the shoulder and knee.

Because the articulating bones of freely movable joints have a variety of shapes, a number of different movements are possible. Examples of these joints, shown in figure 8.39, can be classified as follows:

1. **Ball and socket joints.** A ball and socket joint consists of a bone with a ball-shaped head that articulates with a cup-shaped socket of another bone. Such a joint allows for a wider range of motion than does any other kind. Movements in all planes, as well as rotational movement around a central axis, are possible. The hip and shoulder contain joints of this type.

2. **Condyloid joints.** In a condyloid joint, an oval-shaped condyle of one bone fits into an elliptical cavity of another bone, as in the case of the joints between the metacarpals and phalanges. This type of joint allows a variety of movements in different planes; rotational movements, however, are not possible.

Fig. 8.37 The shoulder joint allows movements in all directions. Note the bursa associated with this joint.

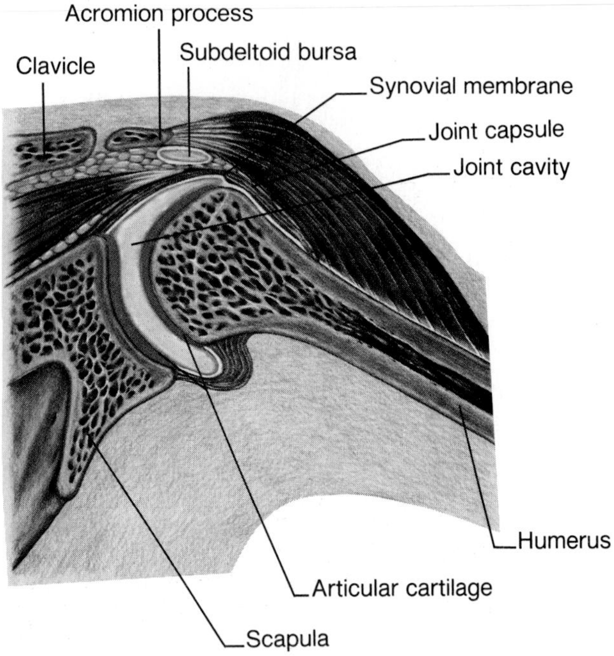

Acromion process
Clavicle
Subdeltoid bursa
Synovial membrane
Joint capsule
Joint cavity
Humerus
Articular cartilage
Scapula

Fig. 8.38 The knee, which is a modified hinge joint, is the most complex of the synovial joints.

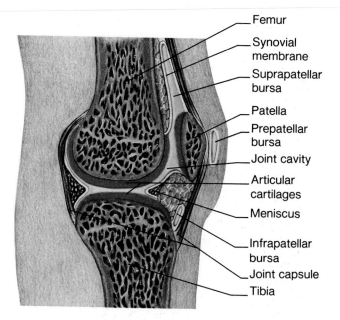

Femur
Synovial membrane
Suprapatellar bursa
Patella
Prepatellar bursa
Joint cavity
Articular cartilages
Meniscus
Infrapatellar bursa
Joint capsule
Tibia

3. **Gliding joints.** The articulating surfaces of gliding joints are nearly flat or only slightly curved. Such joints are found between some wrist bones, some ankle bones, and between the articular processes of adjacent vertebrae. They allow sliding and twisting movements.

4. **Hinge joints.** In a hinge joint, the convex surface of one bone fits into the concave surface of another, as in the case of the elbow and the joints of the phalanges. This type of joint allows movement in one plane only, like the motion of a single-hinged door.

5. **Pivot joints.** In a pivot joint, a cylindrical surface of one bone rotates within a ring formed of bone and fibrous tissue. The movement at such a joint is limited to rotation about a central axis. The joint between the proximal ends of the radius and the ulna is of this type.

6. **Saddle joints.** A saddle joint is formed between bones whose articulating surfaces have both concave and convex regions. The surfaces of one bone fit the complementary surfaces of the other. This arrangement allows a wide variety of movements, as in the case of the joint between a carpal (trapezium) and the metacarpal of the thumb.

Chart 8.3 summarizes the characteristics of the various types of joints.

Arthritis is a condition that causes inflamed, swollen, and painful joints. Although there are several different types of arthritis, the most common forms are *rheumatoid arthritis* and *osteoarthritis*.

In rheumatoid arthritis, which is the most painful and crippling of the arthritic diseases, the synovial membrane of a freely movable joint becomes inflamed and grows thicker. This change is usually followed by damage to the articular cartilages on the ends of the bones and an invasion of the joint by fibrous tissues. These fibrous tissues increasingly interfere with joint movement, and in time the tissues may become ossified so that the articulating bones are fused together. The cause of rheumatoid arthritis is unknown.

Osteoarthritis is a degenerative disease that occurs as a result of aging and affects a large percentage of persons over 60 years of age. In this condition, the articular cartilages soften and disintegrate gradually so that the articular surfaces become roughened. Consequently, the joints are sore and less movement is possible. Osteoarthritis is most likely to affect joints that have received the greatest use over the years, such as those in the knees and the lower regions of the vertebral column.

Fig. 8.39 Types of freely movable joints: (a) hinge joint of the elbow, (b) pivot joint between proximal ends of the radius and ulna, (c) saddle joint between trapezium of the wrist and metacarpal of the thumb, (d) condyloid joint between metacarpals and phalanges of the hand, (e) ball and socket joint of the hip, and (f) gliding joint between tarsals of the ankle.

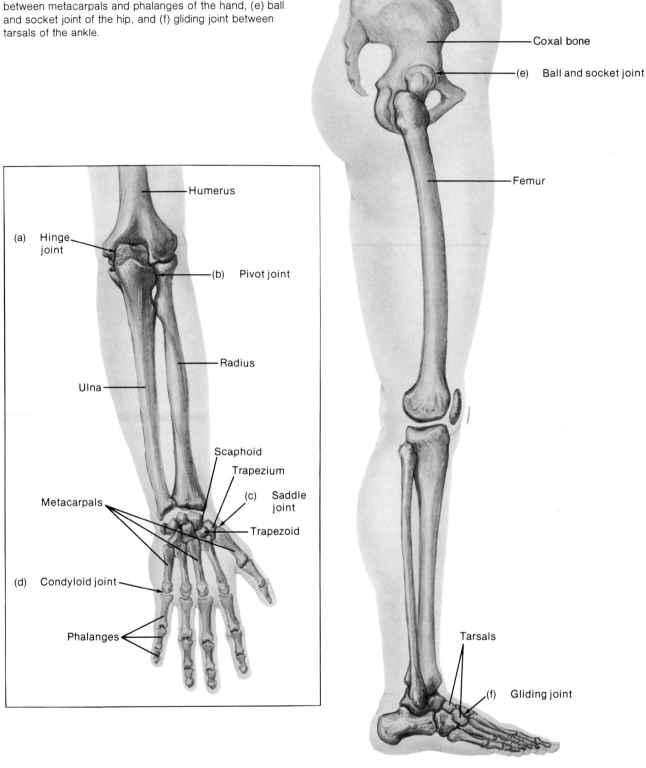

Coxal bone

(e) Ball and socket joint

Femur

Humerus

(a) Hinge joint

(b) Pivot joint

Radius

Ulna

Scaphoid

Trapezium

(c) Saddle joint

Metacarpals

Trapezoid

(d) Condyloid joint

Phalanges

Tarsals

(f) Gliding joint

Chart 8.3 Types of joints

Type	Description	Possible Movements	Example
Immovable	Articulating bones in close contact and separated by a thin layer of fibrous tissue or cartilage	No active movement	Suture between bones of the cranium
Slightly movable	Articulating bones separated by disks of fibrocartilage	Limited movements as when back is bent or twisted	Joints between vertebrae, symphysis pubis, sacroiliac joint
Freely movable	Articulating bones surrounded by joint capsule of ligaments and synovial membranes; ends of articulating bones covered by hyaline cartilage and separated by synovial fluid		
1. Ball and socket	Ball-shaped head of one bone articulates with cup-shaped socket of another	Movements in all planes and rotation	Shoulder, hip
2. Condyloid	Oval-shaped condyle of one bone articulates with elliptical cavity of another	Variety of movements in different planes, but no rotation	Joints between metacarpals and phalanges
3. Gliding	Articulating surfaces are nearly flat or slightly curved	Sliding or twisting	Joints between various bones of wrist and ankle
4. Hinge	Convex surface of one bone articulates with concave surface of another	Up and down motion in one plane	Elbow, joints of phalanges
5. Pivot	Cylindrical surface of one bone articulates with ring of bone and fibrous tissue	Rotation	Joint between proximal ends of radius and ulna
6. Saddle	Articulating surfaces have both concave and convex regions; surface of one bone fits complementary surface of another	Variety of movements	Joint between carpal and metacarpal of thumb

Types of Joint Movements

Movements at synovial joints are produced by actions of skeletal muscles. Typically, one end of a muscle is attached to a relatively immovable or fixed part on one side of a joint, and the other end of the muscle is fastened to a movable part on the other side. When the muscle contracts, fibers within the muscle pull its movable end (insertion) toward its fixed end (origin), and a movement occurs at the joint.

The following terms are used to describe various movements of body parts at joints. (See figs. 8.40, 8.41, and 8.42.)

flexion (flek'shun)—bending parts at a joint so that the angle between them is decreased and the parts come closer together (bending the leg at the knee).

extension (ek-sten'shun)—straightening parts at a joint so that the angle between them is increased and the parts move further apart (straightening the leg at the knee).

hyperextension (hi''per-ek-sten'shun)—excessive extension of the parts at a joint, beyond the anatomical position (bending the head back beyond the upright position).

dorsiflexion (dor''si-flek'shun)—flexing the foot at the ankle (bending the foot upward).

plantar flexion (plan'tar flek'shun)—extending the foot at the ankle (bending the foot downward).

abduction (ab-duk'shun)—moving a part away from the midline (lifting the arm horizontally to form a right angle with the side of the body).

adduction (ah-duk'shun)—moving a part toward the midline (returning the arm from the horizontal position to the side of the body).

rotation (ro-ta'shun)—moving a part around an axis (twisting the head from side to side).

circumduction (ser''kum-duk'shun)—moving a part so that its end follows a circular path (moving the finger in a circular motion without moving the hand).

supination (soo''pi-na'shun)—turning the hand so the palm is upward.

pronation (pro-na'shun)—turning the hand so the palm is downward.

eversion (e-ver'zhun)—turning the foot so the sole is outward.

inversion (in-ver'zhun)—turning the foot so the sole is inward.

protraction (pro-trak'shun)—moving a part forward (thrusting the chin forward).

retraction (re-trak'shun)—moving a part backward (pulling the chin backward).

elevation (el''ĕ-va'shun)—raising a part (shrugging the shoulders).

depression (de-presh'un)—lowering a part (drooping the shoulders).

    Support and Movement

Fig. 8.40 Joint movements illustrating flexion, extension, hyperextension, dorsiflexion, plantar flexion, abduction, and adduction.

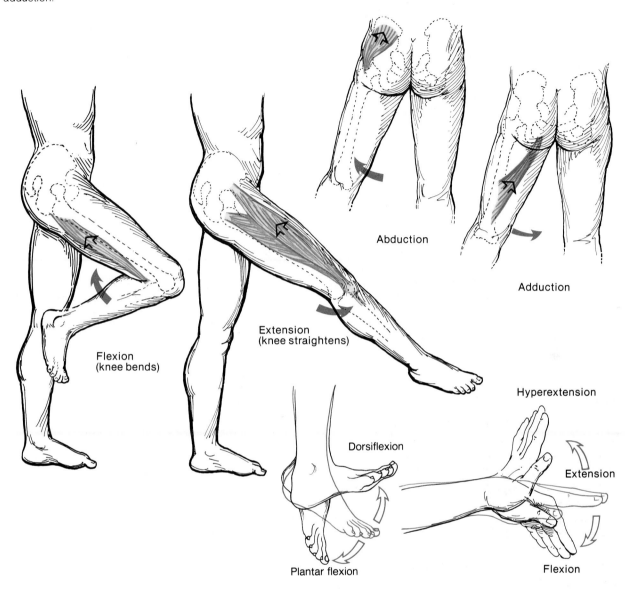

Abduction

Adduction

Flexion
(knee bends)

Extension
(knee straightens)

Hyperextension

Dorsiflexion

Extension

Plantar flexion

Flexion

Fig. 8.41 Joint movements illustrating rotation, circumduction, supination, and pronation.

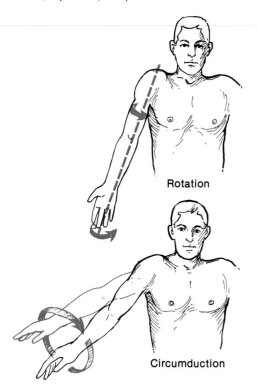

Rotation

Circumduction

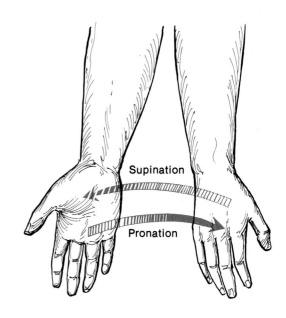

Supination

Pronation

1. Describe the characteristics of the three major types of joints.
2. List six different types of freely movable joints.
3. Describe the movements each type of joint makes possible.
4. What terms are used to describe various movements at joints?

Clinical Terms Related to the Skeletal System

achondroplasia (a-kon″dro-pla′ze-ah)—an inherited condition in which the formation of cartilaginous bone is retarded. The result is a type of dwarfism.

acromegaly (ak″ro-meg′ah-le)—a condition due to an overproduction of growth hormone in adults and characterized by abnormal enlargement of facial features, hands, and feet.

ankylosis (ang″kǐ-lo′sis)—abnormal stiffness of a joint, often due to damage of joint membranes from chronic rheumatoid arthritis.

arthralgia (ar-thral′je-ah)—pain in a joint.

arthrocentesis (ar″thro-sen-te′sis)—puncture and removal of fluid from a joint cavity.

arthrodesis (ar″thro-de′sis)—surgery performed to fuse the bones at a joint.

arthroplasty (ar′thro-plas″te)—surgery performed to make a joint movable.

Colles fracture (kol′ēz frak′tūre)—a fracture at the distal end of the radius in which the smaller fragment is displaced posteriorly.

epiphysiolysis (ep″ǐ-fiz″e-ol′ǐ-sis)—a separation or loosening of the epiphysis from the diaphysis of a bone.

gout (gowt)—a metabolic disease in which excessive uric acid in the blood may be deposited in the joints, causing them to become inflamed, swollen, and painful.

hemarthrosis (hem″ar-thro′sis)—blood in a joint cavity.

laminectomy (lam″ǐ-nek′to-me)—surgical removal of the posterior arch of a vertebra, usually to relieve the symptoms of a ruptured intervertebral disk.

lumbago (lum-ba′go)—a dull ache in the lumbar region of the back.

orthopedics (or″tho-pe′diks)—the science of prevention, diagnosis, and treatment of diseases and abnormalities involving the skeletal and muscular systems.

ostalgia (os-tal′je-ah)—pain in a bone.

ostectomy (os-tek′to-me)—surgical removal of a bone.

osteitis (os″te-i′tis)—inflammation of bone tissue.

osteochondritis (os″te-o-kon-dri′tis)—inflammation of bone and cartilage tissues.

Fig. 8.42 Joint movements illustrating eversion, inversion, elevation, depression, protraction, and retraction.

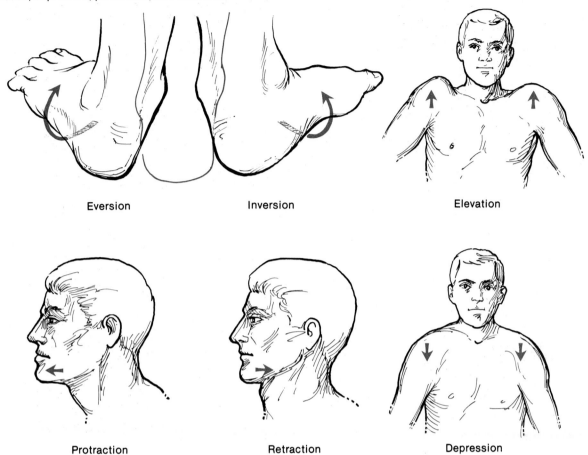

Eversion   Inversion   Elevation

Protraction   Retraction   Depression

osteogenesis (os″te-o-jen′ĕ-sis)—the development of bone.

osteogenesis imperfecta (os″te-o-jen′ĕ-sis im-per-fek′ta)—a congenital condition characterized by the development of deformed and abnormally brittle bones.

osteoma (os″te-o′mah)—a tumor composed of bone tissue.

osteomalacia (os″te-o-mah-la′she-ah)—a softening of adult bone due to a disorder in calcium and phosphorus metabolism, usually caused by a deficiency of vitamin D.

osteomyelitis (os″te-o-mi″ĕ-li′tis)—inflammation of bone caused by the action of bacteria or fungi.

osteonecrosis (os″te-o-ne-kro′sis)—death of bone tissue. This condition occurs most commonly in the head of the femur in elderly persons and may be due to obstructions in arteries that supply the bone.

osteopathology (os″te-o-pah-thol′o-je)—the study of bone diseases.

osteotomy (os″te-ot′o-me)—the cutting of a bone.

Roentgenogram (rent-gen′o-gram″)—an X-ray film.

Chapter Summary

Introduction

Individual bones are the organs of the skeleton system.
Bone contains very active tissues.

Bone Structure

Bone structure reflects its function.

1. Parts of a long bone
   a. Epiphyses are covered with articular cartilage and articulate with other bones.
   b. The shaft of a bone is called the diaphysis.
   c. Except for the articular cartilage, a bone is covered by a periosteum.
   d. Compact bone provides strength and resistance to bending.
   e. Spongy bone provides strength and reduces the weight of bone.
   f. The diaphysis contains a medullary cavity filled with marrow.

2. Microscopic structure
   a. Compact bone contains haversian systems cemented together.
   b. Haversian canals contain blood vessels that nourish the cells of haversian systems.
   c. Cells of spongy bone are nourished by diffusion from the surface of the bony plates.

Bone Development and Growth

1. Intramembranous bones
   a. Intramembranous bones develop from layers of connective tissues.
   b. Bone tissue is formed by osteoblasts within the membranous layers.
   c. Mature bone cells are called osteocytes.

2. Endochondral bones
   a. Endochondral bones develop first as hyaline cartilage that is later replaced by bone tissue.
   b. The primary ossification center appears in the diaphysis, while secondary ossification centers appear in the epiphyses.
   c. An epiphyseal disk remains between the primary and secondary ossification centers.
   d. The epiphyseal disk is responsible for growth in length.
   e. Long bones continue to grow in length until the epiphyseal disks are ossified.
   f. Growth in thickness is due to intramembranous ossification occurring beneath the periosteum.

Functions of Bones

1. Support and protection
   a. Skeletal parts provide shape and form for body structures.
   b. They support and protect softer, underlying tissues.

2. Body movement
   a. Bones and muscles function together as levers.
   b. A lever consists of a rod, pivot (fulcrum), weight that is moved, and a force that supplies energy.

3. Blood cell formation
   a. At different ages, hematopoiesis occurs in the yolk sac, liver and spleen, and red bone marrow.
   b. Red marrow functions in the production of red blood cells, white blood cells, and blood platelets.

4. Storage of inorganic salts
   a. The intercellular material of bone tissue contains large quantities of calcium phosphate.
   b. It also stores lesser amounts of magnesium, potassium, and carbonate ions.

Organization of the Skeleton

1. Number of bones
   Usually there are 206 bones in the human skeleton, but the number may vary.

2. Divisions of the skeleton
   a. The skeleton can be divided into axial and appendicular portions.
   b. The axial skeleton consists of the skull, hyoid bone, vertebral column, and thoracic cage.
   c. The appendicular skeleton consists of the pectoral girdle, upper limbs, pelvic girdle, and lower limbs.

The Skull

The skull consists of 22 bones, which include 8 cranial bones, 13 facial bones, and 1 mandible.

1. The cranium
   a. The cranium encloses and protects the brain.
   b. Some cranial bones contain air-filled sinuses.
   c. Cranial bones include frontal bone, parietal bones, occipital bone, temporal bones, sphenoid bone, and ethmoid bone.

2. The facial skeleton
   a. Facial bones provide the basic shape of the face and attachments for muscles.
   b. Facial bones include maxillary bones, palatine bones, zygomatic bones, lacrimal bones, nasal bones, vomer bone, inferior nasal conchae, and mandible.

3. The infantile skull
   a. Incompletely developed bones are separated by fontanels.
   b. Proportions of the infantile skull are different from those of the adult skull.

The Vertebral Column

The vertebral column extends from the skull to the pelvis and protects the spinal cord.
It is composed of vertebrae separated by intervertebral disks.
It has 4 curvatures—cervical, thoracic, lumbar, and pelvic.

1. A typical vertebra
   a. A typical vertebra consists of a body and a bony arch that surrounds the spinal cord.
   b. Notches on the lower surfaces provide intervertebral foramina through which spinal nerves pass.

2. Cervical vertebrae
   a. Transverse processes bear transverse foramina.
   b. Atlas (first vertebra) supports and balances the head.
   c. Odontoid process of the axis (second vertebra) provides a pivot for the atlas.

3. Thoracic vertebrae
   a. Thoracic vertebrae are larger than cervical vertebrae.
   b. Facets on the sides articulate with the ribs.
4. Lumbar vertebrae
   a. Vertebral bodies are large and strong.
   b. They support more body weight than other vertebrae.
5. The sacrum
   a. The sacrum is a triangular structure formed of five fused vertebrae.
   b. Vertebral foramina form the sacral canal.
6. The coccyx
   a. The coccyx forms the lowest part of the vertebral column.
   b. It is composed of four fused vertebrae.

The Thoracic Cage

The thoracic cage includes the ribs, thoracic vertebrae, sternum, and costal cartilages.
It supports the shoulder girdle and arms, protects visceral organs, and functions in breathing.

1. The ribs
   a. Ribs are attached to the thoracic vertebrae.
   b. Costal cartilages of true ribs join the sternum directly; those of the false ribs join indirectly.
   c. A typical rib bears a shaft, head, and tubercles that articulate with the vertebrae.
2. The sternum
   a. The sternum consists of a manubrium, body, and xiphoid process.
   b. It articulates with the clavicles.

The Pectoral Girdle

The pectoral girdle is composed of two clavicles and two scapulae.
It forms an incomplete ring that supports the arms and provides attachments for muscles.

1. The clavicles
   a. The clavicles are located between the manubrium and scapulae.
   b. They function to hold the shoulders in place and provide attachments for muscles.
2. The scapulae
   a. The scapulae are broad, triangular bones.
   b. They articulate with the humerus and provide attachments for muscles.

The Upper Limb

Bones of the upper limb provide frameworks for arms, wrists, palms, and fingers.
They also provide attachments for muscles and function in levers that move the limb and its parts.

1. The humerus
   a. The humerus extends from the glenoid cavity of the scapula to the elbow.
   b. It articulates with the radius and ulna at the elbow.
2. The radius
   a. The radius extends from the elbow to the wrist.
   b. It articulates with the humerus, ulna, and wrist.
3. The ulna
   a. The ulna overlaps the humerus posteriorly.
   b. It articulates with the radius laterally and with a disk of fibrocartilage inferiorly.
4. The hand
   a. The hand is composed of a wrist, palm, and 5 fingers.
   b. It includes 8 carpals that form a carpus, 5 metacarpals, and 14 phalanges.

The Pelvic Girdle

The pelvic girdle consists of two coxal bones that articulate with each other anteriorly and with the sacrum posteriorly.
The sacrum, coccyx, and pelvic girdle form the pelvis.

1. The coxal bones
   Each coxal bone consists of three bones, which are fused in the region of the acetabulum.
   a. The ilium
      (1) The ilium is the largest portion of the coxal bone.
      (2) It joins the sacrum at the sacroiliac joint.
   b. The ischium
      (1) The ischium is the lowest portion of the coxal bone.
      (2) It supports body weight when sitting.
   c. The pubis
      (1) The pubis is the anterior portion of the coxal bone.
      (2) Pubis bones are fused anteriorly at the symphysis pubis.

The Lower Limb

Bones of the lower limb provide frameworks for the leg, ankle, instep, and toes.

1. The femur
   a. The femur extends from the knee to the hip.
   b. The patella articulates with its anterior surface.
2. The tibia
   a. The tibia is located on the medial side of the lower leg.
   b. It articulates with the talus of the ankle.
3. The fibula
   a. The fibula is located on the lateral side of the tibia.
   b. It articulates with the ankle, but does not bear body weight.
4. The foot
   a. The foot consists of an ankle, instep, and 5 toes.
   b. It includes 7 tarsals that form the tarsus, 5 metatarsals, and 14 phalanges.

The Joints

Joints can be classified on the basis of the amount of movement they make possible.

1. Immovable joints
   a. Bones of immovable joints are in close contact, separated by a thin layer of fibrous tissue or cartilage, as in a suture.
   b. No active movements are possible at these joints.
2. Slightly movable joints
   a. Bones of slightly movable joints are connected by disks of fibrocartilage or by ligaments, as the vertebrae.
   b. Such a joint allows a limited amount of movement.
3. Freely movable joints
   a. Bones of a freely movable joint are covered with hyaline cartilage and held together by a fibrous capsule.
   b. The joint capsule consists of an outer layer of ligaments and an inner lining of synovial membrane.
   c. Bursae are often located between the skin and underlying bony prominences.
   d. Freely movable joints include several types: ball and socket, condyloid, gliding, hinge, pivot, and saddle.
4. Types of joint movements
   a. Movements of synovial joints are produced by muscles that are fastened on either side of the joint.
   b. Movements include flexion, extension, hyperextension, dorsiflexion, plantar flexion, abduction, adduction, rotation, circumduction, supination, pronation, eversion, inversion, elevation, depression, protraction, and retraction.

Application of Knowledge

1. What steps do you think should be taken to reduce the chances of persons accumulating abnormal metallic elements such as lead, radium, and strontium in their bones?
2. When a child's bone is fractured, growth may be stimulated at the epiphyseal disk of that bone. What problems might this extra growth create in an arm or leg before the growth of the other limb compensates for the difference in length?

Review Activities

Part A

1. Sketch a typical long bone and label its epiphyses, diaphysis, medullary cavity, periosteum, and articular cartilages.
2. Distinguish between spongy and compact bone.
3. Explain how haversian and Volkmann's canals are related.
4. Explain how the development of intramembranous bone differs from that of endochondral bone.
5. Distinguish between osteoblasts and osteoclasts.
6. Explain the function of an epiphyseal disk.
7. Explain how a long bone grows in thickness.
8. Provide several examples to illustrate how bones support and protect body parts.
9. Describe a lever.
10. Explain the functions of red and yellow bone marrow.
11. Distinguish between the axial and appendicular skeletons.
12. List the bones that form the pectoral and the pelvic girdles.
13. Name the bones of the cranium and the facial skeleton.
14. Explain the importance of fontanels.
15. Describe a typical vertebra.
16. Explain the differences between cervical, thoracic, and lumbar vertebrae.
17. Name the bones that comprise the thoracic cage.
18. Name the bones of the upper limb.
19. Define *coxal bone.*
20. List the bones of the lower limb.
21. Describe an immovable joint, a slightly movable joint, and a freely movable joint.
22. Define *bursa.*

Part B

Match the parts listed in column I with the bones
  listed in column II.

I

1. Coronoid process
2. Cribriform plate
3. Foramen magnum
4. Mastoid process
5. Palatine process
6. Sella turcica
7. Supraorbital foramen
8. Temporal process

II

A. Ethmoid bone
B. Frontal bone
C. Mandible
D. Maxillary bone
E. Occipital bone
F. Temporal bone
G. Sphenoid bone
H. Zygomatic bone

---

9. Acromion process
10. Deltoid tuberosity
11. Greater trochanter
12. Lateral malleolus
13. Medial malleolus
14. Olecranon process
15. Radial tuberosity
16. Xiphoid process

I. Femur
J. Fibula
K. Humerus
L. Radius
M. Scapula
N. Sternum
O. Tibia
P. Ulna

Part C

Match the terms in column I with the movements in
  column II.

I

17. Rotation
18. Supination
19. Extension
20. Eversion
21. Protraction
22. Flexion
23. Pronation
24. Abduction
25. Depression

II

Q. Turning palm
   upward
R. Decreasing angle
   between parts
S. Moving part
   forward
T. Moving part
   around an axis
U. Turning sole of
   foot outward
V. Increasing angle
   between parts
W. Lowering a part
X. Turning palm
   downward
Y. Moving part
   away from
   midline

The Muscular System

9

Muscles, the organs of the *muscular system,* consist largely of muscle cells. They are specialized to undergo muscular contractions during which the chemical energy of some nutrients is converted into mechanical energy or movement.

When muscle cells contract, they pull on the parts to which they are attached. This action usually causes some movement, as when the joints of the legs are flexed and extended during walking. At other times, muscular contractions resist motion, as when they help to hold body parts in postural positions. Muscles are also responsible for the movement of body fluids such as blood and urine, and they function in heat production, which aids in maintaining body temperature.

Chapter Objectives

After you have studied this chapter, you should be able to

1. Describe how connective tissue is included in the structure of a skeletal muscle.

2. Name the major parts of a skeletal muscle fiber and describe the function of each part.

3. Explain the major events that occur during muscle fiber contraction.

4. Explain how energy is supplied to the muscle fiber contraction mechanism.

5. Describe how oxygen debt develops and how a muscle may become fatigued.

6. Distinguish between a twitch and a sustained contraction.

7. Explain how various types of muscular contractions are used to produce body movements and maintain posture.

8. Describe how skeletal muscles are affected by exercise.

9. Distinguish between the structure and function of a multiunit smooth muscle and a visceral smooth muscle.

10. Compare the fiber contraction mechanisms of skeletal, smooth, and cardiac muscles.

11. Explain how the locations of skeletal muscles are related to the movements they produce and how muscles interact in producing such movements.

12. Identify and describe the location of the major skeletal muscles of each body region and describe the action of each muscle.

13. Complete the review activities at the end of this chapter. Note that the items are worded in the form of specific learning objectives. You may want to refer to them before reading the chapter.

actin (ak′tin)

antagonist (an-tag′o-nist)

aponeurosis (ap″o-nu-ro′sēz)

fascia (fash′e-ah)

insertion (in-ser′shun)

motor neuron (mo′tor nu′ron)

motor unit (mo′tor u′nit)

muscle impulse (mus′el im′puls)

myofibril (mi″o-fi′bril)

myosin (mi′o-sin)

neurotransmitter (nu″ro-trans′mit-er)

origin (or′ĭ-jin)

oxygen debt (ok′sĭ-jen det)

prime mover (prīm moov′er)

recruitment (re-krōōt′ment)

synergist (sin′er-jist)

threshold stimulus (thresh′old stim′u-lus)

calat-, something inserted: inter*calat*ed disk—membranous band that separates adjacent cardiac muscle cells.

erg-, work: syn*erg*ist—muscle that works together with a prime mover to produce a movement.

hyper-, over, more: muscular *hyper*trophy—enlargement of muscle fibers.

inter-, between: *inter*calated disk—membranous band that separates adjacent cardiac muscle cells.

laten-, hidden: *laten*t period—period between the time a stimulus is applied and the beginning of a muscle contraction.

myo-, muscle: *myo*fibril—contractile fiber of a muscle cell.

syn-, together: *syn*ergist—muscle that works together with a prime mover to produce a movement.

tetan-, stiff: *tetan*ic contraction—sustained muscular contraction.

-troph, well fed: muscular hyper*troph*y—enlargement of muscle fibers.

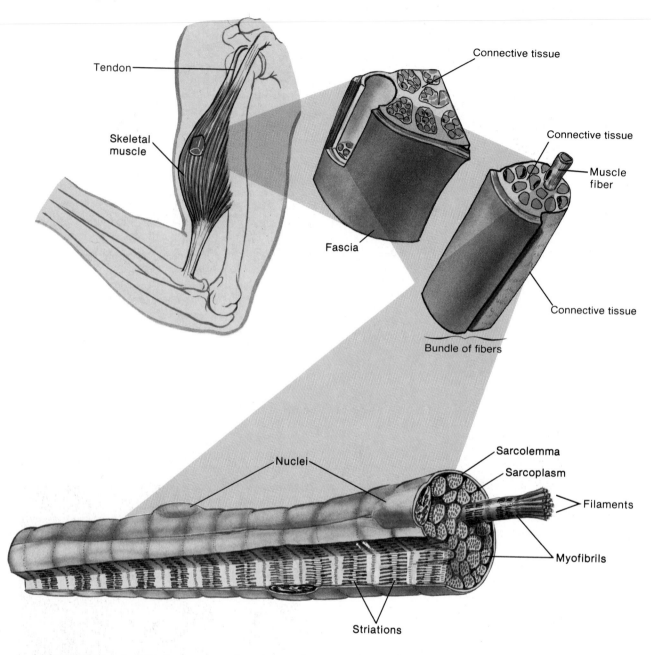

Tendon

Skeletal muscle

Fascia

Connective tissue

Connective tissue

Muscle fiber

Connective tissue

Bundle of fibers

Nuclei

Sarcolemma

Sarcoplasm

Filaments

Myofibrils

Striations

As is described in chapter 5, there are three types of muscle tissue within the body—skeletal muscle, smooth muscle, and cardiac muscle. This chapter, however, is primarily concerned with skeletal muscle, the type found in muscles that are attached to bones and are under conscious control. (See fig. 5.15.)

Structure of a Skeletal Muscle

A skeletal muscle is an organ of the muscular system and is composed of several kinds of tissue. These include skeletal muscle tissue, nerve tissue, blood, and various connective tissues.

Connective Tissue Coverings

An individual skeletal muscle is separated from adjacent muscles and held in position by layers of fibrous connective tissue called **fascia.**

Sometimes the connective tissues surrounding a muscle project beyond the end of the muscle fibers to become part of a cordlike **tendon.** (See fig. 9.1.) Fibers in the tendon may intertwine with those in the periosteum of a bone, thus attaching the muscle to the bone. In other cases, the connective tissues form broad fibrous sheets called **aponeuroses,** which may be attached to the coverings of adjacent muscles. (See fig. 9.13.)

Fig. 9.2 (a) A skeletal muscle fiber contains numerous myofibrils, each consisting of (b) units called sarcomeres. (c) The characteristic striations of a sarcomere are due to the arrangement of actin and myosin filaments.

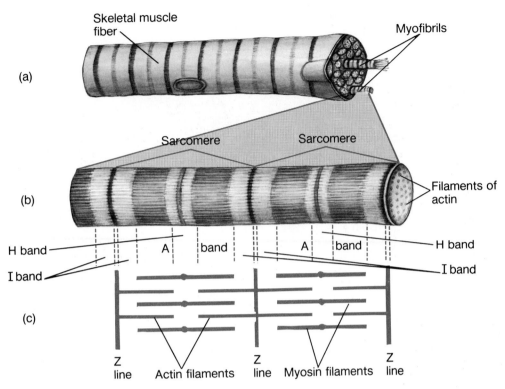

Figure 9.1 illustrates how the connective tissue that closely surrounds a skeletal muscle extends inward and separates the muscle tissue into small compartments. These compartments contain bundles of skeletal muscle fibers, and each muscle fiber within a bundle is surrounded by a thin layer of connective tissue.

Thus, all parts of a skeletal muscle are enclosed in layers of connective tissue that form a network extending throughout the muscular system.

Skeletal Muscle Fibers

A skeletal muscle fiber represents a single cell of a muscle. This fiber is responsive to stimulation, and when it responds it contracts and then relaxes. Each skeletal muscle fiber is a thin, elongated cylinder with rounded ends, and it may extend the full length of the muscle. Just beneath its cell membrane (*sarcolemma*), the cytoplasm (*sarcoplasm*) of the fiber contains many small, oval nuclei and mitochondria. Also within the cytoplasm are numerous threadlike **myofibrils** that lie parallel to one another. (See fig. 9.1.)

The myofibrils play a fundamental role in the muscle contraction mechanism. They contain two kinds of protein filaments—thick ones composed of

Fig. 9.3 Identify the bands of the striations in this electron micrograph of myofibrils.

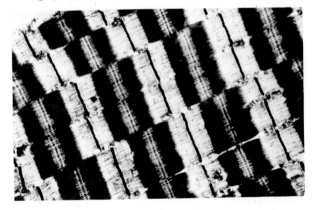

the protein **myosin** and thin ones composed of the protein **actin.** The arrangement of these filaments produces the characteristic alternating light and dark striations of muscle fiber. (See figs. 9.2 and 9.3.)

Myosin filaments are located primarily within the dark portions (*A bands*) of the striations, while actin filaments occur primarily in the light areas (*I bands*). The actin filaments, however, also extend into the A bands, and when the muscle fiber contracts, the actin filaments slide further into these bands.

Fig. 9.4 Within the sarcoplasm of a skeletal muscle fiber are a network of sarcoplasmic reticulum and a system of transverse tubules.

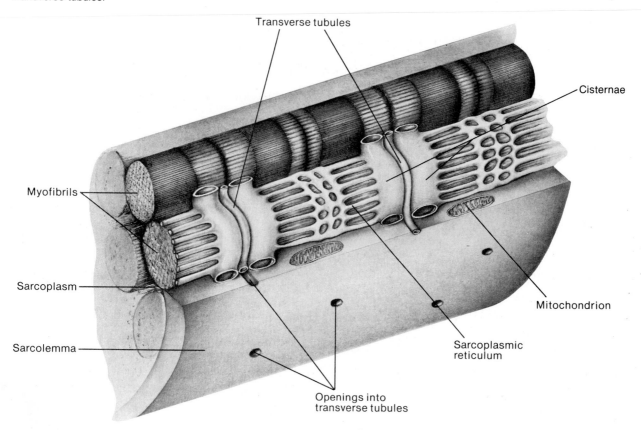

Transverse tubules

Cisternae

Myofibrils

Sarcoplasm

Sarcolemma

Mitochondrion

Sarcoplasmic reticulum

Openings into transverse tubules

The actin filaments are attached to Z lines at the ends of the I bands. These Z lines extend across the muscle fiber so that those of adjacent myofibrils lie side by side. The segment of the myofibril between two successive Z lines is called a **sarcomere,** and the consequent regular arrangement of the sarcomeres causes the muscle fiber to appear striated.

Within the cytoplasm of a muscle fiber is a network of membranous channels that surrounds each myofibril and runs parallel to it. This is the **sarcoplasmic reticulum,** and it is similar to the endoplasmic reticulum of other cells. Another set of membranous channels called **transverse tubules** (T-tubules) extends inward as invaginations from the fiber's membrane and passes all the way through the fiber. Thus, each of the tubules opens to the outside of the muscle fiber and contains extracellular fluid. Furthermore, each transverse tubule lies between two enlarged portions of the sarcoplasmic reticulum called *cisternae* near the region where the actin and myosin filaments overlap. The sarcoplasmic reticulum and transverse tubules function in activating the muscle contraction mechanism when the fiber is stimulated. (See fig. 9.4.)

1. Describe how connective tissue is associated with a skeletal muscle.
2. Describe the general structure of a skeletal muscle fiber.
3. Explain why skeletal muscle fibers appear striated.
4. Explain the relationship between the sarcoplasmic reticulum and transverse tubules.

The Neuromuscular Junction

Each skeletal muscle fiber is connected to a fiber from a nerve cell called a **motor neuron.** This nerve fiber extends outward from the brain or spinal cord, and usually a muscle fiber will contract only when it is stimulated by the action of such a motor neuron.

The site where the nerve fiber and muscle fiber meet is called a **neuromuscular junction.** At this junction the muscle fiber membrane is specialized to form a **motor end plate.** In this region mitochondria are abundant and the cell membrane is highly folded. (See fig. 9.5.)

The end of the motor nerve fiber is branched, and the ends of these branches project into recesses (synaptic clefts) of the muscle fiber membrane. The

Fig. 9.5 A neuromuscular junction includes the end of a motor neuron and the motor end plate of a muscle fiber. The neurotransmitter is stored in vesicles at the end of the nerve fiber.

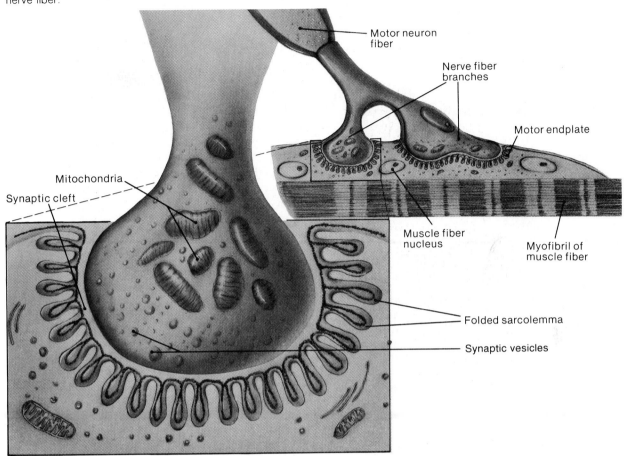

cytoplasm at the ends of the nerve fibers is rich in mitochondria and contains many tiny vesicles (synaptic vesicles) that store chemicals called **neurotransmitters.** *acetyl choline*

When a nerve impulse traveling from the brain or spinal cord reaches the end of a motor nerve fiber, some of the vesicles release a neurotransmitter into the gap between the nerve and the motor end plate. This action stimulates the muscle fiber to contract.

Motor Units

Although a muscle fiber usually has a single motor end plate, the nerve fibers of motor neurons are thickly branched, and one motor fiber may connect to many muscle fibers. Furthermore, when the motor nerve fiber transmits an impulse, all of the muscle fibers it is connected to are stimulated to contract simultaneously. Together, a motor neuron and the muscle fibers that it controls constitute a **motor unit.**

Skeletal Muscle Contraction

A muscle fiber contraction is a complex action involving a number of cell parts and chemical substances. The final result is a sliding movement within the myofibrils in which the filaments of actin and myosin merge. When this happens, the muscle fiber is shortened, and it pulls on its attachments.

Role of Actin and Myosin

A myosin molecule is composed of two twisted protein strands with globular protein parts projecting outward along their lengths. In the presence of calcium ions, these globular parts can react with actin filaments and form cross-bridges with them. This reaction between the actin and myosin filaments somehow generates the force involved with shortening the myofibrils during a contraction.

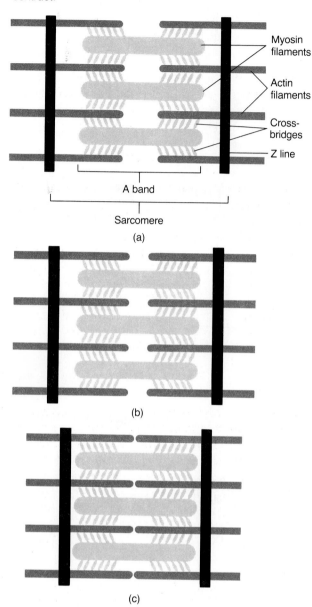

Fig. 9.6 (a) As cross-bridges form between filaments of actin and myosin, (b, c) the actin filaments are pulled toward the center of the A band, causing the myofibril to contract.

Myosin filaments

Actin filaments

Cross-bridges

Z line

A band

Sarcomere

(a)

(b)

(c)

An actin molecule also consists of double, twisted strands of protein, and it has ADP molecules attached to its surface. These ADP molecules serve as active sites for the formation of cross-bridges with the globular parts of myosin molecules.

Stimulus for Contraction

A skeletal muscle fiber normally does not contract until it is stimulated by a motor nerve fiber that releases neurotransmitter at the motor end plate. In skeletal muscle, the neurotransmitter is a compound called **acetylcholine.** This substance is synthesized in the cytoplasm of the motor neuron and stored in vesicles. When a nerve impulse reaches the end of the nerve fiber, many of these vesicles discharge their acetylcholine into the gap between the nerve fiber and the motor end plate. (See fig. 9.5.)

The acetylcholine diffuses rapidly across the gap and combines with certain protein molecules (receptors) in the muscle fiber membrane, thus stimulating it. As a result of this stimulus, a **muscle impulse** very much like a nerve impulse (described in chapter 10), passes in all directions over the surface of the muscle fiber membrane. It also travels through the transverse tubules, deep into the fiber.

The sarcoplasmic reticulum contains a high concentration of calcium ions. In response to a muscle impulse, the membranes of the cisternae become more permeable to these ions, and the ions diffuse into the sarcoplasm of the muscle fiber.

When calcium ions are present in the sarcoplasm in a relatively high concentration, cross-bridges form between the actin and myosin filaments, and a muscle contraction occurs. (See fig. 9.6.) The contraction continues while the calcium ions are present, but they are moved quickly back into the sarcoplasmic reticulum by an active transport mechanism (calcium pump). Consequently, the calcium concentration of the sarcoplasm is lowered, the cross-bridges are broken, and within a fraction of a second, the muscle fiber relaxes.

Meanwhile, the acetylcholine that stimulated the muscle fiber in the first place is rapidly decomposed by the action of an enzyme called **cholinesterase,** which is present at the neuromuscular junction within the membranes of the motor end plate. This action prevents a single nerve impulse from causing a continued stimulation of the muscle fiber.

1. *Describe a neuromuscular junction.*
2. *Define motor unit.*
3. *Explain how the filaments of a myofibril interact during muscle contraction.*
4. *Explain how a motor nerve impulse can trigger a muscle contraction.*

Energy Sources for Contraction

The energy used in muscle fiber contraction comes from ATP molecules, which are supplied by numerous mitochondria positioned close to the myofibrils. The globular portions of the myosin filaments contain an enzyme called **ATPase** that causes ATP to decompose into ADP and phosphate and, at the same time, to release some energy. This energy makes possible the reaction between the actin and myosin filaments. There is only enough ATP in a muscle fiber to operate the contraction mechanism for a very short time, however, so when a fiber is active, ATP must be regenerated.

The primary source of energy available to regenerate ATP from ADP and phosphate is a substance called **creatine phosphate.** Like ATP, creatine phosphate contains high-energy phosphate bonds, and it is actually four to six times more abundant than ATP in muscle fibers. Creatine phosphate, however, cannot directly supply energy for a cell's energy-utilizing reactions. Instead, it acts to store energy released from mitochondria whenever sufficient amounts of ATP are already present. Then, at times when ATP is being decomposed, energy from creatine phosphate can be transferred to ADP molecules, which are quickly converted back into ATP.

In active muscle, the supply of creatine phosphate is quickly exhausted. When this happens, the muscle fibers become dependent upon cellular respiration of glucose as a source of energy for synthesizing ATP.

Oxygen Supply and Cellular Respiration

As is described in chapter 4, the early phase of cellular respiration takes place in the absence of oxygen. On the other hand, the more complete breakdown of glucose occurs in the mitochondria and requires oxygen.

The oxygen needed to support this aerobic respiration is carried from the lungs to body cells by the blood. It is transported within the red blood cells loosely bonded to molecules of **hemoglobin,** the pigment responsible for the red color of blood.

Another pigment, **myoglobin,** is synthesized in the muscle cells and is responsible for the reddish brown color of skeletal muscle tissue. It has properties similar to hemoglobin in that it can combine loosely with oxygen. This ability to store oxygen temporarily reduces a muscle's need for a continuous blood supply during muscular contraction. (See fig. 9.7.)

Oxygen Debt

When a person is resting or is moderately active, the ability of the respiratory and circulatory systems to supply oxygen to the skeletal muscles is usually adequate to support aerobic respiration. When skeletal muscles are used strenuously for even a minute or two, however, these systems usually cannot supply oxygen efficiently enough to meet the needs of aerobic respiration. Consequently, the muscle fibers must depend more and more on the anaerobic phase of respiration to obtain energy.

In anaerobic respiration, glucose molecules are changed to *pyruvic acid.* If the oxygen supply is low, the pyruvic acid is then converted to *lactic acid,* which diffuses out of the muscle fibers and is carried to the liver by the blood. (See fig. 9.7.)

Liver cells change lactic acid into *glucose;* however this conversion also requires energy from ATP. During strenuous exercise, the available oxygen is used primarily to synthesize the ATP needed for the muscle fiber contraction rather than to make ATP for changing lactic acid into glucose. Consequently, as lactic acid accumulates, a person develops an **oxygen debt** that must be paid back at a later time. The

Fig. 9.7 The oxygen needed to support aerobic respiration is carried in the blood and stored in myoglobin.

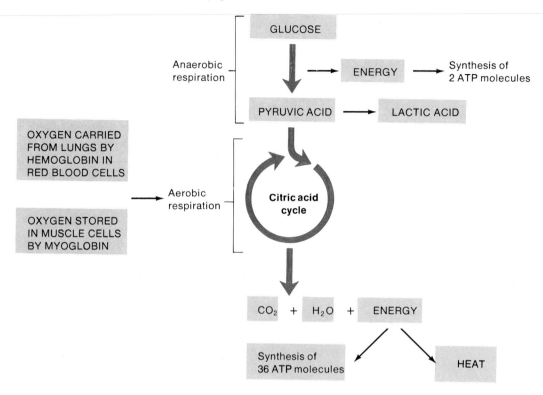

amount of this oxygen debt is equal to the amount of oxygen needed by the liver cells to convert the accumulated lactic acid into glucose plus the amount needed by muscle cells to resynthesize ATP and creatine phosphate, and return them to their original concentrations.

Since the conversion of lactic acid back into glucose is a relatively slow process, it may require several hours to pay back an oxygen debt following heavy exercise.

Muscle Fatigue

If a muscle is exercised strenuously for a prolonged period, it may lose its ability to contract and is said to be *fatigued*. This condition may result from an interruption in the muscle's blood supply or, rarely, from an exhaustion of the supply of acetylcholine in its motor nerve fibers. Muscle fatigue, however, is more likely to arise from an accumulation of lactic acid in the muscle as a result of anaerobic respiration. The lactic acid causes factors, such as pH, to change so that muscle fibers are no longer responsive to stimulation.

Occasionally a muscle becomes fatigued and cramps at the same time. A cramp is a painful condition in which the muscle contracts spasmodically, but does not relax completely. This condition seems

to be due to a lack of ATP, which is needed to move calcium ions back into the sarcoplasmic reticulum and to break the connections between actin and myosin filaments before relaxation in muscle fibers can occur.

A few hours after death, the skeletal muscles undergo a partial contraction that causes the joints to become fixed. This condition, *rigor mortis*, may continue for 72 hours or more. It seems to result from a lack of ATP in the muscle fibers, which prevents relaxation. Thus, the actin and myosin filaments of the muscle fibers remain bound together until the muscles begin to decompose.

Heat Production

Only about 25% of the energy released by cellular respiration is available for use in metabolic processes; the rest is lost as heat. Although heat is generated by all active cells, muscle tissue is a particularly important source because it represents such a large proportion of the total body mass. Thus, whenever muscles are active, large amounts of heat are released. This heat is transported to other tissues by the blood and helps to maintain body temperature.

1. What substances provide energy used to regenerate ATP?
2. What are the sources of oxygen needed for aerobic respiration?
3. What is the relationship between lactic acid and oxygen debt?
4. What is the relationship between cellular respiration and heat production?

Muscular Responses

One way to observe muscle contraction is to remove a single muscle fiber from a skeletal muscle and connect it to a device that records changes in the fiber's length. In such experiments, the muscle fiber is usually stimulated with an electrical stimulator that is capable of producing stimuli of varying strength and frequency.

Threshold Stimulus

By exposing an isolated muscle fiber to a series of stimuli of increasing strength, it can be shown that the fiber remains unresponsive until a certain strength of stimulation is applied. This minimal strength needed to elicit a contraction is called the **threshold stimulus.**

All-or-None Response

When a muscle fiber is exposed to a stimulus of threshold strength (or above), it responds to its fullest extent. Increasing the strength of the stimulus does not affect the degree to which the fiber contracts. In other words, there are no partial contractions of a muscle fiber—if it contracts at all, it will contract completely, even though in some instances it may not shorten completely. This phenomenon is called the **all-or-none response.**

Recruitment of Motor Units

Since the muscle fibers within a muscle are organized into motor units, and each motor unit is controlled by a single motor neuron, all the muscle fibers in a motor unit are stimulated at the same time. Consequently, a motor unit also responds in an all-or-none manner. A whole muscle, however, does not behave like this, because it is composed of many motor units controlled by different motor neurons responding to different thresholds of stimulation. Thus, if only the

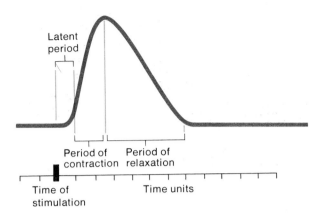

Fig. 9.8 A myogram of a single muscle twitch.

motor neurons with low thresholds of stimulation are stimulated, a relatively small number of motor units contract. At higher intensities of stimulation, other motor neurons respond, and more motor units are activated. Such an increase in the number of motor units being activated is called **recruitment.** As the intensity of stimulation increases, recruitment of motor units continues, until finally all possible motor units are activated, and the muscle is contracting with maximal tension.

Twitch Contraction

To demonstrate how a whole muscle responds to stimulation, a skeletal muscle can be removed from a frog or other laboratory animal. The muscle is stimulated electrically, and when it contracts, its movement is recorded. The resulting pattern is called a **myogram.**

If a muscle is exposed to a single stimulus of sufficient strength to activate some of its motor units, the muscle will contract and then relax. This action— a single contraction that lasts only a fraction of a second—is called a **twitch.** A twitch produces a myogram like that in figure 9.8. It is apparent from this record that the muscle response did not begin immediately following the stimulation. There is a lag between the time that the stimulus was applied and the time that the muscle responded. This time lag is called the **latent period.** In a frog muscle, the latent period lasts for about 0.01 second, and it is even shorter in a human muscle.

The latent period is followed by a **period of contraction** during which the muscle pulls at its attachments, and a **period of relaxation** during which it returns to its former length. (See fig. 9.8.)

If a muscle is exposed to a series of stimuli of increasing frequency, a point is reached when the muscle is unable to complete its relaxation period before the next stimulus in the series arrives. When this happens, the contractions begin to combine and the muscle contraction is sustained.

At the same time that twitches are combining, the strength of the contractions may be increasing. This is due to the recruitment of motor units. The smaller motor units, which have finer fibers, are most easily stimulated and tend to respond earlier in the series of stimuli. The larger motor units, which contain thicker fibers, respond later and produce more forceful contractions. When a sustained, forceful contraction lacks even partial relaxation, it is termed a **tetanic contraction** (tetany). (See fig. 9.9.)

Although twitches may occur occasionally in human skeletal muscles, as when an eyelid twitches, such contractions are of limited use. More commonly, muscular contractions are sustained. When a person lifts a weight or walks, for example, sustained contractions are maintained in arm or leg muscles for varying lengths of time. These contractions are responses to a rapid series of stimuli transmitted from the brain and spinal cord on motor neuron fibers.

Actually, even when a muscle appears to be at rest, a certain amount of sustained contraction is occurring in its fibers. This is called **muscle tone** (tonus). It is a response to nerve impulses originating in the spinal cord from moment to moment that travel to small numbers of muscle fibers.

Muscle tone is particularly important in maintaining posture. If tone is suddenly lost, as happens when a person loses consciousness, the body will collapse.

Use and Disuse of Skeletal Muscles

Skeletal muscles are very responsive to use and disuse. Muscles that are forcefully exercised tend to enlarge. This phenomenon, which is termed **muscular hypertrophy,** involves an increase in size of individual muscle fibers rather than an increase in the number of fibers. The mitochondria within these fibers are stimulated to reproduce, the sarcoplasmic reticula become more extensive, and new filaments of actin and myosin are produced.

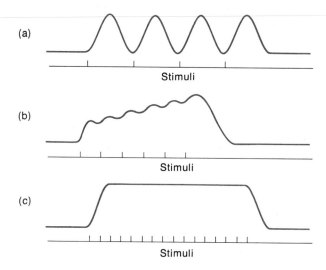

Fig. 9.9 Myograms of (a) a series of twitches; (b) a combining of twitches; and (c) a tetanic contraction.

As a result of hypertrophy a muscle becomes capable of more forceful contractions because the strength of a contraction is directly related to the diameter of the muscle fibers.

When skeletal muscles are contracted very forcefully they may generate up to 50 pounds of pull for each square inch of muscle cross section. Consequently, large muscles, such as those in the thigh, can pull with several hundred pounds of force. Occasionally this force is so great that the tendons of muscles are torn away from their attachments on bones.

A muscle that is not used or is only used to produce weak contractions tends to decrease in size, strength, and number of cellular organelles in its fibers. This condition is called **atrophy,** and it commonly occurs when limbs are immobilized by casts or when accidents or diseases interfere with motor nerve impulses. A muscle that cannot be exercised may decrease to less than one-half its usual size within a few months.

1. *Define* threshold stimulus.
2. *What is meant by an all-or-none response?*
3. *Distinguish between a twitch and a sustained contraction.*
4. *Distinguish between hypertrophy and atrophy.*

Smooth Muscles

Although the contractile mechanisms of smooth and cardiac muscles are essentially the same as that of skeletal muscles, the cells of these tissues have important structural and functional differences.

Smooth Muscle Fibers

As is discussed in chapter 5, smooth muscle cells (fig. 5.16) are elongated with tapering ends. They contain filaments of *actin* and *myosin* in myofibrils that extend the lengths of the cells. However, these filaments are very thin and more randomly arranged than in skeletal muscle. Consequently, smooth muscle cells lack striations.

There are two major types of smooth muscles. In one type, called **multiunit smooth muscle,** the muscle fibers are less well organized and occur as separate fibers rather than in sheets. Smooth muscle of this type is found in the irises of the eyes and in the walls of blood vessels. Typically it contracts only after stimulation by motor nerve impulses.

The second type of smooth muscle is called **visceral.** It is composed of sheets of spindle-shaped cells that are in close contact with one another. Visceral smooth muscle is the more common type and is found in the walls of hollow visceral organs, such as the stomach, intestines, urinary bladder, and uterus.

The fibers of visceral smooth muscles are capable of stimulating each other. Consequently, when one fiber is stimulated, the impulse moving over its surface may excite adjacent fibers, which in turn stimulate still others. Visceral smooth muscles also display *rhythmicity*—a pattern of repeated contractions. This phenomenon is due to the presence of self-exciting fibers from which spontaneous impulses travel periodically into the surrounding muscle tissue.

These two features—transmission of impulses from cell to cell and rhythmicity—are largely responsible for the wavelike motion called **peristalsis** that occurs in various tubular organs, such as the intestines, and helps force the contents along their lengths.

Smooth Muscle Contraction

Smooth muscle contraction resembles skeletal muscle contraction in a number of ways. Both mechanisms involve reactions of actin and myosin, both are triggered by membrane impulses and the release of calcium ions, and both use energy from ATP molecules. There are, however, significant differences between these two types of muscle tissue.

For example, although acetylcholine is the neurotransmitter substance in skeletal muscle, two neurotransmitters affect smooth muscle—*acetylcholine* and *norepinephrine*. Each of these substances stimulates contractions in some smooth muscles and inhibits it in others. (See chapter 10.) Also, smooth muscles are affected by a number of hormones, which stimulate contractions in some cases and alter the degree of response to neurotransmitters in others.

Smooth muscle is slower to contract and slower to relax than skeletal muscle. On the other hand, it can maintain a forceful contraction for a longer time with a given amount of ATP. Also, unlike skeletal muscle, smooth muscle fibers can change length without changing tautness; therefore, smooth muscles in the stomach and intestinal walls can stretch as these organs become filled, while the pressure inside remains unchanged.

---

1. Describe the two major types of smooth muscle.
2. What special characteristics of visceral smooth muscle make peristalsis possible?
3. How does smooth muscle contraction differ from that of skeletal muscle?

Cardiac Muscle

Cardiac muscle (fig. 5.17) occurs only in the heart. It is composed of striated cells that are joined end to end, forming fibers. These fibers are interconnected in branching, three-dimensional networks. Each cell contains numerous filaments of actin and myosin, similar to those in skeletal muscle. It also has a well-developed sarcoplasmic reticulum, many mitochondria, and a system of transverse tubules. The cisternae of cardiac muscle fibers, however, are less well developed and store less calcium than those of skeletal muscle. On the other hand, the transverse tubules of cardiac muscle are larger and they release large quantities of calcium ions into the sarcoplasm in response to muscle impulses. This extra calcium from the transverse tubules is obtained from the fluid outside the muscle fibers, and it makes it possible for cardiac muscle fibers to maintain a contraction for a longer time than skeletal muscle.

Chart 9.1 Types of muscle tissue

	Skeletal	Smooth	Cardiac
Major location	Skeletal muscles	Walls of hollow visceral organs	Wall of heart
Major function	Movement of bones at joints, maintenance of posture	Movement of visceral organs, peristalsis	Pumping action of heart
Cellular characteristics			
Striations	Present	Absent	Present
Nucleus	Many nuclei	Single nucleus	Single nucleus
Special features	Transverse tubule system well developed	Lacks transverse tubules	Transverse tubule system well developed, adjacent cells separated by intercalated disks
Mode of control	Voluntary	Involuntary	Involuntary
Contraction characteristics	Contracts and relaxes relatively rapidly	Contracts and relaxes relatively slowly, self-exciting, rhythmic	Network of fibers contracts as a unit, self-exciting, rhythmic

The opposing ends of cardiac muscle cells are separated by cross-bands called *intercalated disks.* These bands are the result of elaborate junctions of the membranes at the cell's boundary. They help to hold adjacent cells together and to transmit the force of contraction from cell to cell. The intercalated disks also have very low resistance to the passage of impulses, so that cardiac muscle fibers transmit impulses relatively rapidly.

When one portion of the cardiac muscle network is stimulated, the impulse passes to the other fibers of the net, and the whole structure contracts as a unit; that is, the network responds to stimulation in an all-or-none manner. Cardiac muscle is also self-exciting and rhythmic. Consequently, a pattern of contraction and relaxation is repeated again and again and is responsible for the rhythmic contractions of the heart.

Chart 9.1 summarizes the characteristics of the three types of muscles.

---

1. *How is cardiac muscle similar to smooth muscle?*
2. *What is the function of intercalated disks?*

Skeletal Muscle Actions

Skeletal muscles are responsible for a variety of body movements. The action of each muscle depends largely upon the kind of joint it is associated with and the way it is attached on either side of the joint.

Origin and Insertion

As is mentioned in chapter 8, one end of a skeletal muscle usually is fastened to a relatively immovable or fixed part, and the other end is connected to a movable part on the other side of a joint. The immovable end of the muscle is called its **origin,** and the movable one is its **insertion.** When a muscle contracts, its insertion is pulled toward its origin. (See fig. 9.10.)

Some muscles have more than one origin or insertion. The *biceps brachii* in the upper arm, for example, has two origins. This is reflected in its name, *biceps,* which means *two heads.* One head is attached to the coracoid process of the scapula, while the other one arises from a tubercle above the glenoid cavity of the scapula. The muscle extends along the front surface of the humerus and is inserted by means of a tendon on the radial tuberosity of the radius. When the biceps brachii contracts, its insertion is pulled toward its origin, and the arm bends at the elbow. (See fig. 8.5.)

Sometimes the end of a muscle changes its function in different body movements; that is, the origin may become the insertion or vice versa. For instance, the *latissimus dorsi* muscle in the back, shown in figure 9.10, has its origin on the pelvic girdle and vertebral column and its insertion on the proximal end of the humerus. Usually it functions to pull the arm toward the back as in rowing a boat or swimming. During chinning exercises, however, the humerus becomes the fixed origin of the muscle and the end attached to the pelvis and vertebral column moves.

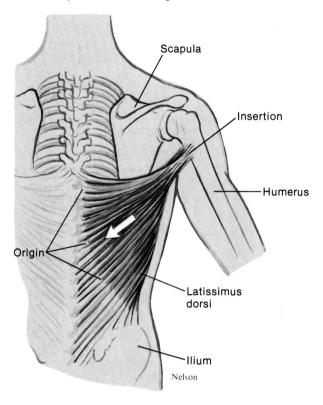

Fig. 9.10 When the fibers of a muscle contract, its insertion is pulled toward its origin.

Scapula

Insertion

Humerus

Origin

Latissimus dorsi

Ilium

Nelson

Interaction of Skeletal Muscles

Skeletal muscles almost always function in groups rather than singly. Consequently, when a particular body movement occurs, a person must do more than command a single muscle to contract; instead the person wills the movement to occur, and the appropriate group of muscles responds to the decision.

By carefully observing body movements, it is possible to determine the particular roles of various muscles. For instance, when the arm is lifted horizontally away from the side, a *deltoid* muscle (fig. 9.11) is responsible for most of the movement, and so it is said to be the **prime mover.** However, while the prime mover is acting, certain nearby muscles also are contracting. They help to hold the shoulder steady and in this way make the action of the prime mover more effective. Muscles that assist the prime mover are called **synergists.**

Still other muscles act as **antagonists** to prime movers. These muscles are capable of resisting a prime mover's action and are responsible for movement in the opposite direction—the antagonist of the prime mover that raises the arm can lower the arm, or the antagonist of the prime mover that bends the arm can straighten it. If both a prime mover and its antagonist contract simultaneously, the part they act upon remains rigid. Consequently, smooth body movements depend upon the antagonists' giving way to the actions of the prime movers whenever the prime movers are contracting. These complex actions are controlled by the nervous system, as is described in chapter 10.

1. *Distinguish between the origin and insertion of a muscle.*
2. *Define prime mover.*
3. *What is the function of a synergist? An antagonist?*

Major Skeletal Muscles

The following section concerns the locations, actions, and attachments of some of the major skeletal muscles of the body. (Figures 9.11 and 9.12 show the locations of the superficial skeletal muscles—those near the surface.)

Muscles of Facial Expression

A number of small muscles that lie beneath the skin of the face and scalp enable us to communicate feelings through facial expression. Many of these muscles are located around the eyes and mouth, and they are responsible for such expressions as surprise, sadness, anger, fear, disgust, and pain. As a group, the muscles of facial expression fasten the bones of the skull to connective tissue in various regions of the overlying skin. They include the following. (See fig. 9.13.)

Epicranius
Orbicularis oculi
Orbicularis oris
Buccinator
Zygomaticus
Platysma

Fig. 9.13 Muscles of expression and mastication.

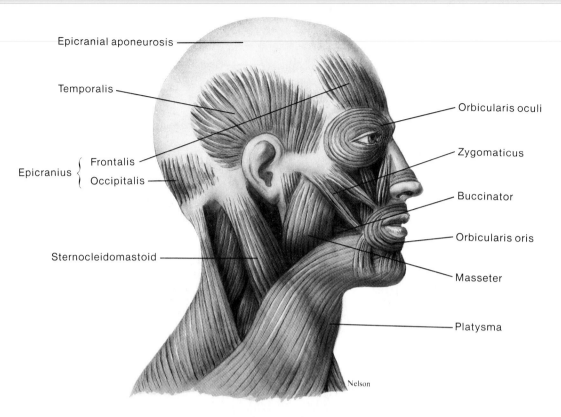

Epicranial aponeurosis

Temporalis

Orbicularis oculi

Zygomaticus

Epicranius { Frontalis
             Occipitalis

Buccinator

Orbicularis oris

Sternocleidomastoid

Masseter

Platysma

Nelson

Chart 9.2 Muscles of facial expression

Muscle	Origin	Insertion	Action
Epicranius	Occipital bone	Skin and muscles around eye	Raises eyebrow
Orbicularis oculi	Maxillary and frontal bones	Skin around eye	Closes eye
Orbicularis oris	Muscles near the mouth	Skin of central lip	Closes lips, protrudes lips
Buccinator	Outer surfaces of maxilla and mandible	Orbicularis oris	Compresses cheeks inward
Zygomaticus	Zygomatic bone	Orbicularis oris	Raises corner of mouth
Platysma	Fascia in upper chest	Lower border of mandible	Draws angle of mouth downward

From Hole, John W., Jr., *Human Anatomy and Physiology 3d ed.* © 1978, 1981, 1984 Wm. C. Brown Publishers, Dubuque, Iowa. All Rights Reserved. Reprinted by permission.

Chart 9.3 Muscles of mastication

Muscle	Origin	Insertion	Action
Masseter	Lower border of zygomatic arch	Lateral surface of mandible	Closes jaw
Temporalis	Temporal bone	Coronoid process and lateral surface of mandible	Closes jaw

From Hole, John W., Jr., *Human Anatomy and Physiology 3d ed.* © 1978, 1981, 1984 Wm. C. Brown Publishers, Dubuque, Iowa. All Rights Reserved. Reprinted by permission.

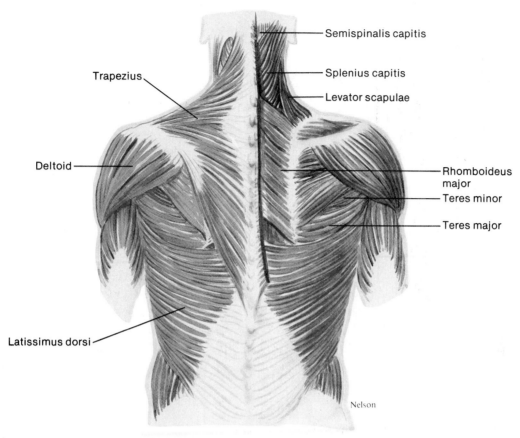

Trapezius

Deltoid

Latissimus dorsi

Semispinalis capitis

Splenius capitis

Levator scapulae

Rhomboideus major

Teres minor

Teres major

Nelson

Chart 9.2 lists the origins, insertions, and actions of the muscles of facial expression. (The muscles that move the eyes are listed in chapter 11.)

Muscles of Mastication

Chewing movements are produced by muscles that are attached to the mandible. Two pairs of these muscles act to close the lower jaw, as in biting. They include the following. (See fig. 9.13.)

> *Masseter*
>
> *Temporalis*

Chart 9.3 lists the origins, insertions, and actions of the muscles of mastication.

Muscles That Move the Head

Head movements result from the actions of paired muscles in the neck and upper back. These muscles are responsible for flexing, extending, and rotating the head. They include the following. (See figs. 9.14 and 9.15.)

> *Sternocleidomastoid*
>
> *Splenius capitis*
>
> *Semispinalis capitis*

Chart 9.4 lists the origins, insertions, and actions of the muscles that move the head.

Fig. 9.15 Muscles of the anterior chest and the abdominal wall.

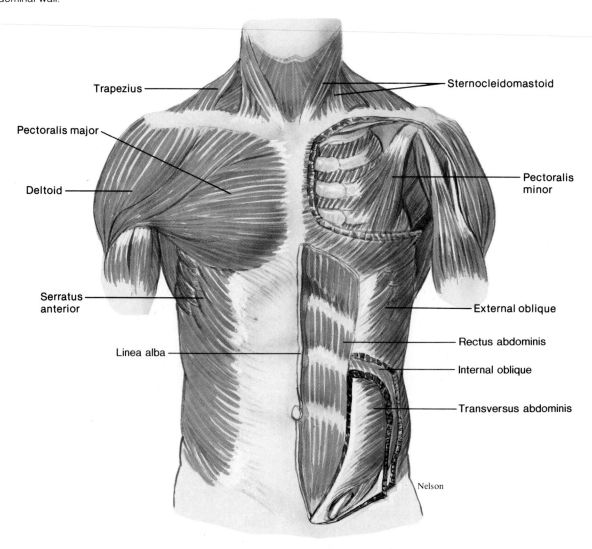

Trapezius

Sternocleidomastoid

Pectoralis major

Pectoralis minor

Deltoid

Serratus anterior

External oblique

Rectus abdominis

Linea alba

Internal oblique

Transversus abdominis

Nelson

Chart 9.4   Muscles that move the head

Muscle	Origin	Insertion	Action
Sternocleidomastoid	Anterior surface of sternum and upper surface of clavicle	Mastoid process of temporal bone	Pulls head to one side, pulls head toward chest, or raises sternum
Splenius capitis	Spinous processes of lower cervical and upper thoracic vertebrae	Mastoid process of temporal bone	Rotates head, bends head to one side, or brings head into upright position
Semispinalis capitis	Processes of lower cervical and upper thoracic vertebrae	Occipital bone	Extends head, bends head to one side, or rotates head

From Hole, John W., Jr., *Human Anatomy and Physiology 3d ed.* © 1978, 1981, 1984 Wm. C. Brown Publishers, Dubuque, Iowa. All Rights Reserved. Reprinted by permission.

Chart 9.5 Muscles that move the pectoral girdle

Muscle	Origin	Insertion	Action
Trapezius	Occipital bone and spines of cervical and thoracic vertebrae	Clavicle, spine, and acromion process of scapula	Rotates scapula and flexes arm; raises scapula; pulls scapula medially; or pulls scapula and shoulder downward
Rhomboideus major	Spines of upper thoracic vertebrae	Medial border of scapula	Raises and adducts scapula
Levator scapulae	Transverse processes of cervical vertebrae	Medial margin of scapula	Elevates scapula
Serratus anterior	Outer surfaces of upper ribs	Ventral surface of scapula	Pulls scapula anteriorly and downward
Pectoralis minor	Sternal ends of upper ribs	Coracoid process of scapula	Pulls scapula forward and downward or raises ribs

From Hole, John W., Jr., *Human Anatomy and Physiology 3d ed.* © 1978, 1981, 1984 Wm. C. Brown Publishers, Dubuque, Iowa. All Rights Reserved. Reprinted by permission.

Muscles That Move the Pectoral Girdle

The muscles that move the pectoral girdle are closely associated with those that move the upper arm. A number of these chest and shoulder muscles connect the scapula to nearby bones and act to move the scapula upward, downward, forward, and backward. They include the following. (See figs. 9.14 and 9.15.)

> *Trapezius*
> *Rhomboideus major*
> *Levator scapulae*
> *Serratus anterior*
> *Pectoralis minor*

Chart 9.5 lists the origins, insertions, and actions of the muscles that move the pectoral girdle.

Muscles That Move the Upper Arm

The upper arm is one of the more freely movable parts of the body. Its many movements are made possible by muscles that connect the humerus to various regions of the pectoral girdle, ribs, and vertebral column. (See figs. 9.14, 9.15, 9.16, and 9.17.) These muscles can be grouped according to their primary actions—flexion, extension, abduction, and rotation—as follows:

Flexors	Abductors
Coracobrachialis	*Supraspinatus*
Pectoralis major	*Deltoid*
Extensors	Rotators
Teres major	*Subscapularis*
Latissimus dorsi	*Infraspinatus*
	Teres minor

Chart 9.6 lists the origins, insertions, and actions of the muscles that move the upper arm.

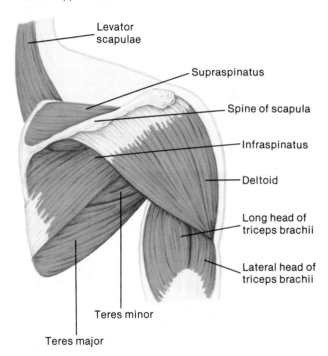

Fig. 9.16 Muscles of the posterior surface of the scapula and the upper arm.

Levator scapulae

Supraspinatus

Spine of scapula

Infraspinatus

Deltoid

Long head of triceps brachii

Lateral head of triceps brachii

Teres minor

Teres major

Fig. 9.17 Muscles of the anterior shoulder and the upper arm.

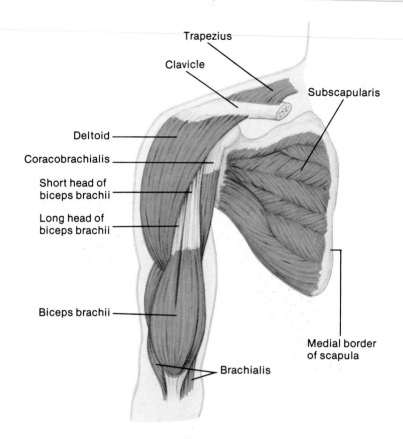

Chart 9.6 Muscles that move the upper arm

Muscle	Origin	Insertion	Action
Coracobrachialis	Coracoid process of scapula	Shaft of humerus	Flexes and adducts the upper arm
Pectoralis major	Clavicle, sternum, and costal cartilages of upper ribs	Intertubercular groove of humerus	Pulls arm forward and across chest, rotates humerus, or adducts arm
Teres major	Lateral border of scapula	Intertubercular groove of humerus	Extends humerus, or adducts and rotates arm medially
Latissimus dorsi	Spines of sacral, lumbar, and lower thoracic vertebrae, iliac crest, and lower ribs	Intertubercular groove of humerus	Extends and adducts the arm and rotates humerus inwardly, or pulls the shoulder downward and back
Supraspinatus	Posterior surface of scapula	Greater tubercle of humerus	Abducts the upper arm
Deltoid	Acromion process, spine of the scapula, and the clavicle	Deltoid tuberosity of humerus	Abducts upper arm, extends humerus, or flexes humerus
Subscapularis	Anterior surface of scapula	Lesser tubercle of humerus	Rotates arm medially
Infraspinatus	Posterior surface of scapula	Greater tubercle of humerus	Rotates arm laterally
Teres minor	Lateral border of scapula	Greater tubercle of humerus	Rotates arm laterally

Fig. 9.18 Muscles of the anterior forearm.

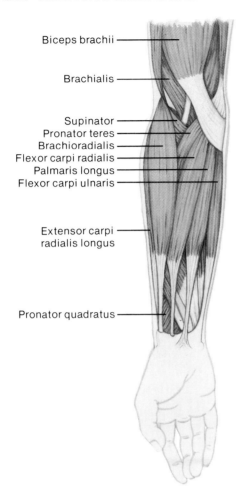

Biceps brachii

Brachialis

Supinator
Pronator teres
Brachioradialis
Flexor carpi radialis
Palmaris longus
Flexor carpi ulnaris

Extensor carpi
radialis longus

Pronator quadratus

Muscles That Move the Forearm

Most forearm movements are produced by muscles that connect the radius or ulna to the humerus or pectoral girdle. A group of muscles located along the anterior surface of the humerus act to flex the elbow, while a single posterior muscle serves to extend this joint. Other muscles cause movements at the radioulnar joint and are responsible for rotating the forearm.

The muscles that move the forearm include the following. (See figs. 9.16, 9.17, and 9.18.)

Flexors

Biceps brachii

Brachialis

Brachioradialis

Extensor

Triceps brachii

Rotators

Supinator

Pronator teres

Pronator quadratus

Chart 9.7 lists the origins, insertions, and actions of the muscles that move the forearm.

Chart 9.7   Muscles that move the forearm

Muscle	Origin	Insertion	Action
Biceps brachii	Coracoid process and tubercle above glenoid cavity of scapula	Radial tuberosity of radius	Flexes arm at elbow and rotates the hand laterally
Brachialis	Anterior shaft of humerus	Coronoid process of ulna	Flexes arm at elbow
Brachioradialis	Distal lateral end of humerus	Lateral surface of radius above styloid process	Flexes arm at elbow
Triceps brachii	Tubercle below glenoid cavity and lateral and medial surfaces of humerus	Olecranon process of ulna	Extends arm at elbow
Supinator	Lateral epicondyle of humerus and crest of ulna	Lateral surface of radius	Rotates forearm laterally
Pronator teres	Medial epicondyle of humerus and coronoid process of ulna	Lateral surface of radius	Rotates arm medially
Pronator quadratus	Anterior distal end of ulna	Anterior distal end of radius	Rotates arm medially

Fig. 9.19 Muscles of the posterior forearm.

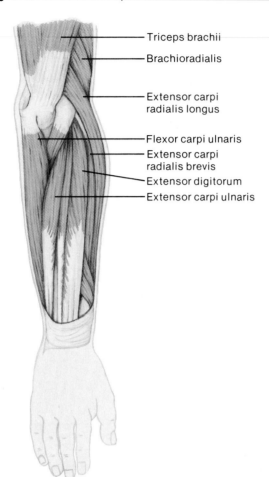

- Triceps brachii
- Brachioradialis
- Extensor carpi radialis longus
- Flexor carpi ulnaris
- Extensor carpi radialis brevis
- Extensor digitorum
- Extensor carpi ulnaris

Muscles That Move the Wrist, Hand, and Fingers

Many muscles are responsible for wrist, hand, and finger movements. They originate from the distal end of the humerus and from the radius and ulna. The two major groups of these muscles are flexors on the anterior side of the forearm and extensors on the posterior side. These muscles include the following. (See figs. 9.18 and 9.19.)

Flexors

Flexor carpi radialis

Flexor carpi ulnaris

Palmaris longus

Flexor digitorum profundus

Extensors

Extensor carpi radialis longus

Extensor carpi radialis brevis

Extensor carpi ulnaris

Extensor digitorum

Chart 9.8 lists the origins, insertions, and actions of the muscles that move the wrist, hand, and fingers.

Muscles of the Abdominal Wall

Although the walls of the chest and pelvic regions are supported directly by bone, those of the abdomen are not. Instead, the anterior and lateral walls of the abdomen are composed of broad, flattened muscles arranged in layers. These muscles connect the rib cage and vertebral column to the pelvic girdle. A band of

Chart 9.8 Muscles that move the wrist, hand, and fingers

Muscle	Origin	Insertion	Action
Flexor carpi radialis	Medial epicondyle of humerus	Base of second and third metacarpals	Flexes and abducts wrist
Flexor carpi ulnaris	Medial epicondyle of humerus and olecranon process	Carpal and metacarpal bones	Flexes and adducts wrist
Palmaris longus	Medial epicondyle of humerus	Fascia of palm	Flexes wrist
Flexor digitorum profundus	Anterior surface of ulna	Bases of distal phalanges in fingers two through five	Flexes distal joints of fingers
Extensor carpi radialis longus	Distal end of humerus	Base of second metacarpal	Extends wrist and abducts hand
Extensor carpi radialis brevis	Lateral epicondyle of humerus	Base of second and third metacarpals	Extends wrist and abducts hand
Extensor carpi ulnaris	Lateral epicondyle of humerus	Base of fifth metacarpal	Extends and adducts wrist
Extensor digitorum	Lateral epicondyle of humerus	Posterior surface of phalanges in fingers two through five	Extends fingers

From Hole, John W., Jr., *Human Anatomy and Physiology 3d ed.* © 1978, 1981, 1984 Wm. C. Brown Publishers, Dubuque, Iowa. All Rights Reserved. Reprinted by permission.

Chart 9.9 Muscles of the abdominal wall

Muscle	Origin	Insertion	Action
External oblique	Outer surfaces of lower ribs	Outer lip of iliac crest and linea alba	Tenses abdominal wall and compresses abdominal contents
Internal oblique	Crest of ilium and inguinal ligament	Cartilages of lower ribs, linea alba, and crest of pubis	Same as above
Transversus abdominis	Costal cartilages of lower ribs, processes of lumbar vertebrae, lip of iliac crest, and inguinal ligament	Linea alba and crest of pubis	Same as above
Rectus abdominis	Crest of pubis and symphysis pubis	Xiphoid process of sternum and costal cartilages	Same as above, also flexes vertebral column

From Hole, John W., Jr., *Human Anatomy and Physiology 3d ed.* © 1978, 1981, 1984 Wm. C. Brown Publishers, Dubuque, Iowa. All Rights Reserved. Reprinted by permission.

Fig. 9.20 Muscles of the male pelvic outlet.

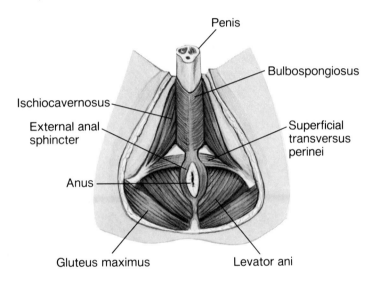

Penis

Bulbospongiosus

Ischiocavernosus

External anal sphincter

Superficial transversus perinei

Anus

Gluteus maximus

Levator ani

tough connective tissue, called the **linea alba,** extends from the xiphoid process of the sternum to the symphysis pubis. It serves as an attachment for some of the abdominal wall muscles.

Contraction of these muscles decreases the size of the abdominal cavity and increases the pressure inside. This action helps to press air out of the lungs during forceful exhalation and also aids in defecation, urination, vomiting, and childbirth.

The abdominal wall muscles (fig. 9.15) include the following:

> *External oblique*
> *Internal oblique*
> *Transversus abdominis*
> *Rectus abdominis*

Chart 9.9 lists the origins, insertions, and actions of the muscles of the abdominal wall.

Muscles of the Pelvic Outlet

The outlet of the pelvis is spanned by two muscular sheets—a deeper **pelvic diaphragm** and a more superficial **urogenital diaphragm.** The pelvic diaphragm forms the floor of the pelvic cavity, and the urogenital diaphragm fills the space within the pubic arch. The muscles of the male and female pelvic outlets include the following (figs. 9.20 and 9.21).

> Pelvic diaphragm
> > *Levator ani*
> > *Coccygeus*
> Urogenital diaphragm
> > *Superficial transversus perinei*
> > *Bulbospongiosus*
> > *Ischiocavernosus*
> > *Sphincter urethrae*

Chart 9.10 lists the origins, insertions, and actions of the pelvic outlet muscles.

The Muscular System 191

Fig. 9.21 Muscles of the female pelvic outlet.

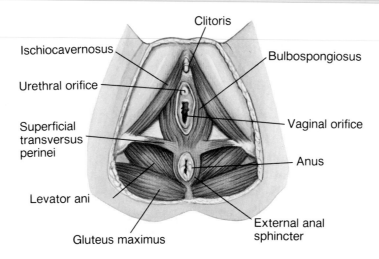

From Hole, John W., Jr., *Human Anatomy and Physiology 3d ed.* © 1978, 1981, 1984 Wm. C. Brown Publishers, Dubuque, Iowa. All Rights Reserved. Reprinted by permission.

Chart 9.10 Muscles of the pelvic outlet

Muscle	Origin	Insertion	Action
Levator ani	Pubic bone and ischial spine	Coccyx	Supports pelvic viscera, and provides sphincterlike action in anal canal and vagina
Coccygeus	Ischial spine	Sacrum and coccyx	Same as above
Superficial transversus perinei	Ischial tuberosity	Central tendon	Supports pelvic viscera
Bulbospongiosus	Central tendon	Males: urogenital diaphragm and fascia of penis	Males: assists emptying urethra
		Females: pubic arch and root of clitoris	Females: constricts vagina
Ischiocavernosus	Ischial tuberosity	Pubic arch	Assists the function of the bulbospongiosus
Sphincter urethrae	Margins of pubis and ischium	Fibers of each unite with those from other side	Opens and closes urethra

Muscles That Move the Thigh

The muscles that move the thigh are attached to the femur and to some part of the pelvic girdle. They can be separated into anterior and posterior groups. The muscles of the anterior group act primarily to flex the thigh; those of the posterior group extend, abduct, or rotate it. The muscles in these groups (figs. 9.22, 9.23, and 9.24) include the following:

> Anterior group
> *Psoas major*
> *Iliacus*

Posterior group
> *Gluteus maximus*
> *Gluteus medius*
> *Gluteus minimus*
> *Tensor fasciae latae*

Still another group of muscles attached to the femur and pelvic girdle function to adduct the thigh. They include the following:

> *Adductor longus*
> *Adductor magnus*
> *Gracilis*

Chart 9.11 lists the origins, insertions, and actions of the muscles that move the thigh.

Fig. 9.22 Muscles of the anterior right thigh.

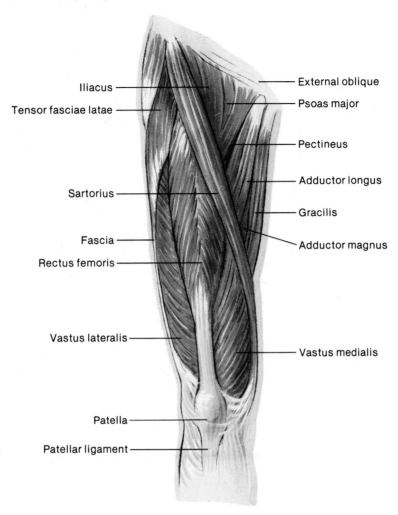

Iliacus

Tensor fasciae latae

Sartorius

Fascia

Rectus femoris

Vastus lateralis

Patella

Patellar ligament

External oblique

Psoas major

Pectineus

Adductor longus

Gracilis

Adductor magnus

Vastus medialis

Muscles That Move the Lower Leg

The muscles that move the lower leg connect the tibia or fibula to the femur or the pelvic girdle. (See figs. 9.22, 9.23, and 9.24.) They can be separated into two major groups—those that cause flexion at the knee and those that cause extension at the knee. The muscles of these groups include the following:

Flexors

Biceps femoris

Semitendinosus

Semimembranosus

Sartorius

Extensor

Quadriceps femoris

Chart 9.12 lists the origins, insertions, and actions of the muscles that move the lower leg.

Fig. 9.23 Muscles of the lateral right thigh.

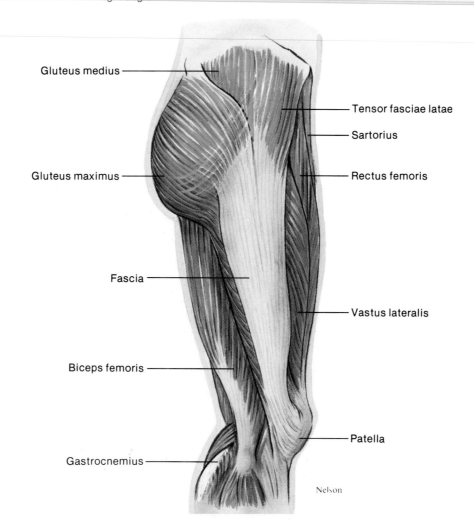

Gluteus medius

Tensor fasciae latae

Sartorius

Gluteus maximus

Rectus femoris

Fascia

Vastus lateralis

Biceps femoris

Patella

Gastrocnemius

Nelson

Chart 9.11 Muscles that move the thigh

Muscle	Origin	Insertion	Action
Psoas major	Lumbar intervertebral disks, bodies and transverse processes of lumbar vertebrae	Lesser trochanter of femur	Flexes thigh
Iliacus	Iliac fossa of ilium	Lesser trochanter of femur	Flexes thigh
Gluteus maximus	Sacrum, coccyx, and posterior surface of ilium	Posterior surface of femur and fascia of thigh	Extends leg at hip
Gluteus medius	Lateral surface of ilium	Greater trochanter of femur	Abducts and rotates thigh medially

Fig. 9.24 Muscles of the posterior right thigh.

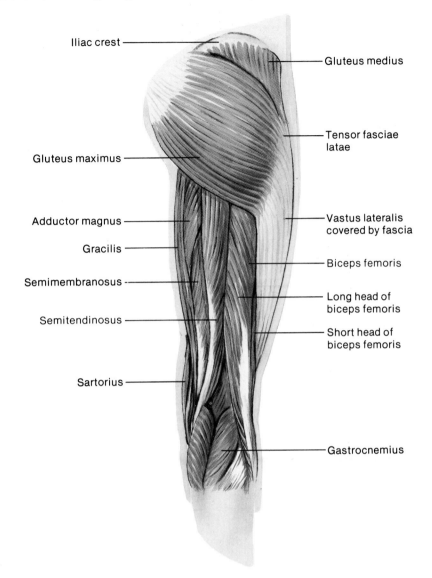

Iliac crest

Gluteus medius

Tensor fasciae latae

Gluteus maximus

Vastus lateralis covered by fascia

Adductor magnus

Gracilis

Biceps femoris

Semimembranosus

Long head of biceps femoris

Semitendinosus

Short head of biceps femoris

Sartorius

Gastrocnemius

Muscle	Origin	Insertion	Action
Gluteus minimus	Lateral surface of ilium	Greater trochanter of femur	Same as gluteus medius
Tensor fasciae latae	Anterior iliac crest	Fascia of thigh	Abducts, flexes, and rotates thigh medially
Adductor longus	Pubic bone near symphysis pubis	Posterior surface of femur	Adducts, flexes, and rotates thigh laterally
Adductor magnus	Ischial tuberosity	Posterior surface of femur	Adducts, extends, and rotates thigh laterally
Gracilis	Lower edge of symphysis pubis	Medial surface of tibia	Adducts thigh, flexes and rotates leg medially at knee

Chart 9.12   Muscles that move the lower leg

Muscle	Origin	Insertion	Action
Hamstring group			
Biceps femoris	Ischial tuberosity and posterior surface of femur	Head of fibula and lateral condyle of tibia	Flexes and rotates leg laterally and extends thigh
Semitendinosus	Ischial tuberosity	Medial surface of tibia	Flexes and rotates leg medially and extends thigh
Semimembranosus	Ischial tuberosity	Medial condyle of tibia	Flexes and rotates leg medially and extends thigh
Sartorius	Spine of ilium	Medial surface of tibia	Flexes leg and thigh, abducts thigh, rotates thigh laterally, and rotates leg medially
Quadriceps femoris group			
Rectus femoris	Spine of ilium and margin of acetabulum	Patella by tendon, which continues as patellar ligament to tibial tuberosity	Extends leg at knee
Vastus lateralis	Greater trochanter and posterior surface of femur	Patella by tendon, which continues as patellar ligament to tibial tuberosity	Extends leg at knee
Vastus medialis	Medial surface of femur	Patella by tendon, which continues as patellar ligament to tibial tuberosity	Extends leg at knee
Vastus intermedius	Anterior and lateral surfaces of femur	Patella by tendon, which continues as patellar ligament to tibial tuberosity	Extends leg at knee

Muscles That Move the Ankle, Foot, and Toes

A number of muscles that function to move the ankle, foot, and toes are located in the lower leg. (See figs. 9.25, 9.26, and 9.27.) They attach the femur, tibia, and fibula to various bones of the foot and are responsible for a variety of movements—moving the foot upward (dorsal flexion) or downward (plantar flexion), and turning the sole of the foot inward (inversion) or outward (eversion). These muscles include the following:

> Dorsal flexors
>> *Tibialis anterior*
>> *Peroneus tertius*
>> *Extensor digitorum longus*
> Plantar flexors
>> *Gastrocnemius*
>> *Soleus*
>> *Flexor digitorum longus*
> Invertor
>> *Tibialis posterior*
> Evertor
>> *Peroneus longus*

Chart 9.13 lists the origins, insertions, and actions of the muscles that move the ankle, foot, and toes.

Fig. 9.25   Muscles of the anterior right lower leg.

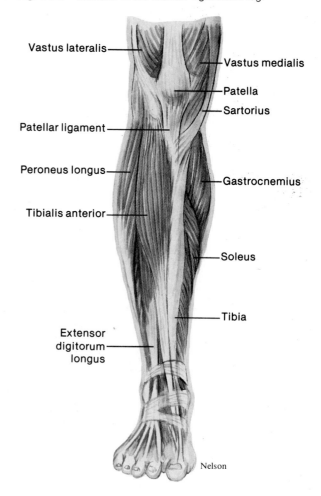

Fig. 9.26 Muscles of the lateral right lower leg.

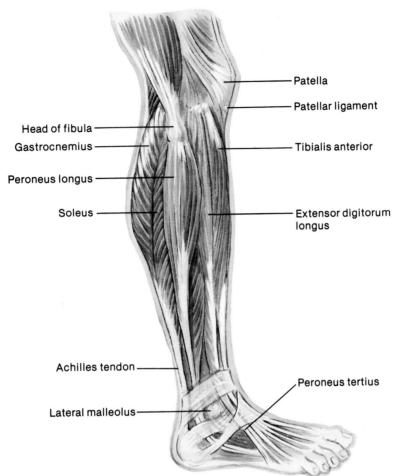

Patella

Patellar ligament

Head of fibula

Gastrocnemius

Tibialis anterior

Peroneus longus

Soleus

Extensor digitorum longus

Achilles tendon

Peroneus tertius

Lateral malleolus

Chart 9.13 Muscles that move the ankle, foot, and toes

Muscle	Origin	Insertion	Action
Tibialis anterior	Lateral condyle and lateral surface of tibia	Tarsal bone (cuneiform) and first metatarsal	Dorsal flexion and inversion of foot
Peroneus tertius	Anterior surface of fibula	Dorsal surface of fifth metatarsal	Dorsal flexion and eversion of foot
Extensor digitorum longus	Lateral condyle of tibia and anterior surface of fibula	Dorsal surfaces of second and third phalanges of four lateral toes	Dorsal flexion and eversion of foot and extension of toes
Gastrocnemius	Lateral and medial condyles of femur	Posterior surface of calcaneus	Plantar flexion of foot and flexion of leg at knee
Soleus	Head and shaft of fibula and posterior surface of tibia	Posterior surface of calcaneus	Plantar flexion of foot
Flexor digitorum longus	Posterior surface of tibia	Distal phalanges of four lateral toes	Plantar flexion and inversion of foot and flexion of four lateral toes
Tibialis posterior	Lateral condyle and posterior surface of tibia and posterior surface of fibula	Tarsal and metatarsal bones	Plantar flexion and inversion of foot
Peroneus longus	Lateral condyle of tibia and head and shaft of fibula	Tarsal and metatarsal bones	Plantar flexion and eversion of foot; also supports arch

Fig. 9.27 Muscles of the medial right lower leg.

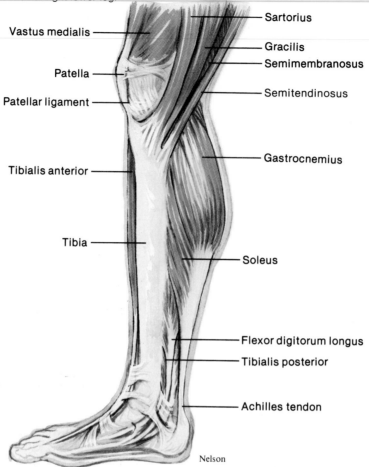

Vastus medialis

Patella

Patellar ligament

Tibialis anterior

Tibia

Sartorius

Gracilis

Semimembranosus

Semitendinosus

Gastrocnemius

Soleus

Flexor digitorum longus

Tibialis posterior

Achilles tendon

Nelson

Clinical Terms Related to the Muscular System

convulsion (kun-vul′shun)—involuntary contraction of muscles.

electromyography (e-lek″tro-mi-og′rah-fe)—technique for recording the electrical changes that occur in muscle tissues.

fibrillation (fi″brĭ-la′shun)—spontaneous contractions of individual muscle fibers, producing rapid and uncoordinated activity within a muscle.

fibrosis (fi-bro′sis)—degenerative disease in which skeletal muscle tissue is replaced by fibrous connective tissue.

fibrositis (fi″bro-si′tis)—inflammatory condition of fibrous connective tissues, especially in the muscle fascia. This disease is also called muscular rheumatism.

muscular dystrophy (mus′ku-lar dis′tro-fe)—a progressively crippling disease of unknown cause in which the muscles gradually weaken and atrophy.

myalgia (mi-al′je-ah)—pain resulting from any muscular disease or disorder.

myasthenia gravis (mi″as-the′ne-ah gra′vis)—chronic disease characterized by muscles that are weak and easily fatigued. It results from a disorder at some of the neuromuscular junctions so that a stimulus is not transmitted from the motor neuron to the muscle fiber.

myokymia (mi″o-ki′me-ah)—persistent quivering of a muscle.

myology (mi-ol′o-je)—the study of muscles.

myoma (mi-o′mah)—a tumor composed of muscle tissue.

myopathy (mi-op′ah-the)—any muscular disease.

myositis (mi″o-si′tis)—inflammation of skeletal muscle tissue.

myotomy (mi-ot′o-me)—the cutting of muscle tissue.

myotonia (mi″o-to′ne-ah)—prolonged muscular spasm.

paralysis (pah-ral′ĭ-sis)—loss of ability to move a body part.

shin splints (shin splints)—a soreness on the front of the lower leg due to straining the flexor digitorum longus, often as a result of walking up and down hills.

torticollis (tor″tĭ-kol′is)—condition in which the neck muscles, such as the sternocleidomastoids, contract involuntarily. It is more commonly called wryneck.

Chapter Summary

Introduction

There are three types of muscle tissue—skeletal, smooth, and cardiac.

Structure of a Skeletal Muscle

Individual muscles are the organs of the muscular system.

They contain skeletal muscle tissue, nerve tissue, blood, and various connective tissues.

1. Connective tissue coverings
   a. Skeletal muscles are covered with fascia.
   b. Other connective tissues attach muscles to bones or to other muscles.

2. Skeletal muscle fibers
   a. Each fiber represents a single muscle cell, which is the unit of contraction.
   b. The cytoplasm contains mitochondria, sarcoplasmic reticulum, and myofibrils composed of actin and myosin.
   c. Striations are produced by the arrangement of the actin and myosin filaments.
   d. Transverse tubules extend inward from the cell membrane and are associated with the sarcoplasmic reticulum.

3. The neuromuscular junction
   a. Muscle fibers are stimulated to contract by motor neurons.
   b. In response to a nerve impulse, the end of the motor nerve fiber secretes a neurotransmitter, which stimulates the muscle fiber.

4. Motor units
   a. One motor neuron and the muscle fibers associated with it constitute a motor unit.
   b. All the fibers of a motor unit contract together.

Skeletal Muscle Contraction

Muscle fiber contraction results from a sliding movement in which actin and myosin filaments merge.

1. Role of actin and myosin
   a. Globular portions of myosin filaments can form cross-bridges with actin filaments.
   b. The reaction between actin and myosin filaments generates the force of contraction.

2. Stimulus for contraction
   a. A skeletal muscle fiber is usually stimulated by acetylcholine released from a motor nerve fiber.
   b. Acetylcholine causes the muscle fiber to conduct an impulse that reaches the deep parts of the fiber by means of the transverse tubules.
   c. A muscle impulse signals the sarcoplasmic reticulum to release calcium ions.

   d. Cross-bridges form between actin and myosin, and the actin filaments move inward.
   e. The muscle fiber relaxes when calcium ions are transported back into the sarcoplasmic reticulum.
   f. Cholinesterase decomposes acetylcholine.

3. Energy sources for contraction
   a. ATP supplies the energy for muscle fiber contraction.
   b. Creatine phosphate stores energy that can be used to synthesize ATP.

4. Oxygen supply and cellular respiration
   a. Aerobic respiration requires the presence of oxygen.
   b. Oxygen is carried to the body cells by red blood cells.
   c. Myoglobin in muscle cells stores oxygen temporarily.

5. Oxygen debt
   a. During rest or moderate exercise, oxygen is supplied to muscles in sufficient concentration to support aerobic respiration.
   b. During strenuous exercise, an oxygen deficiency may develop, and lactic acid may accumulate.
   c. The amount of oxygen needed to convert accumulated lactic acid to glucose and to restore the supplies of ATP and creatine phosphate is called oxygen debt.

6. Muscle fatigue
   a. A fatigued muscle loses its ability to contract.
   b. Muscle fatigue is usually due to an accumulation of lactic acid.

7. Heat production
   a. Most of the energy released by cellular respiration is lost as heat.
   b. Muscles represent an important source of body heat.

Muscular Responses

1. Threshold stimulus is the minimal stimulus needed to elicit a muscular contraction.

2. All-or-none response
   a. If a muscle fiber contracts at all, it will contract completely.
   b. Motor units respond in an all-or-none manner.

3. Recruitment of motor units
   a. At low intensity of stimulation, relatively small numbers of motor units contract.
   b. At higher intensity of stimulation, other motor units are recruited until the muscle contracts with maximal tension.

4. Twitch contraction
   a. A myogram is a recording of a muscular contraction.
   b. The myogram of a twitch includes a latent period, a period of contraction, and a period of relaxation.

5. Sustained contractions
   a. A rapid series of stimuli may produce a combining of twitches and a sustained contraction.
   b. A tetanic contraction is forceful and sustained.
   c. Even when a muscle is at rest, its fibers usually remain partially contracted.
6. Use and disuse of skeletal muscle
   a. Muscles that are forcefully exercised tend to enlarge and become stronger.
   b. Muscles that are not used tend to decrease in size and become weaker.

Smooth Muscles

The contractile mechanisms of smooth and cardiac muscles are similar to that of skeletal muscle.

1. Smooth muscle fibers
   a. Cells contain filaments of actin and myosin.
   b. Types include multiunit smooth muscle and visceral smooth muscle.
   c. Visceral smooth muscle displays rhythmicity and is self-exciting.
2. Smooth muscle contraction
   a. Smooth muscles are affected by two neurotransmitters—acetylcholine and norepinephrine—and are influenced by a number of hormones.
   b. Smooth muscle can maintain a contraction for a longer time with a given amount of energy than can skeletal muscle.
   c. Smooth muscles can change their lengths without changing their tautness.

Cardiac Muscle

1. Cardiac muscle can maintain a contraction for a longer time than skeletal muscle.
2. The ends of adjacent cells are separated by intercalated disks.
3. A network of fibers contracts as a unit and responds to stimulation in an all-or-none manner.
4. Cardiac muscle is self-exciting and rhythmic.

Skeletal Muscle Actions

The type of movement produced by a muscle depends on the way it is attached on either side of a joint.

1. Origin and insertion
   a. The movable end of a muscle is its insertion, and the immovable end is its origin.
   b. Sometimes the end of a muscle changes its function in different body movements.
2. Interaction of skeletal muscles
   a. Skeletal muscles function in groups.
   b. A prime mover is responsible for most of a movement; synergists aid prime movers; antagonists can resist the movement of a prime mover.
   c. Smooth movements depend upon antagonists giving way to the actions of prime movers.

Major Skeletal Muscles

1. Muscles of facial expression
   a. These muscles lie beneath the skin of the face and scalp and are used to communicate feelings through facial expression.
   b. They include the epicranius, orbicularis oculi, orbicularis oris, buccinator, zygomaticus, and platysma.
2. Muscles of mastication
   a. These muscles are attached to the mandible and are used in chewing.
   b. They include the masseter and temporalis.
3. Muscles that move the head
   a. Head movements are produced by muscles in the neck and upper back.
   b. They include the sternocleidomastoid, splenius capitis, and semispinalis capitis.
4. Muscles that move the pectoral girdle
   a. Most of these muscles connect the scapula to nearby bones and are closely associated with muscles that move the upper arm.
   b. They include the trapezius, rhomboideus major, levator scapulae, serratus anterior, and pectoralis minor.
5. Muscles that move the upper arm
   a. These muscles connect the humerus to various regions of the pectoral girdle, ribs, and vertebral column.
   b. They include the coracobrachialis, pectoralis major, teres major, latissimus dorsi, supraspinatus, deltoid, subscapularis, infraspinatus, and teres minor.
6. Muscles that move the forearm
   a. These muscles connect the radius and ulna to the humerus or pectoral girdle.
   b. They include the biceps brachii, brachialis, brachioradialis, triceps brachii, supinator, pronator teres, and pronator quadratus.
7. Muscles that move the wrist, hand, and fingers
   a. These muscles arise from the distal end of the humerus and from the radius and ulna.
   b. They include the flexor carpi radialis, flexor carpi ulnaris, palmaris longus, flexor digitorum profundus, extensor carpi radialis longus, extensor carpi radialis brevis, extensor carpi ulnaris, and extensor digitorum.
8. Muscles of the abdominal wall
   a. These muscles connect the rib cage and vertebral column to the pelvic girdle.
   b. They include the external oblique, internal oblique, transversus abdominus, and rectus abdominis.

9. Muscles of the pelvic outlet
   a. These muscles form the floor of the pelvic cavity and fill the space of the pubic arch.
   b. They include the levator ani, coccygeus, superficial transversus perinei, bulbospongiosus, ischiocavernosus, and sphincter urethrae.
10. Muscles that move the thigh
    a. These muscles are attached to the femur and to some part of the pelvic girdle.
    b. They include the psoas major, iliacus, gluteus maximus, gluteus medius, gluteus minimus, tensor fasciae latae, adductor longus, adductor magnus, and gracilis.
11. Muscles that move the lower leg
    a. These muscles connect the tibia or fibula to the femur or pelvic girdle.
    b. They include the biceps femoris, semitendinosus, semimembranosus, sartorius, and quadriceps femoris.
12. Muscles that move the ankle, foot, and toes
    a. These muscles attach the femur, tibia, and fibula to various bones of the foot.
    b. They include the tibialis anterior, peroneus tertius, extensor digitorum longus, gastrocnemius, soleus, flexor digitorum longus, tibialis posterior, and peroneus longus.

Application of Knowledge

1. Why do you think athletes generally perform better if they warm up by exercising lightly before a competitive event?
2. What steps might be taken to minimize atrophy of skeletal muscles in patients who are confined to bed for prolonged times?
3. As lactic acid and other substances accumulate in an active muscle, they tend to stimulate pain receptors and the muscle may feel sore. How might the application of heat or substances that cause blood vessels to dilate help to relieve such soreness?

Review Activities

Part A

1. List the three types of muscle tissue.
2. Distinguish between a tendon and an aponeurosis.
3. Describe how connective tissue is associated with skeletal muscle.
4. List the major parts of a skeletal muscle fiber and describe the function of each part.
5. Describe a neuromuscular junction and a neurotransmitter substance.

6. Define *motor unit.*
7. Describe the major events that occur when a muscle fiber contracts.
8. Explain how ATP and creatine phosphate are related.
9. Describe how oxygen is supplied to muscles.
10. Describe how oxygen debt may develop.
11. Explain how muscles may become fatigued.
12. Explain how the maintenance of body temperature is related to the actions of skeletal muscles.
13. Define *threshold stimulus.*
14. Explain what is meant by an all-or-none response.
15. Explain what is meant by motor unit recruitment.
16. Sketch a myogram of a single muscular twitch and identify the latent period, period of contraction, and period of relaxation.
17. Explain how a skeletal muscle can be stimulated to produce a sustained contraction.
18. Distinguish between tetanic contraction and muscle tone.
19. Distinguish between hypertrophy and atrophy and explain how each may be caused.
20. Distinguish between multiunit and visceral smooth muscles.
21. Compare the characteristics of smooth and skeletal muscle contractions.
22. Compare the structure of cardiac and skeletal muscles.
23. Distinguish between a muscle's origin and its insertion.
24. Define *prime mover, synergist,* and *antagonist.*

Part B

Match the muscles in column I with the descriptions and functions in column II.

I	II
1. Buccinator	A. Inserted on the coronoid process of the mandible.
2. Epicranius	B. Draws the corner of the mouth upward.
3. Orbicularis oris	
4. Platysma	C. Can raise and adduct the scapula.
5. Rhomboideus major	D. Can pull the head into an upright position.
6. Splenius capitus	E. Raises the eyebrow.
7. Temporalis	F. Compresses the cheeks.
8. Zygomaticus	G. Extends over the neck from the chest to the face.
	H. Closes the lips.

Fig. 9-C (A)

Fig. 9-C (B)

I	II	I	II
9. Biceps brachii	I. Primary extensor of the elbow.	17. Biceps femoris	Q. Inverts the foot.
10. Brachialis	J. Pulls the shoulder back and downward.	18. External oblique	R. A member of the quadriceps femoris group.
11. Deltoid	K. Abducts the arm.	19. Gastrocnemius	
12. Latissimus dorsi	L. Rotates the arm laterally.	20. Gluteus maximus	
13. Pectoralis major	M. Pulls the arm forward and across the chest.	21. Gluteus medius	S. A plantar flexor of the foot.
14. Pronator teres	N. Rotates the arm medially.	22. Gracilis	T. Compresses the contents of the abdominal cavity.
15. Teres minor	O. Strongest flexor of the elbow.	23. Rectus femoris	
16. Triceps brachii	P. Strongest supinator of the forearm.	24. Tibialis anterior	U. Heaviest muscle in the body.
			V. A hamstring muscle.
			W. Adducts the thigh.
			X. Abducts the thigh.

Part C

What muscles can you identify in the bodies of these models whose muscles are enlarged by exercise?

Integration and Coordination

The chapters of unit 3 are concerned with the structures and functions of the nervous and endocrine systems. They describe how the organs of these systems serve to keep the parts of the human body functioning together as a whole and how these organs help to maintain a stable internal environment, which is vital to the survival of the organism.

This unit includes

The Nervous System

10

If a human organism is to survive, the actions of its cells, tissues, organs, and systems must be directed toward a single goal—the maintenance of homeostasis. To accomplish this, the functions of all body parts must be controlled and coordinated so that they work as a unit and respond to changes in ways that help maintain a stable internal environment.

The general task of controlling and coordinating body activities is handled by the nervous and endocrine systems. Of these, the *nervous system* provides the more rapid and precise mode of action. It controls muscular contractions and glandular secretions by means of nerve fibers that extend from the brain and spinal cord to various body parts. Some of these nerve fibers bring information concerning changes occurring inside and outside the body to the brain and spinal cord from sensory receptors. Other fibers carry impulses away from the brain or spinal cord and stimulate muscles or glands to respond.

After you have studied this chapter, you should be able to

1. Describe the general structure of a neuron.

2. Name four types of neuroglial cells and describe the functions of each.

3. Describe the events that lead to the conduction of a nerve impulse.

4. Explain how a nerve impulse is transmitted from one neuron to another.

5. Explain two ways impulses are processed in neuronal pools.

6. Explain how differences in structure and function are used to classify neurons.

7. Name the parts of a reflex arc and describe the function of each part.

8. Describe the coverings of the brain and spinal cord.

9. Describe the structure of the spinal cord and its major functions.

10. Name the major parts of the brain and describe the functions of each part.

11. Distinguish between motor, sensory, and association areas of the cerebral cortex.

12. Describe the formation and the function of cerebrospinal fluid.

13. List the major parts of the peripheral nervous system.

14. Name the cranial nerves and list their major functions.

15. Describe the structure of a spinal nerve.

16. Describe the functions of the autonomic nervous system.

17. Distinguish between the sympathetic and the parasympathetic divisions of the autonomic nervous system.

18. Complete the review activities at the end of this chapter. Note that the items are worded in the form of specific learning objectives. You may want to refer to them before reading the chapter.

action potential   (ak'shun po-ten'shal)

autonomic nervous system
    (aw''to-nom'ik ner'vus sis'tem)

axon   (ak'son)

central nervous system
    (sen'tral ner'vus sis'tem)

convergence   (kon-ver'jens)

dendrite   (den'drīt)

divergence   (di-ver'jens)

effector   (ĕ-fek'tor)

facilitation   (fah-sil''ĭ-ta'shun)

ganglion   (gang'gle-on)

meninges   (mĕ-nin'jēz)

myelin   (mi'ĕ-lin)

neurolemma   (nu''ro-lem'ah)

neurotransmitter   (nu''ro-trans-mit'er)

parasympathetic nervous system
    (par''ah-sim''pah-thet'ik ner'vus sis'tem)

peripheral nervous system
    (pĕ-rif'er-al ner'vus sis'tem)

plexus   (plek'sus)

receptor   (re-sep'tor)

reflex   (re'fleks)

sympathetic nervous system
    (sim''pah-thet'ik ner'vus sis'tem)

synapse   (sin'aps)

ax-, axle: *ax*on—a cylindrical nerve fiber that carries impulses away from a neuron cell body.

dendr-, tree: *dendr*ite—branched nerve fibers that serve as receptor surfaces of a neuron.

funi-, small cord or fiber: *funi*culus—a major nerve tract or bundle of myelinated nerve fibers within the spinal cord.

gangli-, a swelling: *gangli*on—a mass of neuron cell bodies.

-lemm, rind or peel: neuro*lemm*a—a sheath that surrounds the myelin of a nerve fiber.

mening-, membrane: *mening*es—membranous coverings of the brain and spinal cord.

moto-, moving: *moto*r neuron—neuron that stimulates a muscle to contract or a gland to release a secretion.

peri, all around: *peri*pheral nervous system—portion of the nervous system that consists of the nerves branching from the brain and spinal cord.

plex-, interweaving: choroid *plex*us—a mass of specialized capillaries associated with spaces in the brain.

sens-, feeling: *sens*ory neuron—neuron that can be stimulated by a sensory receptor and conducts impulses into the brain or spinal cord.

syn-, together: *syn*apse—junction between two neurons.

ventr-, belly or stomach: *ventr*icle—fluid-filled space within the brain.

The organs of the nervous system, like other organs, are composed of various tissues, including nerve tissue, blood, and connective tissues. These organs can be divided into two groups. One group, consisting of the brain and spinal cord, forms the **central nervous system,** while the other includes the nerves (peripheral nerves) that connect the central nervous system to other body parts and is called the **peripheral nervous system.** Together these parts provide three general functions—a sensory function, an integrative function, and a motor function.

General Functions of the Nervous System

The **sensory function** of the nervous system involves *sensory receptors* at the ends of peripheral nerves. These receptors are specialized to gather information by detecting changes that occur within and around the body. They monitor such factors as light and sound intensities outside the body as well as the temperature, oxygen concentration and other conditions of its internal fluids.

The information gathered is converted into signals in the form of *nerve impulses,* and they are transmitted over peripheral nerves to the central nervous system. There the signals are integrated; that is, they are brought together, creating sensations, adding to memory, or being used to produce thoughts. As a result of this **integrative function,** conscious or subconscious decisions are made and then acted upon using motor functions.

The **motor functions** of the nervous system employ peripheral nerves that carry impulses from the central nervous system to responsive parts called *effectors.* These effectors are outside the nervous system and include muscles that may contract when they are stimulated by nerve impulses, and glands that may produce secretions when they are stimulated.

Thus, the nervous system can detect changes occurring outside and within the body, make decisions on the basis of the information received, and cause muscles or glands to respond. Typically, these responses are directed toward counteracting the effects of the changes that were detected, and in this way the nervous system helps to maintain stable internal conditions.

Nerve Tissue

As is described in chapter 5, nerve tissue contains masses of nerve cells, or *neurons.* These cells are the structural and functional units of the nervous system and are specialized to react to physical and chemical changes occurring in their surroundings. They also function to conduct nerve impulses to other neurons and to cells outside the nervous system.

Between the neurons, there are *neuroglial cells* that function much like the connective tissue cells in other systems. (See fig. 10.4.)

Neuroglial cells are described in more detail in a later section of this chapter.

Neuron Structure

Although neurons vary considerably in size and shape, they have certain features in common. These include a cell body and tubular processes filled with cytoplasm called *nerve fibers,* that conduct nerve impulses to or from the cell body. (See fig. 10.1.)

The neuron **cell body** contains a mass of granular cytoplasm, a cell membrane, and various other organelles usually found in cells. Inside the cell body, for example, are mitochondria, lysosomes, a Golgi apparatus, and a network of fine threads called **neurofibrils,** which extend into the nerve fibers.

Near the center of the cell body there is a large, spherical nucleus with a conspicuous nucleolus. This nucleus, however, apparently does not undergo mitosis after the nervous system is developed, and consequently, mature neurons seem to be incapable of reproduction.

Two kinds of nerve fibers, called **dendrites** and **axons,** extend from the cell bodies of most neurons. Although a neuron usually has many dendrites, it has a single axon.

In most neurons, the dendrites are relatively short and highly branched. These fibers, together with the membrane of the cell body, provide the main receptive surfaces of the neuron to which fibers from other neurons communicate.

The axon usually arises from a slight elevation of the cell body. It is specialized to conduct nerve impulses away from the region of the cell body. Many mitochondria, microtubules, and neurofibrils occur within its cytoplasm. Although it begins as a single fiber, the axon may give off side branches, and at its terminal end it may have many fine branches, each with a specialized ending which contacts the receptive surface of another cell.

Fig. 10.1 Two types of neurons: (a) motor neuron and (b) sensory neuron.

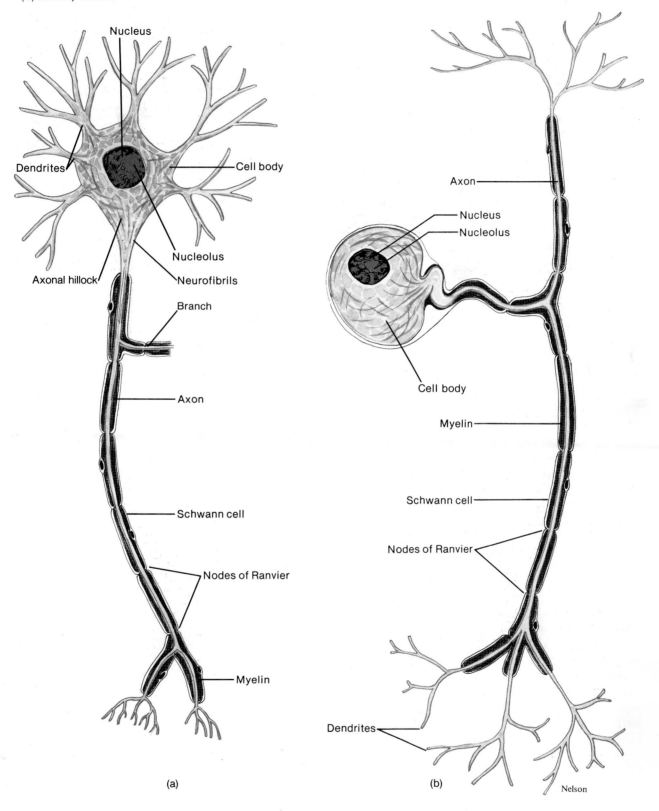

Nucleus

Dendrites

Cell body

Nucleolus

Axonal hillock

Neurofibrils

Branch

Axon

Schwann cell

Nodes of Ranvier

Myelin

(a)

Axon

Nucleus

Nucleolus

Cell body

Myelin

Schwann cell

Nodes of Ranvier

Dendrites

(b)

Nelson

Larger axons passing through peripheral nerves commonly are enclosed in sheaths composed of neuroglial cells called **Schwann cells.** These cells are tightly wound around the axons, somewhat like insulation on an electric wire. As a result, the axons are coated with many layers of cell membrane that have little or no cytoplasm between them. These membrane layers are composed largely of a lipid-protein that has a higher proportion of lipid than other cell surface membranes. This lipid-protein is called **myelin,** and it forms a *myelin sheath* on the outside of an axon. In addition, the portions of the Schwann cells that contain cytoplasm and nuclei remain outside the myelin sheath and comprise a *neurolemma* (neurolemmal sheath) which surrounds the myelin sheath. (See figs. 10.2 and 10.3.)

Smaller axons also are enclosed by Schwann cells, but the cells may not be wound around these axons. Consequently, such axons lack myelin sheaths.

Axons that possess myelin sheaths are called *myelinated* nerve fibers, while those that lack these sheaths are *unmyelinated.* Groups of myelinated fibers appear white, and masses of such fibers are responsible for the *white matter* in the nervous system. Unmyelinated nerve fibers and neuron cell bodies appear as *gray matter.*

When neurons are deprived of oxygen, they undergo a series of irreversible structural changes in which their shapes are altered and their nuclei shrink. This phenomenon is called *ischemic cell change,* and in time the affected cells disintegrate. Such an oxygen deficiency can result from lack of blood flow through nerve tissue (ischemia), an abnormally low blood oxygen concentration, or the presence of toxins that block aerobic respiration.

Neuroglial Cells

Neuroglial cells occur within the organs of the nervous system where they function to fill spaces, support neurons, provide structural frameworks, produce myelin, and carry on phagocytosis.

Within the peripheral nervous system, the neuroglial cells include the *Schwann cells,* previously described. In the central nervous system, where neuroglial cells are several times more numerous than neurons, they include the following types. (See fig. 10.4.)

1. **Astrocytes.** These cells are commonly found between nerve tissues and blood vessels, where they seem to function in providing structural support and in removing cellular debris.

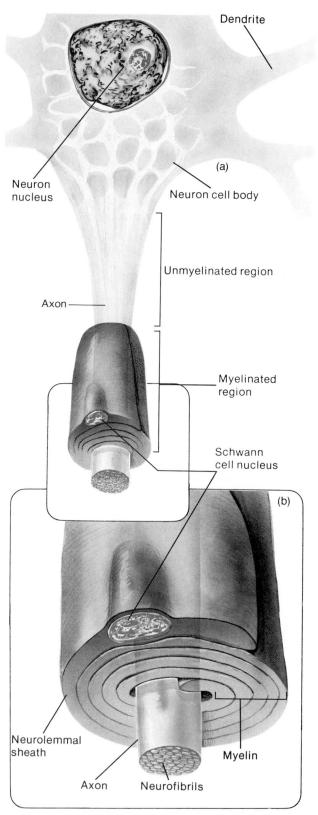

Fig. 10.2 (a) The portion of a Schwann cell that is tightly wound around an axon forms the myelin sheath; (b) the cytoplasm and nucleus of the Schwann cell, remaining on the outside, form the neurolemmal sheath.

Fig. 10.3 An electron micrograph of a myelinated axon in cross section. The arrow indicates the layers of Schwann cell membranes.

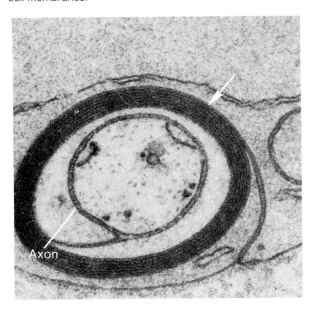

Axon

2. **Oligodendrocytes.** These cells are commonly arranged in rows along nerve fibers, and they function in the formation of myelin within the brain and spinal cord. However, unlike the Schwann cells in the peripheral nervous system, oligodendrocytes fail to form neurolemmal sheaths.

3. **Microglia.** These cells are scattered throughout the central nervous system where they help support neurons and are able to phagocytize bacterial cells and cellular debris.

4. **Ependyma.** These cells form an epitheliallike membrane that covers specialized parts (*choroid plexuses*) and form the lining that encloses spaces (ventricles) within the brain.

1. *Describe a neuron.*
2. *Distinguish between an axon and a dendrite.*
3. *Describe how a myelin sheath is formed.*
4. *Distinguish between a neuron and a neuroglial cell.*

Fig. 10.4 Types of neuroglial cells found within the central nervous system include (a) microglial cell, (b) oligodendrocyte, (c) astrocyte, and (d) ependymal cell.

(a) Microglial cell

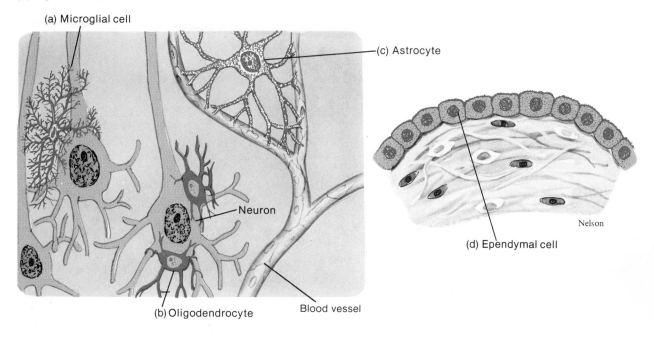

(c) Astrocyte

Neuron

Nelson

(d) Ependymal cell

(b) Oligodendrocyte

Blood vessel

Fig. 10.5 A nerve fiber at rest is polarized as a result of an unequal distribution of ions on either side of its membrane.

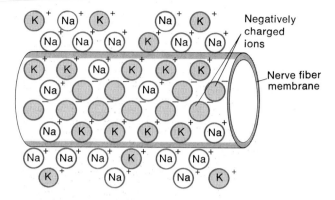

Fig. 10.6 When a polarized nerve fiber is stimulated, sodium channels open, some of the sodium ions diffuse inward, and the membrane is said to depolarize.

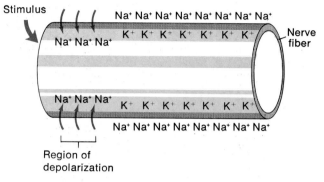

Cell Membrane Potential

The outside of a cell membrane is usually electrically charged or *polarized* with respect to the inside. This is due to an unequal distribution of ions on either side of the membrane and is of particular importance to the functions of nerve and muscle cells.

When nerve cells are at rest (that is, not conducting impulses), their membranes are relatively impermeable to sodium ions (Na^+) and there is a relatively greater concentration of sodium ions on the outside. Similarly, there is a relatively greater concentration of potassium ions (K^+) on the inside. In the cytoplasm of these cells, there are large numbers of negatively charged ions, including those of phosphate, sulfate, and protein, that cannot diffuse through the cell membranes. Consequently, in a resting cell, the outside of the membrane is positively charged with respect to the inside, which is negative. (See fig. 10.5.)

The difference in electrical charge between the inside and the outside of the membrane is called the *resting potential*. As long as the nerve cell membrane is undisturbed, it remains in this polarized state.

The Nerve Impulse

Nerve cells are excitable; that is, they are specialized to detect changes in their surroundings. If a change reaches a certain intensity or *threshold* of stimulation, it may disturb the membrane's resting potential and trigger a nerve impulse. Changes in temperature, pressure, chemical concentration, or electrical condition, for example, can stimulate various nerve cells.

Fig. 10.7 When the potassium channels open, potassium ions diffuse outward, and the membrane is repolarized.

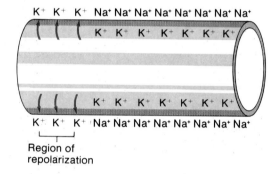

Action Potential

Once the threshold is reached, the portion of the cell membrane that is being stimulated undergoes a sudden change in permeability. Channels in the membrane that are highly selective for sodium ions open and allow the ions to pass inward. This movement is aided by the fact that the sodium ions are attracted by the negative electrical condition on the other side of the membrane. (See fig. 10.6.)

As the sodium ions rush inward, the membrane loses its electrical charge and becomes *depolarized*. At almost the same time, channels open in the membrane that allow some of the potassium ions to pass through, and as they diffuse outward, the outside of the membrane becomes positively charged once more. The membrane, then, is said to become *repolarized*, and it remains in this state until it is stimulated again. (See fig. 10.7.)

This rapid sequence of changes, involving depolarization and repolarization, takes about 1/1000th second or less and is called an **action potential.** Actually, only a small proportion of the sodium and potassium ions present move through the membrane

Fig. 10.8 (a) An action potential in one region stimulates the adjacent region, and (b, c) a wave of action potentials or a nerve impulse moves along the fiber.

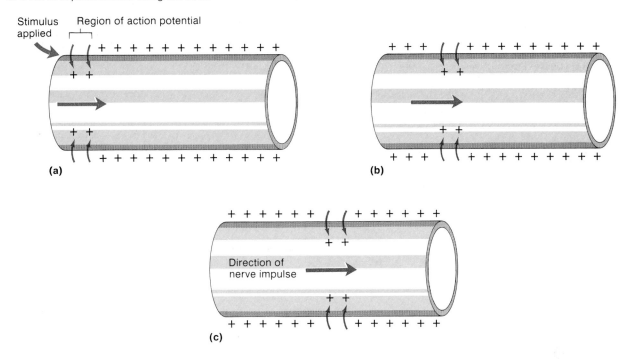

during an action potential, so that many action potentials can occur before the concentrations of sodium ions and potassium ions on either side of the membrane change significantly. Eventually, however, these ion concentrations may change enough that action potentials cease. When this happens, an active transport mechanism in the membrane reestablishes the original concentrations of sodium and potassium, and the resting potential returns.

When an action potential occurs at one point in a nerve cell membrane, it stimulates other action potentials in adjacent portions of the cell membrane. These in turn stimulate still other areas, and a wave of action potentials moves away in all directions from the point of stimulation to the ends of the nerve fiber. This transmission of action potentials along a fiber constitutes a **nerve impulse.** (See fig. 10.8.)

Impulse Conduction

An unmyelinated nerve fiber conducts an impulse over its entire membrane surface. A myelinated fiber functions differently because myelin serves as an insulator that prevents almost all flow of ions through the membrane.

Considering this, it might seem that the myelin sheath would prevent the conduction of a nerve impulse altogether, and this would be true if the sheath were continuous. It is, however, interrupted by constrictions called **nodes of Ranvier,** which occur between adjacent Schwann cells. (See fig. 10.1.) At these nodes the fiber membrane can become especially permeable to sodium and potassium ions, and a nerve impulse traveling along a myelinated fiber appears to jump from node to node. This type of impulse conduction (saltatory conduction), is many times faster than conduction on an unmyelinated fiber.

All-or-None Response

Like muscle fiber contraction, nerve impulse conduction is an *all-or-none response.* In other words, if a nerve fiber responds at all, it responds completely. Thus a nerve impulse is conducted whenever a stimulus of threshold intensity or above is applied to a nerve fiber, and all impulses carried on that fiber will be of the same strength. Greater intensity of stimulation of a nerve fiber does not produce a stronger impulse.

Certain drugs, such as procaine and cocaine, produce effects by decreasing membrane permeability to sodium ions. When one of these drugs is present in the tissue fluids surrounding a nerve fiber, impulses are prevented from passing through the affected region. Consequently, the drugs are useful as local anesthetics because they help keep impulses from reaching the brain and thus prevent the sensations of touch and pain.

---

1. Explain how a nerve fiber becomes polarized.
2. List the major events that occur during an action potential.
3. Explain how impulse conduction differs in myelinated and unmyelinated nerve fibers.

The Synapse

Within the nervous system, nerve impulses travel from neuron to neuron along complex nerve pathways. The junction between the parts of two such neurons is called a **synapse.** Actually, the neurons are not in direct contact at a synapse. There is a gap called a *synaptic cleft* between them. For an impulse to continue along a nerve pathway it must cross this gap. (See fig. 10.9.)

Synaptic Transmission

As was mentioned, a nerve impulse travels in both directions away from the point of stimulation. Within a neuron, however, an impulse often will travel from a dendrite to its cell body and then move along the axon to the end. There it crosses a synapse and continues to a dendrite or cell body of another neuron. The process of crossing the gap at a synapse is called *synaptic transmission.*

The typical one-way transmission from axon to dendrite or cell body is due to the fact that axons usually have rounded *synaptic knobs* at their ends, which dendrites lack. These knobs contain numerous membranous sacs, called *synaptic vesicles,* and when a nerve impulse reaches a knob, some of the vesicles respond by releasing a substance called a **neurotransmitter.**

As figure 10.10 shows, the neurotransmitter diffuses across the synaptic cleft and reacts with specific receptors in the neuron membrane on the other side. If a sufficient amount of neurotransmitter is released, the membrane is stimulated, and a nerve impulse is triggered.

Neurotransmitters usually are destroyed through rapid decomposition by enzymes present in synaptic clefts or are removed. Such destruction or removal of the neurotransmitter is important in preventing continuous stimulation of a neuron on the distal side of a synapse.

Acetylcholine is the neurotransmitter released by most axons outside the brain and spinal cord and by some axons within these organs. This is the same substance that is released from the ends of motor neurons at the motor end plates of muscle fibers. In both cases, acetylcholine is decomposed by the action of the enzyme *cholinesterase.*

Excitatory and Inhibitory Actions

In the brain and spinal cord, some synaptic knobs release neurotransmitters that cause an increase in membrane permeability to sodium ions and thus trigger nerve impulses. This action is said to be *excitatory.* Substances of this type include serotonin, dopamine, and norepinephrine.

Other synaptic knobs release substances that decrease membrane permeability to sodium ions, thus causing the threshold of stimulation to be raised. This action is called *inhibitory,* because it lessens the chance that a nerve impulse will be transferred to an adjoining neuron. Inhibitory substances include gamma-aminobutyric acid (GABA), which is an amino acid that is synthesized primarily in the brain and spinal cord. Another amino acid, glycine, seems to function as an inhibitor in some of the spinal cord synapses.

The synaptic knobs of a thousand or more neurons may communicate with the dendrites and cell body of a particular neuron. Some of these knobs probably have an excitatory action, while others are likely to be inhibitory. The effect on the neuron will depend on which knobs are activated from moment to moment. In other words, if more excitatory than inhibitory knobs are functioning, the neuron's threshold may be exceeded, and a nerve impulse will be triggered to pass over its surface. Conversely, if more inhibitory knobs are active, no impulse will be conducted.

Fig. 10.9 For an impulse to continue from one neuron to another, it must cross the synaptic cleft at a synapse.

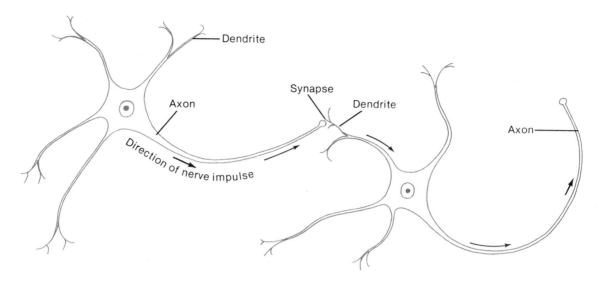

Fig. 10.10 (a) When a nerve impulse reaches the synaptic knob at the end of an axon, (b) synaptic vesicles release a neurotransmitter substance that diffuses across the synaptic cleft.

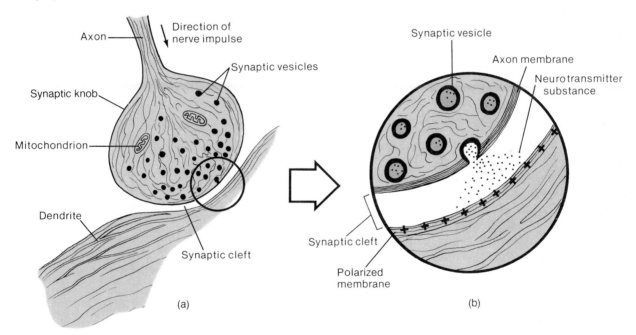

A number of drugs apparently produce their special effects by interfering with the normal actions of neurotransmitters. These include LSD, which seems to counteract the function of serotonin; cocaine, which enhances the effects of norepinephrine by preventing its normal inactivation; and amphetamine, which promotes the excessive release of dopamine. Some of the antianxiety drugs, including diazepam (Valium), seem to produce their effects by increasing the effectiveness of the inhibitory transmitter GABA.

Processing of Impulses

The way the nervous system processes nerve impulses and acts upon them reflects, in part, the organization of neurons and their nerve fibers within the brain and spinal cord.

Neuronal Pools

The neurons within the central nervous system are organized into many groups with varying numbers of cells called *neuronal pools*. Each pool receives impulses from input nerve fibers. These impulses are processed according to the special characteristics of the pool, and any resulting impulses are conducted away on output fibers.

Each input fiber divides many times as it enters, and its branches spread over a certain region of the neuronal pool. The branches give off smaller branches and their terminals form hundreds of synapses with the dendrites and cell bodies of the neurons in the pool.

Facilitation

As a result of incoming impulses, a particular neuron of a pool may receive excitatory and inhibitory stimulation. As before, if the net effect of the stimulation is excitatory, threshold may be reached, and an outgoing impulse will be triggered. If the net effect is excitatory but subthreshold, an impulse will not be triggered. However, in this case the neuron becomes more excitable to incoming stimulation than before, and it is said to be *facilitated*.

Convergence

Any single neuron in a neuronal pool may receive impulses from two or more incoming fibers. Furthermore, these fibers may originate from different parts of the nervous system, and they are said to *converge* when they lead to the same neuron. (See fig. 10.11.)

Convergence makes it possible for impulses arriving from different sources to create an additive effect upon a neuron. For example, if a neuron is facilitated by receiving subthreshold stimulation from one input fiber, its threshold may be reached if it receives additional stimulation from a second input fiber. As a result, an output impulse may travel to a particular effector and cause a special response.

Incoming impulses often represent information from various sensory receptors that have detected changes taking place. Convergence allows the nervous system to bring a variety of kinds of information together, to process it, and to respond to it in a special way.

Divergence

Impulses leaving a neuron of a neuronal pool often *diverge* by passing into several other output fibers. For example, an impulse from one neuron may stimulate two others; each of these, in turn, may stimulate several others, and so forth. This arrangement of diverging nerve fibers can cause an impulse to be *amplified*—that is, to be spread to increasing numbers of neurons within the pool.

As a result of divergence, an impulse originating from a single neuron in the central nervous system may be amplified so that enough impulses reach the motor units within a skeletal muscle to cause a forceful contraction.

Similarly, an impulse originating from a sensory receptor may diverge and reach several different regions of the central nervous system, where the resulting impulses can be processed and acted upon. (See fig. 10.11.)

---

1. Describe the function of a neurotransmitter.
2. Distinguish between excitatory and inhibitory actions of synaptic knobs.
3. Define neuronal pool.
4. Distinguish between convergence and divergence.

Fig. 10.11 (a) Nerve fibers of neurons 1 and 2 converge to the cell body of neuron 3 (b) the nerve fiber of neuron 4 diverges to the cell bodies of neurons 5 and 6.

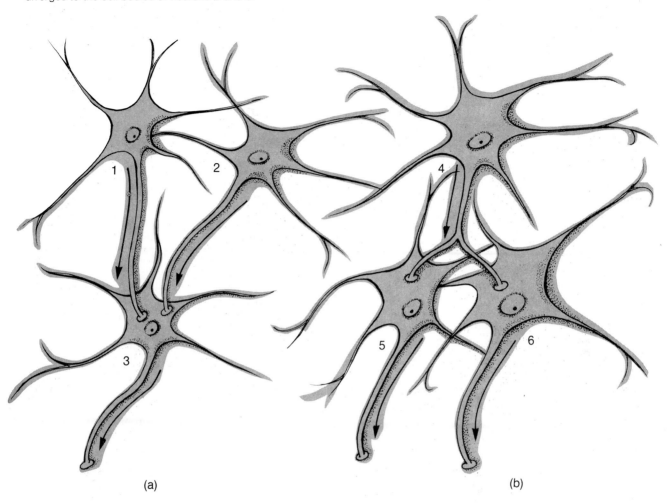

(a)

(b)

Types of Neurons and Nerves

Various neurons differ in the structure, size, and shape of their cell bodies. They also vary in the length and size of their axons and dendrites, and in the number of synaptic knobs by which they communicate with other neurons.

Neurons also vary in function. Some carry impulses into the brain or spinal cord, others carry impulses out from the brain or spinal cord, and still others conduct impulses from neuron to neuron within the brain or spinal cord.

Classification of Neurons

On the basis of *structural* differences, neurons can be classified into three major groups as follows:

1. **Multipolar neurons.** Multipolar neurons have many processes arising from their cell bodies. Only one process of each neuron is an axon; the rest are dendrites. Most of the neurons whose cell bodies lie within the brain or spinal cord are of this type. (See fig. 10.12.)

2. **Bipolar neurons.** The cell body of a bipolar neuron has only two processes, one arising from each end. Although these processes have similar structural characteristics, one serves as an axon and the other as a dendrite. Such neurons are found within specialized parts of the eyes and ears.

Fig. 10.12 Structural types of neurons include the (a) multipolar neuron, (b) bipolar neuron, and (c) unipolar neuron. Where do examples of each of these types of neurons occur in the body?

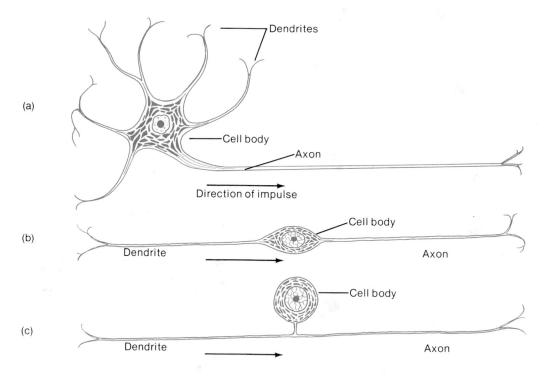

3. **Unipolar neurons.** Unipolar neurons occur in specialized masses of nerve tissue called *ganglia,* which are located outside the brain and spinal cord. Each of these neurons has a single process extending from its cell body. A short distance from the cell body, this process divides into two branches: one branch is connected to some peripheral body part and serves as a dendrite, and the other enters the brain or spinal cord and serves as an axon.

Based upon functional differences, neurons can be grouped as follows:

1. **Sensory neurons.** Sensory neurons (afferent neurons) are those that carry nerve impulses from peripheral body parts into the brain or spinal cord. (See fig. 10.1.) These neurons either have specialized *receptor ends* at the tips of their dendrites, or dendrites that are closely associated with *receptor cells* located in the skin or various sensory organs.

Changes that occur inside or outside the body are likely to stimulate receptor ends or receptor cells, triggering nerve impulses. The impulses travel along the sensory neuron fibers, which lead to the brain or spinal cord, and are processed in these parts by other neurons.

2. **Interneurons.** Interneurons (association neurons) lie within the brain or spinal cord. They form links between other neurons. Interneurons function to transmit impulses from one part of the brain or spinal cord to another. That is, they may direct incoming sensory impulses to appropriate parts for processing and interpreting. (See fig. 10.11.) Other incoming impulses are transferred to motor neurons.

3. **Motor neurons.** Motor neurons (efferent neurons) carry nerve impulses out from the brain or spinal cord to **effectors**—parts of the body capable of responding, such as muscles or glands. (See fig. 10.1.) When motor impulses reach muscles, for example, these effectors are stimulated to contract; when they reach glands, the glands are stimulated to release secretions.

Types of Nerves

While a nerve fiber is an extension of a neuron, a **nerve** is a cordlike bundle (or group of bundles) of nerve fibers held together by layers of connective tissues, as shown in figure 10.13.

Like nerve fibers, nerves that conduct impulses into the brain or spinal cord are called **sensory nerves,** while those that carry impulses out to muscles or

Fig. 10.13 A nerve is composed of bundles of nerve fibers held together by connective tissues.

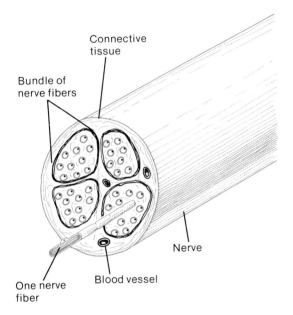

Connective tissue

Bundle of nerve fibers

Nerve

One nerve fiber

Blood vessel

glands are termed **motor nerves.** Most nerves, however, include both sensory and motor fibers, and they are called **mixed nerves.**

---

1. Explain how neurons are classified according to structure or function.
2. How is a neuron related to a nerve?
3. What is a mixed nerve?

Nerve Pathways

The routes followed by nerve impulses as they travel through the nervous system are called *nerve pathways.* The simplest of these pathways includes only a few neurons and is called a reflex arc.

Reflex Arcs

A **reflex arc** begins with the receptor at the end of a sensory nerve fiber. This fiber usually leads to several interneurons within the central nervous system, which serve as a processing or *reflex center.* Fibers from the interneurons may connect with interneurons in other parts of the nervous system. They also communicate with motor neurons, whose fibers pass outward to effectors that respond when they are stimulated. Such a reflex arc represents the *behavioral unit* of the nervous system. That is, it constitutes the structural and functional basis for the simplest acts—the reflexes.

Reflex Behavior

Reflexes are automatic, unconscious responses to changes occurring inside or outside the body. They are mechanisms that help to maintain homeostasis by controlling many of the body's involuntary processes such as heart rate, breathing rate, blood pressure, and digestive activities. Reflexes also are involved in the automatic actions of swallowing, sneezing, coughing, and vomiting.

The *knee-jerk reflex* (patellar reflex) is an example of a simple reflex that may employ only two neurons—a sensory neuron connected directly to a motor neuron (fig. 10.14). This reflex is initiated by striking the patellar ligament just below the patella. As a result, the quadriceps femoris group of muscles, which is attached to the patella by a tendon, is pulled slightly, and stretch receptors located within the muscle are stimulated. These receptors, in turn, trigger impulses that pass along the fibers of a sensory neuron into the spinal cord. Within the spinal cord, the sensory axon forms a synapse with a dendrite of a motor neuron. The impulse then continues along the axon of the motor neuron and travels back to the quadriceps femoris. The muscle group responds by contracting, and the reflex is completed as the lower leg extends.

This reflex is helpful in maintaining an upright posture. For example, if the knee begins to bend as a result of gravity pulling downward when a person is standing still, the quadriceps femoris group is stretched, the reflex is triggered, and the leg straightens again.

Another type of reflex, called a *withdrawal reflex* (fig. 10.15), occurs when a person unexpectedly touches a finger to something hot or painful. As this happens, some of the skin receptors are activated, and sensory impulses travel to the spinal cord. There the impulses pass on to interneurons of the reflex center and are directed to motor neurons. The motor neurons transmit the signals to flexor muscles in the arm, and they contract in response. At the same time, the antagonistic extensor muscles are inhibited, and the hand is rapidly and unconsciously withdrawn from the harmful source of stimulation.

Concurrent with the withdrawal reflex, other interneurons in the spinal cord carry sensory impulses to the brain, and the person becomes aware of the painful experience.

A withdrawal reflex is, of course, protective because it serves to prevent excessive tissue damage when a body part touches something that is potentially harmful.

Chart 10.1 summarizes the parts of a reflex arc.

Fig. 10.14 The knee-jerk reflex involves only two neurons—a sensory neuron and a motor neuron.

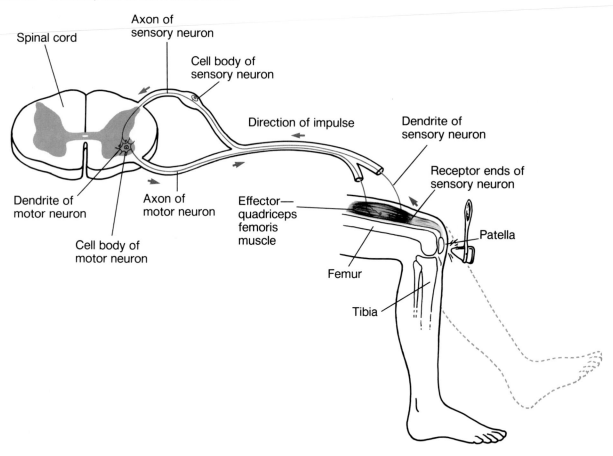

Chart 10.1 Parts of a reflex arc

Part	Description	Function
Receptor	The receptor end of a dendrite or a specialized receptor cell in a sensory organ	Sensitive to a specific type of internal or external change
Sensory neuron	Dendrite, cell body, and axon of a sensory neuron	Transmits nerve impulse from the receptor into the brain or spinal cord
Interneuron	Dendrite, cell body, and axon of a neuron within the brain or spinal cord	Conducts nerve impulse from the sensory neuron to a motor neuron
Motor neuron	Dendrite, cell body, and axon of a motor neuron	Transmits nerve impulse from the brain or spinal cord out to an effector
Effector	A muscle or gland outside the nervous system	Responds to stimulation by the motor neuron and produces the reflex or behavioral action

From Hole, John W., Jr., *Human Anatomy and Physiology 3d ed.* © 1978, 1981, 1984 Wm. C. Brown Publishers, Dubuque, Iowa. All Rights Reserved. Reprinted by permission.

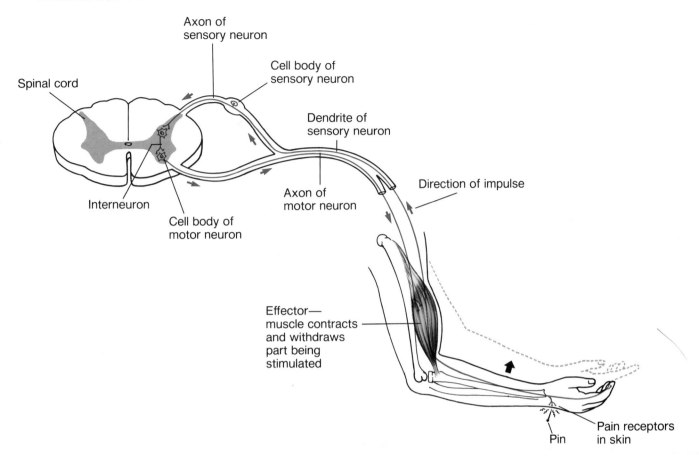

Since normal reflexes depend on normal neuron functions, reflexes are commonly used to obtain information concerning the condition of the nervous system. An anesthesiologist, for instance, may try to initiate a reflex in a patient who is being anesthetized in order to determine how the anesthetic drug is affecting nerve functions. Also, in the case of injury to some part of the nervous system, various reflexes may be tested to discover the location and extent of the damage.

1. What is a nerve pathway?
2. List the parts of a reflex arc.
3. Define reflex.
4. Review the actions that occur during a withdrawal reflex.

Coverings of the Central Nervous System

The organs of the central nervous system (CNS) are surrounded by bones and membranes. More specifically, the brain lies within the cranial cavity of the skull, and the spinal cord occupies the vertebral canal within the vertebral column. Beneath these bony coverings, the brain and spinal cord are protected by membranes called *meninges* that are located between the bone and the soft tissues of the nervous system.

The Meninges

The **meninges** have three layers—a dura mater, an arachnoid mater, and a pia mater (fig. 10.16).

The **dura mater** is the outermost layer. It is composed primarily of tough, white fibrous connective tissue and contains many blood vessels and nerves. It is attached to the inside of the cranial cavity and forms the internal periosteum of the surrounding skull bones.

Fig. 10.16 (a) The brain and spinal cord are enclosed by bone and by membranes called meninges. (b) The meninges include three layers—dura mater, arachnoid mater, and pia mater.

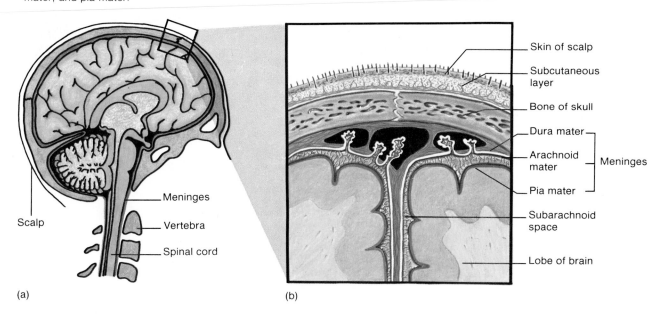

(a)

(b)

Fig. 10.17 The epidural space between the dural sheath and the bone of the vertebra is filled with tissues that provide a protective pad around the spinal cord.

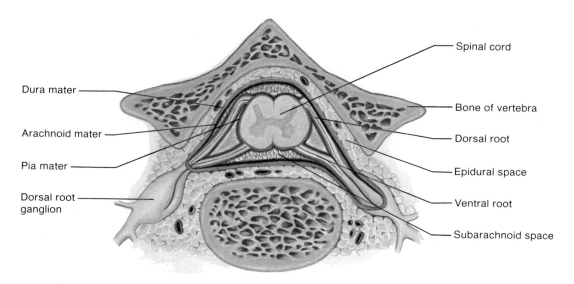

In some regions, the dura mater extends inward between lobes of the brain and forms partitions that support and protect these parts.

The dura mater continues into the vertebral canal as a strong, tubular sheath that surrounds the spinal cord. It terminates as a blind sac below the end of the cord. The membrane around the spinal cord is not attached directly to the bones of the vertebrae, but is separated by the *epidural space,* which lies between the dural sheath and the bony walls (fig. 10.17). This space contains loose connective tissue and fat tissue that provide a protective pad around the delicate tissues of the spinal cord.

The **arachnoid mater** is a thin, netlike membrane that lacks blood vessels and is located between the dura and pia maters. It spreads over the brain and spinal cord, but generally does not dip into the grooves and depressions on their surfaces.

Between the arachnoid and pia maters is a *subarachnoid space* that contains clear, watery **cerebrospinal fluid.**

The **pia mater** is very thin and contains many nerves, as well as blood vessels that aid in nourishing the underlying cells of the brain and spinal cord. This layer is attached to the surfaces of these organs and follows their irregular contours, passing over the high areas and dipping into the depressions.

---

1. Describe the meninges.
2. Name the layers of the meninges.
3. Explain where cerebrospinal fluid occurs.

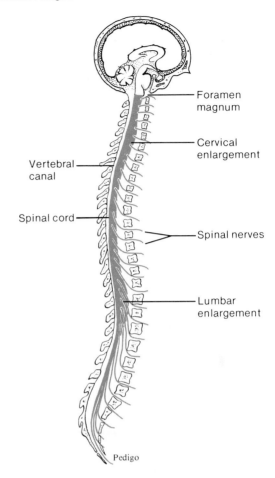

Fig. 10.18 The spinal cord begins at the level of the foramen magnum. At what level does it terminate?

Foramen magnum

Cervical enlargement

Vertebral canal

Spinal cord

Spinal nerves

Lumbar enlargement

Pedigo

The Spinal Cord

The **spinal cord** is a slender nerve column that passes downward from the brain into the vertebral canal. Although it is continuous with the brain, the spinal cord is said to begin where nerve tissue leaves the cranial cavity at the level of the foramen magnum. The cord usually tapers to a point and terminates near the intervertebral disk that separates the first and second lumbar vertebrae. (See fig. 10.18.)

Structure of the Spinal Cord

The spinal cord consists of thirty-one segments, each of which gives rise to a pair of **spinal nerves.** These nerves branch out to various body parts and connect them with the central nervous system.

In the neck region, there is a bulge in the spinal cord, called the *cervical enlargement,* that gives off nerves to the arms. A similar thickening in the lower back, the *lumbar enlargement,* gives off nerves to the legs. (See fig. 10.18.)

Two grooves, a deep *anterior median fissure* and a shallow *posterior median sulcus,* extend the length of the spinal cord, dividing it into right and left halves. A cross section of the cord (fig. 10.19) reveals that it consists of a core of gray matter surrounded by white matter. The pattern produced by the gray matter roughly resembles a butterfly with its wings outspread. The upper and lower wings of gray matter are called the *posterior horns,* and *anterior horns,* respectively. Between them on either side there is a protrusion of gray matter called the *lateral horn.*

Fig. 10.19 A cross section of the spinal cord.

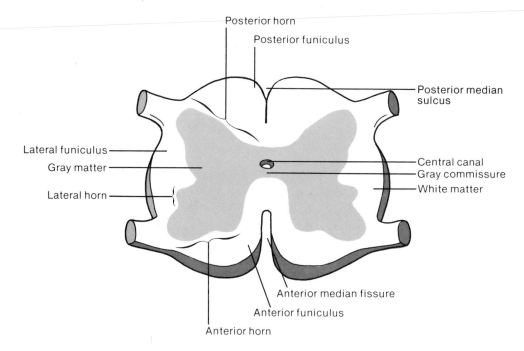

Posterior horn

Posterior funiculus

Posterior median sulcus

Lateral funiculus

Gray matter

Lateral horn

Central canal

Gray commissure

White matter

Anterior median fissure

Anterior funiculus

Anterior horn

A horizontal bar of gray matter in the middle of the spinal cord, the *gray commissure,* connects the wings of the gray matter on the right and left sides. This bar surrounds the **central canal,** which contains cerebrospinal fluid.

Neurons with large cell bodies located in the anterior horns give rise to motor fibers that pass out through spinal nerves and lead to various skeletal muscles. The majority of neurons in the gray matter of the spinal cord, however, are interneurons.

The white matter of the spinal cord is divided into three sections on each side by the gray matter. They are known as the *anterior, lateral,* and *posterior funiculi,* and each consists of longitudinal bundles of myelinated nerve fibers, which comprise major nerve pathways called **nerve tracts.** (See fig. 10.19.)

Functions of the Spinal Cord

The spinal cord has two major functions—to conduct nerve impulses and to serve as a center for spinal reflexes.

The tracts of the spinal cord provide a two-way system of communication between the brain and parts outside the nervous system. Those tracts that conduct impulses from body parts and carry sensory information to the brain are called **ascending tracts;** those that conduct motor impulses from the brain to muscles and glands are **descending tracts.**

The nerve fibers within these tracts are axons, and usually, all the axons within a given tract begin from cell bodies located in the same part of the nervous system and end together in some other part. The names that identify nerve tracts often reflect these common origins and terminations. For example, a *spinothalamic* tract begins in the spinal cord and carries sensory impulses to the thalamus of the brain; a *corticospinal* tract originates in the cortex of the brain and carries motor impulses downward through the spinal cord and spinal nerves to various effectors.

In addition to serving as a pathway for various nerve tracts, the spinal cord functions in many reflexes like the knee-jerk and withdrawal reflexes described previously. Such reflexes are called **spinal reflexes** because the reflex arcs pass through the cord.

Injuries to the spinal cord may be caused indirectly, as by a blow to the head or by a fall, or they may be due to forces applied directly to the cord. The consequences will depend on the amount of damage sustained by the cord. The spinal cord may, for example, be compressed or distorted by a minor injury, and its functions may be disturbed only temporarily. If nerve fibers are severed, however, some of the cord's functions are likely to be lost permanently.

Fig. 10.20 The major portions of the brain include the cerebrum, cerebellum, and brain stem.

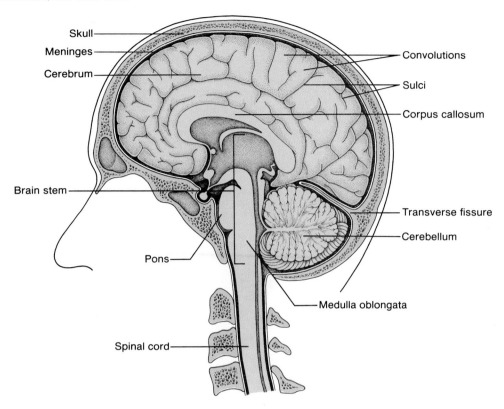

1. Describe the structure of the spinal cord.
2. Distinguish between an ascending and a descending tract.
3. Describe the general functions of the spinal cord.

The Brain

The **brain** is composed of about one hundred billion (10^{11}) neurons and innumerable nerve fibers by which the neurons communicate with one another and with neurons in other parts of the system.

As figure 10.20 shows, the brain can be divided into three major portions—a cerebrum, a cerebellum, and a brain stem. The **cerebrum,** which is the largest part, contains nerve centers associated with sensory and motor functions. It is also concerned with the higher mental functions, including memory and reasoning. The **cerebellum** includes centers associated with the coordination of voluntary muscular movements. The **brain stem** contains nerve pathways by which various parts of the nervous system are interconnected, as well as nerve pathways and centers involved in the regulation of various visceral activities.

Structure of the Cerebrum

The cerebrum consists of two large masses called **cerebral hemispheres,** which are mirror images of each other. These hemispheres are connected by a deep bridge of nerve fibers called the **corpus callosum** and are separated by a layer of dura mater.

The surface of the cerebrum is marked by numerous ridges or **convolutions,** which are separated by grooves. A shallow groove is called a **sulcus,** while a very deep one is a **fissure.** Although the arrangement of these elevations and depressions is complex, they form fairly distinct patterns in all brains. For example, a *longitudinal fissure* separates the right and left cerebral hemispheres, a *transverse fissure* separates the cerebrum from the cerebellum, and various sulci divide each hemisphere into lobes.

The lobes of the cerebral hemispheres (shown in fig. 10.21) are named after the skull bones that they underlie. They include the following:

1. **Frontal lobe.** The frontal lobe forms the anterior portion of each cerebral hemisphere. It is delimited posteriorly by a *central sulcus* that passes out from the longitudinal fissure at a right angle, and inferiorly by a *lateral sulcus* that passes out from the undersurface of the brain along its sides.

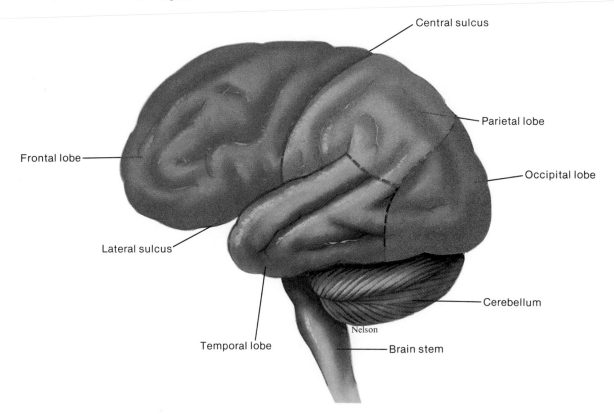

2. **Parietal lobe.** The parietal lobe is posterior to the frontal lobe and is separated from it by the central sulcus.

3. **Temporal lobe.** The temporal lobe lies below the frontal lobe and is separated from it by the lateral sulcus.

4. **Occipital lobe.** The occipital lobe forms the posterior portion of each cerebral hemisphere and is separated from the cerebellum by a shelflike extension of dura mater. There is no distinct boundary between the occipital lobe and the parietal and temporal lobes.

5. **The insula.** The insula is located deep within the lateral sulcus and is covered by parts of the frontal, parietal, and temporal lobes. It is separated from them by a circular sulcus. (See fig. 10.22.)

A thin layer of gray matter called the **cerebral cortex** constitutes the outermost portion of the cerebrum. This layer, which covers the convolutions and dips into the sulci and fissures, is estimated to contain nearly 75% of all the neuron cell bodies in the nervous system.

Just beneath the cerebral cortex are masses of white matter, making up the bulk of the cerebrum. These masses contain bundles of myelinated nerve fibers that connect the neuron cell bodies of the cortex with other parts of the nervous system. Some of these fibers pass from one cerebral hemisphere to the other by way of the corpus callosum, while others carry sensory or motor impulses from portions of the cortex to nerve centers in the brain or spinal cord.

Deep within each hemisphere of the cerebrum are several masses of gray matter called *basal ganglia.* Although the precise function of these parts is poorly understood, it is known that they contain a group of neuron cell bodies. This group serves as a relay station for motor impulses passing between the cerebral cortex, the brain stem, and the spinal cord. Impulses from basal ganglia normally inhibit motor functions and thus aid in the control of muscular activities. (See fig. 10.22.)

Fig. 10.22 A frontal section of the cerebral hemispheres.

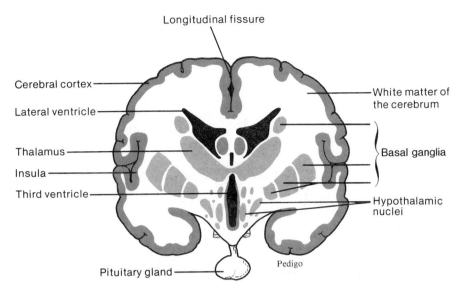

Functions of the Cerebrum

The cerebrum is concerned with the higher brain functions in that it contains centers for interpreting sensory impulses arriving from various sense organs as well as centers for initiating voluntary muscular movements. It stores the information of memory and utilizes this information in the processes associated with reasoning. It also functions in determining a person's intelligence and personality.

Functional Regions of the Cortex. The regions of the cerebral cortex that perform specific functions have been identified. Although there is considerable overlap in these areas, the cortex can be divided into sections known as *motor, sensory,* and *association areas.*

The primary **motor areas** of the cerebral cortex lie in the frontal lobes just anterior to the central sulcus. The nerve tissue in these regions contains numerous, large *pyramidal cells,* so named because of their pyramid-shaped cell bodies.

Impulses from pyramidal cells travel downward through the brain stem and into the spinal cord on the *corticospinal tracts* (pyramidal tracts). Most of the nerve fibers in these tracts cross over from one side of the brain to the other within the brain stem. As a result of this crossing over, the motor area of the right cerebral hemisphere generally controls skeletal muscles on the left side of the body and vice versa. (See fig. 10.23.)

Nerve tracts other than the corticospinal tracts that transmit signals from the cerebral cortex into the spinal cord are called *extrapyramidal tracts.* They function to coordinate and control motor functions involved with the maintenance of balance and posture.

In addition to the primary motor areas, certain other regions of the frontal lobe are involved with higher motor functions. For example, a region called *Broca's area* is located just anterior to the primary motor cortex and above the lateral sulcus. It coordinates the complex muscular actions of the mouth, tongue, and larynx, which make speech possible. Above Broca's area is a region called the *frontal eye field.* The motor cortex in this area controls the voluntary movements of the eyes and eyelids. Another region just in front of the primary motor area controls the muscular movements of the hands and fingers that make skills such as writing possible.

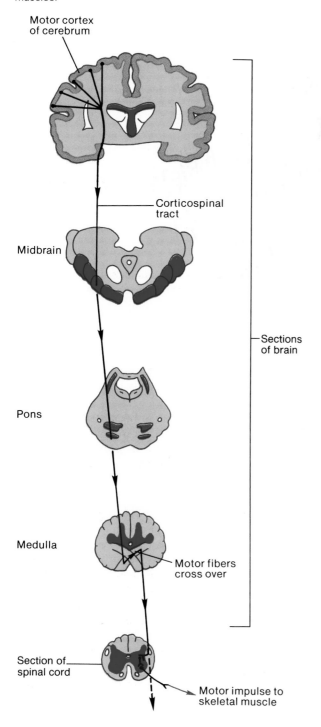

Fig. 10.23 Motor fibers of the corticospinal tract begin in the cerebral cortex, cross over in the medulla, and descend in the spinal cord. There they synapse with neurons that have fibers that lead to spinal nerves supplying skeletal muscles.

Motor cortex of cerebrum

Corticospinal tract

Midbrain

Sections of brain

Pons

Medulla

Motor fibers cross over

Section of spinal cord

Motor impulse to skeletal muscle

Sensory areas, which occur in several lobes of the cerebrum, function in interpreting impulses that arrive from various sensory receptors. These interpretations give rise to feelings or sensations. For example, sensations from all parts of the skin arise in the anterior portions of the parietal lobes along the central sulcus (fig. 10.24). The posterior parts of the occipital lobes are concerned with vision, while the temporal lobes contain the centers for hearing. The sensory areas for taste are located near the bases of the central sulci along the lateral sulci, and the sense of smell arises from centers deep within the cerebrum.

Like motor fibers, the sensory fibers cross over so that centers in the right cerebral hemisphere interpret impulses originating from the left side of the body and vice versa. (See fig. 10.25.)

Association areas function to analyze and interpret sensory experiences and involve memory, reasoning, verbalizing, judgment, and emotional feelings. These areas occupy the anterior portions of the frontal lobes and are widespread in the lateral portions of the parietal, temporal, and occipital lobes. (See fig. 10.26.)

The association areas of the frontal lobes are concerned with a number of higher intellectual processes, including those necessary for concentration, planning, complex problem-solving, and judging the possible consequences of behavior.

The association areas of the parietal lobes aid in understanding speech and choosing words needed to express thoughts and feelings.

The association areas of the temporal lobes and the regions at the posterior ends of the lateral fissures are concerned with the interpretation of complex sensory experiences, such as those needed to understand speech and read printed words. These regions also are involved with the memory of visual scenes, music, and other complex sensory patterns.

The association areas of the occipital lobes that are adjacent to the visual centers are important in analyzing visual patterns and combining visual images with other sensory experiences—necessary, for instance, when one recognizes another person or an object.

Of particular importance is the region where the parietal, temporal, and occipital association areas come together, near the posterior end of the lateral sulcus. This region is called the *general interpretative area,* and it plays the primary role in complex thought processing (fig. 10.26).

Fig. 10.24 Some motor and sensory areas of the cerebral cortex.

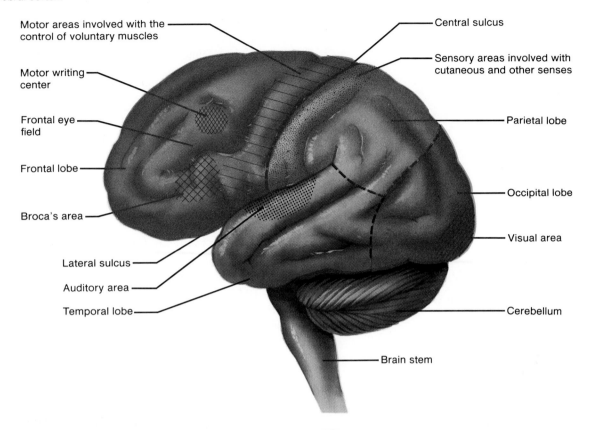

Motor areas involved with the control of voluntary muscles

Motor writing center

Frontal eye field

Frontal lobe

Broca's area

Lateral sulcus

Auditory area

Temporal lobe

Central sulcus

Sensory areas involved with cutaneous and other senses

Parietal lobe

Occipital lobe

Visual area

Cerebellum

Brain stem

The effects of injuries to the cerebral cortex depend on which areas are damaged and to what extent. When particular portions of the cortex are injured, the special functions of these portions are likely to be lost or at least depressed.

It is often possible to deduce the location and extent of a brain injury by determining what abilities the patient is missing. For example, if the motor areas of one frontal lobe have been damaged, the person is likely to be partially or completely paralyzed on the opposite side of the body.

A person with damage to the association areas of the frontal lobes may have difficulty in concentrating on complex mental tasks. Such an individual usually appears disorganized and is easily distracted. A person who suffers damage to association areas of the temporal lobes may have difficulty recognizing printed words or arranging words into meaningful thoughts.

1. List the major divisions of the brain.
2. Describe the location of the cerebral cortex.
3. What are the major functions of the cerebrum?

Hemisphere Dominance. Both cerebral hemispheres participate in basic functions, such as receiving and analyzing sensory impulses, controlling skeletal muscles on opposite sides of the body, and storing memory. However, in most persons, one acts as a **dominant hemisphere** for certain other functions.

In over 90% of the population, the left hemisphere is dominant for the language related activities of speech, writing, and reading. It is also dominant for complex intellectual functions requiring verbal, analytical, and computational skills. In other persons, the right hemisphere is dominant, and in some, the hemispheres are equally dominant.

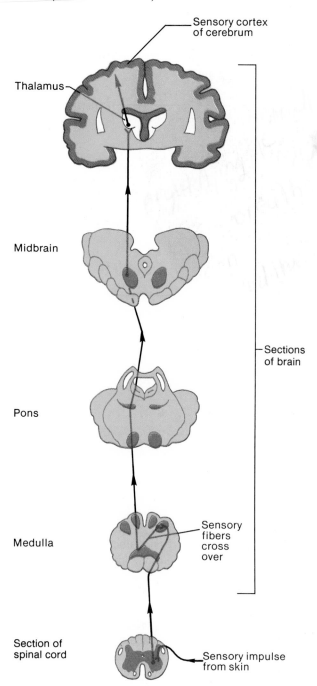

In addition to carrying on basic functions, the nondominant hemisphere seems to specialize in nonverbal functions such as those involving motor tasks that require orientation of the body in the surrounding space, understanding and interpreting musical patterns, and nonverbal visual experiences. It also is concerned with emotional and intuitive thought processes.

Nerve fibers of the *corpus callosum,* which connect the cerebral hemispheres, make it possible for the dominant side to control the motor cortex of the nondominant hemisphere. These fibers also allow sensory information reaching the nondominant hemisphere to be transferred to the dominant one, where it can be used in decision making.

The Ventricles and Cerebrospinal Fluid

Within the cerebral hemispheres and brain stem is a series of interconnected cavities called **ventricles.** (See fig. 10.27.) These spaces are continuous with the central canal of the spinal cord, and like it, they are filled with cerebrospinal fluid.

The largest of the ventricles are the *lateral ventricles* (first and second ventricles), which extend into the cerebral hemispheres and occupy portions of the frontal, temporal, and occipital lobes.

A narrow space that constitutes the *third ventricle* is located in the midline of the brain, beneath the corpus callosum. This ventricle communicates with the lateral ventricles through openings in its anterior end.

The *fourth ventricle* is located in the brain stem just in front of the cerebellum. It is connected to the third ventricle by a narrow canal, the *cerebral aqueduct* which passes lengthwise through the brain stem. This ventricle is continuous with the central canal of the spinal cord and has openings in its roof that lead into the subarachnoid space of the meninges.

Cerebrospinal Fluid. Cerebrospinal fluid is a clear liquid that is secreted by tiny cauliflowerlike masses of specialized capillaries from the pia mater called **choroid plexuses.** These structures project outward from the inner walls of the ventricles. Most of the cerebrospinal fluid seems to arise in the lateral ventricles. From there it circulates slowly into the third and fourth ventricles and into the central canal of the spinal cord. It also enters the subarachnoid space of the meninges by passing through the wall of the fourth ventricle near the cerebellum and completes its circuit by being reabsorbed into the blood.

Fig. 10.26 Some association areas of the cerebral cortex.

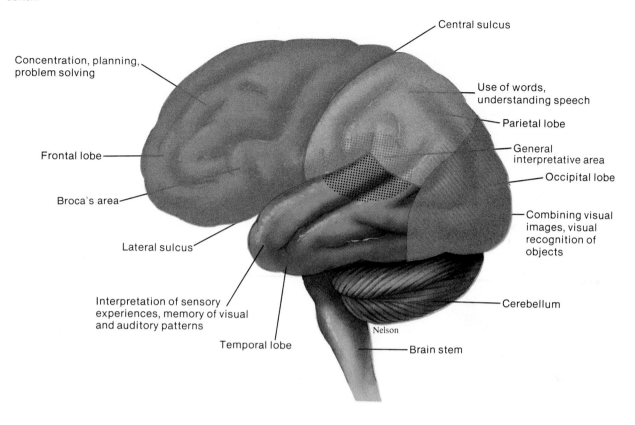

Central sulcus

Concentration, planning, problem solving

Use of words, understanding speech

Parietal lobe

General interpretative area

Occipital lobe

Frontal lobe

Broca's area

Combining visual images, visual recognition of objects

Lateral sulcus

Interpretation of sensory experiences, memory of visual and auditory patterns

Cerebellum

Nelson

Temporal lobe

Brain stem

Fig. 10.27 Anterior view of the ventricles within the cerebral hemispheres and brain stem.

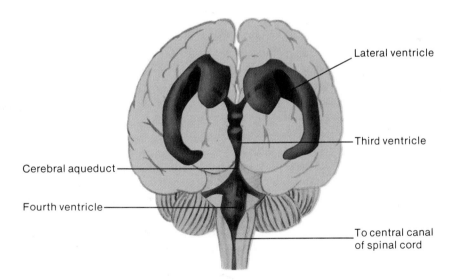

Lateral ventricle

Third ventricle

Cerebral aqueduct

Fourth ventricle

To central canal of spinal cord

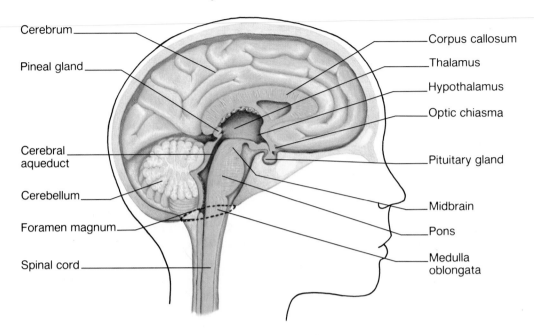

Cerebrum
Pineal gland
Cerebral aqueduct
Cerebellum
Foramen magnum
Spinal cord

Corpus callosum
Thalamus
Hypothalamus
Optic chiasma
Pituitary gland
Midbrain
Pons
Medulla oblongata

Since it occupies the subarachnoid space of the meninges, cerebrospinal fluid completely surrounds the brain and spinal cord. In effect, these organs float in the fluid, which supports and protects them by absorbing shocks and other forces that might otherwise jar and damage their delicate tissues. Cerebrospinal fluid also aids in maintaining a stable ionic concentration in the central nervous system and provides a pathway to the blood for waste substances.

Because cerebrospinal fluid is secreted and reabsorbed continuously, the fluid pressure in the ventricles normally remains relatively constant. Sometimes, however, an infection, a tumor, or a blood clot interferes with the circulation of this fluid, and the pressure within the ventricles increases. When this happens, there is a danger that the brain tissues may be injured by being forced against the skull.

1. What is meant by hemisphere dominance?
2. What are the major functions of the dominant hemisphere? The nondominant one?
3. Where are the ventricles of the brain located?
4. Describe the pattern of cerebrospinal fluid circulation.

The Brain Stem

The **brain stem** is a bundle of nerve tissue that connects the cerebrum to the spinal cord. It consists of numerous tracts of nerve fibers and several masses of gray matter called *nuclei*. The parts of the brain stem include the diencephalon, midbrain, pons, and medulla oblongata (fig. 10.28).

The Diencephalon. The **diencephalon** is located between the cerebral hemispheres and above the midbrain. It generally surrounds the third ventricle and is composed largely of gray matter organized into nuclei. Among these, a dense nucleus, called the *thalamus,* bulges into the third ventricle from each side. Another region of the diencephalon that includes many nuclei is called the *hypothalamus*. It lies below the thalamic nuclei and forms the lower walls and the floor of the third ventricle. (See fig. 10.22.)

The **thalamus** serves as a central relay station for sensory impulses ascending from other parts of the nervous system to the cerebral cortex. It receives all sensory impulses (except those associated with the sense of smell) and channels them to appropriate regions of the cortex for interpretation. In addition, all regions of the cerebral cortex can communicate with the thalamus by means of descending fibers, so these parts function closely together.

Although the cerebral cortex pinpoints the origin of sensory stimulation, the thalamus seems to produce a general awareness of certain sensations such as pain, touch, and temperature.

The **hypothalamus** is interconnected by nerve fibers to the cerebral cortex, thalamus, and other parts of the brain stem so that it can receive impulses from them and send impulses to them. The hypothalamus plays key roles in maintaining homeostasis by regulating a variety of visceral activities and by serving as a link between the nervous and endocrine systems.

Among the many important functions of the hypothalamus are the following:

1. Regulation of heart rate and arterial blood pressure.

2. Regulation of body temperature.

3. Regulation of water and electrolyte balance.

4. Control of hunger and regulation of body weight.

5. Control of movements and glandular secretions of the stomach and intestines.

6. Production of neurosecretory substances that stimulate the pituitary gland to release various hormones.

7. Regulation of sleep and wakefulness.

Structures in the general region of the diencephalon also play important roles in the control of emotional responses. For example, portions of the cerebral cortex in the medial parts of the frontal and temporal lobes are interconnected with the hypothalamus, thalamus, basal ganglia, and other deep nuclei. Together these structures comprise a complex called the limbic system.

The **limbic system** can modify the way a person acts because it functions to produce such emotional feelings as fear, anger, pleasure, and sorrow. More specifically, the limbic system seems to recognize upsets in a person's physical or psychological condition that might threaten survival. By causing pleasant or unpleasant feelings about experiences, the limbic system guides the person into behavior that is likely to increase the chance of survival.

The diencephalon also includes the **pineal gland** which is a small, cone-shaped structure attached to the upper portion of the thalamus. Its possible function as an endocrine gland is discussed in chapter 12.

The Midbrain. The **midbrain** (fig. 10.28) is a short section of the brain stem located between the diencephalon and the pons. It contains bundles of myelinated nerve fibers that join lower parts of the brain stem and spinal cord with higher parts of the brain. The midbrain includes several masses of gray matter that serve as reflex centers. For example, it contains the centers for certain visual reflexes, such as those responsible for moving the eyes to view something as the head is turned. It also contains the auditory reflex centers that operate when it is necessary to move the head so that sounds can be heard more distinctly.

Two prominent bundles of nerve fibers on the underside of the midbrain include the corticospinal tracts, and are the main motor pathways between the cerebrum and lower parts of the nervous system.

The Pons. The **pons** (fig. 10.28) appears as a rounded bulge on the underside of the brain stem, where it separates the midbrain from the medulla oblongata. The dorsal portion of the pons consists largely of longitudinal nerve fibers, which relay impulses to and from the medulla oblongata and the cerebrum. Its ventral portion contains large bundles of transverse nerve fibers, which transmit impulses from the cerebrum to centers within the cerebellum.

Several nuclei of the pons relay sensory impulses from peripheral nerves to higher brain centers. Other nuclei function with centers of the medulla oblongata in regulating the rate and depth of breathing.

Medulla Oblongata. The **medulla oblongata** (fig. 10.28) is an enlarged continuation of the spinal cord extending from the level of the foramen magnum to the pons. Its dorsal surface is flattened to form the floor of the fourth ventricle, and its ventral surface is marked by the corticospinal tracts, most of whose fibers cross over at this level.

Because of its location, all ascending and descending nerve fibers connecting the brain and spinal cord must pass through the medulla oblongata. As in the spinal cord, the white matter of the medulla surrounds a central mass of gray matter. Here, however, the gray matter is broken up into nuclei that are separated by nerve fibers. Some of these nuclei relay ascending impulses to the other side of the brain stem and then onto higher brain centers.

Other nuclei within the medulla oblongata function as control centers for vital visceral activities. These centers include the following:

1. **Cardiac center.** Impulses originating in the cardiac center are transmitted to the heart on peripheral nerves. Impulses on these nerves can cause the heart to beat more slowly or more rapidly.

2. **Vasomotor center.** Certain cells of the vasomotor center initiate impulses that travel to smooth muscles in the walls of certain blood vessels and stimulate them to contract. This action causes constriction of the blood vessels (vasoconstriction) and a rise in blood pressure. Other cells of the vasomotor center produce the opposite effect—a dilation of blood vessels and a consequent drop in the blood pressure.

3. **Respiratory center.** The respiratory center functions with centers in the pons to regulate the rate, rhythm, and depth of breathing.

Still other nuclei within the medulla oblongata function as centers for reflexes associated with coughing, sneezing, swallowing, and vomiting.

Reticular Formation. Scattered throughout the medulla oblongata, pons, and midbrain is a complex network of nerve fibers associated with tiny islands of gray matter. This network, the **reticular formation** (reticular activating system), extends from the upper portion of the spinal cord into the diencephalon. Its intricate system of nerve fibers interconnects centers of the hypothalamus, basal ganglia, cerebellum, and cerebrum with fibers in all the major ascending and descending tracts.

When sensory impulses reach the reticular formation, it responds by signaling the cerebral cortex, activating it into a state of wakefulness. Without this arousal, the cortex remains unaware of stimulation and cannot interpret sensory information or carry on thought processes. Thus, sleep results from decreased activity in the reticular formation. If the reticular formation ceases to function, as in certain injuries, the person remains unconscious and cannot be aroused, even with strong stimulation.

---

1. List the structures of the brain stem.
2. What are the major functions of the thalamus? The hypothalamus?
3. How may the limbic system influence a person's behavior?
4. What vital reflex centers are located in the brain stem?
5. What is the function of the reticular formation?

The Cerebellum

The **cerebellum** (fig. 10.28) is a large mass of tissue located below the occipital lobes of the cerebrum and posterior to the pons and medulla oblongata. It consists of two lateral hemispheres partially separated by a layer of dura mater and connected in the midline by a structure called the *vermis.*

Like the cerebrum, the cerebellum is composed primarily of white matter with a thin layer of gray matter, the **cerebellar cortex,** on its surface.

The cerebellum communicates with other parts of the central nervous system by means of three pairs of nerve tracts called *cerebellar peduncles.* Some of these tracts travel upward from the midbrain to the motor areas of the cerebral cortex, while others pass downward through the pons, medulla oblongata, and spinal cord.

The cerebellum functions mainly as a reflex center in the coordination of skeletal muscle movements. The sensory impulses involved in these reflexes come from receptors, called **proprioceptors,** that are found in muscles, tendons, and joints and from special sense organs such as the eyes and ears.

As a result of the sensory information it receives, the cerebellum becomes aware of the conditions of muscles, the attitudes of joints, and the positions of various body parts. When this information is analyzed, the cerebellum acts on it to stimulate or inhibit muscles at appropriate times, making complex muscular activities possible. The cerebellum also helps to maintain equilibrium.

Damage to the cerebellum is likely to result in tremors, inaccurate movements of voluntary muscles, loss of tone, reeling walk, and loss of equilibrium.

---

1. Where is the cerebellum located?
2. What are the major functions of the cerebellum?
3. What is the function of the proprioceptors?

The Peripheral Nervous System

The **peripheral nervous system** consists of the nerves that branch out from the central nervous system (CNS) and connect it to other body parts. It includes the *cranial nerves* that arise from the brain and the *spinal nerves* that arise from the spinal cord.

This portion of the nervous system also can be subdivided into somatic and autonomic nervous systems. Generally, the **somatic system** consists of the cranial and spinal nerve fibers that connect the CNS to the skin and skeletal muscles. The **autonomic system** includes those fibers that connect the CNS to visceral organs such as the heart, stomach, intestines, and various glands. Chart 10.2 outlines the subdivisions of the nervous system.

The Cranial Nerves

Twelve pairs of **cranial nerves** (fig. 10.29) arise from various locations on the underside of the brain. With the exception of the first pair, which springs from the cerebrum, these nerves originate from the brain stem. They pass from their sites of origin through various foramina of the skull and lead to parts of the head, neck, and trunk.

Fig. 10.29 Except for the first pair, the cranial nerves arise from the brain stem. They are identified either by numbers indicating their order, or by names describing their function or the general distribution of their fibers.

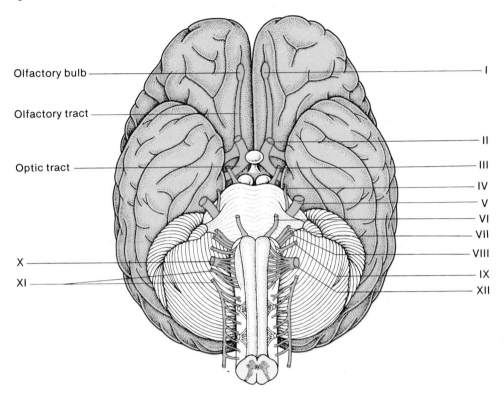

Olfactory bulb — I
Olfactory tract — II
Optic tract — III
— IV
— V
— VI
— VII
— VIII
X — IX
XI — XII

Although most cranial nerves are mixed, some of those associated with special senses, such as smell and vision, contain only sensory fibers. Others that are closely involved with the activities of muscles and glands are composed primarily of motor fibers and have limited sensory functions.

When sensory fibers are present in cranial nerves, the neuron cell bodies to which the fibers are attached are located outside the brain and are usually in groups called *ganglia*. On the other hand, motor neuron cell bodies are typically located within the gray matter of the brain.

Cranial nerves are designated either by a number or a name. The numbers indicate the order in which the nerves arise from the front to the back of the brain, and the names describe their primary functions or the general distribution of their fibers.

The names, types, and functions of the cranial nerves are listed in chart 10.3.

1. Define the peripheral nervous system.
2. Distinguish between somatic and autonomic nerve fibers.
3. Name the cranial nerves and list the major functions of each.

Chart 10.3   Functions of cranial nerves

Nerve	Type	Function
I Olfactory	Sensory	Sensory fibers transmit impulses associated with the sense of smell
II Optic	Sensory	Sensory fibers transmit impulses associated with the sense of vision
III Oculomotor	Primarily motor	Motor fibers transmit impulses to muscles that raise eyelids, move eyes, adjust amount of light entering eyes, and focus lenses
		Some sensory fibers transmit impulses associated with the condition of muscles
IV Trochlear	Primarily motor	Motor fibers transmit impulses to muscles that move the eyes
		Some sensory fibers transmit impulses associated with the condition of muscles
V Trigeminal	Mixed	
Ophthalmic divison		Sensory fibers transmit impulses from the surface of the eyes, the tear glands, scalp, forehead, and upper eyelids
Maxillary division		Sensory fibers transmit impulses from upper teeth, upper gum, upper lip, lining of the palate, and skin of the face
Mandibular division		Sensory fibers transmit impulses from the scalp, skin of the jaw, lower teeth, lower gum, and lower lip
		Motor fibers transmit impulses to muscles of mastication and muscles in floor of the mouth
VI Abducens	Primarily motor	Motor fibers transmit impulses to muscles that move the eyes
		Some sensory fibers transmit impulses associated with the condition of muscles
VII Facial	Mixed	Sensory fibers transmit impulses associated with taste receptors of the anterior tongue
		Motor fibers transmit impulses to muscles of facial expression, tear glands, and salivary glands
VIII Vestibulo-cochlear	Sensory	
Vestibular branch		Sensory fibers transmit impulses associated with sense of equilibrium
Cochlear branch		Sensory fibers transmit impulses associated with the sense of hearing
IX Glosso-pharyngeal	Mixed	Sensory fibers transmit impulses from the pharynx, tonsils, posterior tongue, and carotid arteries
		Motor fibers transmit impulses to muscles of pharynx used in swallowing and to salivary glands
X Vagus	Mixed	Motor fibers transmit impulses to muscles associated with speech and swallowing, the heart, and smooth muscles of visceral organs in the thorax and abdomen
		Sensory fibers transmit impulses from the pharynx, larynx, esophagus, and visceral organs of the thorax and abdomen
XI Accessory	Motor	
Cranial branch		Motor fibers transmit impulses to muscles of soft palate, pharynx, and larynx
Spinal branch		Motor fibers transmit impulses to muscles of neck and back
XII Hypoglossal	Motor	Motor fibers transmit impulses to muscles that move the tongue

From Hole, John W., Jr., *Human Anatomy and Physiology 3d ed.* © 1978, 1981, 1984 Wm. C. Brown Publishers, Dubuque, Iowa. All Rights Reserved. Reprinted by permission.

Fig. 10.30 Spinal nerves and plexuses.

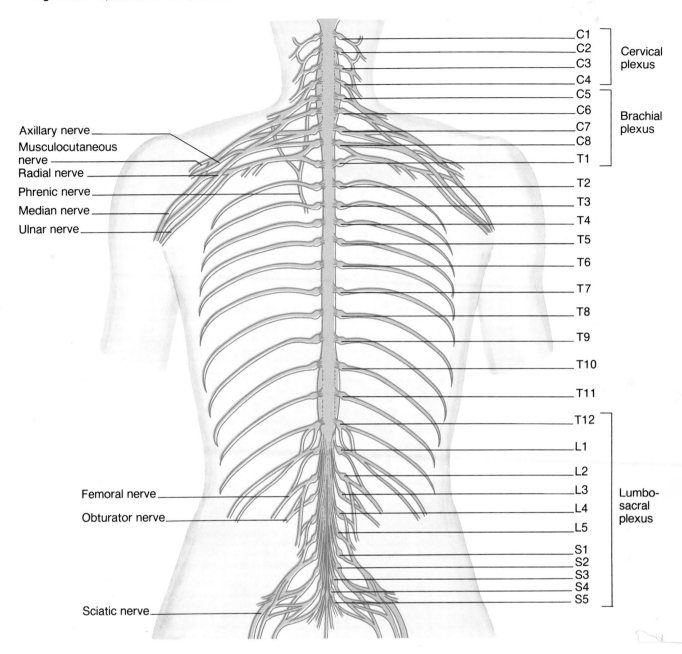

Axillary nerve
Musculocutaneous nerve
Radial nerve
Phrenic nerve
Median nerve
Ulnar nerve

Femoral nerve
Obturator nerve

Sciatic nerve

C1
C2
C3
C4
— Cervical plexus

C5
C6
C7
C8
T1
— Brachial plexus

T2
T3
T4
T5
T6
T7
T8
T9
T10
T11
T12
L1
L2
L3
L4
L5
S1
S2
S3
S4
S5
— Lumbo-sacral plexus

The Spinal Nerves

Thirty-one pairs of **spinal nerves** originate from the spinal cord. They are all mixed nerves, and they provide a two-way communication system between the spinal cord and parts in the arms, legs, neck, and trunk.

Although spinal nerves are not named individually, they are grouped according to the level from which they arise, and each nerve is numbered in sequence. (See fig. 10.30.) Thus, there are eight pairs of *cervical nerves* (numbered C1 to C8), twelve pairs of *thoracic nerves* (numbered T1 to T12), five pairs of *lumbar nerves* (numbered L1 to L5), five pairs of *sacral nerves* (numbered S1 to S5), and one pair of *coccygeal nerves.*

The adult spinal cord ends at the level between the first and second lumbar vertebrae, so the lumbar, sacral, and coccygeal nerves descend to their exits beyond the end of the cord. These descending nerves form a structure called *cauda equina,* which is shaped somewhat like a horse's tail.

Each spinal nerve emerges from the cord by two short branches, or *roots,* which lie within the vertebral column. The **dorsal root** (sensory root) can be identified by the presence of an enlargement called the *dorsal root ganglion.* This ganglion contains the cell bodies of the sensory neurons whose dendrites conduct impulses inward from peripheral body parts. The axons of these neurons extend through the dorsal root and into the spinal cord, where they form synapses with dendrites of other neurons. (See fig. 10.17.)

The **ventral root** (motor root) of each spinal nerve consists of axons from motor neurons whose cell bodies are located within the gray matter of the cord.

A ventral root and a dorsal root unite to form a spinal nerve, which passes outward from the vertebral canal through an *intervertebral foramen.* (See fig. 8.15.) Just beyond its foramen, each spinal nerve divides into several parts.

Except in the thoracic region, the main portions of the spinal nerves combine to form complex networks, called **plexuses,** instead of continuing directly to the peripheral body parts. In a plexus, the fibers of various spinal nerves are sorted and recombined, so that fibers associated with a particular peripheral part reach it in the same nerve, even though the fibers originate from different spinal nerves. (See fig. 10.30.)

Cervical Plexuses. The **cervical plexuses** lie deep in the neck on either side. They are formed by branches of the first four cervical nerves, and fibers from these plexuses supply the muscles and skin of the neck. In addition, fibers from the third, fourth, and fifth cervical nerves pass into the right and left **phrenic nerves,** which conduct motor impulses to the muscle fibers of the diaphragm.

Brachial Plexuses. Branches of the lower four cervical nerves and the first thoracic nerve give rise to the **brachial plexuses.** These nets of nerve fibers are located deep within the shoulders between the neck and the axillae (armpits). The major branches emerging from the brachial plexuses supply muscle and skin in the arm, forearm, and hand, and include the **musculocutaneous, ulnar, median, radial,** and **axillary nerves.**

Lumbosacral Plexuses. The **lumbosacral plexuses** are formed on either side by the last thoracic and the lumbar, sacral, and coccygeal nerves. These networks of nerve fibers extend from the lumbar region of the back into the pelvic cavity, giving rise to a number of motor and sensory fibers associated with muscles and skin of the lower abdominal wall, external genitalia, buttock, thighs, legs, and feet. The major branches of these plexuses include the **obturator, femoral,** and **sciatic nerves.**

Spinal nerves may be injured in a variety of ways including stabs, gunshot wounds, birth injuries, dislocations and fractures of vertebrae, and pressure from tumors in surrounding tissues. The nerves of the cervical plexuses are sometimes compressed by a sudden bending of the neck, called whiplash, that may occur during rear end automobile collisions. A victim of such an injury may suffer continuing headache and pain in the neck and skin, which are supplied by the cervical nerves.

1. How are spinal nerves grouped?
2. Describe the way a spinal nerve joins the spinal cord.
3. Name and locate the major nerve plexuses.

The Autonomic Nervous System

The *autonomic nervous system* is the portion of the nervous system that functions independently (autonomously) without conscious effort. This system controls the actions of smooth muscles, cardiac muscle, and most glands. It is concerned with regulating heart rate, blood pressure, breathing rate, body temperature, and other visceral functions that aid in the maintenance of homeostasis. Portions of the autonomic system also are responsive during times of emotional stress, and they serve to prepare the body to meet the demands of strenuous physical activity.

General Characteristics

Autonomic functions operate largely by reflexes in which sensory signals originate from receptors in visceral organs and skin. These signals are received by nerve centers within the hypothalamus, brain stem, or spinal cord. Motor impulses then travel out from these centers on peripheral nerve fibers within cranial and spinal nerves to various muscles or glands, which respond by contracting, secreting, or being inhibited.

The autonomic nervous system is composed of two sections called the **sympathetic** and **parasympathetic subdivisions,** which act together. For example, some visceral organs are supplied with nerve fibers from each of the subdivisions. Impulses on one set of fibers tend to activate such an organ, while impulses on the other set inhibit it. Thus, the actions of the organ are regulated by alternately being activated or inhibited.

Fig. 10.31 Visceral organs usually are innervated by fibers from both the parasympathetic and sympathetic divisions.

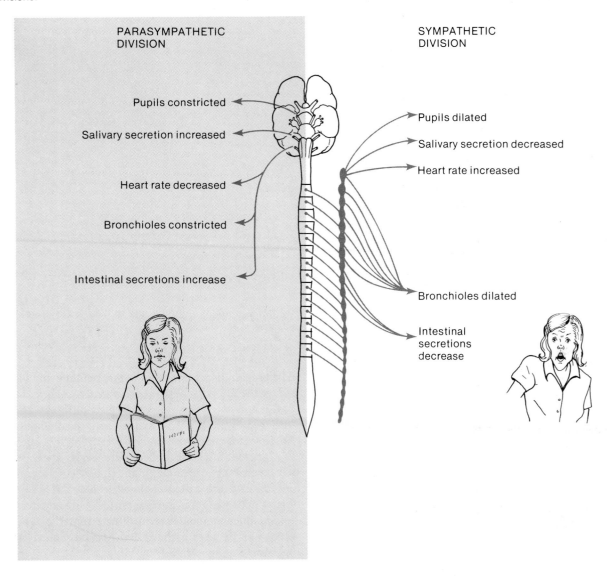

PARASYMPATHETIC DIVISION

Pupils constricted

Salivary secretion increased

Heart rate decreased

Bronchioles constricted

Intestinal secretions increase

SYMPATHETIC DIVISION

Pupils dilated

Salivary secretion decreased

Heart rate increased

Bronchioles dilated

Intestinal secretions decrease

Although the functions of the subdivisions are mixed—each activates some organs and inhibits others—they have general functional differences. The sympathetic division is concerned primarily with preparing the body for energy-expending, stressful, or emergency situations. The parasympathetic division is most active under ordinary, restful conditions. It also counterbalances the effects of the sympathetic division and restores the body to a resting state following a stressful experience. For example, during an emergency, the sympathetic division will cause the heart and breathing rates to increase, and following the emergency, the parasympathetic division will slow these activities. (See fig. 10.31.)

Autonomic Nerve Fibers

The nerve fibers of the autonomic nervous system are primarily motor fibers. Unlike the motor pathways of the somatic nervous system, which usually include a single neuron between the brain or spinal cord and an effector, those of the autonomic system involve two neurons, as shown in figure 10.32. The cell body of one neuron is located in the brain or spinal cord. Its axon, the **preganglionic fiber,** leaves the CNS and forms a synapse with one or more nerve fibers whose cell bodies are housed within an autonomic ganglion outside the brain or spinal cord. The axon of such a second neuron is called a **postganglionic fiber,** and it extends to a visceral effector.

Fig. 10.32 (a) Somatic neuron pathways usually have a single neuron between the central nervous system and an effector. (b) Autonomic neuron pathways involve two neurons between the central nervous system and an effector.

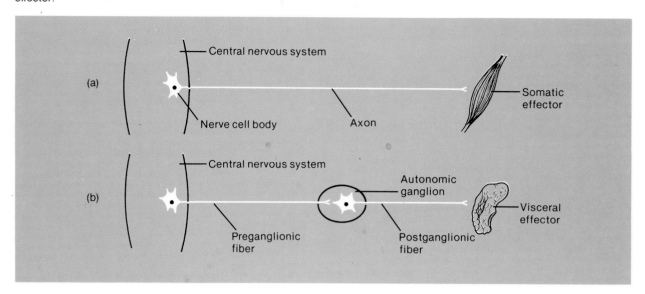

In the sympathetic division, the preganglionic fibers begin from neurons in the gray matter of the spinal cord. (See fig. 10.33.) They leave the cord with the ventral roots of spinal nerves in the first thoracic through the second lumbar segments. After traveling a short distance, these fibers leave the spinal nerves, and each enters a member of a chain of *sympathetic ganglia.* One of these chains extends longitudinally along each side of the vertebral column.

Within a sympathetic ganglion, a preganglionic fiber forms a synapse with a second neuron. The axon of this neuron, the postganglionic fiber, typically returns to a spinal nerve and extends with it to a visceral effector.

The preganglionic fibers of the parasympathetic division arise from the *brain stem* and the *sacral region* of the spinal cord. From there, they lead outward on cranial or sacral nerves to ganglia located near or within various visceral organs. The relatively short postganglionic fibers continue from the ganglia to specific muscles or glands within these visceral organs. (See fig. 10.34.)

---

1. *What parts of the nervous system are included in the autonomic nervous system?*
2. *How are the subdivisions of the autonomic system distinguished?*
3. *Describe a sympathetic nerve pathway. A parasympathetic nerve pathway.*

Autonomic Transmitter Substances

The preganglionic fibers of the sympathetic and parasympathetic divisions all secrete *acetylcholine;* their postganglionic fibers, however, use different transmitter substances. Most sympathetic postganglionic fibers secrete *norepinephrine* (noradrenalin), and for this reason they are called **adrenergic fibers.** The parasympathetic postganglionic fibers secrete *acetylcholine* and are called **cholinergic fibers.** These different postganglionic transmitter substances are responsible for the different effects that the sympathetic and parasympathetic divisions have on visceral organs. (See fig. 10.35.)

Although each division can activate some effectors and inhibit others, most visceral organs are controlled primarily by one division. In other words, the divisions usually are not actively antagonistic. For example, the diameter of most blood vessels, which lack parasympathetic innervation, is regulated by the sympathetic division. Smooth muscles in the walls of these vessels are continuously stimulated and thus maintained in a state of partial contraction (tone). The diameter of a vessel can be increased (dilated) by decreasing the degree of sympathetic stimulation, which allows the muscular wall to relax. Conversely, the vessel can be constricted by increasing the amount of sympathetic stimulation.

SYMPATHETIC DIVISION

PARASYMPATHETIC DIVISION

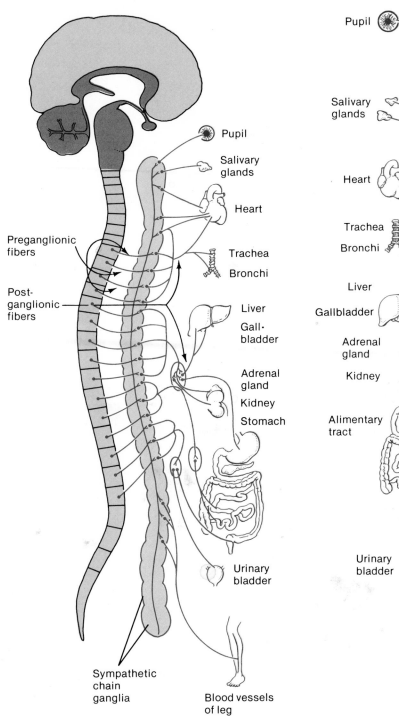

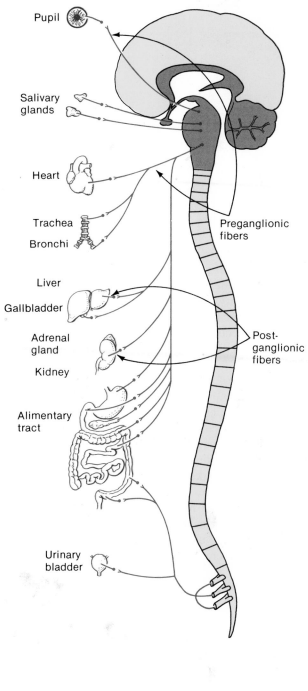

Fig. 10.35 (a) Sympathetic fibers are adrenergic and secrete norepinephrine at the ends of their postganglionic fibers; (b) parasympathetic fibers are cholinergic and secrete acetylcholine at the ends of their postganglionic fibers.

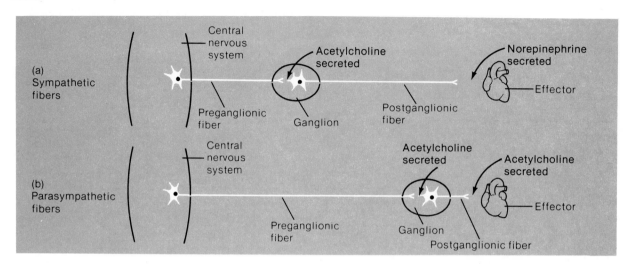

Similarly, the parasympathetic division is dominant in controlling movements in the digestive system. Parasympathetic impulses stimulate stomach and intestinal motility, and when these impulses decrease, movement is reduced.

The effects of adrenergic and cholinergic fibers on some visceral effectors are summarized in chart 10.4.

The acetylcholine released by cholinergic fibers is rapidly decomposed by the action of *cholinesterase.* Thus, acetylcholine usually produces an effect for only a fraction of a second. Most of the norepinephrine released from adrenergic fibers, however, is taken back into the nerve endings by an active transport mechanism. This may take a few seconds, and during that time some molecules may diffuse into nearby tissues and be decomposed by enzymes. On the other hand, some norepinephrine molecules may enter the blood and remain active until they diffuse into tissues containing the decomposing enzymes. For this reason, norepinephrine is likely to produce a more prolonged effect than acetylcholine.

1. *What neurotransmitters are used in the autonomic nervous system?*
2. *How do the divisions of the autonomic system function to regulate visceral activities?*

Chart 10.4 Some effects of transmitter substances upon visceral effectors or actions

Visceral Effector or Action	Response to Adrenergic Fibers (sympathetic)	Response to Cholinergic Fibers (parasympathetic)
Pupil of the eye	Dilation	Constriction
Heart rate	Increases	Decreases
Bronchioles of lungs	Dilation	Constriction
Muscles of intestinal wall	Slows peristaltic action	Speeds peristaltic action
Intestinal glands	Secretion decreases	Secretion increases
Blood distribution	More blood to skeletal muscles; less blood to digestive organs	More blood to digestive organs; less blood to skeletal muscles
Blood glucose concentration	Increases	Decreases
Salivary glands	Secretion decreases	Secretion increases
Tear glands	No action	Secretion
Muscles of gallbladder	Relaxation	Contraction
Muscles of urinary bladder	Relaxation	Contraction

Clinical Terms Related to the Nervous System

analgesia (an″al-je′ze-ah)—loss or reduction in the ability to sense pain, but without loss of consciousness.

analgesic (an″al-je′sik)—a pain-relieving drug.

anesthesia (an″es-the′ze-ah)—a loss of feeling.

aphasia (ah-fa′ze-ah)—a disturbance or loss in the ability to use words or to understand them, usually due to damage to cerebral association areas.

apraxia (ah-prak′se-ah)—an impairment in a person's ability to make correct use of objects.

ataxia (ah-tak′se-ah)—a partial or complete inability to coordinate voluntary movements.

cerebral palsy (ser′ĕ-bral pawl′ze)—a condition characterized by partial paralysis and lack of muscular coordination.

coma (ko′mah)—an unconscious condition in which there is an absence of responses to stimulation.

cordotomy (kor-dot′o-me)—a surgical procedure in which a nerve tract within the spinal cord is severed, usually to relieve intractable pain.

craniotomy (kra″ne-ot′o-me)—a surgical procedure in which part of the skull is opened.

electroencephalogram (EEG) (e-lek″tro-en-sef′ah-lo-gram″)—a recording of the electrical activity of the brain.

encephalitis (en″sef-ah-li′tis)—an inflammation of the brain and meninges characterized by drowsiness and apathy.

epilepsy (ep′ĭ-lep″se)—a disorder of the central nervous system that is characterized by temporary disturbances in normal brain impulses; it may be accompanied by convulsive seizures and loss of consciousness.

hemiplegia (hem″ĭ-ple′je-ah)—paralysis on one side of the body and the limbs on that side.

Huntington's chorea (hunt′ing-tunz ko-re′ah)—a rare hereditary disorder of the brain characterized by involuntary convulsive movements and mental deterioration.

laminectomy (lam″ĭ-nek′to-me)—surgical removal of the posterior arch of a vertebra, usually to relieve the symptoms of a ruptured intervertebral disk.

monoplegia (mon″o-ple′je-ah)—paralysis of a single limb.

multiple sclerosis (mul′tĭ-pl skle-ro′sis)—a disease of the central nervous system characterized by loss of myelin and the appearance of scarlike patches throughout the brain and spinal cord or both.

neuralgia (nu-ral′je-ah)—a sharp, recurring pain associated with a nerve, usually caused by inflammation or injury.

neuritis (nu-ri′tis)—an inflammation of a nerve.

paraplegia (par″ah-ple′je-ah)—paralysis of both legs.

quadriplegia (kwod″rĭ-ple′je-ah)—paralysis of all four limbs.

vagotomy (va-got′o-me)—severing of a vagus nerve.

Chapter Summary

Introduction

Organs of the nervous system are divided into the central and peripheral nervous systems.

These parts provide sensory, integrative, and motor functions.

General Functions of the Nervous System

1. Sensory functions employ receptors that detect internal and external changes in body conditions.

2. Integrative functions bring sensory information together and make decisions that are acted upon using motor functions.

3. Motor functions make use of effectors that respond when they are stimulated by motor impulses.

Nerve Tissue

1. Nerve tissue includes neurons, which are the structural and functional units of the nervous system, and neuroglial cells.

2. Neuron structure
   a. A neuron includes a cell body, nerve fibers, and other organelles usually found in cells.
   b. Dendrites and the cell body provide receptive surfaces.
   c. A single axon arises from the cell body and may be enclosed in a myelin sheath and a neurolemma.

3. Neuroglial cells
   a. Neuroglial cells fill spaces, support neurons, hold nerve tissue together, produce myelin, and carry on phagocytosis.
   b. They include Schwann cells, astrocytes, oligodendrocytes, microglia, and ependyma.

Cell Membrane Potential

1. A cell membrane is usually polarized as a result of unequal distribution of ions.

2. Nerve cell at rest
   a. There is a high concentration of sodium ions on the outside of the membrane and a high concentration of potassium ions on the inside.
   b. There are large numbers of negatively charged ions on the inside of the cell.
   c. In a resting cell, the outside develops a positive charge with respect to the inside.

The Nerve Impulse

1. Action potential
   a. If the membrane is disturbed by a stimulus of threshold intensity, a nerve impulse is triggered.

b. An action potential involves changes in membrane permeability and the movement of sodium and potassium ions through the membrane.

c. When ion concentrations have changed as a result of action potentials, an active transport mechanism in the membrane reestablishes the resting potential.

d. A wave of action potentials is a nerve impulse.

2. Impulse conduction
   a. Unmyelinated fibers conduct impulses that travel over their entire surfaces.
   b. Myelinated fibers conduct impulses more rapidly.

3. All-or-none response
   a. A nerve impulse is conducted in an all-or-none manner whenever a stimulus of threshold intensity is applied to a fiber.
   b. All the impulses conducted on a fiber are of the same strength.

The Synapse

A synapse is a junction between two neurons.

1. Synaptic transmission
   a. Impulses usually travel from dendrite to cell body, then along the axon to a synapse.
   b. Axons have synaptic knobs at their ends that secrete neurotransmitters.
   c. The neurotransmitter is released when a nerve impulse reaches the end of an axon.
   d. When the neurotransmitter reaches the nerve fiber on the distal side of the cleft, a nerve impulse is triggered.
   e. Neurotransmitters usually are decomposed by enzymes or otherwise are removed.

2. Excitatory and inhibitory actions
   a. Neurotransmitters that trigger nerve impulses are excitatory; those that inhibit impulses are inhibitory.
   b. The effect of the synaptic knobs communicating with a neuron will depend upon which knobs are activated from moment to moment.

Processing of Impulses

The way impulses are processed reflects the organization of the neurons in the brain and spinal cord.

1. Neuronal pools
   a. Neurons are organized into pools within the central nervous system.
   b. Each pool receives impulses, processes them, and conducts impulses away.

2. Facilitation
   a. Each neuron in a pool may receive excitatory and inhibitory stimuli.
   b. A neuron is facilitated when it receives subthreshold stimuli and becomes more excitable.

3. Convergence
   a. Impulses from two or more incoming fibers may converge on a single neuron.
   b. Convergence makes it possible for impulses from different sources to create an additive effect on a neuron.

4. Divergence
   a. Impulses leaving a pool may diverge by passing onto several output fibers.
   b. Divergence allows impulses to be amplified.

Types of Neurons and Nerves

1. On the basis of structure, neurons can be classified as multipolar, bipolar, or unipolar.

2. On the basis of function, neurons can be classified as sensory neurons, interneurons, or motor neurons.

3. Types of nerves
   a. Nerves are cordlike bundles of nerve fibers.
   b. Nerves can be classified as sensory, motor, or mixed.

Nerve Pathways

A nerve pathway is a route followed by an impulse as it travels through the nervous system.

1. Reflex arcs
   a. A reflex arc includes a sensory neuron, a reflex center composed of interneurons, and a motor neuron.
   b. The reflex arc is the behavioral unit of the nervous system.

2. Reflex behavior
   a. Reflexes are automatic, unconscious responses to changes.
   b. They help in the maintenance of homeostasis.
   c. The knee-jerk reflex may employ only two neurons.
   d. Withdrawal reflexes are protective actions.

Coverings of the Central Nervous System

1. The brain and spinal cord are surrounded by bone and protective membranes called meninges.

2. The meninges
   a. The meninges consist of a dura mater, arachnoid mater, and pia mater.
   b. Cerebrospinal fluid occupies the space between the arachnoid and pia maters.

The Spinal Cord

The spinal cord is a nerve column that extends from the brain into the vertebral canal.

1. Structure of the spinal cord
   a. The spinal cord is composed of thirty-one segments, each of which gives rise to a pair of spinal nerves.
   b. It is characterized by a cervical enlargement and a lumbar enlargement.
   c. It has a central core of gray matter that is surrounded by white matter.
   d. The white matter is composed of bundles of myelinated nerve fibers.

2. Functions of the spinal cord
   a. The cord provides a two-way communication system between the brain and other body parts.
   b. Ascending tracts carry sensory impulses to the brain; descending tracts carry motor impulses to muscles and glands.

The Brain

The brain is divided into a cerebrum, cerebellum, and brain stem.

1. Structure of the cerebrum
   a. The cerebrum consists of two cerebral hemispheres connected by the corpus callosum.
   b. The cerebral cortex is a thin layer of gray matter near the surface.
   c. White matter consists of myelinated nerve fibers that interconnect neurons within the nervous system and with other body parts.

2. Functions of the cerebrum
   a. The cerebrum is concerned with higher brain functions.
   b. The cerebral cortex can be divided into sensory, motor, and association areas.
   c. In most persons, one cerebral hemisphere is dominant for certain intellectual functions.

3. The ventricles and cerebrospinal fluid
   a. Ventricles are interconnected cavities within the cerebral hemispheres and brain stem.
   b. These spaces are filled with cerebrospinal fluid.
   c. Cerebrospinal fluid is secreted by choroid plexuses in the walls of the ventricles.

4. The brain stem
   a. The brain stem consists of the diencephalon, midbrain, pons, and medulla oblongata.
   b. The diencephalon contains the thalamus, which serves as a central relay station for incoming sensory impulses, and the hypothalamus, which plays important roles in maintaining homeostasis.
   c. The limbic system functions to produce emotional feelings and to modify behavior.
   d. The midbrain contains reflex centers associated with eye and head movements.
   e. The pons transmits impulses between the cerebrum and other parts of the nervous system and contains centers that help to regulate the rate and depth of breathing.
   f. The medulla oblongata transmits all ascending and descending impulses and contains several vital and nonvital reflex centers.
   g. The reticular formation acts to filter incoming sensory impulses, arousing the cerebral cortex into wakefulness whenever significant impulses are received.

5. The cerebellum
   a. The cerebellum consists of two hemispheres.
   b. It functions primarily as a reflex center in the coordination of skeletal muscle movements and the maintenance of equilibrium.

The Peripheral Nervous System

The peripheral nervous system consists of cranial and spinal nerves that branch out from the brain and spinal cord to all body parts.
It can be divided into somatic and autonomic portions.

1. The cranial nerves
   a. Twelve pairs of cranial nerves connect the brain to parts in the head, neck, and trunk.
   b. The names of cranial nerves indicate their primary functions or the general distributions of their fibers.

2. The spinal nerves
   a. Thirty-one pairs of spinal nerves originate from the spinal cord.
   b. These mixed nerves provide a two-way communication system between the spinal cord and parts in the arms, legs, neck and trunk.
   c. Spinal nerves are grouped according to the levels from which they arise, and they are numbered in sequence.
   d. Each nerve emerges by a dorsal and a ventral root.
   e. Just beyond its foramen, each spinal nerve divides into several branches.
   f. Most spinal nerves combine to form plexuses in which nerve fibers are sorted and recombined so that those fibers associated with a particular part reach it together.

The Autonomic Nervous System

The autonomic nervous system consists of the portions of the nervous system that function without conscious effort.

It is concerned primarily with the regulation of visceral activities that aid in maintaining homeostasis.

1. General characteristics
   a. Autonomic functions operate as reflex actions controlled from centers in the hypothalamus, brain stem, and spinal cord.
   b. The autonomic nervous system consists of two divisions—sympathetic and parasympathetic.
   c. The sympathetic division functions in meeting stressful and emergency conditions.
   d. The parasympathetic division is most active under ordinary conditions.

2. Autonomic nerve fibers
   a. The autonomic fibers are largely motor.
   b. Sympathetic fibers leave the spinal cord and synapse in chain ganglia.
   c. Parasympathetic fibers begin in the brain stem and sacral region of the spinal cord and synapse in ganglia near visceral organs.

3. Autonomic transmitter substances
   a. Preganglionic sympathetic and parasympathetic fibers secrete acetylcholine.
   b. Postganglionic sympathetic fibers secrete norepinephrine; postganglionic parasympathetic fibers secrete acetylcholine.
   c. The different effects of the autonomic divisions are due to different transmitter substances released by the postganglionic fibers.
   d. Most visceral organs are controlled mainly by one division.

Application of Knowledge

1. What functional losses would you expect to observe in a patient who has suffered injury to the right occipital lobe of the cerebral cortex? The right temporal lobe?

2. Based on your knowledge of the cranial nerves, devise a set of tests to assess the normal functions of each of these nerves.

3. Multiple sclerosis is a disease in which nerve fibers in the central nervous system lose their myelin. Why would this loss be likely to affect the person's ability to control skeletal muscles?

Review Activities

1. Explain the relationship between the central nervous system and the peripheral nervous system.

2. List three general functions of the nervous system.

3. Distinguish between neurons and neuroglial cells.

4. Describe the generalized structure of a neuron and explain the functions of its parts.

5. Distinguish between myelinated and unmyelinated nerve fibers.

6. Discuss the functions of each type of neuroglial cell.

7. Explain how a membrane may become polarized.

8. Explain how nerve impulses are related to action potentials.

9. Define *synapse*.

10. Explain how a nerve impulse is transmitted from one neuron to another.

11. Distinguish between excitatory and inhibitory actions of neurotransmitters.

12. Describe a neuronal pool.

13. Distinguish between convergence and divergence.

14. Explain how neurons can be classified on the basis of their structure or function.

15. Distinguish between sensory, motor, and mixed nerves.

16. Define *reflex*.

17. Describe a reflex arc that consists of two neurons.

18. Name the layers of the meninges and explain their functions.

19. Describe the structure of the spinal cord.

20. Distinguish between ascending and descending tracts of the spinal cord.

21. Name the three major portions of the brain and describe the general functions of each.

22. Describe the general structure of the cerebrum.

23. Distinguish between the *cerebral cortex* and *basal ganglia*.

24. Describe the location of the motor, sensory, and association areas of the cerebral cortex and describe the general functions of each.

25. Define *hemisphere dominance*.

26. Explain the function of the corpus callosum.

27. Describe the location of the ventricles of the brain.

28. Explain how cerebrospinal fluid is produced and how it functions.

29. Name the parts of the brain stem and describe the general functions of each part.

30. Define the limbic system and explain its functions.

31. Name the parts of the midbrain and describe the general functions of each part.

32. Describe the pons and its functions.

33. Describe the medulla oblongata and its functions.

34. Describe the functions of the cerebellum.

35. Name, locate, and describe the major functions of each pair of cranial nerves.

36. Explain how the spinal nerves are grouped and numbered.

37. Describe the structure of a spinal nerve.

38. Define *plexus* and locate the major plexuses of the spinal nerves.

39. Distinguish between the sympathetic and the parasympathetic divisions of the autonomic nervous system.

40. Distinguish between a preganglionic and a postganglionic nerve fiber.

41. Explain why the effects of the sympathetic and parasympathetic divisions differ.

Somatic and Special Senses

Before parts of the nervous system can act to control body functions, they must detect what is happening inside and outside the body. This information is gathered by sensory receptors that are sensitive to changes occurring in their surroundings.

Although receptors vary greatly in their individual characteristics, they can be grouped into two major categories. The members of one group are widely distributed throughout the skin and deeper tissues and generally have simple forms. These receptors are associated with the *somatic senses* of touch, pressure, temperature, and pain. Members of the second group function as parts of complex, specialized sensory organs that are responsible for the *special senses* of smell, taste, hearing, equilibrium, and vision.

11

After you have studied this chapter, you should be able to

1. Name five kinds of receptors and explain the function of each kind.

2. Explain how a sensation is produced.

3. Describe the somatic senses.

4. Describe the receptors associated with the senses of touch, pressure, temperature, and pain.

5. Describe how the sense of pain is produced.

6. Explain the relationship between the senses of smell and taste.

7. Name the parts of the ear and explain the function of each part.

8. Distinguish between static and dynamic equilibrium.

9. Name the parts of the eye and explain the function of each part.

10. Explain how light is refracted by the eye.

11. Describe the visual nerve pathway.

12. Complete the review activities at the end of this chapter. Note that the items are worded in the form of specific learning objectives. You may want to refer to them before reading the chapter.

accommodation   (ah-kom″o-da′shun)

chemoreceptor   (ke″mo-re-sep′tor)

cochlea   (kok′le-ah)

dynamic equilibrium   (di-nam′ik e″kwĭ-lib′re-um)

labyrinth   (lab′i-rinth)

mechanoreceptor   (mek″ah-no-re-sep′tor)

olfactory   (ol-fak′to-re)

optic   (op′tik)

photoreceptor   (fo″to-re-sep′tor)

projection   (pro-jek′shun)

referred pain   (re-furd′ pān)

refraction   (re-frak′shun)

sensory adaptation   (sen′so-re ad″ap-ta′shun)

static equilibrium   (stat′ik e″kwĭ-lib′re-um)

choroid, skinlike: *choroid* coat—middle, vascular layer of the eye.

cochlea, snail: *cochlea*—coiled tube within the inner ear.

iris, rainbow: *iris*—colored, muscular part of the eye.

labyrinth, maze: *labyrinth*—complex system of interconnecting chambers and tubes of the inner ear.

lacri-, tears: *lacri*mal gland—tear gland.

macula, spot: *macula* lutea—yellowish spot on the retina.

olfact-, to smell: *olfact*ory—pertaining to the sense of smell.

scler-, hard: *scler*a—tough, outer protective layer of the eye.

tympan-, drum: *tympan*ic membrane—the eardrum.

vitre-, glass: *vitre*ous humor—clear, jellylike substance within the eye.

As changes occur within the body and its surroundings, *sensory receptors* are stimulated, and they trigger nerve impulses. These impulses travel on sensory pathways into the central nervous system to be processed and interpreted. As a result, the person often experiences a particular type of feeling or sensation.

Receptors and Sensations

Although there are many kinds of sensory receptors, they have some features in common. For example, each type of receptor is particularly sensitive (that is, has a low threshold) to a distinct kind of environmental change and is much less sensitive to other forms of stimulation.

Types of Receptors

On the basis of their sensitivities, it is possible to identify five general groups of sensory receptors. They are: those stimulated by changes in the chemical concentration of substances (chemoreceptors); those stimulated by tissue damage (pain receptors); those stimulated by changes in temperature (thermoreceptors); those stimulated by changes in pressure or movement in fluids (mechanoreceptors); and those stimulated by light energy (photoreceptors).

Sensations

A **sensation** is a feeling that occurs when sensory impulses are interpreted by the brain. Because all the nerve impulses that travel from sensory receptors to the central nervous system are alike, different kinds of sensations must be due to the way the brain interprets the impulses rather than to differences in the receptors. In other words, when a receptor is stimulated, the resulting sensation depends on what region of the brain receives the impulse. For example, impulses reaching one region are always interpreted as sounds, and those reaching another portion are always sensed as touch. Impulses reaching still other regions of the brain are always interpreted as pain.

At the same time a sensation is created, the brain causes the feeling to seem to come from the receptors being stimulated. This process if called **projection,** because the brain projects the sensation back to its apparent source. Projection allows the person to pinpoint the region of stimulation. Thus the eyes appear to see, the ears appear to hear, and so forth.

Sensory Adaptation

When subjected to continuous stimulation, many receptors undergo an adjustment called **sensory adaptation.** As the receptors adapt, impulses leave them at lower and lower rates, until finally they may fail completely to send signals. Once receptors have adapted, impulses can be triggered only if the strength of the stimulus is changed.

Sensory adaptation is experienced when a person enters a room where there is a strong odor. At first the scent seems intense, but it becomes less and less noticeable as the smell receptors adapt.

---

1. List five general types of sensory receptors.
2. Explain how a sensation occurs.
3. What is meant by sensory adaptation?

Somatic Senses

The somatic senses include the senses associated with the skin, as well as those associated with muscles, joints, and visceral organs.

Touch and Pressure Senses

The senses of touch and pressure employ several kinds of receptors (fig. 11.1). As a group, these receptors are sensitive to mechanical forces that cause tissues to be deformed or displaced. They include the following:

1. **Free ends of sensory nerve fibers.** These receptors occur commonly in epithelial tissues, where they end between epithelial cells. They are associated with the sensations of touch and pressure.

2. **Meissner's corpuscles.** These structures consist of small, oval masses of flattened connective tissue cells surrounded by connective tissue sheaths. Two or more nerve fibers branch into each corpuscle and end within it as tiny knobs.

Meissner's corpuscles are especially numerous in the hairless portions of the skin, such as the lips, fingertips, palms, soles, nipples, and external genital organs. They are sensitive to the motion of objects that barely contact the skin, and impulses from them involve the sensation of light touch.

3. **Pacinian corpuscles.** These sensory bodies are relatively large structures composed of connective tissue fibers and cells. They are commonly found in the deeper subcutaneous tissues and occur in tendons and the ligaments of joints.

Fig. 11.1 Touch and pressure receptors include (a) free ends of sensory nerve fibers, (b) Meissner's corpuscles, and (c) Pacinian corpuscles.

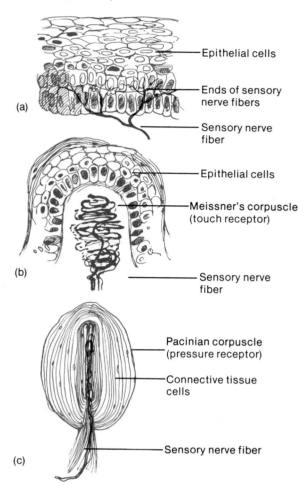

(a)
— Epithelial cells
— Ends of sensory nerve fibers
— Sensory nerve fiber

(b)
— Epithelial cells
— Meissner's corpuscle (touch receptor)
— Sensory nerve fiber

(c)
— Pacinian corpuscle (pressure receptor)
— Connective tissue cells
— Sensory nerve fiber

Pacinian corpuscles are stimulated by heavy pressure and are associated with the sensation of deep pressure.

Temperature Senses

The temperature senses employ two kinds of skin receptors. Although there is some question about the identity of these receptors, they seem to include two types of *free nerve endings* called *heat receptors* and *cold receptors*. The heat receptors are most sensitive to temperatures above 25°C (77°F) and become unresponsive at temperatures above 45°C (113°F). As 45°C is approached, pain receptors also are triggered, producing a burning sensation.

Cold receptors are most sensitive to temperatures between 10°C (50°F) and 20°C (68°F). If the temperature drops below 10°C, pain receptors are stimulated, and the person feels a freezing sensation.

Both heat and cold receptors demonstrate rapid adaptation, so that within about a minute following stimulation, the sensation of hot or cold begins to fade.

Sense of Pain

The sense of pain involves receptors that also consist of *free nerve fiber endings*. These receptors are widely distributed throughout the skin and internal tissues, except for the nerve tissue in the brain, which lacks pain receptors.

Pain receptors have a protective function in that they are stimulated whenever tissues are being damaged. The pain sensation is usually perceived as unpleasant, and it serves as a signal that something should be done to remove the source of the stimulation.

Pain receptors adapt poorly, if at all, and once such a receptor has been activated, even by a single stimulus, it may continue to send impulses into the central nervous system for some time.

The exact way that tissue damage excites pain receptors is unknown. It is thought that injuries promote the release of certain chemicals and that sufficient quantities of these chemicals may stimulate pain receptors. A deficiency of oxygen-rich blood (ischemia) in a tissue or the stimulation of certain mechanical-sensitive receptors also triggers pain sensations. Pain elicited during a muscle cramp, for example, seems to be related to an interruption of blood flow that occurs as the sustained contraction squeezes capillaries and reduces blood flow, as well as to the stimulation of mechanical-sensitive pain receptors.

Visceral Pain. As a rule, pain receptors are the only receptors in visceral organs whose stimulation produce sensations. Pain receptors in these organs seem to respond differently to stimulation than those associated with surface tissues. For example, localized damage to intestinal tissue, as may occur during surgical procedures, may not elicit any pain sensations even in a conscious person. When visceral tissues are subjected to more widespread stimulation however, as when intestinal tissues are stretched, or when the smooth muscles in the intestinal walls undergo spasms, a strong pain sensation may follow. Once again, the resulting pain seems to be related to the stimulation of mechanical-sensitive receptors and to a decreased blood flow accompanied by lower tissue oxygen concentration and an accumulation of pain-stimulating chemicals.

Another characteristic of visceral pain is that it may feel as if it is coming from some part of the body other than the part being stimulated—a phenomenon called **referred pain.** For example, pain originating from the heart may be referred to the left shoulder or the inside of the left arm.

Fig. 11.2 Surface regions to which some visceral pain may be referred.

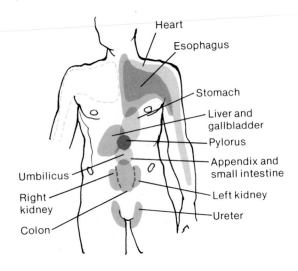

Figure 11.2 shows some other surface areas to which pain originating in various visceral organs may be referred.

Probably the most common pain sensations are those associated with headaches. Although brain tissue lacks pain receptors, nearly all the other tissues of the head, including blood vessel walls and meninges, are well supplied with them.

Most headaches seem to be related to stressful life situations that result in fatigue, emotional tension, anxiety, or frustration. These conditions are reflected in physiological changes. The changes that appear most likely to result in headache are dilations of cranial blood vessels, accompanied by edema in surrounding tissues, or spasms in skeletal muscles of the face, scalp, and neck.

Even though the pain of some headaches originates inside the cranium, the sensation is often referred to the outside.

1. Describe three types of touch and pressure receptors.
2. Describe the receptors involved in temperature senses.
3. What types of stimuli excite pain receptors?
4. What is referred pain?

Sense of Smell

The sense of smell, like the other special senses, is associated with complex sensory structures in the head.

Olfactory Receptors

The smell or olfactory receptors are similar to those for taste (described in a subsequent section) in that they are chemoreceptors, stimulated by chemicals dissolved in liquids. These two senses function closely together and aid in food selection, since food is often smelled at the same time that it is tasted.

Olfactory Organs

The **olfactory organs,** which contain the olfactory receptors, appear as yellowish-brown masses that cover the upper parts of the nasal cavity, the superior nasal conchae, and a portion of the nasal septum.

The **olfactory receptor cells** are neurons surrounded by columnar epithelial cells. The neurons have tiny knobs at the distal ends of their dendrites that are covered by hairlike cilia. The cilia project into the nasal cavity and are thought to be the sensitive portions of the receptors (fig. 11.3).

Chemicals that stimulate olfactory receptors enter the nasal cavity as gases; but they must dissolve at least partially in the watery fluids that surround the cilia before they can be detected.

Olfactory Nerve Pathway

Once the olfactory receptors have been stimulated, nerve impulses are triggered, and they travel along the axons of the receptor cells that are the fibers of the olfactory nerves. These fibers lead to neurons located in enlargements called **olfactory bulbs,** which lie on either side of the crista galli of the ethmoid bone. (See fig. 8.11.) From the olfactory bulbs, the impulses travel along the **olfactory tracts** to interpreting centers located at the bases of the cerebral frontal lobes.

Olfactory Stimulation

The way various substances stimulate the olfactory receptors is poorly understood. One hypothesis suggests the shapes of gaseous molecules may fit receptor sites on the cilia that have complementary shapes. A nerve impulse, according to this idea, is triggered when a molecule binds to its particular receptor site.

Since the olfactory organs are located high in the nasal cavity above the usual pathway of inhaled air, a person may have to sniff and force air over the

Fig. 11.3 The olfactory receptor cells, which have cilia at their distal ends, are supported by columnar epithelial cells.

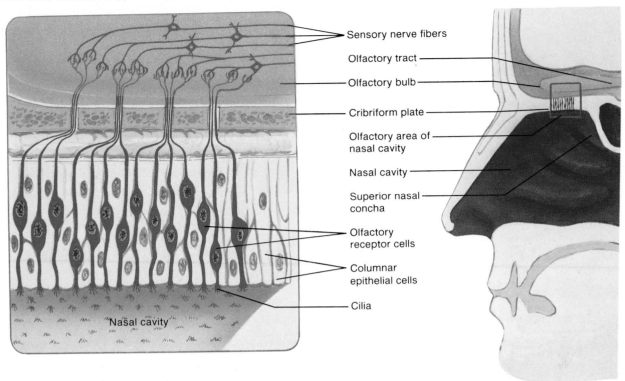

Sensory nerve fibers
Olfactory tract
Olfactory bulb
Cribriform plate
Olfactory area of nasal cavity
Nasal cavity
Superior nasal concha
Olfactory receptor cells
Columnar epithelial cells
Cilia
Nasal cavity

receptor areas to smell something that has a faint odor. Also, olfactory receptors undergo sensory adaptation rather rapidly, but even though they have adapted to one scent, their sensitivity to other odors remains unchanged.

Partial or complete loss of smell is called *anosmia*. This condition may be caused by a variety of factors including inflammation of the nasal cavity lining, as occurs during a head cold or from excessive tobacco smoking, or as a result of the use of certain drugs such as adrenalin or cocaine.

1. *Where are the olfactory receptors located?*
2. *Trace the pathway of an olfactory impulse from a receptor to the cerebrum.*

Sense of Taste

Taste buds are the special organs of taste. They occur primarily on the surface of the tongue and are associated with tiny elevations called *papillae.* They also are found in smaller numbers in the roof of the mouth and the walls of the pharynx (fig. 11.4).

Taste Receptors

Each taste bud consists of a group of modified epithelial cells, the **taste cells** (gustatory cells), which function as receptors, along with a number of epithelial supporting cells. The entire taste bud is somewhat spherical with an opening, the **taste pore,** on its free surface. Tiny projections, called **taste hairs,** protrude from the outer ends of the taste cells and jut out through the taste pore. It is believed that these taste hairs are the sensitive parts of the receptor cells.

Interwoven among the taste cells, and wrapped around them, is a network of nerve fibers. When a receptor cell is stimulated, an impulse is triggered on a nearby nerve fiber and is carried into the brain.

Before the taste of a particular chemical can be detected, it must be dissolved in the watery fluid surrounding the taste buds. This fluid is supplied by the salivary glands.

The mechanism by which various substances stimulate taste cells is not well understood. One possible explanation holds that various substances combine with specific receptor sites on the taste hair surfaces. Such a combination is thought to be responsible for the generation of sensory impulses on nearby nerve fibers.

Fig. 11.4 (a) Taste buds on the surface of the tongue are associated with nipplelike elevations called papillae. (b) A taste bud contains taste cells and has an opening, the taste pore, at its free surface.

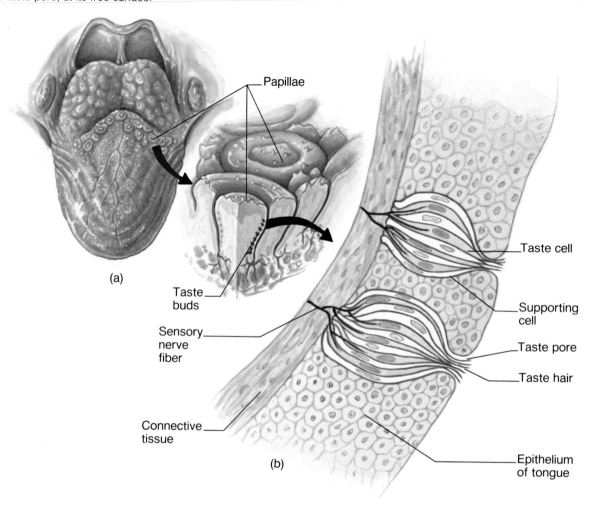

Papillae

(a)

Taste buds

Sensory nerve fiber

Connective tissue

(b)

Taste cell

Supporting cell

Taste pore

Taste hair

Epithelium of tongue

Although the taste cells in all taste buds appear very much alike microscopically, there are at least four types. Each type is most sensitive to a particular kind of chemical stimulus, and consequently, there are at least four primary taste (gustatory) sensations.

Taste Sensations

The four *primary taste sensations* are:
1. *Sweet,* as produced by table sugar.
2. *Sour,* as produced by vinegar.
3. *Salty,* as produced by table salt.
4. *Bitter,* as produced by caffeine or quinine.

In addition to these four, some investigators recognize two other taste sensations, which they call *alkaline* and *metallic.*

Each of the four major types of taste receptors is most highly concentrated in certain regions of the tongue's surface. Figure 11.5 illustrates this special distribution of receptors.

Each of the many flavors we experience daily is believed to result from one of the primary sensations or from some combination of two or more of them. The way we experience flavors also may involve the concentration of chemicals as well as the sensations of odor, texture (touch), and temperature. Furthermore, chemicals in some foods—chili peppers and ginger, for instance—may stimulate pain receptors that cause the tongue to burn.

Taste receptors, like olfactory receptors, undergo sensory adaptation relatively rapidly. The resulting loss of taste can be circumvented by moving bits of food over the surface of the tongue to stimulate different receptors at different times.

Taste Nerve Pathway

Sensory impulses from taste receptors located in various regions of the tongue travel on fibers of the facial, glossopharyngeal, and vagus nerves into the

Fig. 11.5 Patterns of taste receptor distribution are indicated by color in these diagrams. (a) Sweet receptors, (b) sour receptors, (c) salt receptors, (d) bitter receptors.

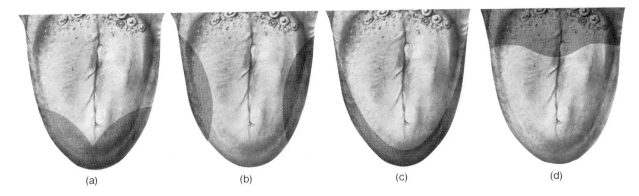

(a)   (b)   (c)   (d)

medulla oblongata. From there the impulses ascend to the thalamus and are directed to the gustatory cortex, which is located in the parietal lobe of the cerebrum along a deep portion of the lateral sulcus.

1. Why is saliva necessary for the sense of taste?
2. Name the four primary taste sensations.
3. Trace a sensory impulse from a taste receptor to the cerebral cortex.

Sense of Hearing

The organ of hearing, the **ear,** has external, middle, and inner parts. In addition to making hearing possible, the ear functions in the sense of equilibrium, which is discussed in a subsequent section of this chapter.

External Ear

The external ear consists of an outer, funnellike structure called the **auricle** and an S-shaped tube, called the **external auditory meatus,** that leads into the temporal bone for about 2.5 centimeters (1 inch). (See fig. 11.6.)

Sounds generally are created by vibrations of objects, which are transmitted through matter in the form of sound waves. For example, the sounds of some musical instruments are produced by vibrating strings or reeds, and the sounds of the voice are created by vibrating vocal folds in the larynx. The auricle of the ear helps to collect sound waves traveling through air and directs them into the auditory meatus.

Middle Ear

The middle ear consists of an air-filled space in the temporal bone called the **tympanic cavity,** an eardrum or **tympanic membrane,** and three small bones called **auditory ossicles.**

The eardrum is a semitransparent membrane covered by a thin layer of skin on its outer surface and by mucous membrane on the inside. It has an oval margin and is cone-shaped with the apex of the cone directed inward. Its cone shape is maintained by the attachment of one of the auditory ossicles.

Sound waves that enter the auditory meatus cause pressure changes on the eardrum, which moves back and forth in response, and thus reproduces the vibrations of the sound wave source.

The **auditory ossicles**—the *malleus* (hammer), *incus* (anvil), and *stapes* (stirrup)—are attached to the wall of the tympanic cavity by tiny ligaments and are covered by mucous membrane. These bones form a bridge connecting the eardrum to the inner ear and function to transmit vibrations between these parts. Specifically, the malleus is attached to the eardrum, and when the eardrum vibrates, the malleus vibrates in unison. The malleus causes the incus to vibrate, and it passes the movement onto the stapes. The stapes is held by ligaments to an opening in the wall of the tympanic cavity. This opening, called the **oval window,** leads into the inner ear. Vibration of the stapes at the oval window causes motion in a fluid within the inner ear. These vibrations are responsible for stimulating the receptors of hearing. (See fig. 11.6.)

In addition to transmitting vibrations, the auditory ossicles help to increase the force of the vibrations as they are passed from the eardrum to the oval window. For example, because the ossicles transmit

Fig. 11.6 The ear consists of external, middle, and inner portions.

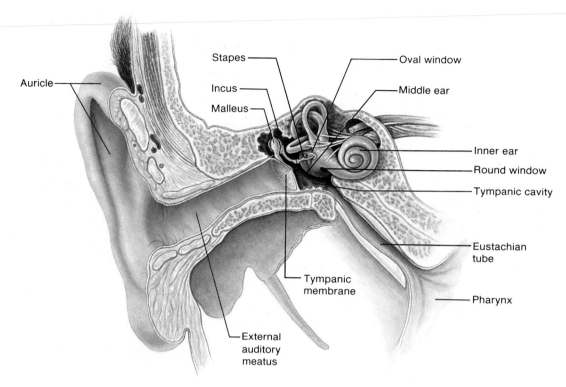

Labels: Auricle · Stapes · Incus · Malleus · Oval window · Middle ear · Inner ear · Round window · Tympanic cavity · Eustachian tube · Pharynx · Tympanic membrane · External auditory meatus

vibrations from the relatively large surface of the eardrum to a much smaller area at the oval window, the vibration force becomes concentrated as it travels from the external to the inner ear. As a result, the pressure (per mm^2) applied by the stapes at the oval window is many times greater than that exerted on the eardrum by sound waves.

Eustachian Tube. An **eustachian tube** (fig. 11.6) connects each middle ear to the throat. This tube allows air to pass between the tympanic cavity and the outside of the body by way of the throat and mouth. It is important in maintaining equal air pressure on both sides of the eardrum, which is necessary for normal hearing.

The function of the eustachian tube can be experienced during rapid altitude change. For example, as a person moves from a high altitude to a lower one, the air pressure on the outside of the membrane becomes greater and greater. The eardrum may be pushed inward, out of its normal position, and hearing may be impaired.

When the air pressure difference is great enough, some air may force its way up through the eustachian tube into the middle ear. At the same time, the pressure on both sides of the eardrum is equalized, and the drum moves back into its regular position. The person usually hears a popping sound at this moment, and normal hearing is restored.

The mucous membranes that line the eustachian tubes are continuous with the linings of the middle ears. Consequently, the tubes provide a route by which mucous membrane infections of the throat may pass to the ear and cause an infection of the middle ear. For this reason, it is poor practice to pinch one nostril when blowing the nose, because pressure in the nasal cavity may force material from the throat up the eustachian tube and into the middle ear.

Inner Ear

The inner ear (fig. 11.6) consists of a complex system of intercommunicating chambers and tubes called a **labyrinth;** in fact, there are two such structures in each ear—the osseous and membranous labyrinths.

The *osseous labyrinth* is a bony canal in the temporal bone; the *membranous labyrinth* lies within the osseous one and has a similar shape (fig. 11.7). Between the osseous and membranous labyrinths is a fluid called *perilymph* that is secreted by cells in the wall of the bony canal. The membranous labyrinth contains another fluid, called *endolymph.*

The parts of the labyrinths include a **cochlea** that functions in hearing and three **semicircular canals** that function in providing a sense of equilibrium.

The **cochlea** contains a bony core and a thin bony shelf that winds around the core like the threads of a

Fig. 11.7 The osseous labyrinth of the inner ear is separated from the membranous labyrinth by perilymph.

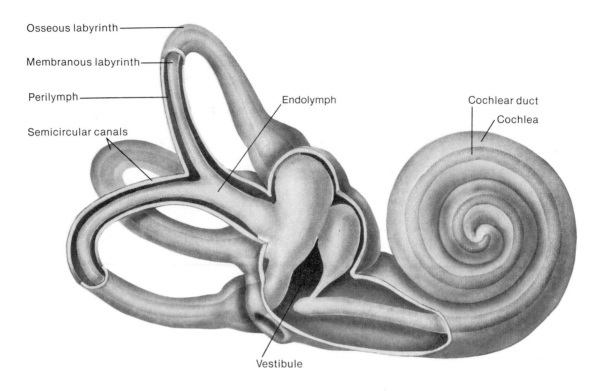

Osseous labyrinth

Membranous labyrinth

Perilymph

Semicircular canals

Endolymph

Cochlear duct

Cochlea

Vestibule

Fig. 11.8 (a) The cochlea consists of a coiled, bony canal with a membranous tube inside. (b) If the cochlea is unwound, the membranous tube is seen ending as a closed sac at the apex of the bony canal.

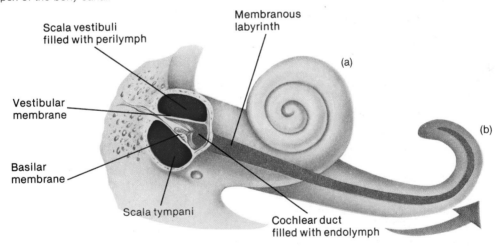

Scala vestibuli filled with perilymph

Membranous labyrinth

(a)

(b)

Vestibular membrane

Basilar membrane

Scala tympani

Cochlear duct filled with endolymph

screw. The shelf divides the bony labyrinth of the cochlea into upper and lower compartments. The upper compartment, called the *scala vestibuli,* leads from the oval window to the apex of the spiral. The lower compartment, the *scala tympani,* extends from the apex of the cochlea to a membrane-covered opening in the wall of the inner ear called the **round window.** (See figs. 11.6 and 11.7.)

The membranous labyrinth of the cochlea is represented by the *cochlear duct.* It lies between the two bony compartments and ends as a closed sac at the apex of the cochlea. The cochlear duct is separated from the scala vestibuli by a *vestibular membrane* (Reissner's membrane) and from the scala tympani by a *basilar membrane.* (See fig. 11.8.)

Fig. 11.9 (a) Cross section of the cochlea; (b) organ of Corti.

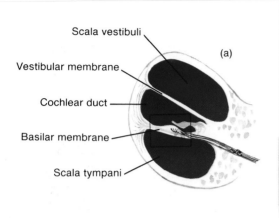

Scala vestibuli

Vestibular membrane

Cochlear duct

Basilar membrane

Scala tympani

(a)

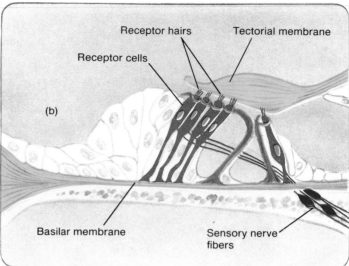

Receptor hairs

Tectorial membrane

Receptor cells

(b)

Basilar membrane

Sensory nerve fibers

Fig. 11.10 A scanning electron micrograph of hair cells in the organ of Corti.

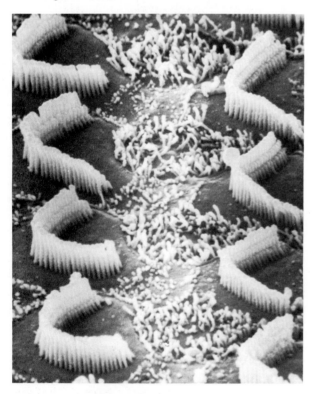

The basilar membrane contains many thousands of stiff, elastic fibers whose lengths vary, becoming progressively longer from the base of the cochlea to its apex. It is believed that vibrations entering the perilymph at the oval window travel along the scala vestibuli and pass through the vestibular membrane to enter the endolymph of the cochlear duct, where they cause movements in the basilar membrane.

After passing through the basilar membrane, the sound vibrations enter the perilymph of the scala tympani, and their forces are dissipated to the air in the tympanic cavity by movements of the membrane covering the round window.

An **organ of Corti,** which contains the hearing receptors, is located on the upper surface of the basilar membrane. Its receptor cells are arranged in rows, and they possess numerous hairlike processes that project into the endolymph of the cochlear duct. A *tectorial membrane,* attached to the bony shelf of the cochlea, passes over the receptor cells and makes contact with the tips of their hairs (fig. 11.9).

Because of the way these parts are arranged, different frequencies of vibrations cause greater movements in particular fibers of the basilar membrane. These fibers, in turn, cause specific receptor cells to move, and as the hairs of the receptors sheer back and forth against the tectorial membrane, sensory impulses are initiated in nerve fibers that are clustered at the bases of the cells. (See fig. 11.10.)

Although the human ear is able to detect sound waves with frequencies varying from about 20 to 20,000 vibrations per second, the range of greatest sensitivity is between 2,000 and 3,000 vibrations per second.

Auditory Nerve Pathway

The nerve fibers associated with the hearing receptors enter the auditory nerve pathways that pass into the auditory cortices of the temporal lobes of the cerebrum. On the way, some of these fibers cross over, so that impulses arising from each ear are interpreted on both sides of the brain. Consequently, damage to a temporal lobe on one side of the brain is not necessarily accompanied by complete hearing loss in the ear on that side.

Partial or complete hearing loss can be caused by a variety of factors, including interference with the transmission of vibrations to the inner ear (conductive deafness) or damage to the cochlea, auditory nerve, or auditory nerve pathways (sensorineural deafness).

Conductive deafness may be due to plugging of the external auditory meatus or to changes in the eardrum or the auditory ossicles. The eardrum, for example, may harden as a result of disease and thus be less responsive to sound waves, or it may be torn or perforated by disease or injury.

Sensorineural deafness can be caused by exposure to excessively loud sounds, by tumors in the central nervous system, by brain damage as a result of vascular accidents, or by the use of certain drugs. Some of the antibiotic drugs in the streptomycin group, for example, are known to be able to cause damage to inner ear parts.

---

1. *How are sound waves transmitted through the external, middle, and inner ears?*
2. *Distinguish between the osseous and the membranous labyrinths.*
3. *Describe the organ of Corti.*

Sense of Equilibrium

The sense of equilibrium actually involves two senses—a sense of *static equilibrium* and a sense of *dynamic equilibrium*—that result from the actions of different sensory organs.

Static Equilibrium

The organs of **static equilibrium** function in maintaining the stability of the head and body when these parts are motionless. They are located within the **vestibule,** the bony chamber between the semicircular canals and the cochlea. The membranous labyrinth inside the vestibule consists of two expanded chambers—an **utricle** and a **saccule.** (See fig. 11.11.)

On the anterior wall of the utricle is a tiny structure, called a **macula,** that contains numerous hair cells and supporting cells. When the head is upright, the hairs of the hair cells project upward into a mass of gelatinous material that has grains of calcium carbonate (otoliths) embedded in it. These particles increase the weight of the gelatinous structure, and the hair cells serve as the sensory receptors.

The usual stimulus to the hair cells occurs when the head is bent forward, backward, or to one side. Such movements may cause the gelatinous masses of the maculae to be tilted, and as they sag in response to gravity, the hairs projecting into them are bent. This action stimulates the hair cells, and they signal the nerve fibers associated with them. The resulting nerve impulses travel into the central nervous system, informing the brain as to the position of the head. The brain may act on this information by sending motor impulses to skeletal muscles, and they may contract or relax appropriately so that balance is maintained (fig. 11.12).

Dynamic Equilibrium

The three **semicircular canals** aid in balancing the head and body when they are moved suddenly. These canals lie at right angles to each other, and each occupies a different plane in space.

Suspended in the perilymph of the bony portion of each semicircular canal is a membranous canal that ends in a swelling, called an **ampulla.** (See fig. 11.11.)

The sensory organs of the semicircular canals are located within the ampullae. Each of these organs, called a **crista ampullaris,** contains a number of sensory hair cells and supporting cells. As in a macula, the hair cells have hairs that extend upward into a dome-shaped gelatinous mass, called the *cupula.* (See fig. 11.13.)

The hair cells of the crista are ordinarily stimulated by rapid turns of the head or body. At such times, the semicircular canals move with the head or body, but the fluid inside the membranous canals remains stationary. This causes the cupula in one or more of the canals to be bent in a direction opposite to that of the head or body movement, and the hairs embedded in it also are bent. This stimulates the hair cells to signal their associated nerve fibers, and as a result, impulses travel to the brain.

Parts of the cerebellum are particularly important in interpreting impulses from the semicircular canals. Analysis of such information allows the nervous system to predict the consequences of rapid body movements. The brain then can send motor impulses to stimulate appropriate skeletal muscles so that loss of balance can be prevented.

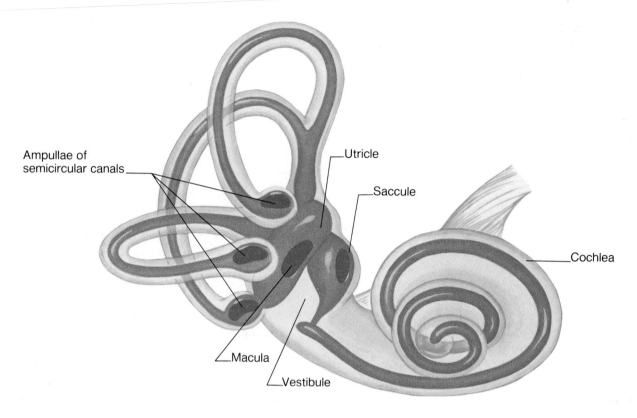

Fig. 11.11 The saccule and utricle, which are expanded portions of the membranous labyrinth, are located within the bony chamber of the vestibule.

Ampullae of semicircular canals

Utricle

Saccule

Cochlea

Macula

Vestibule

Fig. 11.12 The macula is responsive to changes in the position of the head. (a) Macula with the head in an upright position; (b) with the head bent forward.

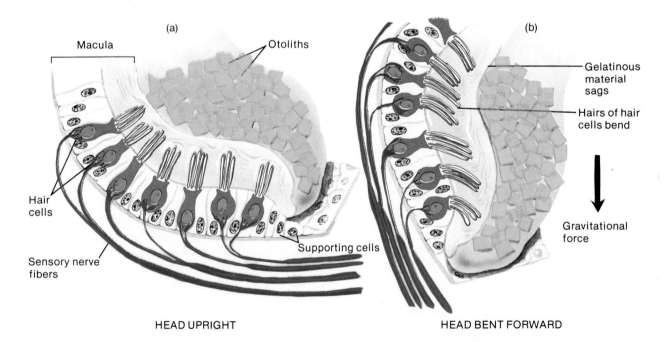

(a)

Macula

Otoliths

(b)

Gelatinous material sags

Hairs of hair cells bend

Hair cells

Sensory nerve fibers

Supporting cells

Gravitational force

HEAD UPRIGHT

HEAD BENT FORWARD

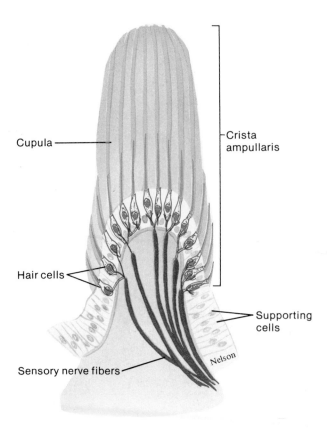

Cupula

Crista ampullaris

Hair cells

Supporting cells

Sensory nerve fibers

Nelson

Other sensory structures also aid in the maintenance of equilibrium. For example, certain mechanoreceptors (proprioceptors), particularly those associated with the joints of the neck, supply the brain with information concerning the position of certain body parts. In addition, the eyes can detect changes in posture that result from body movements. Visual information is so important that a person who has suffered damage to the organs of equilibrium may be able to maintain normal balance as long as the eyes remain open and body movements are performed slowly.

---

1. Distinguish between the senses of static and dynamic equilibrium.
2. What structures function to provide the sense of static equilibrium? Dynamic equilibrium?
3. How does sensory information from other receptors help maintain equilibrium?

Sense of Sight

Although the eye contains the visual receptors and is the primary organ of sight, its functions are assisted by a number of *accessory organs*. These include the eyelids and the lacrimal apparatus, which aid in protecting the eye, and a set of extrinsic muscles, which move it.

Visual Accessory Organs

The eye, lacrimal gland, and the extrinsic muscles of the eye are housed within the pear-shaped orbital cavity of the skull. (See fig. 8.7.) This orbit, which is lined with the periosteums of various bones, also contains fat, blood vessels, nerves, and a variety of connective tissues.

Each **eyelid** is composed of four layers—skin, muscle, connective tissue, and conjunctiva. The skin of the eyelid, which is the thinnest skin of the body, covers the lid's outer surface and fuses with its inner lining, the *conjunctiva,* near the margin of the lid. (See fig. 11.14.)

The eyelids are moved by the *orbicularis oculi* muscle (fig. 9.13), which acts as a sphincter and closes the lids when it contracts, and by the *levator palpebrae superioris* muscle, which raises the upper lid and thus opens the eye. (See fig. 11.14.)

The **conjunctiva** is a mucous membrane that lines the inner surfaces of the eyelids and folds back to cover the anterior surface of the eyeball, except for its central portion (cornea).

The *lacrimal apparatus* consists of the **lacrimal gland,** which secretes tears, and a series of *ducts,* which carry the tears into the nasal cavity. The gland is located in the orbit and secretes tears continuously. The tears pass out through tiny tubules and flow downward and medially across the eye.

Tears are collected by two small ducts (superior and inferior canaliculi) and flow first into the *lacrimal sac,* which lies in a deep groove of the lacrimal bone, and then into the *nasolacrimal duct* that empties into the nasal cavity. (See fig. 11.15.)

The secretion of the lacrimal gland keeps the surface of the eye and the lining of the lids moist and lubricated. Also, tears contain an enzyme (*lysozyme*) that functions as an antibacterial agent, reducing the chance of eye infections.

Fig. 11.14 Sagittal section of the eyelids and the anterior portion of the eye.

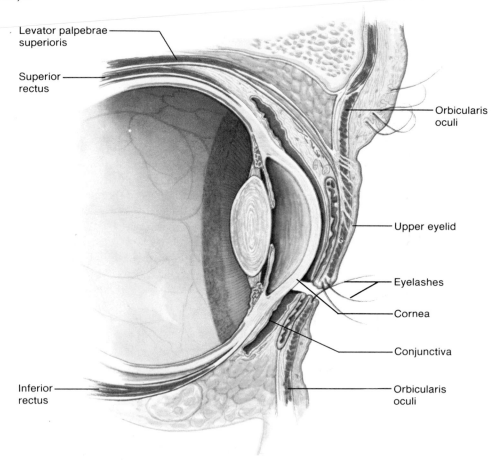

Levator palpebrae superioris

Superior rectus

Orbicularis oculi

Upper eyelid

Eyelashes

Cornea

Conjunctiva

Orbicularis oculi

Inferior rectus

Fig. 11.15 The lacrimal apparatus consists of a tear-secreting gland and a series of ducts.

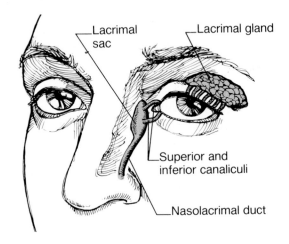

Lacrimal sac

Lacrimal gland

Superior and inferior canaliculi

Nasolacrimal duct

Chart 11.1 Skeletal muscles associated with the eye

Name	Innervation	Function
Orbicularis oculi	Facial nerve (VII)	Closes eye
Levator palpebrae superioris	Oculomotor nerve (III)	Opens eye
Superior rectus	Oculomotor nerve (III)	Rotates eye upward and toward midline
Inferior rectus	Oculomotor nerve (III)	Rotates eye downward and toward midline
Medial rectus	Oculomotor nerve (III)	Rotates eye toward midline
Lateral rectus	Abducens nerve (VI)	Rotates eye away from midline
Superior oblique	Trochlear nerve (IV)	Rotates eye downward and away from midline
Inferior oblique	Oculomotor nerve (III)	Rotates eye upward and away from midline

From Hole, John W. Jr., *Human Anatomy and Physiology 3d ed.* © 1978, 1981, 1984 Wm. C. Brown Publishers, Dubuque, Iowa. All Rights Reserved. Reprinted by permission.

Fig. 11.16 The extrinsic muscles of the eye.

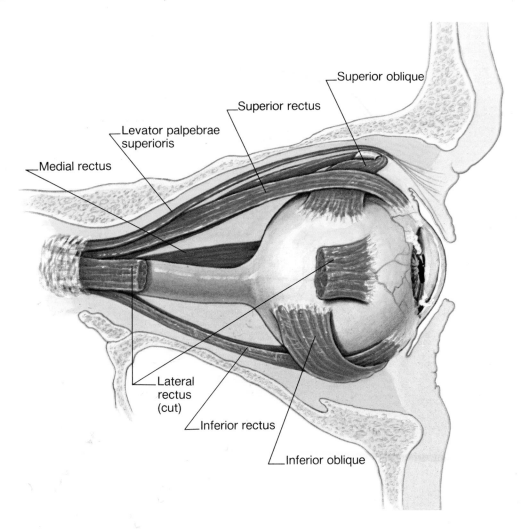

Levator palpebrae superioris

Medial rectus

Superior rectus

Superior oblique

Lateral rectus (cut)

Inferior rectus

Inferior oblique

The **extrinsic muscles** of the eye arise from the bones of the orbit and are inserted by broad tendons on the eye's tough outer surface. There are six such muscles that function to move the eye in various directions. Although any given eye movement may involve more than one of them, each muscle is associated with one primary action. Figure 11.16 illustrates the locations of these muscles, and chart 11.1 lists their functions.

When one eye deviates from the line of vision, the person may have double vision (diplopia). If this condition persists, there is danger that changes will occur in the brain to suppress the image from the deviated eye. As a result, the turning eye may become blind (suppression amblyopia). Such monocular blindness often can be prevented if the eye deviation is treated early in life with exercises, glasses, and surgery.

1. Explain how the eyelid is moved.
2. Describe the conjunctiva.
3. What is the function of the lacrimal apparatus?

Structure of the Eye

The eye is a hollow, spherical structure about 2.5 centimeters (one inch) in diameter. Its wall has three distinct layers—an outer *fibrous tunic,* a middle *vascular tunic,* and an inner *nervous tunic.* The spaces within the eye are filled with fluids that provide support for its wall and internal parts and help maintain its shape. Figure 11.17 shows the major parts of the eye.

Fig. 11.17 · Sagittal section of the eye.

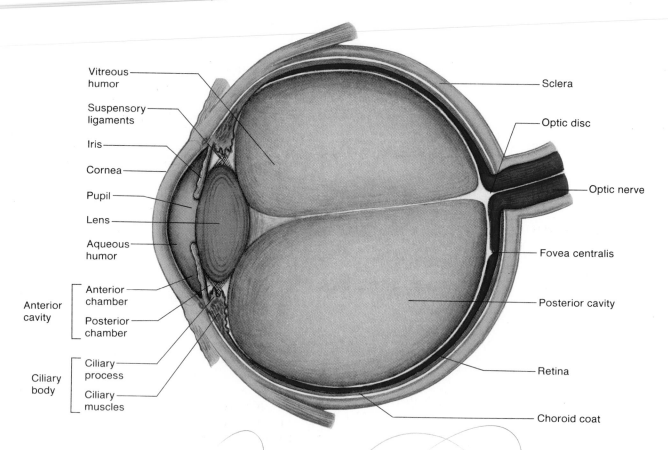

Vitreous humor
Suspensory ligaments
Iris
Cornea
Pupil
Lens
Aqueous humor
Anterior cavity — Anterior chamber / Posterior chamber
Ciliary body — Ciliary process / Ciliary muscles

Sclera
Optic disc
Optic nerve
Fovea centralis
Posterior cavity
Retina
Choroid coat

The Outer Tunic. The anterior ⅙ of the outer tunic bulges forward as the transparent **cornea,** which serves as the window of the eye and helps focus entering light rays. It is composed largely of connective tissue with a thin layer of epithelium on the surface. The transparency of the cornea is due to the fact that it contains relatively few cells and no blood vessels, and its cells and collagenous fibers are arranged in unusually regular patterns.

Along its circumference, the cornea is continuous with the **sclera,** the white portion of the eye. This part makes up the posterior ⅚ of the outer tunic and is opaque due to the presence of many large, haphazardly arranged, collagenous and elastic fibers. The sclera provides protection and serves as an attachment for the extrinsic muscles.

In the back of the eye, the sclera is pierced by the **optic nerve** and certain blood vessels. It is attached to the dura mater that encloses the optic nerve and other parts of the central nervous system.

The Middle Tunic. The middle or vascular tunic of the eyeball includes the choroid coat, the ciliary body, and the iris.

The **choroid coat,** in the posterior ⅚ of the globe, is loosely joined to the sclera and is honeycombed with blood vessels that provide nourishment to surrounding tissues. The choroid also contains numerous pigment-producing melanocytes. The melanin of these cells functions to absorb excess light and thus helps to keep the inside of the eye dark.

The **ciliary body,** which is the thickest part of the middle tunic, extends forward from the choroid and forms an internal ring around the front of the eye. Within the ciliary body there are many radiating folds called *ciliary processes* and groups of muscle fibers that constitute the *ciliary muscles.* These structures are shown in figure 11.17.

The transparent **lens** is held in position by a large number of strong but delicate fibers, called *suspensory ligaments,* that extend inward from the ciliary processes. The distal ends of these fibers are attached along the margin of a thin capsule that surrounds the lens. The body of the lens lies directly behind the iris and is composed of "fibers" that arise from epithelial cells. In fact, the cytoplasm of these cells makes up the transparent substance of the lens.

Fig. 11.18 (a) How is the focus of the eye affected when the lens becomes thin as the ciliary muscle fibers relax? (b) How is it affected when the lens thickens as the ciliary muscle fibers contract?

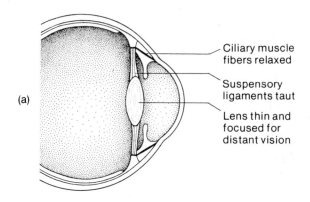

(a)

Ciliary muscle fibers relaxed

Suspensory ligaments taut

Lens thin and focused for distant vision

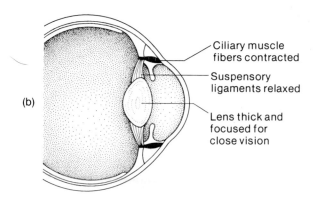

(b)

Ciliary muscle fibers contracted

Suspensory ligaments relaxed

Lens thick and focused for close vision

A relatively common eye disorder, particularly in older people, is called *cataract*. In this condition, the lens or its capsule slowly loses its transparency and becomes cloudy and opaque. In time, the person may become blind.

Cataract usually is treated by surgically removing the lens. The lens may then be replaced by an artificial one, or the loss of refractive power of the eye may be corrected with eyeglasses or contact lenses.

The lens capsule is a clear, membranelike structure composed largely of intercellular material. It is quite elastic, and this quality keeps it under constant tension. As a result, the lens can assume a globular shape. The suspensory ligaments attached to the margin of the capsule are also under tension, however, and as they pull outward, the capsule and the lens inside are kept somewhat flattened.

If the tension on the suspensory ligaments is relaxed, the elastic capsule rebounds, and the lens surface becomes more convex. Such a change occurs in the lens of an eye when it is focused to view a close object. This adjustment is called **accommodation.**

The relaxation of the suspensory ligaments during accommodation is a function of the ciliary muscles. For example, one set of these muscle fibers extends back from fixed points in the sclera to the choroid coat. When the fibers contract, the choroid is pulled forward and the ciliary body is shortened. This action causes the suspensory ligaments to become relaxed, and the lens thickens in response. (See fig. 11.18.) In this thickened state the lens is focused for viewing objects closer than before. To focus on more distant objects, the ciliary muscles are relaxed, tension on the suspensory ligaments increases, and the lens becomes thinner again.

1. Describe the three layers of the eye.
2. What factors contribute to the transparency of the cornea?
3. How is the shape of the lens changed during accommodation?

The iris is a thin diaphragm, composed largely of connective tissue and smooth muscle fibers, that is seen from the outside as the colored portion of the eye. It extends forward from the periphery of the ciliary body and lies between the cornea and the lens. The iris divides the space separating these parts into an *anterior chamber* (between the cornea and the iris) and a *posterior chamber* (between the iris and the lens). (See fig. 11.17.)

The epithelium on the inner surface of the ciliary body secretes a watery fluid called **aqueous humor** into the posterior chamber. The fluid circulates from this chamber through the **pupil,** a circular opening in the center of the iris, and into the anterior chamber. It fills the space between the cornea and the lens, helps to nourish these parts, and aids in maintaining the shape of the front of the eye. The aqueous humor subsequently leaves the anterior chamber through a special drainage canal located in its wall.

Fig. 11.19 The retina consists of several cell layers.

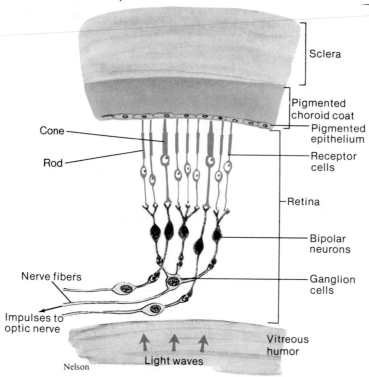

Labels:
Sclera
Pigmented choroid coat
Pigmented epithelium
Cone
Receptor cells
Rod
Retina
Bipolar neurons
Nerve fibers
Ganglion cells
Impulses to optic nerve
Vitreous humor
Nelson
Light waves

A disorder called *glaucoma* sometimes develops in the eyes as a person ages. This condition occurs when the rate of aqueous humor formation exceeds the rate of its removal. As a result, fluid accumulates in the anterior chamber of the eye, and the fluid pressure rises.

Since liquids cannot be compressed, the increasing pressure from the anterior chamber is transmitted to all parts of the eye, and in time the blood vessels that supply the receptor cells of the retina may be squeezed closed. If this happens, cells that fail to receive needed nutrients and oxygen may die and a form of permanent blindness may result.

The smooth muscle fibers of the iris are arranged into two groups, a circular set and a radial set. These muscles function to control the size of the pupil, which is the opening that light passes through as it enters the eye. The circular set acts as a sphincter. When it contracts, the pupil gets smaller, and the intensity of the light entering decreases. When the radial muscle fibers contract, the diameter of the pupil increases, and the intensity of the light entering increases.

The Inner Tunic. The inner tunic of the eye consists of the **retina,** which contains the visual receptor cells (photoreceptors). This nearly transparent sheet of tissue is continuous with the optic nerve in the back of the eye and extends forward as the inner lining of the eyeball. It ends just behind the margin of the ciliary body.

Although the retina is thin and delicate, its structure is quite complex. It has a number of distinct layers as figure 11.19 illustrates.

In the central region of the retina there is a yellowish spot (macula lutea) that has a depression in its center called the **fovea centralis.** This depression is in the region of the retina that produces the sharpest vision.

Just medial to the fovea centralis is an area called the **optic disk.** (See fig. 11.20.) Here the nerve fibers from the retina leave the eye and become parts of the optic nerve. A central artery and vein also pass through at the optic disk. These vessels are continuous with capillary networks of the retina, and together with vessels in the underlying choroid coat, they supply blood to the cells of the inner tunic.

Because there are no receptor cells in the region of the optic disk, it is commonly referred to as the blind spot of the eye.

The space bounded by the lens, ciliary body, and retina is the largest compartment of the eye and is called the *posterior cavity.* It is filled with a clear, jellylike fluid called **vitreous humor.** This fluid functions to support the internal parts of the eye and helps maintain its shape.

Fig. 11.20 Nerve fibers leave the eye in the area of the optic disk (arrow) to form the optic nerve.

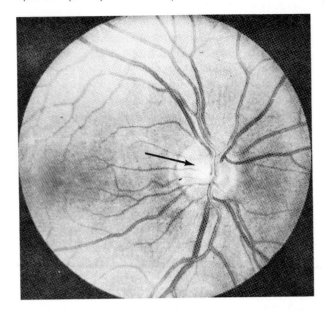

Fig. 11.21 A lens with a convex surface causes light waves to converge.

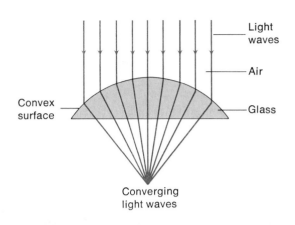

Fig. 11.22 The image of an object formed on the retina is upside down.

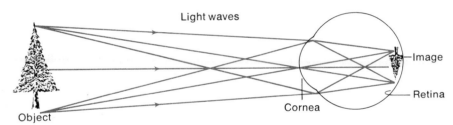

1. Explain the origin of aqueous humor and trace its path through the eye.
2. How is the size of the pupil regulated?
3. Describe the structure of the retina.

Refraction of Light

When a person sees something, it is either giving off light or light waves are being reflected from it. These waves enter the eye, and an image of what is seen is focused upon the retina. This focusing process involves a bending of the light waves—a phenomenon called **refraction.**

Refraction occurs when light waves pass at an oblique angle from a medium of one optical density into a medium of a different optical density. For example, when light passes obliquely from a less dense medium such as air into a denser medium such as glass, or from air into the cornea of the eye, the light is bent toward a line perpendicular to the surface between these substances. When the surface between such refracting media is curved, a lens is formed. A lens with a *convex* surface causes light waves to converge. (See fig. 11.21.)

When light arrives from objects outside the eye, the light waves are refracted primarily by the convex surface of the cornea. Then, the light is refracted again by the convex surface of the lens and, to a lesser extent, by the surfaces of the fluids within the chambers of the eye.

If the shape of the eye is normal, light waves are focused sharply upon the retina, much as a motion picture image is focused on a screen for viewing. Unlike the motion picture image, however, the one formed on the retina is upside down and reversed from left to right (fig. 11.22). When the visual cortex interprets such an image, it somehow corrects this, and things are seen in their proper positions.

1. What is meant by refraction?
2. What parts of the eye provide refracting surfaces?

Visual Receptors

The receptor cells of the eye are actually modified neurons, and there are two distinct kinds, as illustrated in figure 11.19. One group of receptors have long, thin projections at their terminal ends and are called **rods.** The cells of the other group have short, blunt projections and are called **cones.**

Instead of being located in the surface layer of the retina, the rods and cones are found in a deep portion, closely associated with a layer of pigmented epithelium. The projections from the receptors, which are loaded with visual pigments, extend into the pigmented layer.

The epithelial pigment of the retina functions to absorb light waves that are not absorbed by the receptor cells, and with the pigment of the choroid coat, it serves to keep light from reflecting about the inside of the eye.

The visual receptors are stimulated only when light reaches them. Thus when a light image is focused on an area of the retina, some receptors are stimulated, and impulses travel away from them to the brain. The impulse leaving each activated receptor, however, provides only a fragment of the information needed for the brain to interpret a total scene.

Rods and cones each function differently. For example, rods are hundreds of times more sensitive to light than cones, and as a result, they enable persons to see in relatively dim light. In addition, the rods produce colorless vision, while cones can detect colors.

Still another difference involves visual acuity—the sharpness of the images perceived. Cones allow a person to see sharp images, while rods enable one to see more general outlines of objects. This characteristic is related to the fact that nerve fibers from many rods may converge, and their impulses may be transmitted to the brain on the same nerve fiber. Thus, if a point of light stimulates a rod, the brain cannot tell which one of many receptors has actually been stimulated. Such a convergence of impulses occurs to a much lesser degree among cones, so when a cone is stimulated, the brain is able to pinpoint the stimulation more accurately. (See fig. 11.19.)

As was mentioned, the area of sharpest vision is the fovea centralis. This area lacks rods, but contains densely packed cones with few or no converging fibers. Also, the overlying layers of the retina, as well as the retinal blood vessels, are displaced to the sides in the fovea. This displacement more fully exposes the receptors to incoming light. Consequently, to view something in detail one moves the eye so the important part of an image falls upon the fovea.

Visual Pigments. Both rods and cones contain light-sensitive pigments that decompose when they absorb light energy. The light-sensitive substance in rods is called **rhodopsin** (visual purple). In the presence of light, rhodopsin molecules break down into molecules of a colorless protein called *scotopsin* and a yellowish substance called *retinene,* which is synthesized from vitamin A.

At the same time that rhodopsin molecules decompose, some energy is released and it triggers a nerve impulse to travel away from the retina, through the optic nerve, and into the brain.

In bright light, nearly all of the rhodopsin in the rods of the retina is decomposed, and the sensitivity of these receptors is greatly reduced. In dim light, however, rhodopsin can be regenerated from scotopsin and retinene faster than it is broken down. This regeneration process requires cellular energy, which is provided by energy-carrying molecules of ATP.

Many people, particularly children, suffer from vitamin A deficiency due to improper diet. In such cases, the quantity of retinene available for the manufacture of rhodopsin may be reduced, and consequently, the sensitivity of their rods may be low. This condition, called night blindness, is characterized by poor vision in dim light. Fortunately, the problem is usually easy to correct by adding vitamin A to the diet or by providing the victim with injections of vitamin A.

The light-sensitive pigments of cones are similar to rhodopsin in that they are composed of retinene combined with a protein; the protein, however, differs from that in the rods. In fact, there are apparently three different sets of cones, each containing an abundance of one of the three different visual pigments.

The wavelength of a particular kind of light determines the color perceived from it. For example, the shortest wavelengths of visible light are perceived as violet, while the longest visible wavelengths are sensed as red. As far as the cone pigments are concerned, one type (erythrolabe) is thought to be most sensitive to red light waves, another (chlorolabe) to green light waves, and a third (cyanolabe) to blue light waves. The color a person perceives depends upon which set of cones or combination of sets is stimulated by the light in a given image. If all three sets of cones are stimulated, the person senses the light as white, and if none are stimulated, the person senses black.

Fig. 11.23 Stereoscopic vision results from the formation of two slightly different retinal images.

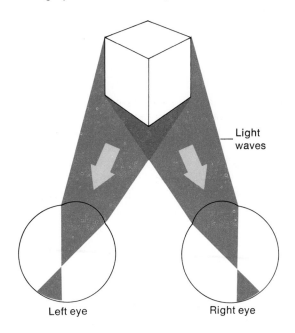

Fig. 11.24 The visual pathway includes the optic nerve, optic chiasma, optic tract, and optic radiations.

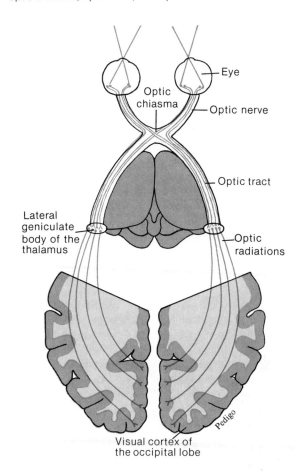

Stereoscopic Vision

Stereoscopic vision is vision that involves the perception of distance and depth as well as the height and width of objects. Such vision is due largely to the fact that the pupils of the eyes are 6–7 centimeters (about 2.5 inches) apart. Consequently, objects that are relatively close (less than 6 meters or 20 feet away) produce slightly different retinal images. That is, the right eye sees a little more of one side of an object, while the left eye sees a little more of the other side. These two images are somehow superimposed and interpreted by the visual cortex of the brain, and the result is the perception of a single object in three dimensions. (See fig. 11.23.)

Visual Nerve Pathway

The axons of the retinal neurons leave the eyes to form the *optic nerves.* Just anterior to the pituitary gland, these nerves give rise to the X-shaped *optic chiasma,* and within the chiasma, some of the fibers cross over. More specifically, the fibers from the nasal half of each retina cross over, while those from the temporal sides do not. Thus, fibers from the nasal half of the left eye and the temporal half of the right eye form the right *optic tract;* and fibers from the nasal half of the right eye and the temporal half of the left eye form the left optic tract.

The nerve fibers continue in the optic tracts, and just before they reach the thalamus, a few of them leave to enter nuclei that function in various visual reflexes. Most of the fibers, however, enter the thalamus and synapse in its posterior portion (lateral geniculate body). From this region the visual impulses enter nerve pathways called *optic radiations,* and they lead to the visual cortex of the occipital lobes. (See fig. 11.24.)

---

1. Distinguish between the rods and the cones of the retina.
2. Explain the roles of visual pigments.
3. What factors make stereoscopic vision possible?

Clinical Terms Related to the Senses

amblyopia (am″ble-o′pe-ah)—dimness of vision due to a cause other than a refractive disorder or lesion.

ametropia (am″ĕ-tro′pe-ah)—an eye condition characterized by inability to focus images sharply on the retina.

anopia (an-o′pe-ah)—absence of the eye.

audiometry (aw″de-om′ĕ-tre)—the measurement of auditory acuity for various frequencies of sound waves.

blepharitis (blef″ah-ri′tis)—an inflammation of the margins of the eyelids.

causalgia (kaw-zal′je-ah)—a persistent, burning pain usually associated with injury to a limb.

conjunctivitis (kon-junk″ti-vi′tis)—an inflammation of the conjunctiva.

diplopia (di-plo′pe-ah)—double vision, or the sensation of seeing two objects when only one is viewed.

emmetropia (em″ĕ-tro′pe-ah)—normal condition of the eyes; eyes with no refractive defects.

enucleation (e-nu″kle-a′shun)—removal of the eyeball.

exophthalmos (ek″sof-thal′mos)—condition in which the eyes protrude abnormally.

hemianopia (hem″e-ah-no′-pe-ah)—defective vision affecting half of the visual field.

hyperalgesia (hi″per-al-je′ze-ah)—an abnormally increased sensitivity to pain.

iridectomy (ir″i-dek′to-me)—the surgical removal of part of the iris.

iritis (i-ri′tis)—an inflammation of the iris.

keratitis (ker″ah-ti′tis)—an inflammation of the cornea.

labyrinthectomy (lab″i-rin-thek′to-me)—the surgical removal of the labyrinth.

labyrinthitis (lab″i-rin-thi′tis)—an inflammation of the labyrinth.

Ménière's disease (men″e-ārz′ di-zez)—an inner ear disorder characterized by ringing in the ears, increased sensitivity to sounds, dizziness, and loss of hearing.

neuralgia (nu-ral′je-ah)—pain resulting from inflammation of a nerve or a group of nerves.

neuritis (nu-ri′tis)—an inflammation of a nerve.

nystagmus (nis-tag′mus)—an involuntary oscillation of the eyes.

otitis media (o-ti′tis me′de-ah)—an inflammation of the middle ear.

otosclerosis (o″to-skle-ro′sis)—a formation of spongy bone in the inner ear, which often causes deafness by fixing the stapes to the oval window.

pterygium (tĕ-rij′e-um)—an abnormally thickened patch of conjunctiva that extends over part of the cornea.

retinitis pigmentosa (ret″i-ni′tis pig″men-to′sa)—a progressive retinal sclerosis characterized by deposits of pigment in the retina and atrophy of the retina.

retinoblastoma (ret″i-no-blas-to′mah)—an inherited, highly malignant tumor arising from immature retinal cells.

tinnitus (ti-ni′tus)—a ringing or buzzing noise in the ears.

tonometry (to-nom′ĕ-tre)—the measurement of fluid pressure within the eyeball.

trachoma (trah-ko′mah)—a virus-caused disease of the eye, characterized by conjunctivitis, that may lead to blindness.

tympanoplasty (tim″pah-no-plas′te)—the surgical reconstruction of the middle ear bones and establishment of continuity from the tympanic membrane to the oval window.

uveitis (u″ve-i′tis)—an inflammation of the uvea, which includes the iris, ciliary body, and the choroid coat.

vertigo (ver′ti-go)—a sensation of dizziness.

Chapter Summary

Introduction

Sensory receptors are sensitive to changes occurring in their surroundings.

Receptors and Sensations

1. Types of receptors
   a. Each type of receptor is most sensitive to a distinct type of stimulus.
   b. The major types of receptors include chemoreceptors, pain receptors, thermoreceptors, mechanoreceptors, and photoreceptors.

2. Sensations
   a. Sensations are feelings resulting from sensory stimulation.
   b. A particular part of the sensory cortex always interprets impulses reaching it in the same way.
   c. The brain projects a sensation back to the region of stimulation.

3. Sensory adaptations are adjustments made by sensory receptors to continuous stimulation in which impulses leave them at lower and lower rates.

Somatic Senses

Somatic senses are those involved with receptors in skin, muscles, joints, and visceral organs.

1. Touch and pressure senses
   a. Free ends of sensory nerve fibers are responsible for the sensations of touch and pressure.

b. Meissner's corpuscles are responsible for the sensation of light touch.

c. Pacinian corpuscles are responsible for the sensation of heavy pressure.

2. Temperature receptors include two sets of free nerve endings that serve as heat and cold receptors.

3. Sense of pain
a. Pain receptors are free nerve endings stimulated by tissue damage.
b. Pain receptors are the only receptors in visceral organs that provide sensations.
c. The sensations produced from visceral receptors are likely to feel as if they were coming from some other part.

Sense of Smell

1. Olfactory receptors
a. Olfactory receptors are chemoreceptors that are stimulated by chemicals dissolved in liquid.
b. They function together with taste receptors and aid in food selection.

2. Olfactory organs
a. The olfactory organs consist of receptors and supporting cells in the nasal cavity.
b. Olfactory receptors are neurons with cilia.

3. Nerve impulses travel from the olfactory receptors through the olfactory nerves, olfactory bulbs, and olfactory tracts to interpreting centers in the cerebrum.

4. Olfactory stimulation
a. Olfactory impulses may result when various gaseous molecules combine with specific sites on the cilia of the receptor cells.
b. Olfactory receptors adapt rapidly.

Sense of Taste

1. Taste receptors
a. Taste buds consist of receptor cells and supporting cells.
b. Taste cells have taste hairs.
c. Taste hair surfaces seem to have receptor sites to which chemicals combine.

2. Taste sensations
a. The four primary taste sensations are sweet, sour, salty, and bitter.
b. Various taste sensations result from the stimulation of one or more sets of taste receptors.

3. Taste nerve pathway
a. Sensory impulses from taste receptors travel on fibers of the facial, glossopharyngeal, and vagus nerves.
b. These impulses are carried to the medulla and then ascend to the thalamus from which they travel to the gustatory cortex in the parietal lobes.

Sense of Hearing

1. The external ear collects sound waves created by vibrating objects.

2. Middle ear
a. Auditory ossicles of the middle ear conduct sound waves from the tympanic membrane to the oval window of the inner ear.
b. Eustachian tubes connect the middle ears to the throat and function to help maintain equal air pressure on both sides of the eardrums.

3. Inner ear
a. The inner ear consists of a complex system of interconnected tubes and chambers—the osseous and membranous labyrinths.
b. The organ of Corti contains the hearing receptors that are stimulated by vibrations in the fluids of the inner ear.
c. Different frequencies of vibrations are thought to stimulate different receptor cells.

4. Auditory nerve pathway
a. The auditory nerves carry impulses to the auditory cortices of the temporal lobes.
b. Some auditory nerve fibers cross over, so impulses arising from each ear are interpreted on both sides of the brain.

Sense of Equilibrium

1. Static equilibrium is concerned with maintaining the stability of the head and body when these parts are motionless.

2. Dynamic equilibrium is concerned with balancing the head and body when they are moved or rotated suddenly.

3. Other parts that help with the maintenance of equilibrium include the eyes and mechanoreceptors associated with certain joints.

Sense of Sight

1. Visual accessory organs include the eyelids, lacrimal apparatus, and extrinsic muscles of the eyes.

2. Structure of the eye
a. The wall of the eye has an outer, a middle, and an inner layer that function as follows:
(1) The outer layer (sclera) is protective, and its transparent anterior portion (cornea) refracts light entering the eye.
(2) The middle layer (choroid) is vascular and contains pigments.
(3) The inner layer (retina) contains the visual receptor cells.
b. The lens is a transparent, elastic structure whose shape is controlled by the action of ciliary muscles.
c. The iris is a muscular diaphragm that controls the amount of light entering the eye.
d. Spaces within the eye are filled with fluids that help to maintain its shape.

3. Refraction of light
   a. Light waves are refracted primarily by the cornea and lens.
   b. The lens must be thickened to focus on close objects.
4. Visual receptors
   a. The visual receptors are called rods and cones.
   b. Rods are responsible for colorless vision in relatively dim light, and cones are responsible for color vision.
   c. Visual pigments
      (1) A light-sensitive pigment in rods decomposes in the presence of light and triggers nerve impulses.
      (2) Color vision seems to be related to the presence of three sets of cones containing different light-sensitive pigments.
5. Stereoscopic vision
   a. Stereoscopic vision involves the perception of distance and depth.
   b. Stereoscopic vision occurs because of the formation of two slightly different retinal images that the brain superimposes and interprets as one image in three dimensions.
6. Visual nerve pathway
   a. Nerve fibers from the retina form the optic nerves.
   b. Some fibers cross over in the optic chiasma.
   c. Most of the fibers enter the thalamus and synapse with others that continue to the visual cortex.

Application of Knowledge

1. How would you explain the following observation? A person enters a tub of water and reports that it is uncomfortably warm, yet a few moments later says the water feels comfortable, even though the water temperature remains unchanged.
2. How would you explain the fact that some serious injuries, such as those produced by a bullet entering the abdomen, may be relatively painless, while others such as those involving a crushing of the skin may produce considerable discomfort?
3. Labyrinthitis is a condition in which the tissues of the inner ear are inflamed. What symptoms would you expect to observe in a patient with this disorder?

Review Activities

1. List five groups of sensory receptors and name the kind of change to which each is sensitive.
2. Define sensation.

3. Explain what is meant by the projection of a sensation.
4. Define sensory adaptation and provide an example of this phenomenon.
5. Describe the functions of free nerve endings, Meissner's corpuscles, and Pacinian corpuscles.
6. Define referred pain and provide an example of this phenomenon.
7. Describe the olfactory organ and its function.
8. Trace a nerve impulse from the olfactory receptor to the interpreting center of the cerebrum.
9. Explain how the salivary glands aid the function of the taste receptors.
10. Name the four primary taste sensations.
11. Trace the pathway of a taste impulse from a receptor to the cerebral cortex.
12. Distinguish between the external, middle, and inner ears.
13. Trace the path of a sound wave from the tympanic membrane to the hearing receptors.
14. Describe the functions of the auditory ossicles.
15. Explain the function of the eustachian tube.
16. Distinguish between the osseous and the membranous labyrinths.
17. Describe the cochlea and its function.
18. Trace a nerve impulse from the organ of Corti to the interpreting centers of the cerebrum.
19. Describe the organs of static and dynamic equilibrium and their functions.
20. List the visual accessory organs and describe the functions of each organ.
21. Name the three layers of the eye wall and describe the functions of each layer.
22. Describe how accommodation is accomplished.
23. Explain how the iris functions.
24. Distinguish between aqueous and vitreous humor.
25. Distinguish between the fovea centralis and the optic disk.
26. Explain how light waves are focused on the retina.
27. Distinguish between rods and cones.
28. Explain why cone vision is generally more acute than rod vision.
29. Describe the function of rhodopsin.
30. Describe the relationship between light wavelengths and color vision.
31. Define stereoscopic vision.
32. Trace a nerve impulse from the retina to the visual cortex.

The Endocrine System

The endocrine system consists of a variety of loosely related cells, tissues, and organs that act together with parts of the nervous system to control body activities and maintain homeostasis. Each of these systems provides a means by which body parts can communicate with one another and adjust to changing needs. Whereas the parts of the nervous system communicate by means of nerve impulses carried on nerve fibers, the parts of the endocrine system make use of hormones, which are carried in body fluids and act as chemical messengers.

After you have studied this chapter, you should be able to

1. Distinguish between endocrine and exocrine glands.

2. Explain how hormones are thought to exert influences on target tissues.

3. Discuss how hormone secretions are regulated by feedback mechanisms.

4. Explain how hormone secretions may be controlled by the nervous system.

5. Name and describe the location of the major endocrine glands of the body, and list the hormones they secrete.

6. Describe the general functions of these hormones.

7. Explain how the secretion of each of these hormones is regulated.

8. Complete the review activities at the end of this chapter. Note that the items are worded in the form of specific learning objectives. You may want to refer to them before reading the chapter.

adrenal cortex   (ah-dre′nal kor′teks)

adrenal medulla   (ah-dre′nal me-dul′ah)

anterior pituitary   (an-ter′e-or pĭ-tu′ĭ-tar″e)

negative feedback   (neg′ah-tiv fēd′bak)

pancreas   (pan′kre-as)

parathyroid gland   (par″ah-thi′roid gland)

pineal gland   (pin′e-al gland)

posterior pituitary   (pos-ter′e-or pĭ-tu′ĭ-tar″e)

prostaglandin   (pros″tah-glan′din)

releasing factor   (re-le′sing fak′tor)

target cell   (tar′get sel)

thymus gland   (thi′mus gland)

thyroid gland   (thi′roid gland)

-crin, to secrete: endo*crine*—pertaining to internal secretions.

diuret-, to pass urine: *diuret*ic—a substance that promotes the production of urine.

endo-, within: *endo*crine gland—a gland that releases its secretion internally into a body fluid.

exo-, outside: *exo*crine gland—a gland that releases its secretion to the outside through a duct.

hyper-, above: *hyper*thyroidism—condition resulting from above normal secretion of thyroid hormone.

hypo-, below: *hypo*thyroidism—condition resulting from below normal secretion of thyroid hormone.

para-, beside: *para*thyroid glands—a set of glands located on the surface of the thyroid gland.

toc-, birth: oxy*toc*in—a hormone that stimulates uterine muscles to contract during childbirth.

tropic-, influencing: adrenocortico*tropic* hormone—a hormone secreted by the anterior pituitary gland that stimulates the adrenal cortex.

The term *endocrine* is used to describe cells, tissues, and organs that secrete hormones into body fluids. In contrast, the term *exocrine* refers to parts whose secretory products are carried by tubes or ducts to some internal or external body surface. Thus, the thyroid and parathyroid glands which secrete hormones into the blood are endocrine glands (ductless glands), while sweat glands and salivary glands are exocrine.

General Characteristics of the Endocrine System

As a group, endocrine structures function to help regulate metabolic processes. They control the rates of certain chemical reactions, aid in the transport of substances through membranes, and help regulate water and electrolyte balances. They also play vital roles in reproductive processes and in development and growth.

Although many hormones have localized effects, those described in this chapter have more widespread actions and are produced by larger endocrine glands. These glands include the pituitary gland, thyroid gland, parathyroid glands, adrenal glands, and pancreas. (See fig. 12.1.) Several other hormone-secreting glands and tissues, such as those involved in the processes of digestion and reproduction, are discussed in subsequent chapters.

Hormones and Their Actions

A **hormone** is an organic substance secreted by a cell that has an effect on the metabolic activity of another cell or tissue. Such substances are released into the extracellular spaces surrounding the endocrine cells. Some hormones travel only short distances and produce their effects locally. Others are transported in the blood to all parts of the body and may produce rather general effects. In either case, the physiological action of a particular hormone is restricted to its *target cells*—those that possess specific receptors for the hormone molecules.

Hormones are very potent and, thus, can produce their effects when they are present in extremely low concentrations. Chemically, most of them are either amines, peptides, proteins, or glycoproteins that are synthesized from amino acids, or they are steroids or steroidlike substances that are synthesized from cholesterol. (See chapter 2.)

Fig. 12.1 Locations of major endocrine glands.

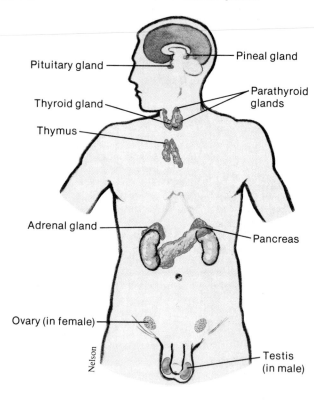

Pituitary gland
Pineal gland
Thyroid gland
Parathyroid glands
Thymus
Adrenal gland
Pancreas
Ovary (in female)
Testis (in male)

Nelson

1. Distinguish between an endocrine and an exocrine gland.
2. Describe the general function of the endocrine system.
3. What is a hormone?

Actions of Hormones

Different kinds of hormones exert their effects in different ways. For example, they may change the rate at which proteins or enzymes are synthesized, alter the rates of enzyme activities, or alter the rates at which molecules are transported through cell membranes. The ability of a hormone to influence a particular kind of cell depends upon the presence of receptor molecules in the cell or on its membranes. In other words, a hormone's target cells possess certain receptors that other cells lack.

The molecular structure formed when a hormone combines with a receptor is responsible for causing the cellular changes that are recognized as the hormone's actions. For instance, a steroid hormone is soluble in lipids such as those found in cell membranes, and for this reason it can enter a cell relatively easily. If such a hormone enters a cell containing a specific cytoplasmic protein—the receptor

Fig. 12.2 (a) A steroid hormone passes through a cell membrane and (b) combines with a receptor protein in the cytoplasm. (c) The steroid-protein complex enters the nucleus and (d) activates the synthesis of messenger RNA. (e) The messenger RNA leaves the nucleus and (f) functions in the manufacture of specific kinds of protein molecules.

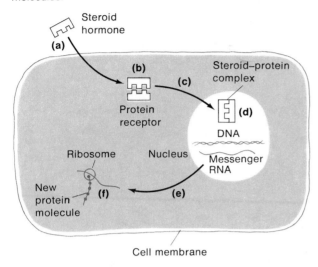

Cell membrane

Fig. 12.3 (a) Nonsteroid hormone molecules reach the target cell by means of body fluids and (b) combine with receptor sites on the cell membrane. (c) As a result, molecules of adenylate cyclase are activated and (d) cause the change of ATP into cyclic AMP. (e) Cyclic AMP brings about various cellular changes.

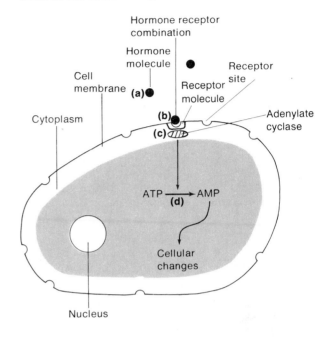

Nucleus

molecule—a steroid-protein complex forms. This complex may then enter the cell's nucleus and activate certain genes, bringing about the synthesis of particular kinds of messenger RNA.

As is described in chapter 4, messenger RNA molecules can leave the nucleus and enter the cytoplasm, where they function in the manufacture of specific protein molecules. Thus, steroid hormones influence cells by causing special proteins to be synthesized—proteins that may act as enzymes, which in turn alter the rates of cellular processes, function in membrane transport systems, or cause other enzymes to be activated or inhibited. (See fig. 12.2.)

On the other hand, many of the protein, peptide, and amine hormones seem to act by combining with receptors on the outer surfaces of target cell membranes. The formation of a hormone-receptor complex activates an enzyme called *adenylate cyclase,* which is bound to the other side of the membrane. This enzyme then causes ATP molecules on the inside of the cell to become molecules of *cyclic AMP* (adenosine monophosphate), and the cyclic AMP, in turn, activates another set of enzymes. These enzymes act upon various protein substrate molecules and in this way cause changes in cellular processes. Thus, the response of any particular cell is determined by the kinds of protein substrate molecules it contains.

To summarize, such a hormone acts as a messenger that stimulates its target cells by combining with a membrane receptor. Cyclic AMP, which is produced in the cytoplasm as a result of this process, serves as a "second messenger." It delivers the stimulus to the inside of the cell and triggers the changes leading to the cellular response. (See fig. 12.3.)

Prostaglandins

Another group of substances, called **prostaglandins,** also have regulating effects on cells. These substances are lipids and are synthesized from fatty acids found in cell membranes. They occur in a wide variety of cells, including those of the liver, kidneys, heart, lungs, thymus gland, pancreas, brain, and various reproductive organs.

Like hormones, prostaglandins are very potent compounds and are present in very small quantities. They are not stored in cells, but instead are synthesized just before they are released and are then inactivated rapidly.

Fig. 12.4 An example of a negative feedback system: (1) gland A secretes a hormone that stimulates gland B to release another hormone; (2) the hormone from gland B inhibits the activity of gland A.

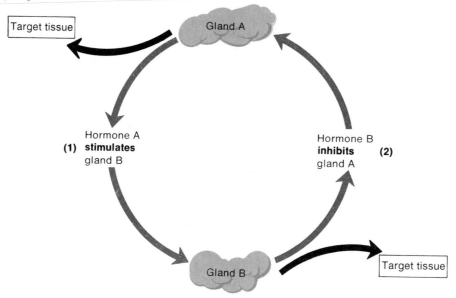

Target tissue ← Gland A

Hormone A
(1) stimulates gland B

Hormone B **inhibits** gland A **(2)**

Gland B → Target tissue

Although their modes of action are not well understood, prostaglandins produce a variety of effects. For example, they can cause the smooth muscles in the airways of the lungs and in the blood vessels to relax; however, they can also cause smooth muscles in the walls of the uterus and intestines to contract. They stimulate the secretions of hormones from the adrenal cortex and inhibit the secretion of hydrochloric acid from the wall of the stomach. They also influence the movement of sodium ions and water molecules in the kidneys, help regulate blood pressure, and have a powerful effect on both male and female reproductive physiology.

---

1. *Explain how a steroid hormone may promote cellular changes.*
2. *Explain what is meant by a "second messenger."*
3. *What are prostaglandins?*
4. *What kinds of effects do prostaglandins produce?*

Control of Hormonal Secretions

Since hormones are very potent substances that function to regulate metabolic activities, the amounts of hormones released by endocrine cells must be regulated. One mechanism commonly employed for this purpose is a *feedback system.*

Negative Feedback Systems

In a feedback system, the hormone-secreting tissue continuously receives information (feedback) concerning the cellular processes it controls. It receives this information in the form of chemical signals. If conditions change, the tissue can act upon the information and adjust its rate of secretion.

Commonly, the control of hormonal secretions involves a **negative feedback system.** In such a system, an endocrine gland is sensitive either to the concentration of a substance it regulates, or to the concentration of a product of a process it regulates. Whenever this concentration reaches a certain level, the endocrine gland is inhibited (a negative effect) and its secretory activity decreases. Then, as the concentration of its hormone drops, the concentration of the regulated substances drops also, and the inhibition on the gland is released. When the gland is no longer inhibited, it begins to secrete its hormone again.

As a result of such negative feedback systems, the concentrations of some hormones remain relatively stable, although they may fluctuate slightly within normal ranges. (See figs. 12.4 and 12.5.)

Fig. 12.5 As a result of negative feedback systems, hormone concentrations remain relatively stable although they may fluctuate slightly above and below the average concentrations.

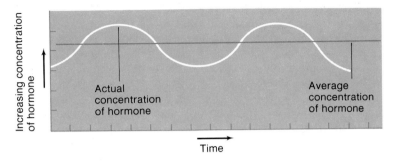

Fig. 12.6 This diagram represents a positive feedback system. How does it differ from a negative feedback system?

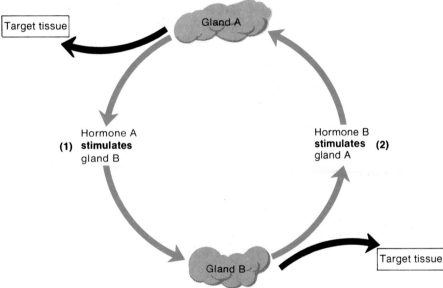

Positive Feedback Systems

A **positive feedback system** is operating when an endocrine gland is stimulated to increase its rate of hormonal secretion (a positive effect) by a substance it causes to be produced. Consequently, still more product forms, and then more hormone is secreted. (See fig. 12.6.)

Such a system is unstable because it tends to produce extreme changes in conditions, and positive feedback systems occur only rarely in organisms. One example of this type of system involves the sex hormone called estrogen. This steroid is secreted by cells in the ovary in response to a hormone from the anterior pituitary gland (FSH). When a certain blood concentration of estrogen occurs, it causes an increase in the secretion of some anterior pituitary hormones (including FSH), so that a positive feedback exists for a short time. (See chapter 20.)

Fig. 12.7 (a) In some cases, an endocrine gland secretes its hormones in response to nerve impulses; (b) the anterior pituitary gland secretes hormones in response to releasing hormones secreted by the hypothalamus.

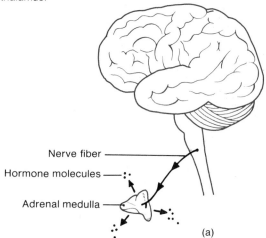

(a)

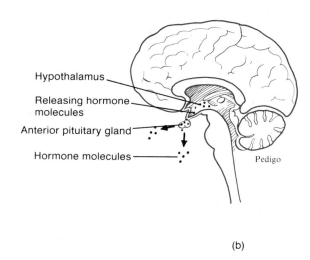

Hypothalamus

Releasing hormone molecules

Anterior pituitary gland

Hormone molecules

Pedigo

(b)

Nerve Control

Still other hormone control mechanisms involve the nervous system. For example, some endocrine glands, such as the adrenal medulla, secrete their hormones in response to nerve impulses.

Another kind of control system involves interaction between an endocrine gland and the hypothalamus of the brain. In this system, neurosecretory cells in the hypothalamus secrete substances, called **releasing** (or inhibiting) **hormones,** whose target tissues are in the anterior pituitary gland. The gland responds to the releasing hormone by secreting its own hormone. Then, as the gland's hormone reaches a certain concentration in the body fluids, a negative feedback system operates to inhibit the hypothalamus, and its secretion of the releasing hormone decreases. (See fig. 12.7.)

1. Describe a negative feedback system.
2. Describe a positive feedback system.
3. Explain two mechanisms involving the nervous system that help to control hormonal secretions.

The Pituitary Gland

The **pituitary gland** (fig. 10.28) is located at the base of the brain, where it is attached to the hypothalamus by the pituitary stalk, or *infundibulum.* The gland is about 1 centimeter (0.4 inch) in diameter and consists of two distinct portions—an anterior lobe and a posterior lobe.

Most of the pituitary activities are controlled by the brain. The release of hormones from the posterior lobe, for example, occurs when nerve impulses from the hypothalamus signal the axon ends of neurosecretory cells in this lobe. Secretions from the anterior lobe are controlled by releasing hormones produced by the hypothalamus. These releasing hormones are transmitted by blood in the vessels of a capillary net in the region of the hypothalamus. These vessels merge to form the **hypophyseal portal veins** that pass downward along the pituitary stalk and give rise to a capillary net in the anterior lobe of the pituitary gland. Thus, substances released into the blood from the hypothalamus are carried directly to the anterior lobe. (See fig. 12.8.)

1. Where is the pituitary gland located?
2. Explain how the hypothalamus is related to the actions of the anterior lobe of the pituitary gland. Of the posterior lobe.

Fig. 12.8 Axons in the posterior lobe of the pituitary gland are stimulated to release hormones by nerve impulses originating in the hypothalamus; cells of the anterior lobe are stimulated by releasing hormones secreted by hypothalamic neurons.

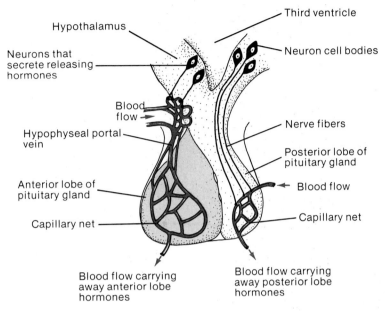

Anterior Lobe Hormones

The anterior lobe of the pituitary gland consists largely of epithelial cells arranged in blocks around many thin-walled blood vessels. There are two groups of secretory cells within this tissue. The cells of one group (acidophils) secrete growth hormone (GH) and a hormone called prolactin (PRL); the members of the second group (basophils) produce thyroid-stimulating hormone (TSH), adrenocorticotropic hormone (ACTH), follicle-stimulating hormone (FSH), and luteinizing hormone (LH).

Growth hormone (GH) generally stimulates body cells to increase in size and undergo more rapid cell division than usual. It also enhances the movement of amino acids through cell membranes and causes an increase in the rate at which cells utilize carbohydrates and fats. The hormone's effect on amino acids, however, seems to be the more important one.

Although the exact mechanism for controlling growth hormone secretion is unknown, it appears to involve two substances from the hypothalamus (*growth hormone-releasing hormone* and *growth hormone release-inhibiting hormone*). A person's nutritional state also seems to play a role in the control of GH, for more of it is released during periods of protein deficiency and of abnormally low blood glucose concentration. Conversely, when blood protein and glucose levels are increased, there is a resulting decrease in growth hormone secretion.

If growth hormone is not secreted in sufficient amounts during childhood, body growth is limited, and a type of *dwarfism* (hypopituitary dwarfism) results. In this condition, body parts are usually correctly proportioned and mental development is normal. However, an abnormally low secretion of growth hormone is usually accompanied by lessened secretions from other anterior lobe hormones, leading to additional hormone deficiency symptoms. For example, a hypopituitary dwarf often fails to develop adult sexual features unless hormone therapy is provided.

An oversecretion of growth hormone during childhood may result in *gigantism*—a condition in which the person's height may exceed 8 feet. Gigantism, which is relatively rare, is usually accompanied by a tumor of the pituitary gland. In such cases, various pituitary hormones in addition to GH are likely to be secreted excessively, so that a giant often suffers from a variety of metabolic disturbances and has a shortened life expectancy.

Prolactin (PRL) functions in milk production in females. (It has little effect in males.) More specifically, prolactin stimulates and sustains milk production following the birth of an infant. This action is discussed in chapter 20.

Thyroid-stimulating hormone (TSH) has as its major function the control of the secretions from the thyroid gland, which are described in a later section of this chapter.

TSH secretion is partially regulated by the hypothalamus, which secretes *thyrotropin-releasing hormone* (TRH). TSH secretion also is regulated by circulating thyroid hormones that exert an inhibiting effect on the release of TRH and TSH; therefore, as the blood concentration of thyroid hormones increases, the secretions of TRH and TSH are reduced. (See fig. 12.9.)

---

1. *How does growth hormone affect the synthesis of proteins?*
2. *What is the function of prolactin?*
3. *How is TSH secretion regulated?*

Adrenocorticotropic hormone (ACTH) controls the manufacture and secretion of certain hormones from the outer layer, or *cortex,* of the adrenal gland. These adrenal hormones are discussed in a later section of this chapter.

The secretion of ACTH may be regulated in part by a *corticotropin-releasing hormone* (CRH), which some investigators believe is released from the hypothalamus in response to decreased concentrations of adrenal cortical hormones. Also, various forms of stress serve to stimulate the secretion of ACTH.

Follicle-stimulating hormone (FSH) and **luteinizing hormone** (LH) are called *gonadotropins,* which means they exert their actions on the gonads or reproductive organs. The functions of these gonadotropins and the ways they interact with each other are discussed in chapter 20.

---

1. *What is the function of ACTH?*
2. *What is a gonadotropin?*

Posterior Lobe Hormones

Unlike the anterior lobe of the pituitary gland, which is composed primarily of glandular epithelial cells, the posterior lobe consists largely of nerve fibers and neuroglial cells. The neuroglial cells function to support the nerve fibers, which originate in the hypothalamus.

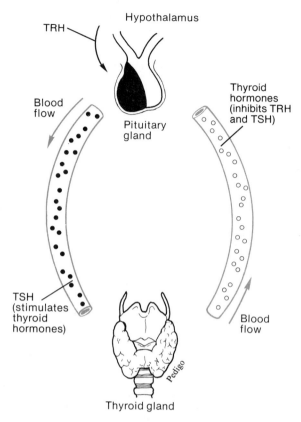

Fig. 12.9 TRH from the hypothalamus stimulates the anterior pituitary gland to release TSH. TSH stimulates the thyroid gland to release hormones, which in turn cause the hypothalamus to reduce its secretion of TRH.

The two hormones associated with the posterior lobe—antidiuretic hormone (ADH) and oxytocin—are actually produced by specialized neurons in the hypothalamus. These substances travel down axons through the pituitary stalk to the posterior lobe and are stored in granules near the ends of the axons. Then the hormones are released into the blood in response to nerve impulses coming from the hypothalamus. (See fig. 12.8.)

A *diuretic* is a substance that acts to increase urine production. An *antidiuretic,* then, is a chemical that inhibits urine formation. **Antidiuretic hormone** (ADH) produces an antidiuretic effect by acting on the kidneys and causing them to reduce the amount of water they excrete. In this way, ADH is important in regulating the water concentration of body fluids.

The secretion of ADH is regulated by the hypothalamus. Certain neurons in this part of the brain called *osmoreceptors* are sensitive to changes in the water concentration of body fluids. If, for example, a person is dehydrating due to lack of water intake, the solutes in blood become more and more concentrated. The osmoreceptors can sense this change and signal

Chart 12.1 Hormones of the pituitary gland

Anterior Lobe

Hormone	Action	Source of Control
Growth hormone (GH)	Stimulates increase in size and rate of reproduction of body cells; enhances movement of amino acids through membranes	Growth hormone-releasing hormone and growth hormone release-inhibiting hormone from the hypothalamus
Prolactin (PRL)	Sustains milk production after birth	Secretion restrained by prolactin release-inhibiting factor and stimulated by prolactin-releasing factor from the hypothalamus
Thyroid-stimulating hormone (TSH)	Controls secretion of hormones from the thyroid gland	Thyrotropin-releasing hormone (TRH) from the hypothalamus
Adrenocorticotropic hormone (ACTH)	Controls secretion of certain hormones from the adrenal cortex	Corticotropin-releasing hormone (CRH) from the hypothalamus
Follicle-stimulating hormone (FSH)	Responsible for the development of egg-containing follicles in ovaries; stimulates follicle cells to secrete estrogen; in male stimulates the production of sperm cells	Gonadotropin-releasing hormone from the hypothalamus
Luteinizing hormone (LH)	Promotes secretion of sex hormones; plays role in release of egg cell in females	Gonadotropin-releasing hormone from the hypothalamus

Posterior Lobe

Hormone	Action	Source of Control
Antidiuretic hormone (ADH)	Causes kidneys to reduce water excretion; in high concentration, causes blood pressure to rise	Hypothalamus in response to changes in blood water concentration
Oxytocin	Causes contractions of muscles in uterine wall; causes muscles associated with milk-secreting glands to contract	Hypothalamus in response to stretch in uterine and vaginal walls and stimulation of breasts

From Hole, John W., Jr., *Human Anatomy and Physiology 3d ed.* © 1978, 1981, 1984 Wm. C. Brown Publishers, Dubuque, Iowa. All Rights Reserved. Reprinted by permission.

the posterior lobe to release ADH. The ADH is transmitted by the blood to the kidneys, and as a result of its effects, less urine is produced. This action conserves water.

If, on the other hand, a person drinks an excess amount of water, the body fluids become more dilute and the release of ADH is inhibited. Consequently, the kidneys excrete more dilute urine until the water concentration of the body fluids returns to normal.

If any parts involved in the ADH regulating mechanism are damaged due to an injury or a tumor, the hormone may not be synthesized or released normally. The resulting ADH deficiency produces a condition called *diabetes insipidus*. This disease is characterized by an output of as much as 25 to 30 liters of very dilute urine per day and a rise in the concentration of dissolved substances in body fluids. It also is accompanied by a sensation of great thirst.

Oxytocin also has an antidiuretic action, but it is weaker in this respect than ADH. In addition, it can cause contractions of the smooth muscles in the uterine wall and may play a role in the later stages of childbirth by stimulating uterine contractions. The mechanism that triggers the release of oxytocin is not clearly understood.

Oxytocin also has an effect upon the breasts, causing contractions in certain cells associated with milk-producing glands and their ducts. In lactating breasts, this action forces liquid from the milk glands into the milk ducts—an effect that is necessary before milk can be removed.

Chart 12.1 reviews the hormones of the pituitary gland.

1. *What is the function of ADH?*
2. *What effects does oxytocin produce?*

The Thyroid Gland

The **thyroid gland** (fig. 12.10) is a very vascular structure that consists of two large lobes connected by a broad isthmus. It is located just below the larynx on either side and in front of the trachea.

Structure of the Gland

The thyroid gland is covered by a capsule of connective tissue and is made up of many secretory parts called *follicles*. The cavities of the follicles are lined with a single layer of cuboidal epithelial cells and are filled with a clear, viscous substance called *colloid*. The follicle cells produce and secrete hormones that may be stored in the colloid or released into the blood of nearby capillaries. (See fig. 12.10.)

Thyroid Hormones

The thyroid gland produces several hormones. Some, which are synthesized in the follicles, have marked effects on the metabolic rates of body cells. Another hormone, produced by the extrafollicular cells of the gland, influences the levels of blood calcium and phosphate.

Of the hormones that affect cellular metabolic rates, the most important are **thyroxine and triiodothyronine.** They help regulate the metabolism of carbohydrates, lipids, and proteins. For example, they increase the rate at which cells release energy from carbohydrates; they enhance the rate of protein synthesis; and they stimulate the breakdown and mobilization of lipids. As a result of their actions, these hormones are needed for normal growth and development of children. They also are essential for the maturation of the nervous system.

Before follicle cells can produce thyroxine and triiodothyronine, they must be supplied with iodine salts (iodides). Such salts are normally obtained from foods, and after they have been absorbed from the intestine, they are carried by the blood to the thyroid gland. An efficient active transport mechanism moves the iodides into the follicle cells, where they are used in the synthesis of the hormones. As was explained previously, the release of these hormones is controlled by the hypothalamus and pituitary gland.

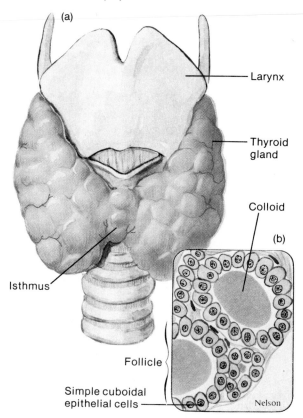

Fig. 12.10 The thyroid gland consists of two lobes connected anteriorly by an isthmus.

(a)

Larynx

Thyroid gland

Colloid

(b)

Isthmus

Follicle

Simple cuboidal epithelial cells

Nelson

The thyroid hormone that influences blood calcium and phosphate levels is called **calcitonin.** This substance helps regulate the concentrations of these ions by inhibiting the rates at which they leave the bones and enter the extracellular fluids.

This is accomplished by inhibiting the bone resorbing activity of osteoclasts. (See chapter 8.) At the same time, calcitonin causes an increase in the rate at which calcium and phosphate are deposited in bone matrix by stimulating the activity of osteoblasts. It also increases the excretion of calcium and phosphate ions by the kidneys. Thus, calcitonin acts to lower the blood calcium and phosphate levels.

The secretion of calcitonin is thought to be controlled directly by the blood calcium level. As this level increases, so does the secretion of calcitonin.

Chart 12.2 reviews the actions and controls of the thyroid hormones.

Chart 12.2 Hormones of the thyroid gland

Hormone	Action	Source of Control
Thyroxine	Increases rate of energy release from carbohydrates; increases rate of protein synthesis; accelerates growth; stimulates activity in the nervous system	TSH from the anterior pituitary gland
Triiodothyronine	Same as above	Same as above
Calcitonin	Lowers blood calcium and phosphate levels by inhibiting the release of calcium and phosphate ions from bones	Blood calcium level

From Hole, John W., Jr., *Human Anatomy and Physiology 3d ed.* © 1978, 1981, 1984 Wm. C. Brown Publishers, Dubuque, Iowa. All Rights Reserved. Reprinted by permission.

Most functional disorders of the thyroid gland are characterized by *overactivity* (hyperthyroidism) or *underactivity* (hypothyroidism) of the glandular cells. One form of *hypothyroidism* appears in infants when their thyroid glands fail to function normally. An affected child may appear normal at birth because it has received an adequate supply of thyroid hormones from its mother during pregnancy. When its own thyroid gland fails to produce sufficient quantities of hormones, the child soon develops a condition called *cretinism.* Cretinism is characterized by severe symptoms including stunted growth, abnormal bone formation, retarded mental development, low body temperature, and sluggishness. Without treatment with thyroid hormones within a month or so following birth, the child is likely to suffer from permanent mental retardation.

Hyperthyroidism is characterized by an elevated metabolic rate, restlessness, and overeating. Also, the person's eyes are likely to protrude (exophthalmos) because of edematous swelling in the tissues behind them. At the same time, the thyroid gland is likely to enlarge, producing a bulge in the neck called a *goiter.*

1. Where is the thyroid gland located?
2. What hormones of the thyroid gland affect carbohydrate metabolism and protein synthesis?
3. How does the thyroid gland influence the concentration of blood calcium?

The Parathyroid Glands

The **parathyroid glands** are located on the posterior surface of the thyroid gland, as shown in figure 12.11. Usually there are four of them—two associated with each of the thyroid's lateral lobes.

Fig. 12.11 The parathyroid glands are embedded in the posterior surface of the thyroid gland.

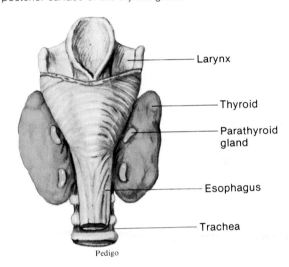

Larynx

Thyroid

Parathyroid gland

Esophagus

Trachea

Pedigo

Structure of the Glands

Each parathyroid gland is a small, yellowish brown structure covered by a thin capsule of connective tissue. The body of the gland consists of numerous tightly packed secretory cells that are closely associated with capillary networks.

Parathyroid Hormone

The only hormone known to be secreted by the parathyroid glands is called **parathyroid hormone** (PTH). This substance causes an increase in the blood calcium concentration and a decrease in the blood phosphate level. It apparently does this by influencing actions in bones, kidneys, and the intestine.

Fig. 12.12 The parathyroid and thyroid glands function to control the concentration of blood calcium.

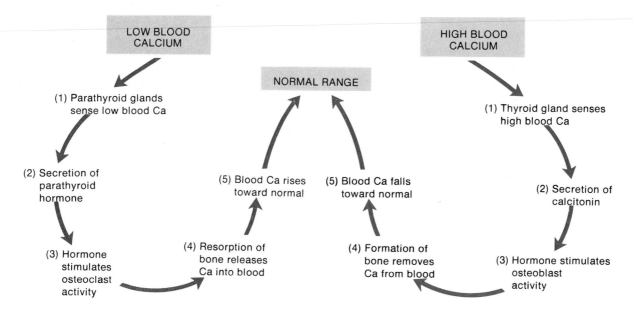

As is described in chapter 8, the intercellular matrix of bone tissue contains a considerable amount of mineral salts, including calcium phosphate. PTH seems to stimulate bone resorption by osteocytes and osteoclasts and to inhibit the activity of osteoblasts. (See chapter 8.) As a result of this increased resorption, calcium and phosphate ions are released from bone, and the blood concentrations of these substances increase. At the same time, PTH causes the kidneys to conserve blood calcium and to increase the excretion of phosphate ions in the urine. It also stimulates the absorption of calcium from food in the intestine so that the blood calcium level rises still more.

The parathyroid glands seem to be free of control by the hypothalamus. Instead, the secretion of PTH is regulated by a *negative feedback mechanism* operating between the glands and the blood calcium level. As the level of blood calcium rises, less PTH is secreted; as the level of blood calcium drops, more PTH is released. (See fig. 12.12.)

Hyperparathyroidism is most often caused by a tumor associated with a parathyroid gland. The resulting increase in PTH secretion stimulates excessive osteoclastic activity, and as bone tissue is resorbed, the bones become soft, deformed, and subject to spontaneous fractures.

The excessive calcium and phosphate released into the body fluids as a result of extra PTH secretion may be deposited in abnormal places, causing new problems such as kidney stones.

Hypoparathyroidism can result from an injury to the parathyroids or from surgical removal of these glands. In either case, decreased PTH secretion is reflected in reduced osteoclastic activity, and although the bones remain strong, the blood calcium level drops. At the same time, the nervous system may become abnormally excitable, and impulses may be triggered spontaneously. As a result, muscles may undergo tetanic contractions, and the person may die due to a failure of respiratory movements.

1. Where are the parathyroid glands located?
2. How does parathyroid hormone help to regulate the concentrations of blood calcium and phosphate?

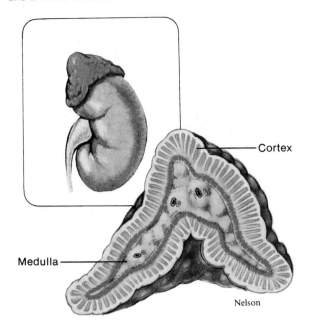

Cortex

Medulla

Nelson

The Adrenal Glands

The **adrenal glands** are located in close association with the kidneys. A gland sits atop each kidney like a cap and is embedded in the mass of fat that encloses the kidney.

Structure of the Glands

Each gland is very vascular and consists of two parts, as shown in figure 12.13. The central portion is termed the adrenal medulla, and the outer part is the adrenal cortex. Although these regions are not sharply divided, they represent distinct glands that secrete different hormones.

The **adrenal medulla** consists of irregularly shaped cells that are arranged in groups around blood vessels. These cells are intimately connected with the sympathetic division of the autonomic nervous system. They are modified postganglionic cells, and preganglionic autonomic nerve fibers lead to them from the central nervous system. (See chapter 10.)

The **adrenal cortex,** which makes up the bulk of the gland, is composed of closely packed masses of epithelial cells that are arranged in layers. These layers form an outer, a middle, and an inner zone of the cortex. As in the case of the medulla, the cells of the adrenal cortex are well supplied with blood vessels.

1. *Where are the adrenal glands located?*
2. *Describe the two portions of an adrenal gland.*

Hormones of the Adrenal Medulla

Cells of the adrenal medulla secrete two closely related hormones, **epinephrine** and **norepinephrine.** These substances have similar molecular structures and physiological functions. In fact, epinephrine, which makes up 80% of the medullary secretion, is produced from norepinephrine by the action of an enzyme.

The effects of the medullary hormones generally resemble those occurring when sympathetic nerve fibers stimulate their effectors. The hormonal effects, however, last up to ten times longer because the hormones are removed from the tissues relatively slowly. These effects include increased heart rate and increased force of cardiac muscle contraction, elevated blood pressure, increased breathing rate, and decreased activity in the digestive system.

Impulses arriving by way of sympathetic nerve fibers stimulate the adrenal medulla to release its hormones at the same time other effectors are being stimulated by sympathetic impulses. As a rule, these impulses originate in the hypothalamus in response to various types of stress. Thus the medullary secretions function together with the sympathetic division of the autonomic nervous system in preparing the body for energy-expending action—fight or flight.

Although a lack of medullary hormones produces no significant effects, tumors in the adrenal medulla are sometimes accompanied by excessive hormone secretions. Affected persons show signs of prolonged sympathetic responses—high blood pressure, increased heart rate, elevated blood sugar, and so forth. Treatment for this condition generally involves surgical removal of the tumorous growth.

1. *Name the hormones secreted by the adrenal medulla.*
2. *What effects are produced by these hormones?*
3. *What is the usual stimulus for their release?*

The cells of the adrenal cortex produce over thirty different steroids, among which are several hormones. Unlike the adrenal medullary hormones, which a person can live without, some of those released by the cortex are vital. In fact, in the absence of cortical secretions, a person usually dies within a week unless extensive electrolyte therapy is provided.

The most important of the cortical hormones are aldosterone, cortisol, and sex hormones.

Aldosterone. **Aldosterone** is synthesized by cells in the outer zone of the adrenal cortex. It is called a *mineralocorticoid* because it helps to regulate the concentration of mineral electrolytes, such as sodium and potassium. More specifically, aldosterone causes sodium ions to be conserved and potassium ions to be excreted by the kidneys. At the same time, it promotes the conservation of water and reduces urine output. By promoting the conservation of sodium ions and water, aldosterone causes the blood sodium concentration and the blood volume to increase. This in turn increases the blood pressure. (See fig. 12.14.)

The cells that secrete aldosterone are stimulated by a decrease in the blood concentration of sodium ions or an increase in the concentration of potassium ions.

Cortisol. **Cortisol** (hydrocortisone) is a *glucocorticoid*, which means it affects glucose metabolism. It is produced in the middle zone of the adrenal cortex and has a molecular structure similar to aldosterone. In addition to affecting glucose, cortisol influences protein and fat metabolism.

Among the more important actions of this hormone are the following:

1. It inhibits the synthesis of protein in various tissues, thus causing an increase in the concentration of amino acids in the blood.

2. It promotes the release of fatty acids from adipose tissue, thus causing an increase in the use of fatty acids as an energy source and a decrease in the use of glucose for this purpose.

3. It stimulates the liver cells to form glucose from noncarbohydrates, such as circulating amino acids and glycerol, thus promoting an increase in the blood glucose concentration.

These actions help to keep the blood glucose concentration within the normal range between meals, since the supply of glycogen stored within the liver, (that is, used to provide glucose) can be exhausted in a few hours without food.

Fig. 12.14 Aldosterone secreted by the adrenal cortex causes the kidney to excrete potassium.

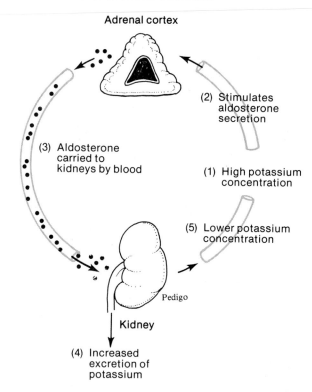

Adrenal cortex

(2) Stimulates aldosterone secretion

(3) Aldosterone carried to kidneys by blood

(1) High potassium concentration

(5) Lower potassium concentration

Pedigo

Kidney

(4) Increased excretion of potassium

The release of cortisol is controlled by a *negative feedback mechanism*, which involves the brain, anterior pituitary gland, and adrenal cortex. More specifically, the hypothalamus of the brain secretes CRH (corticotropin-releasing hormone) into the hypophyseal portal veins that were described previously. These vessels carry the CRH to the anterior pituitary gland, stimulating it to secrete ACTH. In turn, the ACTH causes the adrenal cortex to release cortisol. Cortisol has an inhibiting effect on the release of both CRH and ACTH, and as these substances decrease in concentration, less cortisol is produced.

Such a mechanism might act to maintain a relatively stable blood concentration of hormone, but in the case of cortisol, the set-point of the mechanism is changed from time to time. In this way, the output of hormone can be altered to meet the demands of changing conditions. For example, when the body is subjected to stressful conditions—injury, disease, extreme temperatures, extreme emotional feelings, and so forth—nerve impulses provide the brain with information concerning the stressful factors. In response, brain centers acting through the hypothalamus can cause the release of more cortisol, and can maintain a higher level of the hormone until the cause of the stress is reduced.

Chart 12.3 Hormones of the adrenal cortical glands

Hormone	Action	Regulation
Aldosterone	Helps regulate the concentration of extracellular electrolytes	Electrolyte concentrations in body fluids
Cortisol	Influences the metabolism of carbohydrates, proteins, and fats	CRH from hypothalamus and ACTH from the anterior pituitary gland
Adrenal sex hormones	Supplement the sex hormones from the gonads	

From Hole, John W., Jr., *Human Anatomy and Physiology 3d ed.* © 1978, 1981, 1984 Wm. C. Brown Publishers, Dubuque, Iowa. All Rights Reserved. Reprinted by permission.

Sex Hormones. Adrenal sex hormones are produced by cells in the inner zone of the cortex. Although the hormones of this group are primarily male types (adrenal androgens), small quantities of female hormones (adrenal estrogens and progesterone) are also present. The normal functions of these hormones are not clear, but they may supplement the supply of sex hormones from the gonads and stimulate early development of the reproductive organs. Also, there is some evidence that the adrenal androgens play a role in stimulating the female sex drive.

Hyposecretion of cortical hormones leads to *Addison's disease,* a condition characterized by decreased blood sodium, increased blood potassium, low blood glucose concentration (hypoglycemia), dehydration, low blood pressure, and increased skin pigmentation. Treatment for Addison's disease involves the use of mineralocorticoids and glucocorticoids. An untreated person is likely to live only a few days because of severe disturbances in electrolyte balance.

Hypersecretion of cortical hormones, which may be associated with an adrenal tumor or with an oversecretion of ACTH by the anterior pituitary gland, results in *Cushing's disease.* A person with this condition suffers from changes in carbohydrate and protein metabolism and in electrolyte balance. For example, when mineralocorticoids and glucocorticoids are excessive, the blood glucose concentration remains high, and there is a great decrease in tissue protein. Also, sodium is retained abnormally, and as a result, tissue fluids tend to increase, and the skin becomes puffy. At the same time, an increase in adrenal sex hormones may produce masculinizing effects in a female, such as growth of a beard or development of a deeper voice.

---

1. *Name the most important hormones of the adrenal cortex.*
2. *What is the function of aldosterone?*
3. *What actions does cortisol produce?*

Chart 12.3 summarizes the characteristics of the hormones produced by the adrenal cortex.

The Pancreas

The **pancreas** contains two major types of secretory tissues, reflecting its dual function as an exocrine gland that secretes digestive juice, and an endocrine gland that releases hormones.

Structure of the Gland

The pancreas is an elongated, somewhat flattened organ that lies back of the stomach and behind the parietal peritoneum. (See fig. 13.18.) It is attached to the first section of the small intestine (duodenum) by a duct that transports its digestive juice to the intestine.

The endocrine portion of the pancreas consists of cells arranged in groups closely associated with blood vessels. These groups, called *islets of Langerhans,* include two distinct types of cells—alpha cells that secrete the hormone glucagon and beta cells that secrete the hormone insulin. (See fig. 12.15.)

The digestive functions of the pancreas are discussed in chapter 13.

Hormones of the Islets of Langerhans

Glucagon stimulates the liver to convert glycogen and certain noncarbohydrates such as amino acids into glucose. This action causes the blood glucose concentration to rise. Although its effect is similar to the action of epinephrine, glucagon is many times more effective in elevating blood sugar.

The secretion of glucagon is regulated by a negative feedback system in which a low concentration of blood sugar stimulates the release of the hormone from the alpha cells. When the blood sugar concentration rises, glucagon is no longer secreted. This mechanism prevents hypoglycemia from occurring when the glucose concentration is relatively low, such as between meals, or when glucose is being used rapidly—during periods of exercise, for example.

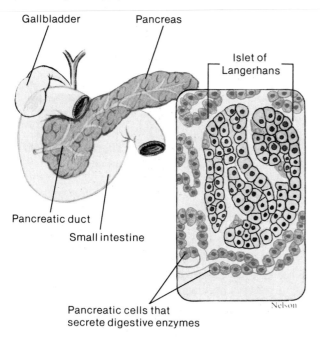

Fig. 12.15 The hormone-secreting cells of the pancreas are arranged in clusters or islets surrounded by cells that secrete digestive enzymes.

Gallbladder

Pancreas

Islet of Langerhans

Pancreatic duct

Small intestine

Pancreatic cells that secrete digestive enzymes

Nelson

Chart 12.4 Hormones of the islets of Langerhans

Hormone	Action	Source of Control
Glucagon	Stimulates the liver to convert glycogen and noncarbohydrates into glucose	Blood glucose concentration
Insulin	Promotes formation of glycogen from glucose; inhibits the conversion of noncarbohydrates into glucose; enhances movement of glucose into adipose and muscle cells	Blood glucose concentration

At the same time that insulin is decreasing, glucagon secretion is increasing, so that these hormones function together to help maintain a relatively stable blood glucose concentration, despite great variation in the amount of carbohydrates eaten. (See fig. 12.16.)

Chart 12.4 summarizes the hormones of the islets of Langerhans.

A deficiency of insulin (hypoinsulinism) results in a disease called *diabetes mellitus*. This condition is characterized by severe upsets in carbohydrate metabolism as well as by disturbances in protein and fat metabolism. More specifically, there is a decrease in movement of glucose into adipose and muscle cells, a decrease in glycogen formation, and a consequent rise in the concentration of blood sugar (hyperglycemia). When the blood sugar reaches a certain high concentration, the kidneys begin to excrete the excess, and glucose appears in the urine (glycosuria). The sugar raises the osmotic pressure of the urine, normal kidney functions are upset, and more water and electrolytes (particularly sodium ions) are excreted than usual. As a result of excessive urine output (diuresis or polyuria), the affected person becomes dehydrated and experiences great thirst (polydipsia).

Meanwhile, because many body cells are unable to obtain adequate amounts of glucose, they begin to utilize abnormal quantities of fat.

Abnormal fat metabolism may lead to an increase in the concentration of ketone bodies in the blood. If they accumulate excessively these substances can cause a condition called *acidosis*. In this instance the pH of the body fluids drops, and the affected person may lose consciousness (*diabetic coma*) and die.

The main effect of the hormone **insulin** is exactly opposite to that of glucagon. It acts on the liver to stimulate the formation of glycogen from glucose and inhibits the conversion of noncarbohydrates into glucose. Insulin also has a special effect in promoting the facilitated diffusion (see chapter 3) of glucose through the membranes of certain cells, including those of adipose and muscle tissues. As a result of these actions, insulin causes a decrease in the concentration of blood glucose. In addition, it promotes the transport of amino acids into cells, increases the synthesis of proteins, and stimulates adipose cells to synthesize and store fat.

As in the case of glucagon, insulin secretion is regulated by a negative feedback system which is sensitive to the concentration of blood glucose. When glucose is relatively high, as may occur following a meal, insulin is released from the beta cells. By promoting the formation of glycogen in the liver and the entrance of glucose into adipose and muscle cells, the hormone helps to prevent an excessive rise in the blood glucose concentration. Then, when the glucose concentration is relatively low, between meals or during the night, less insulin is released.

As insulin decreases, less and less glucose enters the adipose and muscle cells, and the remaining glucose in the blood is available for use by nerve cells. Neurons are dependent upon a continuous supply of glucose for ATP production, and they obtain it from the blood by simple diffusion.

Fig. 12.16 Insulin and glucagon function together to help maintain a relatively stable blood glucose concentration.

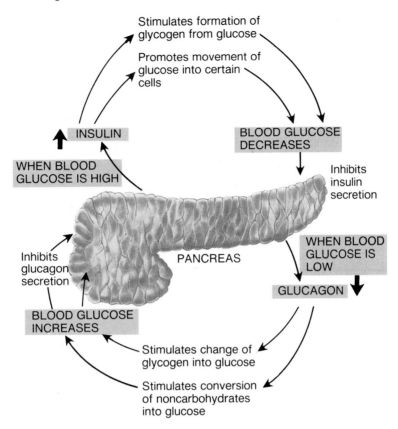

Stimulates formation of glycogen from glucose

Promotes movement of glucose into certain cells

INSULIN

WHEN BLOOD GLUCOSE IS HIGH

BLOOD GLUCOSE DECREASES

Inhibits insulin secretion

Inhibits glucagon secretion

PANCREAS

WHEN BLOOD GLUCOSE IS LOW

GLUCAGON

BLOOD GLUCOSE INCREASES

Stimulates change of glycogen into glucose

Stimulates conversion of noncarbohydrates into glucose

---

1. *What is the name of the endocrine portion of the pancreas?*
2. *What is the function of glucagon?*
3. *What is the function of insulin?*
4. *How are the secretions of these hormones controlled?*

Other Endocrine Glands

Other organs that produce hormones and thus are parts of the endocrine system include the pineal gland, thymus gland, reproductive glands, and certain glands of the digestive tract.

The Pineal Gland

The **pineal gland** is a small structure located deep between the cerebral hemispheres where it is attached to the upper portion of the thalamus near the roof of the third ventricle. Its body consists largely of specialized *pineal cells* and *neuroglial cells* that provide support. (See fig. 10.28.)

Although the pineal gland does not seem to have any direct nerve connections to other brain parts, its functional cells are innervated by postganglionic sympathetic nerve fibers. Norepinephrine released from these fibers apparently stimulates the pineal cells.

While the function of the pineal gland remains in doubt, there is evidence that it secretes at least one hormone, **melatonin.** This substance is thought to inhibit the secretion of gonadotropins from the anterior pituitary gland.

The Thymus Gland

The **thymus gland,** which lies in the mediastinum behind the sternum and between the lungs, is relatively large in young children but diminishes in size with age. It is believed that this gland secretes a hormone, called **thymosin,** that affects the production of certain white blood cells (lymphocytes). This gland is discussed in chapter 17.

The reproductive organs that secrete important hormones include the **ovaries,** which produce estrogens and progesterone; the **placenta,** which produces estrogens, progesterone, and gonadotropin; and the **testes,** which produce testosterone. These glands and their secretions are discussed in chapter 20.

The Digestive Glands

The digestive glands that secrete hormones are generally associated with the linings of the stomach and small intestine. These structures and their secretions are described in chapter 13.

---

1. *Where is the pineal gland located?*
2. *What seems to be the function of the pineal gland?*
3. *Where is the thymus gland located?*
4. *Which reproductive organs secrete hormones?*

Clinical Terms Related to the Endocrine System

adrenalectomy (ah-dre″nah-lek′to-me)—surgical removal of the adrenal glands.

adrenogenital syndrome (ah-dre″no-jen′i-tal sin′drōm)—a group of symptoms associated with changes in sexual characteristics as a result of increased secretion of adrenal androgens.

diabetes insipidus (di″ah-be′tēz in-sip′idus)—a metabolic disorder characterized by a large output of dilute urine containing no sugar, and caused by a decreased secretion of ADH by the posterior pituitary gland.

exophthalmos (ek″sof-thal′mos)—an abnormal protrusion of the eyes.

hirsutism (her′sūt-izm)—excessive growth of hair.

hyperglycemia (hi″per-gli-se′me-ah)—an excess of blood glucose.

hypocalcemia (hi″po-kal-se′me-ah)—a deficiency of blood calcium.

hypoglycemia (hi″po-gli-se′me-ah)—a deficiency of blood glucose.

hypophysectomy (hi-pof″i-sek′to-me)—surgical removal of the pituitary gland.

parathyroidectomy (par″ah-thi″roi-dek′to-me)—surgical removal of the parathyroid glands.

pheochromocytoma (fe-o-kro″mo-si-to′-mah)—a type of tumor found in the adrenal medulla and usually accompanied by high blood pressure.

polyphagia (pol″e-fa′je-ah)—excessive eating.

Simmond's disease (sim′ondz di-zēz′)—a disorder characterized by excessive weight loss and caused by destruction of the pituitary gland.

thymectomy (thi-mek′to-me)—surgical removal of the thymus gland.

thyroidectomy (thi″roi-dek′to-me)—surgical removal of the thyroid gland.

thyroiditis (thi″roi-di′tis)—inflammation of the thyroid gland.

virilism (vir′ī-lizm)—masculinization of a female.

Chapter Summary

Introduction

Endocrine glands secrete their products into body fluids; exocrine glands secrete into ducts that lead to a body surface.

General Characteristics of the Endocrine System

1. As a group, endocrine glands are concerned with the regulation of metabolic processes.

2. Some hormones have localized effects while others produce more general actions.

Hormones and Their Actions

Endocrine glands secrete hormones that have effects on specific target cells.

Hormones are very potent substances.
1. Actions of hormones
   a. Hormones may alter the rates of protein synthesis, enzyme action, or membrane transport.
   b. Hormones combine with receptor molecules in cells or on their membranes.
   c. Steroid hormones enter cells and activate genes.
   d. Protein, peptide, and amine hormones cause cyclic AMP to be produced, and it acts to trigger cellular changes.

2. Prostaglandins
   Prostaglandins are substances present in small quantities that have powerful hormonelike effects.

Control of Hormone Secretions

The concentration of each hormone in the body fluids is regulated.

1. Negative feedback systems
   a. When an imbalance in hormone concentration occurs, information is fed back to some part that acts to correct this imbalance.
   b. Negative feedback systems maintain relatively stable hormone concentrations.

2. Positive feedback systems
   a. In positive feedback systems, a gland is stimulated to increase its rate of secretion by a substance it causes to be produced.
   b. Positive feedback systems produce unstable conditions.

3. Nerve control
   a. Some endocrine glands secrete their hormones in response to nerve impulses.
   b. Other glands secrete hormones in response to releasing hormones.

The Pituitary Gland

The pituitary gland has an anterior lobe and a posterior lobe.

Most pituitary secretions are controlled by the hypothalamus.

1. Anterior lobe hormones
   a. The anterior lobe secretes GH, PRL, TSH, ACTH, FSH, and LH.
   b. Growth hormone (GH)
      (1) Growth hormone stimulates body cells to increase in size and undergo an increased rate of reproduction.
      (2) The secretion of GH is controlled by growth hormone-releasing hormone and growth hormone release-inhibiting hormone from the hypothalamus.
   c. Prolactin promotes breast development and stimulates milk production.
   d. Thyroid-stimulating hormone (TSH)
      (1) TSH controls the secretion of hormones from the thyroid gland.
      (2) TSH secretion is regulated by the hypothalamus, which secretes TRH.
   e. Adrenocorticotropic hormone (ACTH)
      (1) ACTH controls the secretion of hormones from the adrenal cortex.
      (2) The secretion of ACTH is regulated by the hypothalamus, which secretes CRH.
   f. Follicle-stimulating hormone (FSH) and luteinizing hormone (LH) are gonadotropins.

2. Posterior lobe hormones
   a. This lobe consists largely of neuroglial cells and nerve fibers.
   b. The hormones are produced in the hypothalamus.
   c. Antidiuretic hormone (ADH)
      (1) ADH causes the kidneys to reduce the amount of water they excrete.
      (2) The secretion of ADH is regulated by the hypothalamus.
   d. Oxytocin
      (1) Oxytocin can cause muscles in the uterine wall to contract.
      (2) It also causes contraction of certain cells associated with the production and ejection of milk.

The Thyroid Gland

The thyroid gland is located in the neck and has a special ability to remove iodine from the blood.

1. Structure of the gland
   a. The thyroid gland consists of many follicles.
   b. The follicles are fluid-filled and store the hormones.

2. Thyroid hormones
   a. The thyroid hormones thyroxine and triiodothyronine cause the metabolic rate of cells to increase, enhance protein synthesis, and stimulate utilization of lipids.
   b. The thyroid hormone calcitonin helps to regulate the concentrations of blood calcium and phosphate.

The Parathyroid Glands

The parathyroid glands are located on the posterior surface of the thyroid.

1. Structure of the glands
   Each gland consists of secretory cells that are well supplied with capillaries.

2. Parathyroid hormone (PTH)
   a. PTH causes an increase in blood calcium and a decrease in blood phosphate concentrations.
   b. The parathyroid glands seem to be regulated by a negative feedback mechanism that operates between the glands and the blood.

The Adrenal Glands

The adrenal glands are located atop the kidneys.

1. Structure of the glands
   a. Each gland consists of a medulla and a cortex.
   b. These parts represent separate glands.

2. Hormones of the adrenal medulla
   a. The adrenal medulla secretes epinephrine and norepinephrine that have similar effects.
   b. The secretion of these hormones is stimulated by sympathetic impulses.

3. Hormones of the adrenal cortex
   a. The adrenal cortex produces several steroid hormones.
   b. Aldosterone is a mineralocorticoid that causes the kidneys to conserve sodium and water and to excrete potassium.
   c. Cortisol is a glucocorticoid that affects carbohydrate, protein, and fat metabolism.
   d. Adrenal sex hormones
      (1) These hormones are primarily of the male type.
      (2) They are thought to supplement the sex hormones produced by the gonads.

The Pancreas

The pancreas secretes digestive juices as well as hormones.

1. Structure of the gland
   a. The pancreas is attached to the small intestine.
   b. The islets of Langerhans secrete glucagon and insulin.

2. Hormones of the islets of Langerhans
   a. Glucagon stimulates the liver to produce glucose from glycogen and noncarbohydrates.
   b. Insulin promotes the movement of glucose through some cell membranes, stimulates the storage of glucose, promotes the synthesis of proteins, and stimulates the storage of fats.

Other Endocrine Glands

1. The pineal gland
   a. The pineal gland is attached to the thalamus.
   b. It secretes melatonin, which seems to inhibit the secretion of gonadotropins from the anterior pituitary gland.

2. The thymus gland
   a. The thymus lies behind the sternum and between the lungs.
   b. It secretes thymosin that affects the production of lymphocytes.

3. The reproductive glands
   a. The ovaries secrete estrogens and progesterone.
   b. The placenta secretes estrogens, progesterone, and gonadotropin.
   c. The testes secrete testosterone.

4. The digestive glands
   Certain glands of the stomach and small intestine secrete hormones.

Application of Knowledge

1. What hormones would need to be administered to an adult whose anterior pituitary gland had been removed? Why?

2. How might the environment of a patient with hyperthyroidism be modified to minimize the drain on body energy resources?

Review Activities

1. Explain what is meant by an endocrine gland.
2. Define *hormone* and *target cell*.
3. Explain how hormones produce their effects.
4. Distinguish between a negative and a positive feedback system.
5. Define *releasing hormone* and provide an example of such a substance.
6. Describe the location and structure of the pituitary gland.
7. List the hormones secreted by the anterior lobe of the pituitary gland.
8. Explain how pituitary gland activity is controlled by the brain.
9. Explain how growth hormone produces its effects.
10. List the major factors that affect the secretion of growth hormone.
11. Summarize the function of prolactin.
12. Describe the mechanism that regulates the concentrations of circulating thyroid hormones.
13. Explain how the secretion of ACTH is controlled.
14. Compare the cellular structures of the anterior and posterior lobes of the pituitary gland.
15. Describe the function of the posterior pituitary hormones.
16. Explain how the release of ADH is regulated.
17. Describe the location and structure of the thyroid gland.
18. Name the hormones secreted by the thyroid gland and list the general functions of each hormone.
19. Describe the location and structure of the parathyroid glands.
20. Explain the general functions of parathyroid hormone.
21. Describe the mechanism that regulates the secretion of parathyroid hormone.
22. Distinguish between the adrenal medulla and the adrenal cortex.
23. List the hormones produced by the adrenal medulla and describe their general functions.
24. Name the most important hormones of the adrenal cortex and describe the general functions of each.
25. Describe how the pituitary gland controls the secretion of adrenal cortical hormones.
26. Describe the location and structure of the pancreas.
27. List the hormones secreted by the pancreas and describe the general functions of each.
28. Summarize how the secretion of hormones from the pancreas is regulated.
29. Describe the location and general function of the pineal gland.
30. Describe the location and general function of the thymus gland.

Processing and Transporting

The chapters of unit 4 are concerned with the digestive, respiratory, circulatory, lymphatic, and urinary systems. They describe how organs of these systems obtain nutrients and oxygen from outside the body, how the nutrients are altered chemically and absorbed into body fluids, and how the nutrients and oxygen are transported to body cells. They also discuss processes by which these substances are utilized by cells, how resulting wastes are transported and excreted, and how stable concentrations of various substances in body fluids are maintained.

This unit includes

Digestion and Nutrition

13

Most food substances are composed of chemicals whose molecules cannot pass easily through cell membranes and so cannot be absorbed efficiently by cells. The *digestive system* functions to solve this problem. Its parts are adapted to ingest foods, to break large particles into smaller ones, to secrete enzymes that decompose food molecules, to absorb the products of this digestive action, and to eliminate the unused residues.

The substances in foods that are needed for the maintenance of health are called *nutrients,* and they include various carbohydrates, lipids, proteins, vitamins, and minerals. The processes involved with the ingestion and utilization of these nutrients is known as *nutrition.*

Chapter Objectives

After you have studied this chapter, you should be able to

1. Name and describe the location of the organs of the digestive system and their major parts.

2. Describe the general functions of each digestive organ and the liver.

3. Describe the structure of the wall of the alimentary canal.

4. Explain how the contents of the alimentary canal are mixed and moved.

5. List the enzymes secreted by the various digestive organs and describe the function of each enzyme.

6. Describe how digestive secretions are regulated.

7. Explain how the products of digestion are absorbed.

8. List the major sources of carbohydrates, lipids, and proteins.

9. Describe how carbohydrates, lipids, and proteins are utilized by cells.

10. List the fat-soluble and water-soluble vitamins, and summarize the general functions of each vitamin.

11. List the major minerals and trace minerals and summarize the general functions of each mineral.

12. Describe an adequate diet.

13. Complete the review activities at the end of this chapter. Note that the items are worded in the form of specific learning objectives. You may want to refer to them before reading the chapter.

absorption   (ab-sorp′shun)

alimentary canal   (al″ĭ-men′tar-e kah-nal′)

bile   (bīl)

cellulose   (sel′u-lōs)

chyme   (kīm)

deciduous   (de-sid′u-us)

gastric juice   (gas′trik jōōs)

intestinal juice   (in-tes′tĭ-nal jōōs)

intrinsic   (in-trin′sik)

mineral   (min′er-al)

mucous membrane   (mu′kus mem′brān)

nutrient   (nu′tre-ent)

nutrition   (nu-trish′un)

pancreatic juice   (pan″kre-at′ik jōōs)

peristalsis   (per″ĭ-stal′sis)

sphincter muscle   (sfingk′ter mus′el)

villi; singular, villus   (vil′i, vil′us)

vitamin   (vi′tah-min)

aliment-, food: *aliment*ary canal—the tubelike portion of the digestive system.

chym-, juice: *chym*e—semifluid paste of food particles and gastric juice formed in the stomach.

decidu-, falling off: *decidu*ous teeth—those that are shed during childhood.

gastr-, stomach: *gastr*ic gland—portion of the stomach that secretes gastric juice.

hepat-, liver: *hepat*ic duct—duct that carries bile from the liver to the common bile duct.

lingu-, the tongue: *lingu*al tonsil—mass of lymphatic tissue at the root of the tongue.

nutri-, nourishing: *nutri*ent—chemical substance needed to nourish body cells.

peri-, around: *peri*stalsis—wavelike ring of contraction that moves material along the alimentary canal.

pylor-, gatekeeper: *pylor*ic sphincter—muscle that serves as a valve between the stomach and small intestine.

vill-, hairy: *vill*i—tiny projections of mucous membrane in the small intestine.

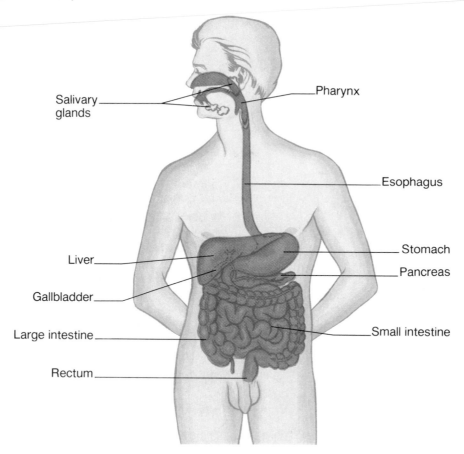

Salivary glands

Pharynx

Esophagus

Liver

Stomach

Pancreas

Gallbladder

Large intestine

Small intestine

Rectum

Digestion is the process by which food substances are changed into forms that can be absorbed through cell membranes. The **digestive system** includes the organs that promote this process. It consists of an *alimentary canal* that extends from the mouth to the anus and several *accessory organs* that release secretions into the canal. The alimentary canal includes the mouth, pharynx, esophagus, stomach, small intestine, and large intestine, while the accessory organs include the salivary glands, liver, gallbladder, and pancreas. Major organs of this system are shown in figure 13.1.

General Characteristics of the Alimentary Canal

The **alimentary canal** is a muscular tube about 9 meters (30 feet) long, which passes through the body's ventral cavity. Although it is specialized in various regions to carry on particular functions, the structure of its wall and the method by which it moves food are similar throughout its length.

Structure of the Wall

The wall of the alimentary canal consists of four distinct layers, although the degree to which they are developed varies from region to region. Beginning with the innermost tissues, these layers, shown in figure 13.2, include the following:

1. **Mucous membrane** or **mucosa.** This layer is formed of surface epithelium, underlying connective tissue, and a small amount of smooth muscle. In some regions, it develops folds and tiny projections which extend into the lumen of the digestive tube and increase its absorptive surface area. It also may contain glands, which are tubular invaginations into which lining cells secrete mucus and digestive enzymes. Thus, the mucosa functions to protect the tissues beneath it and to carry on absorption and secretion.

2. **Submucosa.** The submucosa contains considerable loose connective tissue as well as blood vessels, lymphatic vessels, and nerves. Its vessels serve to nourish surrounding tissues and to carry away absorbed materials.

Fig. 13.2 The wall of the alimentary canal includes four layers: mucous membrane, submucosa, muscular layer, and serous layer.

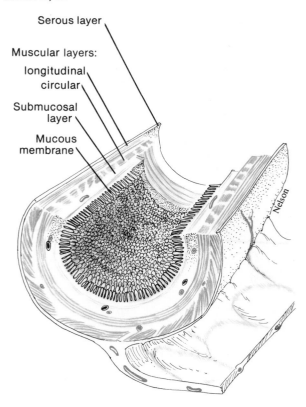

Serous layer

Muscular layers:
longitudinal
circular

Submucosal layer

Mucous membrane

Nelson

3. **Muscular layer.** This layer consists of two coats of smooth muscle tissue. The fibers of the inner coat are arranged so that they encircle the tube, and when these *circular fibers* contract, the diameter of the tube is decreased. The fibers of the outer muscular coat run lengthwise, and when these *longitudinal fibers* contract, the tube is shortened.

4. **Serous layer** (*serosa*). The serous, or outer, covering of the tube is composed of the *visceral peritoneum*. The cells of the serosa secrete serous fluid that keeps the tube's outer surface moist and lubricates it so that the organs within the abdominal cavity can slide freely against one another.

Movements of the Tube

The motor functions of the alimentary canal are of two basic types—mixing movements and propelling movements. Mixing occurs when smooth muscles in relatively small segments of the tube undergo rhythmic contractions. When the stomach is full, for example, waves of muscular contractions move through its walls from one end to the other. These waves tend to mix food substances with digestive juices secreted by the mucosa.

Propelling movements include a wavelike motion called **peristalsis.** When peristalsis occurs, a ring of contraction appears in the wall of the tube. At the same time, the muscular wall just ahead of the ring relaxes. As the peristaltic wave moves along, it pushes the tubular contents ahead of it.

---

1. *What organs constitute the digestive system?*
2. *Describe the wall of the alimentary canal.*
3. *Name the two types of movements that occur in the alimentary canal.*

The Mouth

The mouth is adapted to receive food and to prepare it for digestion by mechanically reducing the size of solid particles and mixing them with saliva. It is surrounded by the lips, cheeks, tongue, and palate, and includes a chamber between the palate and tongue, called the *oral cavity,* as well as a narrow space between the teeth, cheeks, and lips, called the *vestibule.* (See fig. 13.3.)

The Cheeks and Lips

The cheeks consist of outer layers of skin, pads of subcutaneous fat, certain muscles associated with expression and chewing, and inner linings of stratified squamous epithelium.

The lips are highly mobile structures. They contain skeletal muscles and a variety of sensory receptors that are useful in judging the temperature and texture of foods. Their normal reddish color is due to an abundance of blood vessels near their surfaces.

The Tongue

The **tongue** nearly fills the oral cavity when the mouth is closed. It is covered by mucous membrane and is anchored in the midline to the floor of the mouth by a membranous fold called the **frenulum.**

The body of the tongue is composed largely of skeletal muscle. These muscles aid in mixing food particles with saliva during chewing and in moving food toward the pharynx during swallowing. Rough projections, called **papillae,** on the surface of the tongue provide friction that is useful in handling food. These papillae also contain taste buds. (See chapter 11 and fig. 11.4.)

The posterior region, or *root,* of the tongue is anchored to the hyoid bone and is covered with rounded masses of lymphatic tissue called **lingual tonsils.** (See fig. 13.4.)

Fig. 13.3 The mouth is adapted for ingesting food and preparing it for digestion.

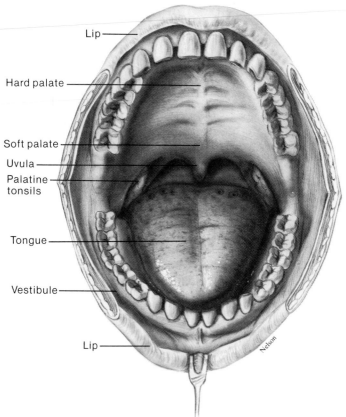

Lip

Hard palate

Soft palate

Uvula

Palatine tonsils

Tongue

Vestibule

Lip

Fig. 13.4 A sagittal section of the mouth and nasal cavity.

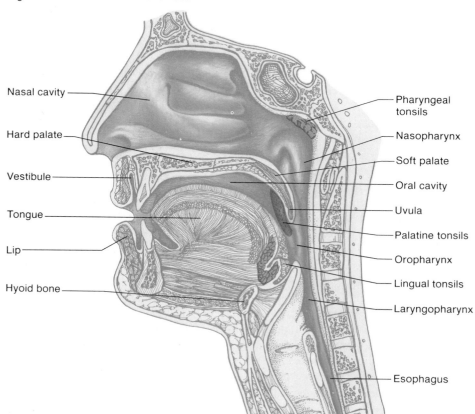

Nasal cavity

Hard palate

Vestibule

Tongue

Lip

Hyoid bone

Pharyngeal tonsils

Nasopharynx

Soft palate

Oral cavity

Uvula

Palatine tonsils

Oropharynx

Lingual tonsils

Laryngopharynx

Esophagus

The Palate

The **palate** forms the roof of the oral cavity and consists of a hard anterior part (*hard palate*) and a soft posterior part (*soft palate*). The soft palate forms a muscular arch that extends posteriorly and downward as a cone-shaped projection called the **uvula.**

During swallowing, muscles draw the soft palate and the uvula upward. This action closes the opening between the nasal cavity and the pharynx, preventing food from entering the nasal cavity.

In the back of the mouth, on either side of the tongue and closely associated with the palate, are masses of lymphatic tissue called **palatine tonsils.** (See fig. 13.4.) These structures lie beneath the epithelial lining of the mouth, and like other lymphatic tissues, help protect the body against infections.

The palatine tonsils are common sites of infections, and if they become inflamed, the condition is termed *tonsillitis.* Infected tonsils may become so swollen that they block the passageways of the pharynx and interfere with breathing and swallowing. Since the mucous membranes of the pharynx, eustachian tubes, and middle ears are continuous, there is always danger that such an infection may travel from the throat into the middle ears.

Still other masses of lymphatic tissue, called **pharyngeal tonsils** or *adenoids* (fig. 13.4), occur on the posterior wall of the pharynx, above the border of the soft palate.

---

1. *How does the tongue contribute to the function of the digestive system?*
2. *What is the role of the soft palate in swallowing?*
3. *Where are the tonsils located?*

The Teeth

The teeth are unique structures in that two different sets form during development. The members of the first set, the *primary* or **deciduous teeth,** usually erupt through the gums at regular intervals between the ages of six months and 2½ years. There are twenty deciduous teeth—ten in each jaw.

The deciduous teeth usually are shed in the same order they appeared. Before this happens, their roots are resorbed. Then the teeth are pushed out of their sockets by pressure from the developing *secondary* or **permanent** teeth. This second set consists of thirty-two teeth—sixteen in each jaw. (See fig. 13.5.)

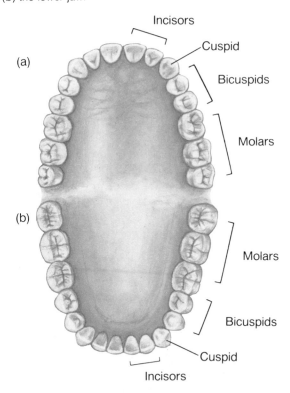

Fig. 13.5 (a) The permanent teeth of the upper jaw and (b) the lower jaw.

The permanent teeth usually begin to appear at about age six years, but the set may not be completed until the third molars appear between seventeen and twenty-five years of age.

Teeth function mechanically to break pieces of food into smaller pieces. This action increases the surface area of the food particles and thus makes it possible for digestive enzymes to react more effectively with food molecules.

Different teeth are adapted to handle food in different ways. *Incisors,* or front teeth, are chisel-shaped and their sharp edges are used to bite off relatively large pieces of food. The *cuspids* (canine teeth) are cone-shaped, and they are useful in grasping or tearing food. The *bicuspids* and *molars* have somewhat flattened surfaces and are specialized for grinding food particles. (See fig. 13.5.)

Chart 13.1 summarizes the number and kinds of teeth that appear during development.

Each tooth consists of two main portions called the *crown* that projects beyond the gum (gingiva), and the *root* that is anchored to the alveolar bone of the jaw. The region where these portions meet is called the *neck* of the tooth.

Fig. 13.6 A section of a typical tooth.

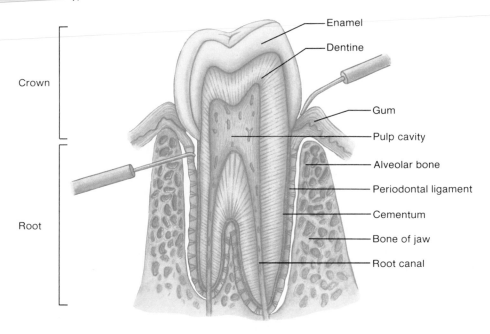

Chart 13.1 Primary and secondary teeth

Primary Teeth (Deciduous)		Secondary Teeth (Permanent)	
Type	*Number*	*Type*	*Number*
Incisor		Incisor	
central	4	central	4
lateral	4	lateral	4
Cuspid	4	Cuspid	4
		Bicuspid	
		first	4
		second	4
Molar		Molar	
first	4	first	4
second	4	second	4
		third	4
Total	20	Total	32

From Hole, John W., Jr., *Human Anatomy and Physiology 3d ed.* © 1978, 1981, 1984 Wm. C. Brown Publishers, Dubuque, Iowa. All Rights Reserved. Reprinted by permission.

The crown is covered by glossy, white *enamel* that consists mainly of calcium salts and is the hardest substance in the body. Unfortunately, if damaged by abrasive action or injury, enamel is not replaced.

The bulk of a tooth beneath the enamel is composed of *dentine,* a substance much like bone but somewhat harder. The dentine, in turn, surrounds the tooth's central cavity (pulp cavity), which contains blood vessels, nerves, and connective tissue (pulp). The blood vessels and nerves reach this cavity through tubular *root canals* that extend into the root.

The root is enclosed by a thin layer of bonelike material called *cementum,* which is surrounded by a *periodontal ligament.* This ligament contains bundles of thick collagenous fibers that pass between the cementum and the alveolar bone, thus firmly attaching the tooth to the jaw. It also contains blood vessels and nerves. (See fig. 13.6.)

Dental caries (decay) involves decalcification of tooth enamel and usually is followed by destruction of the enamel and its underlying dentine. The result is a cavity, which must be cleaned and filled to prevent further erosion of the tooth.

Although the cause or causes of dental caries are not well understood, lack of dental cleanliness and a diet high in sugar and starch seem to promote the problem. Accumulations of food particles on the surfaces and between the teeth are thought to aid the growth of certain kinds of bacteria. These microorganisms apparently utilize carbohydrates in food particles and produce acid by-products. The acids then begin the process of destroying tooth enamel.

Preventing dental caries requires brushing the teeth at least once a day, using dental floss or tape regularly to remove debris from between the teeth, and limiting the intake of sugar and starch, especially between meals. The use of fluoridated drinking water or the application of fluoride solution to children's teeth also helps prevent dental decay.

Fig. 13.7 Locations of the major salivary glands.

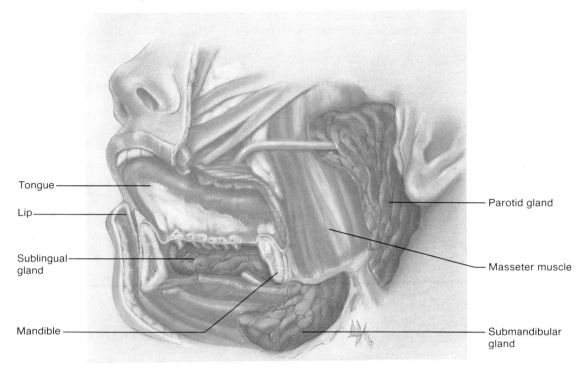

Tongue

Lip

Sublingual gland

Mandible

Parotid gland

Masseter muscle

Submandibular gland

---

1. How do deciduous teeth differ from permanent teeth?
2. Describe the structure of a tooth.
3. Explain how a tooth is attached to the bone of the jaw.

The Salivary Glands

The **salivary glands** function to secrete saliva. This fluid moistens food particles, helps bind them together, and begins the digestion of carbohydrates. Saliva also acts as a solvent by dissolving various food chemicals—a process that is necessary before they can be tasted—and by helping to cleanse the mouth and teeth.

Salivary Secretions

Within a salivary gland there are two types of secretory cells, called *serous* and *mucous* cells. They occur in varying proportions within different glands. The serous cells produce a watery fluid that contains a digestive enzyme, called **amylase.** This enzyme functions to split starch and glycogen molecules into disaccharides—the first step in the digestion of carbohydrates. Mucous cells secrete the thick, stringy liquid called **mucus,** that binds food particles together and acts as a lubricant during swallowing.

When a person sees, smells, tastes, or even thinks about pleasant food, parasympathetic nerve impulses elicit the secretion of a large volume of watery saliva. Conversely, if food looks, smells, or tastes unpleasant, parasympathetic activity is inhibited so that less saliva is produced, and swallowing may become difficult. (See chapter 10.)

Major Salivary Glands

There are three major pairs of salivary glands—the parotid, submandibular, and sublingual glands—and many minor ones that are associated with the mucous membrane of the tongue, palate, and cheeks. (See fig. 13.7.)

The **parotid glands** are the largest of the major salivary glands. One lies in front of and somewhat below each ear, between the skin of the cheek and the masseter muscle. These glands secrete a clear, watery fluid that is rich in amylase.

A person who has *mumps,* which is caused by a virus infection, typically develops swellings in one or both of the parotid glands. Occasionally this disease also affects the submandibular glands as well as the pancreas and the gonads.

The **submandibular** (submaxillary) **glands** are located in the floor of the mouth on the inside surface of the jaw. The secretory cells of these glands are predominantly serous, but some mucous cells are present. Consequently, the submandibular glands secrete a more viscous fluid than the parotid glands.

The **sublingual glands** are the smallest of the major salivary glands. They are found on the floor of the mouth under the tongue. Their cells are primarily the mucous type, and as a result, their secretions tend to be thick and stringy.

---

1. *What is the function of saliva?*
2. *What stimulates the salivary glands to secrete saliva?*
3. *Where are the major salivary glands located?*

The Pharynx and Esophagus

The pharynx is a cavity behind the mouth from which the tubular esophagus leads to the stomach. Although neither of these organs contributes to the digestive process, they are important passageways, and their muscular walls function in swallowing.

Structure of the Pharynx

The **pharynx** connects the nasal and oral cavities with the larynx and esophagus. It can be divided into the nasopharynx, oropharynx, and laryngopharynx. (See fig. 13.4.)

The **nasopharynx** communicates with the nasal cavity and provides a passageway for air during breathing.

The **oropharynx** opens behind the soft palate into the nasopharynx. It functions as a passageway for food moving downward from the mouth and for air moving to and from the nasal cavity.

The **laryngopharynx** is located just below the oropharynx. It opens into the larynx and esophagus.

The Swallowing Mechanism

The act of swallowing involves a set of complex reflexes and can be divided into three stages. In the first, which is initiated voluntarily, food is chewed and mixed with saliva. Then it is rolled into a mass and forced into the pharynx by the tongue.

The second stage begins as the food reaches the pharynx and stimulates sensory receptors located around the pharyngeal opening. This triggers the swallowing reflex, which includes the following actions:

1. The soft palate is raised, preventing food from entering the nasal cavity.
2. The hyoid bone and the larynx are elevated, so that food is less likely to enter the trachea.
3. Muscles in the lower portion of the pharynx relax, opening the esophagus.
4. A peristaltic wave begins in the pharyngeal muscles, and this wave forces the food into the esophagus.

During the third stage of swallowing, the food enters the esophagus and is transported to the stomach by peristalsis.

The Esophagus

The **esophagus** is a straight, collapsible tube about 25 centimeters (10 inches) long that provides a passageway for substances between the pharynx and stomach. It begins at the base of the pharynx and descends behind the trachea, passing through the mediastinum. It penetrates the diaphragm and is continuous with the stomach on the abdominal side of the diaphragm. (See fig. 13.1.)

Just above the point where the esophagus joins the stomach, some of the circular muscle fibers in its wall are thickened. These fibers usually are contracted and function to close the entrance to the stomach. In this way, they help prevent regurgitation of the stomach contents into the esophagus. When peristaltic waves reach the stomach, these muscle fibers relax and allow the food to enter.

There are mucous glands scattered throughout the mucosa of the esophagus, and their secretions keep the inner lining of the tube moist and lubricated.

If regurgitation of stomach contents does occur, gastric secretions may irritate the esophageal wall, producing discomfort or pain. This condition is commonly called *heartburn* because it seems to originate from behind the sternum in the region of the heart.

---

1. *Describe the regions of the pharynx.*
2. *List the major events that occur during swallowing.*
3. *What is the function of the esophagus?*

Fig. 13.8 Regions of the stomach.

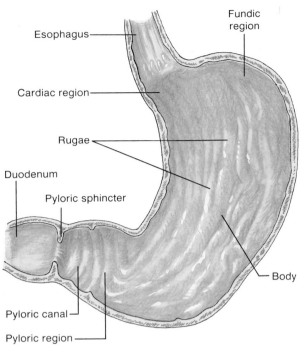

The Stomach

The stomach (fig. 13.1) is a J-shaped, pouchlike organ that hangs under the diaphragm in the upper left portion of the abdominal cavity. It has a capacity of about one liter or more, and its inner lining is marked by thick folds (rugae), which tend to disappear as its wall is distended. The stomach's functions include receiving food from the esophagus, mixing it with gastric juice, initiating the digestion of proteins, carrying on a limited amount of absorption, and moving food into the small intestine.

Parts of the Stomach

The stomach, shown in figure 13.8, can be divided into cardiac, fundic, body, and pyloric regions. The *cardiac region* is a small area near the esophageal opening. The *fundic region,* which balloons above the cardiac portion, acts as a temporary storage area. The dilated *body region* is the main part of the stomach and is located between the fundic and pyloric portions, while the *pyloric region* narrows and becomes the *pyloric canal* as it approaches the junction with the small intestine.

At the end of the pyloric canal, the muscular wall is thickened, forming a powerful circular muscle called the **pyloric sphincter** (pylorus). This muscle serves as a valve that prevents regurgitation of food from the intestine back into the stomach.

Gastric Secretions

The mucous membrane that forms the inner lining of the stomach is relatively thick, and its surface is studded with many small openings. These openings, called *gastric pits,* are located at the ends of tubular **gastric glands** (fig. 13.9).

The gastric glands generally contain three types of secretory cells. One type, the *mucous cell,* occurs in the necks of the glands near the openings of the gastric pits. The other types, *chief cells* and *parietal cells,* are found in the deeper parts of the glands. The chief cells secrete *digestive enzymes,* and the parietal cells release *hydrochloric acid.* The products of the mucous cells, chief cells, and parietal cells together form **gastric juice.**

Although gastric juice contains several digestive enzymes, **pepsin** is by far the most important. It is secreted by the chief cells as an inactive substance called **pepsinogen.** When pepsinogen contacts gastric juice, however, it is changed rapidly into pepsin.

Pepsin is a protein-splitting enzyme capable of beginning the digestion of nearly all types of dietary protein. This enzyme is most active in an acid environment, and the hydrochloric acid in gastric juice seems to function mainly to provide such an environment.

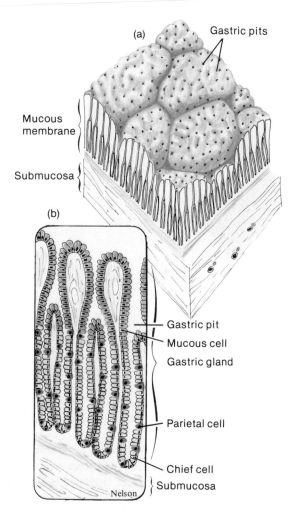

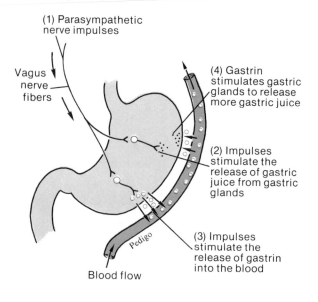

(1) Parasympathetic nerve impulses

Vagus nerve fibers

(4) Gastrin stimulates gastric glands to release more gastric juice

(2) Impulses stimulate the release of gastric juice from gastric glands

(3) Impulses stimulate the release of gastrin into the blood

Blood flow

Pedigo

---

1. *What is secreted by the chief cells of the gastric glands? By the parietal cells?*
2. *What is the most important digestive enzyme in gastric juice? Why?*
3. *How is the stomach prevented from digesting itself?*

The mucous cells of the gastric glands secrete large quantities of mucus. In addition, the cells of the mucous membrane between these glands release a more viscous and alkaline secretion that is thought to form a protective coating on the inside of the stomach wall. This coating is especially important because pepsin is capable of digesting the proteins of the stomach tissues as well as those in foods. Thus the coating normally prevents the stomach from digesting itself.

Still another component of gastric juice is **intrinsic factor.** This substance, which is secreted by the parietal cells of the gastric glands, aids in absorption of vitamin B_{12} from the small intestine.

The substances in gastric juice are summarized in chart 13.2.

Regulation of Gastric Secretions. Although gastric juice is produced continuously, the rate of production varies considerably from time to time and is under the control of neural and hormonal mechanisms. More specifically, when a person tastes, smells, or even sees pleasant food, or when food enters the stomach, parasympathetic impulses stimulate gastric glands to secrete large amounts of gastric juice that is rich in hydrochloric acid and pepsin. These impulses also stimulate certain stomach cells to release a hormone, called **gastrin,** that causes the gastric glands to increase their secretory activity. (See fig. 13.10.)

Later the food leaves the stomach and enters the small intestine. When it first contacts the intestinal wall, the food stimulates intestinal cells to release a hormone that again enhances gastric gland secretions. Although the actual nature of this hormone is unknown, some investigators believe it is identical to gastrin.

As more food moves into the small intestine, the secretion of gastric juice from the stomach wall is inhibited. Apparently this inhibition is due to sympathetic nerve impulses that are triggered by the presence of acid substances in the upper part of the

Chart 13.2 Composition of gastric juice

Component	Source	Function
Pepsinogen	Chief cells of the gastric glands	An inactive form of pepsin
Pepsin	Formed from pepsinogen in the presence of gastric juice	A protein-splitting enzyme capable of digesting nearly all types of protein
Hydrochloric acid	Parietal cells of the gastric glands	Provides acid environment needed for the action of pepsin
Mucus	Goblet cells and mucous glands	Provides viscous, alkaline protective layer on the stomach wall
Intrinsic factor	Parietal cells of the gastric glands	Aids in absorption of vitamin B_{12}

From Hole, John W., Jr., *Human Anatomy and Physiology 3d ed.* © 1978, 1981, 1984 Wm. C. Brown Publishers, Dubuque, Iowa. All Rights Reserved. Reprinted by permission.

small intestine. Also, the presence of fats in this region causes the release of a hormone called *cholecystokinin* from the intestinal wall, and it brings about a decrease in gastric motility as the small intestine fills with food.

An *ulcer* is an open sore in the mucous membrane resulting from a localized breakdown of the tissues. *Gastric ulcers* are most likely to develop in the wall of the stomach near the liver. Ulcers are also common in the first portion of the small intestine. These *duodenal ulcers* occur in regions that are exposed to gastric juice as the contents of the stomach enter the intestine. Thus, both of these types of ulcers are caused by the digestive action of pepsin, and for this reason, are known as *peptic ulcers*.

The cause of peptic ulcers is poorly understood. It is believed however, that they may result either from failure of the mechanism that normally protects the stomach lining from being digested, or from excessive secretion of gastric juice.

Gastric Absorption

Although gastric enzymes begin the breakdown of proteins, the stomach wall is not well adapted to carry on the absorption of digestive products. Small quantities of water, glucose, certain salts, alcohol, and various lipid-soluble drugs may be absorbed by the stomach however.

1. How is the secretion of gastric juice stimulated?
2. How is the secretion of gastric juice inhibited?
3. What substances may be absorbed in the stomach?

Mixing and Emptying Actions

Following a meal, the mixing movements of the stomach wall aid in producing a semifluid paste of food particles and gastric juice called **chyme.** Peristaltic waves push the chyme toward the pyloric region of the stomach, and as it accumulates near the pyloric sphincter, this muscle begins to relax allowing the chyme to move a little at a time into the small intestine.

The rate at which the stomach empties depends on several factors, including the fluidity of the chyme and the type of food present. For example, liquids usually pass through the stomach quite rapidly, while solids remain until they are well mixed with gastric juice. Fatty foods may remain in the stomach from three to six hours; foods high in proteins tend to be moved through more quickly; and carbohydrates usually pass through more rapidly than either fats or proteins.

As the stomach contents enter the duodenum, accessory organs add their secretions to the chyme. These organs include the pancreas, liver, and gallbladder.

Vomiting is a complex reflex that empties the stomach in another way. This action is usually triggered by irritation or distension in some part of the alimentary canal, such as the stomach or intestines. Sensory impulses travel from the site of stimulation to the *vomiting center* in the medulla oblongata, and a number of motor responses follow. These include taking a deep breath, raising the soft palate and thus closing the nasal cavity, closing the opening to the trachea (glottis), relaxing the sphincterlike muscle at the base of the esophagus, contracting the diaphragm so it moves downward over the stomach, and contracting the abdominal wall muscles so the pressure inside the abdominal cavity is increased. As a result, the stomach is squeezed from all sides, and its contents are forced upward and out through the esophagus, pharynx, and mouth.

The Pancreas

The shape of the **pancreas** and its general location are described in chapter 12, as are its endocrine functions. (See fig. 13.11.) The pancreas also has an exocrine function—the secretion of digestive juice.

Structure of the Pancreas

The pancreas is closely associated with the small intestine. It extends horizontally across the posterior abdominal wall in the C-shaped curve of the duodenum.

The cells that produce pancreatic juice make up the bulk of the pancreas. These cells (pancreatic acinar cells) are clustered around tiny tubes into which they release their secretions. The smaller tubes unite to form larger ones, which in turn give rise to a *pancreatic duct* extending the length of the pancreas. This duct usually connects with the duodenum at the same place where the bile duct from the liver and gallbladder joins the duodenum. (See fig. 13.11.)

Pancreatic Juice

Pancreatic juice contains enzymes capable of digesting carbohydrates, fats, proteins, and nucleic acids.

The carbohydrate-digesting enzyme is called *pancreatic amylase*. It splits molecules of starch or glycogen into double sugars (disaccharides); the fat-digesting enzyme, *pancreatic lipase,* breaks fat molecules into fatty acids and glycerol.

The protein-splitting enzymes (proteinases) are *trypsin, chymotrypsin,* and *carboxypeptidase.* Each of these acts to split the bonds between particular combinations of amino acids in proteins. Since no single enzyme can split all possible combinations, the presence of several enzymes is necessary for the complete digestion of protein molecules.

These protein-splitting enzymes are stored in inactive forms within tiny cellular structures called *zymogen granules.* They, like gastric pepsin, are secreted in inactive forms and must be activated by other enzymes after they reach the small intestine. For example, the pancreatic cells release inactive *trypsinogen.* This substance becomes active trypsin when it contacts an enzyme called *enterokinase,* which is secreted by the mucosa of the small intestine.

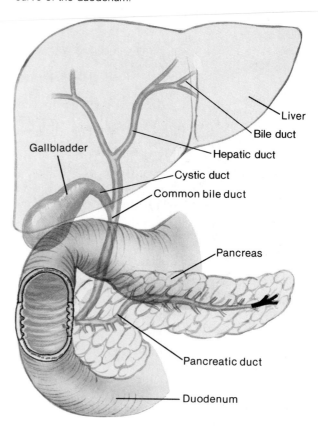

Fig. 13.11 The pancreas is located in the C-shaped curve of the duodenum.

Liver
Bile duct
Hepatic duct
Gallbladder
Cystic duct
Common bile duct
Pancreas
Pancreatic duct
Duodenum

If something happens to block the release of pancreatic juice, it may accumulate in the duct system of the pancreas and the trypsinogen may become activated. As a result, portions of the pancreas may become digested, causing a painful condition called *acute pancreatitis.*

In addition to the above enzymes, pancreatic juice contains two **nucleases** that break down nucleic acid molecules into nucleotides.

Regulation of Pancreatic Secretion. As in the case of gastric and small intestinal secretions, the release of pancreatic juice is regulated by nerve actions as well as hormones. For example, when the secretion of gastric juice is being stimulated, parasympathetic impulses also travel to the pancreas and stimulate it to begin releasing digestive enzymes. Also, as acidic chyme enters the duodenum, a hormone called **secretin** is released into the blood from the duodenal mucous membrane. This hormone stimulates the secretion of pancreatic juice that has a high concentration of bicarbonate ions. These ions function to

Fig. 13.12 Acidic chyme entering the duodenum from the stomach stimulates the release of secretin, which in turn stimulates the release of pancreatic juice.

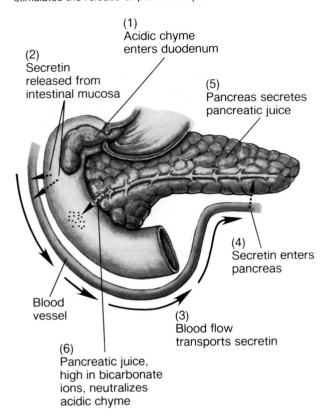

(1)
Acidic chyme
enters duodenum

(2)
Secretin
released from
intestinal mucosa

(5)
Pancreas secretes
pancreatic juice

(4)
Secretin enters
pancreas

Blood
vessel

(3)
Blood flow
transports secretin

(6)
Pancreatic juice,
high in bicarbonate
ions, neutralizes
acidic chyme

neutralize the acid of chyme and to provide a favorable environment for digestive enzymes in the intestine. (See fig. 13.12.)

The presence of chyme in the duodenum also stimulates the release of **cholecystokinin** from the intestinal wall. As before, this hormone reaches the pancreas by way of the blood; however, it causes the secretion of pancreatic juice with a high concentration of digestive enzymes.

---

1. List the enzymes found in pancreatic juice.
2. What are the functions of these enzymes?
3. How is the secretion of pancreatic juice regulated?

The Liver

The liver is located in the upper right portion of the abdominal cavity, just below the diaphragm. It is partially surrounded by the ribs and extends from the level of the fifth intercostal space to the lower margin of the ribs. It has a reddish brown color and is well supplied with blood vessels. (See fig. 13.1.)

Functions of the Liver

The liver is the largest gland in the body and carries on many important metabolic activities. For example, it plays a key role in carbohydrate metabolism by helping to maintain the normal concentration of blood glucose. As described in chapter 12, liver cells responding to various hormones can decrease blood glucose by converting glucose to glycogen; they also can increase blood glucose by changing glycogen to glucose or by converting noncarbohydrates into glucose.

The liver's effects on lipid metabolism include the oxidation of fatty acids at an especially high rate (see chapter 4); the synthesis of lipoproteins, phospholipids, and cholesterol; and the conversion of carbohydrates into proteins and fats. Fats synthesized in the liver are transported by the blood to adipose tissue for storage.

The most vital liver functions probably are those related to protein metabolism. They include the deamination of amino acids; the formation of urea (see chapter 4); the synthesis of various blood proteins, including several that are necessary for blood clotting (see chapter 15); and the conversion of various amino acids to other nonessential types of amino acids.

The liver also stores a variety of substances, including glycogen, vitamins A, D, and B_{12}, and iron. In addition, various liver cells (macrophages) help to destroy damaged red blood cells and foreign substances by phagocytosis. The liver also alters the composition of toxic substances (detoxification) in body fluids, and secretes bile.

Since many of these functions are not directly related to the digestive system, they are discussed in other chapters. Bile secretion, however, is important to digestion and is explained in a subsequent section of this chapter.

Chart 13.3 summarizes the major functions of the liver.

Structure of the Liver

The liver is enclosed in a fibrous capsule and is divided by connective tissue into *lobes*—a large right lobe and a smaller left lobe. (See fig. 13.13.) Each lobe is separated into numerous tiny **hepatic lobules,** which are the functional units of the gland. (See fig. 13.14.) A lobule consists of numerous hepatic cells that radiate outward from a *central vein.* Platelike groups of these cells are separated from each other by vascular channels called **hepatic sinusoids.** Blood from the digestive tract that is carried in portal veins (see chapter 16) brings newly absorbed nutrients into the sinusoids and nourishes the hepatic cells.

Fig. 13.13 (a) Lobes of the liver viewed from the front and (b) from below.

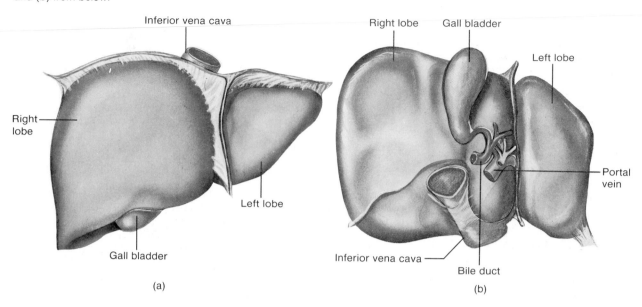

Inferior vena cava

Right lobe

Left lobe

Gall bladder

(a)

Right lobe　　Gall bladder

Left lobe

Portal vein

Inferior vena cava

Bile duct

(b)

Chart 13.3 Major functions of the liver

General Function	Specific Function
Carbohydrate metabolism	Conversion of glucose to glycogen, conversion of glycogen to glucose, conversion of noncarbohydrates to glucose
Lipid metabolism	Oxidation of fatty acids; synthesis of lipoproteins, phospholipids, and cholesterol; conversion of carbohydrates and proteins into fats
Protein metabolism	Deamination of amino acids, synthesis of urea, synthesis of blood proteins, interconversion of amino acids
Storage	Stores glycogen; vitamins A, D, and B_{12}; and iron
Blood filtering	Removes damaged red blood cells and foreign substances by phagocytosis
Detoxification	Alters composition of toxic substances
Secretion	Secretes bile

From Hole, John W., Jr., *Human Anatomy and Physiology 3d ed.* © 1978, 1981, 1984 Wm. C. Brown Publishers, Dubuque, Iowa. All Rights Reserved. Reprinted by permission.

Fig. 13.14 A cross section of a hepatic lobule, which is the functional unit of the liver.

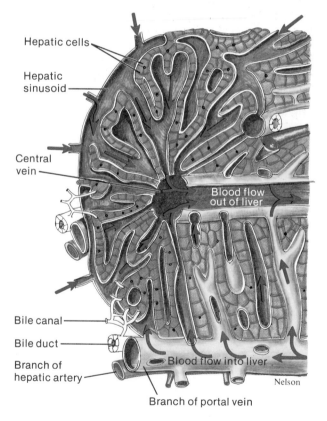

Hepatic cells

Hepatic sinusoid

Central vein

Blood flow out of liver

Bile canal

Bile duct

Branch of hepatic artery

Blood flow into liver

Branch of portal vein

Nelson

Usually the blood in the portal veins contains a large number of bacterial cells that enter through the intestinal wall. Large *Kupffer cells* that are fixed to the inner lining of the hepatic sinusoids remove most of these bacteria by phagocytosis. Then, the blood passes into the central veins of the hepatic lobules and moves out of the liver.

Within the liver lobules, there are many fine *bile canals* that receive secretions from the hepatic cells. The canals of neighboring lobules unite to form larger ducts, and these converge to become the **hepatic ducts.** They, in turn, merge to form the **common bile duct.**

---

1. Describe the location of the liver.
2. Review the functions of the liver.
3. Describe a hepatic lobule.

Composition of Bile

Bile is a yellowish green liquid that is secreted continuously by hepatic cells. In addition to water, it contains *bile salts, bile pigments* (bilirubin and biliverdin), cholesterol, and various electrolytes. Of these, the bile salts are the most abundant, and they are the only substances in bile that have a digestive function.

The bile pigments are products of red blood cell breakdown and are normally excreted in the bile. (See chapter 15.)

If the excretion of bile pigments is prevented due to obstruction of ducts, they tend to accumulate in the blood and tissues, causing a yellowish tinge in the skin and other body parts. This condition is called *obstructive jaundice.*

The skin begins to appear yellow when the concentration of bile pigments in body fluid reaches a level about three times normal. In addition to being caused by obstructions in ducts, jaundice may occur with liver diseases in which liver cells are unable to secrete ordinary amounts of pigment (hepatocellular jaundice) or with excessive destruction of red blood cells accompanied by a rapid release of pigments (hemolytic jaundice).

Fig. 13.15 X-ray film of a gallbladder that contains gallstones. What other anatomical features can you identify?

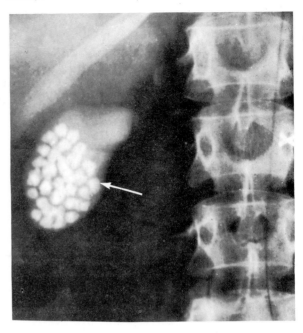

The Gallbladder and Its Functions

The **gallbladder** is a pear-shaped sac attached to the ventral surface of the liver by the **cystic duct,** which in turn joins the hepatic duct. The gallbladder is lined with epithelial cells and has a strong muscular layer in its wall. It stores bile between meals, concentrates bile by reabsorbing water, and releases bile into the small intestine.

The **common bile duct** is formed by the union of the common hepatic and cystic ducts. It leads to the duodenum, where its exit is guarded by a sphincter muscle (sphincter of Oddi). This sphincter normally remains contracted, so that bile collects in the duct and backs up to the cystic duct. When this happens, the bile flows into the gallbladder and is stored there.

Although the cholesterol in bile normally remains in solution, it may precipitate and form crystals under certain conditions. The resulting solids are called *gallstones,* and if they get into the bile duct, they may block the flow of bile into the small intestine and cause considerable pain.

Generally gallstones that cause obstructions are surgically removed. At the same time, the gallbladder is removed by a surgical procedure called cholecystectomy. (See fig. 13.15.)

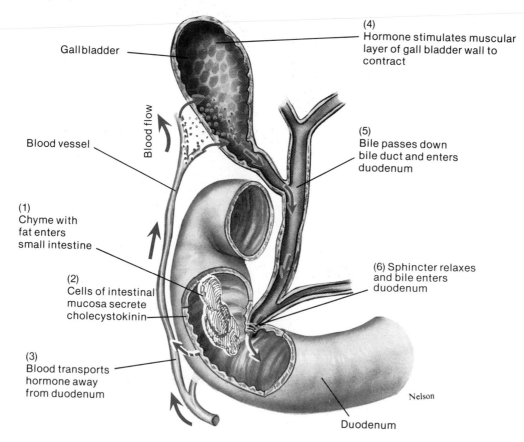

Gallbladder

(4) Hormone stimulates muscular layer of gall bladder wall to contract

Blood flow

Blood vessel

(5) Bile passes down bile duct and enters duodenum

(1) Chyme with fat enters small intestine

(2) Cells of intestinal mucosa secrete cholecystokinin

(6) Sphincter relaxes and bile enters duodenum

(3) Blood transports hormone away from duodenum

Nelson

Duodenum

Regulation of Bile Release.   Normally bile does not enter the duodenum until the gallbladder is stimulated to contract by the hormone cholecystokinin. The usual stimulus involves the presence of fat in the small intestine, which triggers the release of the hormone from the mucosa of the small intestine. The sphincter at the base of the common bile duct remains contracted until a peristaltic wave in the duodenal wall passes by. Then the sphincter relaxes slightly, and a squirt of bile enters the small intestine. (See fig. 13.16.)

Functions of Bile Salts.   Although they do not act as digestive enzymes, bile salts aid the actions of enzymes and enhance the absorption of fatty acids and certain fat-soluble vitamins.

Molecules of fats tend to clump together, forming masses called *fat globules*. Bile salts affect fat globules much as a soap or detergent would affect them. That is, they cause fat globules to break up into smaller droplets, an action called **emulsification.** As a result, the total surface area of the fatty substance is greatly increased, and the tiny droplets mix with water. The fat-splitting enzymes (lipases) can then act on the fat molecules more effectively.

Bile salts also aid in the absorption of fatty acids and cholesterol. Along with these lipids, various fat-soluble vitamins, such as vitamins A, D, E, and K, also are absorbed. Thus if bile salts are lacking, lipids may be poorly absorbed, and the person is likely to develop vitamin deficiencies.

The hormones that help control digestive functions are summarized in chart 13.4.

1. *Explain how bile originates.*
2. *Describe the functions of the gallbladder.*
3. *How is the secretion of bile regulated?*
4. *How does bile function in digestion?*

Chart 13.4 Hormones of the digestive tract

Hormone	Source	Function
Gastrin	Gastric cells, in response to the presence of food	Causes gastric glands to increase their secretory activity
Intestinal gastrin	Cells of small intestine, in response to the presence of chyme	Causes gastric glands to increase their secretory activity
Cholecystokinin	Intestinal wall cells, in response to the presence of fats in the small intestine	Causes gastric glands to decrease their secretory activity and inhibits gastric motility; stimulates pancreas to secrete fluid with a high digestive enzyme concentration; stimulates gallbladder to contract and release bile
Secretin	Cells in the duodenal wall, in response to chyme entering the small intestine	Stimulates pancreas to secrete fluid with a high bicarbonate ion concentration

From Hole, John W., Jr., *Human Anatomy and Physiology 3d ed.* © 1978, 1981, 1984 Wm. C. Brown Publishers, Dubuque, Iowa. All Rights Reserved. Reprinted by permission.

Fig. 13.17 The small intestine includes the duodenum, jejunum, and ileum.

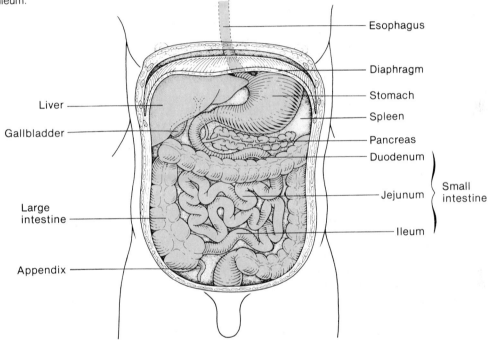

The Small Intestine

The small intestine is a tubular organ that extends from the pyloric sphincter to the beginning of the large intestine. With its many loops and coils, it fills much of the abdominal cavity. (See fig. 13.1.)

As was mentioned, this portion of the alimentary canal receives secretions from the pancreas and liver. It also completes the digestion of the nutrients in chyme, absorbs the various products of digestion, and transports the remaining residues to the large intestine.

Parts of the Small Intestine

The small intestine, shown in figure 13.17, consists of three portions, the duodenum, jejunum, and ileum.

The **duodenum,** which is about 25 centimeters (10 inches) long and 5 centimeters (2 inches) in diameter, lies behind the parietal peritoneum and is the most fixed portion of the small intestine. It follows a C-shaped path as it passes in front of the right kidney and the upper three lumbar vertebrae.

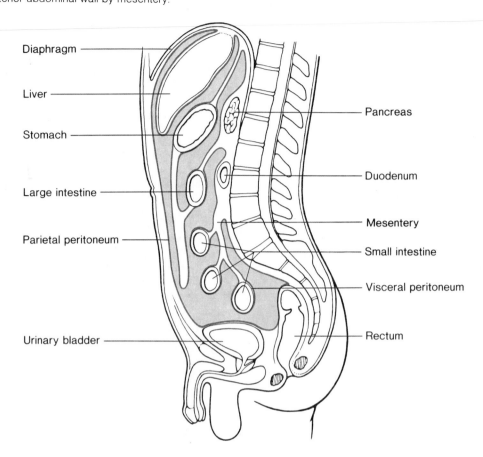

The rest of the small intestine lies free in the peritoneal cavity and is mobile. The proximal 2/5ths of this portion is called the **jejunum,** and the remainder is the **ileum.** These portions are suspended from the posterior abdominal wall by a double-layered fold of peritoneum called **mesentery** (fig. 13.18). This supporting tissue contains the blood vessels, nerves, and lymphatic vessels that supply the intestinal wall.

Although there is no distinct separation between the jejunum and ileum, the diameter of the jejunum tends to be greater, and its wall is thicker, more vascular, and more active than that of the ileum.

Structure of the Small Intestinal Wall

Throughout its length, the inner wall of the small intestine has a velvety appearance. This is due to the presence of innumerable tiny projections of mucous membrane called **intestinal villi** (figs. 13.19 and 13.20). These structures project into the passageway, or **lumen,** of the alimentary canal where they come in contact with the intestinal contents. They serve to increase the surface area of the intestinal lining and play an important role in the absorption of digestive products.

Each villus consists of a layer of simple columnar epithelium and a core of connective tissue containing blood capillaries, a lymphatic capillary (called a *lacteal*), and nerve fibers.

The blood and lymph capillaries function to carry away substances absorbed by the villus, while the nerve fibers act to stimulate or inhibit its activities.

Between the bases of the villi are tubular **intestinal glands** that extend downward into the mucous membrane. (See fig. 13.19.)

Secretions of the Small Intestine

In addition to mucus-secreting goblet cells that occur extensively throughout the mucosa of the small intestine, there are many specialized *mucus-secreting*

Fig. 13.19 Tiny projections, called villi, extend from the mucosa into the lumen of the small intestine. What is the function of these projections?

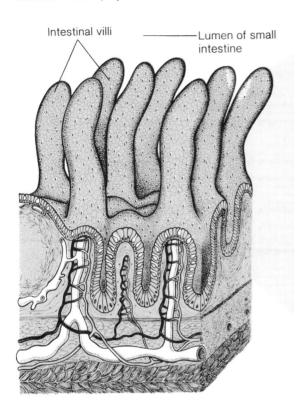

Intestinal villi

Lumen of small intestine

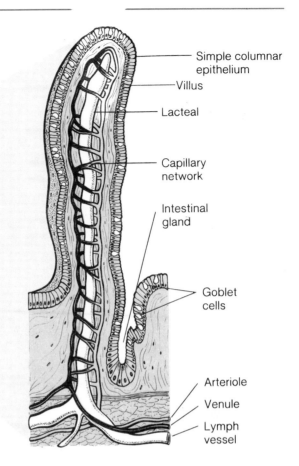

Simple columnar epithelium

Villus

Lacteal

Capillary network

Intestinal gland

Goblet cells

Arteriole

Venule

Lymph vessel

glands in the first part of the duodenum. These glands secrete large quantities of thick, alkaline mucus in response to various stimuli.

The intestinal glands at the bases of the villi secrete great amounts of a watery fluid. The villi rapidly reabsorb this fluid, and it provides a vehicle for moving digestive products into the villi. The fluid secreted by the intestinal glands has a pH that is nearly neutral (6.5–7.5), and seems to lack digestive enzymes. The epithelial cells of the intestinal mucosa have digestive enzymes embedded in the surfaces of their microvilli however, and the enzymes can break down food molecules just before absorption takes place. These enzymes include **peptidases,** which split peptides into amino acids; **sucrase, maltase,** and **lactase,** which split the double sugars (disaccharides) sucrose, maltose, and lactose respectively into the simple sugars (monosaccharides) glucose, fructose, and galactose; and **intestinal lipase,** which splits fats into fatty acids and glycerol.

Chart 13.5 summarizes the sources and actions of the major digestive enzymes.

Fig. 13.20 A scanning electron micrograph of the small intestine of a rat. What anatomical features can you identify?
From *Tissues and Organs: A Text-Atlas of Scanning Electron Microscopy* by Richard G. Kessel and Randy H. Kardon. W. H. Freeman and Company. Copyright © 1979.

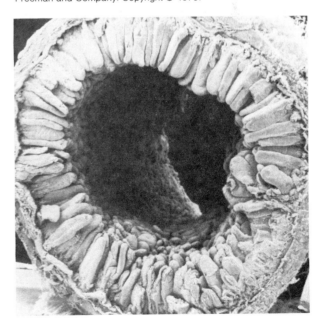

Chart 13.5 A summary of the digestive enzymes

Enzyme	Source	Digestive Action
Salivary enzyme		
Amylase	Salivary glands	Begins carbohydrate digestion by converting starch and glycogen to disaccharides
Gastric enzyme		
Pepsin	Gastric glands	Begins the digestion of proteins
Intestinal enzymes		
Peptidase	Mucosal cells	Converts peptides into amino acids
Sucrase, maltase, lactase	Mucosal cells	Converts disaccharides into monosaccharides
Lipase	Mucosal cells	Converts fats into fatty acids and glycerol
Enterokinase	Mucosal cells	Activates trypsin
Pancreatic enzymes		
Amylase	Pancreas	Converts starch and glycogen into disaccharides
Lipase	Pancreas	Converts fats into fatty acids and glycerol
Proteinases	Pancreas	Converts proteins or partially digested proteins into peptides
a. Trypsin b. Chymotrypsin c. Carboxypeptidase		
Nuclease	Pancreas	Converts nucleic acids into nucleotides

From Hole, John W., Jr., *Human Anatomy and Physiology 3d ed.* © 1978, 1981, 1984 Wm. C. Brown Publishers, Dubuque, Iowa. All Rights Reserved. Reprinted by permission.

The amount of lactase produced in the small intestine usually reaches a maximum shortly after birth and thereafter tends to decrease. As a result, some adults produce insufficient quantities of lactase to break down the lactose or milk sugar in their diets. When this happens, the lactose from milk or certain milk products remains undigested and causes an increase in the osmotic pressure of the intestinal contents. Consequently, water is drawn from the tissues into the intestine. At the same time, intestinal bacteria may act upon the undigested sugar and produce organic acids and gases. As a result, the person may feel bloated and suffer from intestinal cramps and diarrhea.

Regulation of Small Intestinal Secretions. Secretions from goblet cells and intestinal glands are stimulated by direct contact with chyme, which provides both chemical and mechanical stimuli, and by reflexes triggered by distension of the intestinal wall. The reflex actions involve parasympathetic motor impulses that cause secretory cells to increase their activities.

1. Describe the parts of the small intestine.
2. Define intestinal villi.
3. What is the function of the intestinal glands?
4. List the digestive enzymes formed by intestinal cells.

Absorption in the Small Intestine

Because villi greatly increase the surface area of the intestinal mucosa, the small intestine is the most important absorbing organ of the alimentary canal. In fact, the small intestine is so effective at absorbing digestive products, water, and electrolytes that very little absorbable material reaches its distal end.

Carbohydrate digestion begins in the mouth as a result of the activity of salivary amylase, and it is completed in the small intestine by enzymes from the intestinal mucosa and pancreas. The resulting monosaccharides are absorbed by the villi and enter blood capillaries. Even though small quantities of these simple sugars may pass into the villi by diffusion, most of them are absorbed by active transport or facilitated diffusion. (See chapter 3.)

Protein digestion begins in the stomach as a result of pepsin activity and is completed in the small intestine by enzymes from the intestinal mucosa and the pancreas. During this process, large protein molecules are converted into amino acids. These smaller particles then are absorbed into the villi by active transport and are carried away by the blood.

Fig. 13.21 Fatty acid absorption involves several steps.

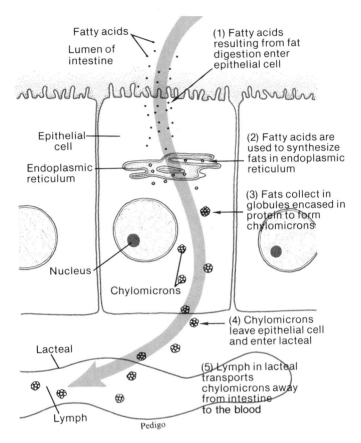

Fat molecules are digested almost entirely by enzymes from the intestinal mucosa and pancreas. The resulting fatty acids and glycerol molecules diffuse into the epithelial cells of the villi and are resynthesized into fat molecules similar to those previously digested. These fats form tiny droplets, called *chylomicrons,* that make their ways to the lacteal of the villus. The lymph carries the chylomicrons to the blood. (See fig. 13.21.)

On the other hand, some fatty acids with relatively short carbon chains may be absorbed directly into the blood capillary of the villus without being converted back into fat.

In addition to absorbing the products of carbohydrate, protein, and fat digestion, the intestinal villi function in the absorption of various electrolytes and water.

---

1. *What substances resulting from the digestion of carbohydrate, protein, and fat molecules are absorbed by the small intestine?*
2. *Describe how fatty acids are absorbed.*

Movements of the Small Intestine

Like the stomach, the small intestine carries on mixing movements and peristalsis. The mixing movements include small, ringlike contractions that occur periodically, cutting the chyme into segments over and over again.

Chyme is propelled through the small intestine by peristaltic waves. These waves are usually weak, and they stop after pushing the chyme a short distance. Consequently, food materials move relatively slowly through the small intestine, taking from 3 to 10 hours to travel its length.

Sometimes however, stimulation of the small intestinal wall by overdistension or by severe irritation may elicit a strong *peristaltic rush* that passes along its entire length. This type of movement serves to sweep the contents of the small intestine into the large intestine relatively rapidly and helps relieve the small intestine of its problem.

Fig. 13.22 Major parts of the large intestine.

The Large Intestine

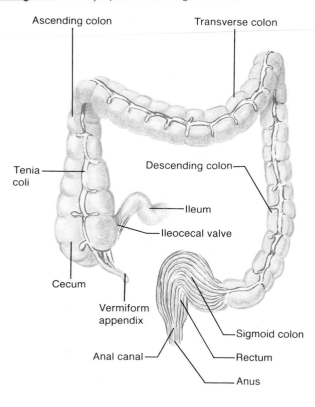

The large intestine is so named because its diameter is greater than that of the small intestine. This portion of the alimentary canal is about 1.5 meters (five feet) long and begins in the lower right side of the abdominal cavity, where the ileum joins the cecum. From there the large intestine (colon) travels upward on the right side, crosses obliquely to the left, and descends into the pelvis. At its distal end it opens to the outside of the body as the anus. (See fig. 13.1.)

The large intestine reabsorbs water and electrolytes from the chyme remaining in the alimentary canal. It also forms and stores the feces until defecation occurs.

Parts of the Large Intestine

The large intestine, shown in figures 13.22 and 13.23, consists of the cecum, colon, rectum, and anal canal.

The **cecum,** which represents the beginning of the large intestine, is a dilated, pouchlike structure that hangs slightly below the ileocecal opening. Projecting downward from it is a narrow tube with a closed end called the **vermiform appendix.** Although the human appendix has no known digestive function, it does contain lymphatic tissue that may serve to resist infections.

Rapid movement of chyme through the intestine prevents the normal absorption of water, nutrients, and electrolytes from the intestinal contents. The result is *diarrhea,* a condition in which defecation becomes more frequent and the stools are watery. If the diarrhea continues for a prolonged time, problems in water and electrolyte balance are likely to develop.

Occasionally the appendix may become infected and inflamed, causing the condition called *appendicitis.* When this happens, the appendix often is removed surgically to prevent its rupture. If it does break open, the contents of the large intestine may enter the abdominal cavity and cause a serious infection of the peritoneum called *peritonitis.*

At the distal end of the small intestine, where the ileum joins the cecum of the large intestine, there is a sphincter muscle called the **ileocecal valve** (fig. 13.22). Normally this sphincter remains constricted, preventing the contents of the small intestine from entering. At the same time, it prevents the contents of the large intestine from backing up into the ileum. After a meal however, a reflex is elicited, and peristalsis in the ileum is increased. This action forces some of the contents of the small intestine into the cecum.

The **colon** can be divided into four portions—the ascending, transverse, descending, and sigmoid colons. The **ascending colon** begins at the cecum and travels upward against the posterior abdominal wall to a point just below the liver. There it turns sharply to the left and becomes the **transverse colon.** The transverse colon is the longest and the most mobile part of the large intestine. It is suspended by a fold of peritoneum and tends to sag in the middle below the stomach. As the transverse colon approaches the spleen, it turns abruptly downward and becomes the **descending colon.** At the brim of the pelvis, the descending colon makes an S-shaped curve, called the **sigmoid colon,** and then becomes the rectum.

1. *Describe the movements of the small intestine.*
2. *What stimulus causes the ileocecal valve to relax?*

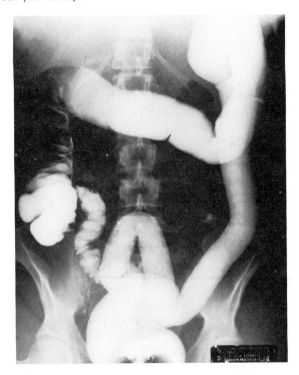

Fig. 13.24 The rectum and anal canal are located at the distal end of the alimentary canal. Muscle fibers exert tension on the wall of the large intestine and create a series of pouches called haustra.

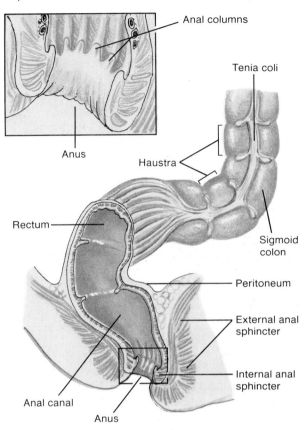

The **rectum** lies next to the sacrum and generally follows its curvature. It is firmly attached to the sacrum by the peritoneum, and it ends about two inches below the tip of the coccyx, where it becomes the anal canal (fig. 13.24).

The **anal canal** is formed by the last 2.5–4.0 centimeters of the large intestine. The mucous membrane in the canal is folded into a series of six to eight longitudinal *anal columns*. At its distal end, the canal opens to the outside as the **anus**. This opening is guarded by two sphincter muscles, an *internal anal sphincter* composed of smooth muscle and an *external anal sphincter* of skeletal muscle.

1. *What is the general function of the large intestine?*
2. *Describe the parts of the large intestine.*

Each anal column contains a branch of the rectal vein, and if something interferes with the blood flow in these vessels, the anal columns may become enlarged and inflamed. This condition, called *hemorrhoids,* may be aggravated by bowel movements and may be accompanied by discomfort and bleeding.

Structure of the Large Intestinal Wall

Although the wall of the large intestine includes the same types of tissues found in other parts of the alimentary canal, it has some unique features. For example, it lacks the villi that are characteristic of the small intestine. Also, the layer of longitudinal muscle fibers does not cover its wall uniformly. Instead, the fibers are arranged in three distinct bands (teniae coli) that extend the entire length of the colon. These bands exert tension on the wall, creating a series of pouches (haustra) (fig. 13.24).

Functions of the Large Intestine

Unlike the small intestine which functions to secrete digestive enzymes and to absorb the products of digestion, the large intestine has little or no digestive function. On the other hand, the mucous membrane that forms the inner lining of the large intestine contains many tubular glands. Structurally these glands are similar to those in the small intestine, but they are composed almost entirely of goblet cells. Consequently, mucus is the only significant secretion of the large intestine.

The mucus secreted into the large intestine functions to protect the intestinal wall against the abrasive action of the material passing through. It also aids in holding particles of the fecal matter together, and because it is alkaline, helps to control the pH of the large intestinal contents.

The chyme entering the large intestine contains materials that could not be digested or absorbed by the small intestine, as well as water, various electrolytes, and bacteria. The proximal half of the large intestine functions to reabsorb some of the water and electrolytes. The substances that remain in the tube become feces, and it is stored for a time in the distal portion of the large intestine.

---

1. How does the structure of the large intestine differ from that of the small intestine?
2. What substances can be absorbed by the large intestine?

Movements of the Large Intestine

The movements of the large intestine—mixing and peristalsis—are similar to those of the small intestine, although they are usually more sluggish. Also, instead of occurring frequently, peristaltic waves in the large intestine come only two or three times each day. These waves produce *mass movements* in which a relatively large section of the colon constricts vigorously, forcing its contents to move toward the rectum. Typically, mass movements occur following a meal, as a result of a reflex that is initiated in the small intestine. Abnormal irritations of the mucosa also can trigger such movements. For instance, a person suffering from an inflamed colon (colitis) may experience frequent mass movements.

When it is appropriate to defecate, a person usually can initiate a *defecation reflex* by holding a deep breath and contracting the abdominal wall muscles. This action increases the internal abdominal pressure and forces feces into the rectum. As the rectum fills, its wall is distended and the defecation reflex is triggered. As a result, peristaltic waves in the descending colon are stimulated, and the internal anal sphincter relaxes. At the same time, other reflexes involving the sacral region of the spinal cord cause the peristaltic waves to strengthen, the diaphragm to lower, the glottis to close, and the abdominal wall muscles to contract. These actions cause an additional increase in the internal abdominal pressure and assist in squeezing the rectum. The external anal sphincter is signaled to relax, and the feces are forced to the outside. A person, of course, can inhibit defecation voluntarily by keeping the external anal sphincter contracted.

The Feces

As was mentioned, **feces** are composed largely of materials that could not be digested, together with water, electrolytes, mucus, and bacteria. Usually they are about 75% water, and the color is normally due to the presence of bile pigments that have been altered somewhat by bacterial actions.

The pungent odor of feces results from a variety of compounds produced by bacteria acting upon the residues.

---

1. How does peristalsis in the large intestine differ from peristalsis in the small intestine?
2. List the major events that occur during defecation.
3. Describe the composition of feces.

Nutrition and Nutrients

Nutrition is the process by which necessary food substances are taken in and utilized by the body. These food substances, or **nutrients,** include the carbohydrates, lipids, proteins, vitamins, and minerals that are discussed in chapters 2 and 4. Some of them, such as certain amino acids and fatty acids, are of particular importance because they cannot be synthesized in adequate amounts by human cells. Since it is essential for health that they be provided in the diet, they are known as *essential nutrients.*

Carbohydrates

As is discussed in chapters 2 and 4, carbohydrates are organic compounds such as sugars and starch that are used primarily to supply energy for cellular processes.

Sources of Carbohydrates. Carbohydrates are ingested in a variety of forms, including starch from grains and certain vegetables, glycogen from meats and seafoods, disaccharides from cane sugar, beet sugar, and molasses, and simple sugars from honey and various fruits. During digestion the more complex carbohydrates, such as starch and glycogen, are changed to monosaccharides—a form in which they can be absorbed and transported to body cells. (See figs. 2.10 and 2.11.)

Fig. 13.25 Monosaccharides from foods are used to supply energy, are stored as glycogen, or are converted to fat.

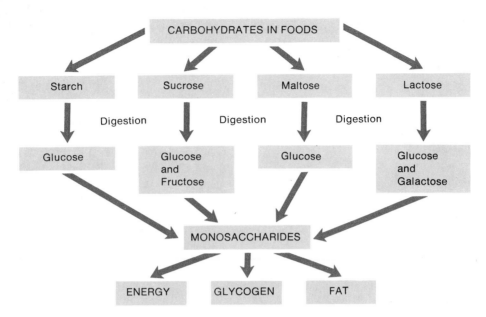

An exception to this is provided by the polysaccharide called *cellulose* that is very abundant in plant foods. Although it is composed of glucose units arranged in chains, somewhat like plant starch, cellulose cannot be broken down by human digestive enzymes. Consequently, it passes through the alimentary canal largely unchanged. The presence of cellulose in foods, however, is important to the function of the digestive system. It provides bulk (sometimes called fiber or roughage) in the digestive tube against which the muscular wall can push. Thus, cellulose facilitates the movement of food substances through the digestive tract.

Utilization of Carbohydrates. The monosaccharides that are absorbed from the digestive tract include *fructose, galactose,* and *glucose.* Fructose and galactose are normally converted into glucose by the action of the liver, and glucose is the form of carbohydrate that is most commonly oxidized by cells as fuel.

If glucose is present in excessive amounts, some of it is converted to *glycogen* and stored in the liver and muscles. Glucose can be mobilized rapidly from glycogen when it is needed to supply energy. Only a certain amount of glycogen can be stored, however, and excess glucose is usually converted into fat and stored in adipose tissue. (See fig. 13.25.)

Carbohydrate Requirements. Although most carbohydrates are used to supply energy, some are used to produce vital cellular substances. These include the 5-carbon sugars *ribose* and *deoxyribose* that are needed for the synthesis of the nucleic acids RNA and DNA, as well as the disaccharide *lactose* (milk sugar) that is produced when the breasts are actively secreting milk.

Many cells also can obtain energy by oxidizing fatty acids. Some cells, however, such as the neurons of the central nervous system, seem to be dependent on a continuous supply of glucose for survival. Even a temporary drop in the glucose concentration may produce a serious functional disorder of the nervous system or the death of nerve cells. Consequently, the presence of some carbohydrates in the body is essential; if an adequate supply is not received from foods, the liver may convert noncarbohydrates, such as amino acids from proteins, into glucose. Thus the need for glucose has priority over the need to manufacture proteins from available amino acids.

Since carbohydrates provide the primary source of fuel for cellular processes, the need for carbohydrates varies with individual energy requirements. Persons who are physically active require more fuel than those who are sedentary. The minimal requirement for carbohydrates in the human diet is unknown. It is estimated, however, that an intake of at least 100 grams daily is necessary to avoid excessive breakdown of protein and to avoid metabolic disorders that sometimes accompany excessive utilization of fats.

---

1. List several common sources of carbohydrates.
2. Explain the importance of cellulose in the diet.
3. Name two uses of carbohydrates other than supplying energy.

Lipids

As is described in chapter 2, **lipids** comprise a group of organic compounds. This group includes fats, oils, and fatlike substances that are used to supply energy for cellular processes and to build structures such as cell membranes. Although lipids include fats, phospholipids, and cholesterol, the most common dietary lipids are fats called *triglycerides.* (See fig. 2.12.)

Sources of Lipids.   Triglycerides are contained in foods of both plant and animal origin. They occur, for example, in meats, eggs, milk, and lard, as well as in various nuts and plant oils such as corn oil, peanut oil, and olive oil.

Cholesterol is obtained in relatively high concentrations from such foods as liver, egg yolk, and brain, and is present in lesser amounts in whole milk, butter, cheese, and meats. It does not occur in foods of plant origin.

Utilization of Lipids.   During digestion, triglycerides are broken down into fatty acids and glycerol, and after being absorbed, these products are transported by the lymph and blood to various tissues. The metabolism of these substances is controlled mainly by the liver and the adipose tissues.

The liver can convert fatty acids from one form to another, but it cannot synthesize one type of fatty acid called *linoleic acid.* Thus, this substance is an **essential fatty acid.** It is needed for the production of certain phospholipids, which in turn are necessary for the formation of cell membranes and the transport of circulating lipids. Good sources of linoleic acid include corn oil, cottonseed oil, and soy oil.

The liver uses free fatty acids in the synthesis of triglycerides, phospholipids, and lipoproteins that may then be released into the blood. Thus the liver is largely responsible for the control of circulating lipids. In addition, it is thought to regulate the total amount of cholesterol in the body by synthesizing cholesterol and releasing it into the blood, or by removing cholesterol from the blood and excreting it into the bile. The liver also uses cholesterol in the production of bile salts. (See fig. 13.26.)

Cholesterol is not used as an energy source, but it does provide structural material for a variety of cell parts and furnishes molecular components for the synthesis of various sex hormones and hormones produced by the adrenal cortex.

Excessive triglycerides are stored in adipose tissue, and if the blood lipid concentration drops (in response to fasting, for example), some of these triglycerides are hydrolyzed into free fatty acids and glycerol and released into the blood.

Lipid Requirements.   As in the case of carbohydrates, humans ingest lipids in a variety of forms.

The amounts and types of fats needed for health are unknown. Since linoleic acid is an essential fatty acid, however, nutritionists recommend that infants receive formulas in which 3% of the energy intake is in the form of linoleic acid. Fatty acid deficiencies have not been observed in adults, so it has been concluded that a typical adult diet consisting of a variety of foods provides an adequate supply of this essential nutrient.

Since fats contain fat-soluble vitamins, the intake of fats also must be sufficient to supply needed amounts of these nutrients.

---

1. Which fatty acid is an essential nutrient?
2. What is the role of the liver in the utilization of lipids?
3. What is the function of cholesterol?

Proteins

As is described in chapter 2, **proteins** are organic compounds that serve as structural materials in cells, act as enzymes that regulate metabolic reactions, and often are used to supply energy.

Sources of Proteins.   Foods that are rich in proteins include meats, fish, poultry, cheese, nuts, milk, eggs, and cereals. Various legumes such as beans and peas contain lesser amounts.

During digestion, proteins are broken down into their component amino acids, and these smaller molecules are absorbed and transported to the body cells in the blood.

Fig. 13.26 Fatty acids are used by the liver to synthesize a variety of lipids.

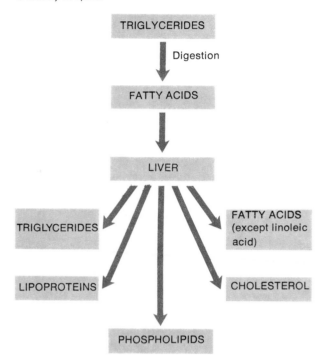

Although human body cells can synthesize many amino acids (nonessential amino acids), eight amino acids needed by the adult body (ten needed by growing children) cannot be synthesized, or are not synthesized in adequate amounts. For this reason they are called **essential amino acids.**

All of these essential amino acids must be present in the body at the same time if growth and repair of tissues are to occur. If one essential molecule is missing, the process of protein synthesis cannot take place.

Chart 13.6 lists the amino acids found in foods and indicates those that are essential.

On the basis of the kinds of amino acids that they provide, proteins can be classified as complete or incomplete. The **complete proteins,** which include those available in milk, meats, fish, poultry, and eggs, contain adequate amounts of the essential amino acids. **Incomplete proteins,** such as *zein* in corn, which lacks the essential amino acids tryptophan and lysine, and *gelatin,* which lacks tryptophan, are unable to support tissue maintenance or normal growth and development.

A protein called *gliadin,* which occurs in wheat, is an example of a **partially complete protein.** It is deficient in the essential amino acid lysine, and although it does not contain enough lysine to promote growth, it does contain enough to maintain life.

Plant proteins typically contain less than adequate amounts of one or more of the essential amino acids. Protein-containing plant foods, however, can be combined in a meal so that one supplies the amino acids another lacks. For example, bean are deficient in the essential amino acid *methionine,* they contain adequate amounts of *lysine.* R deficient in *lysine* but contains adequate amoun *methionine.* Thus, if beans and rice ar together, they complement one anothe he meal provides a suitable combination of ess mino acids.

1. *What foods provide rich sources of proteins?*
2. *Why are some amino acids called essential?*
3. *Distinguish between a complete protein and an incomplete protein.*

Utilization of Amino Acids. Amino acids are used by body cells in a variety of ways. For example, some amino acids are incorporated into protein molecules that provide cellular structure as in the *actin* and *myosin* of muscle fibers, the *collagen* of connective tissue fibers, and the *keratin* of skin. Other amino acids are used to synthesize proteins that function in various body processes: hemoglobin, which transports oxygen; plasma proteins, which help to regulate water balance and control pH; enzymes, which catalyze metabolic reactions; and hormones, which help to coordinate body activities.

Fig. 13.27 Amino acids resulting from the digestion of proteins are used to synthesize proteins and to supply energy.

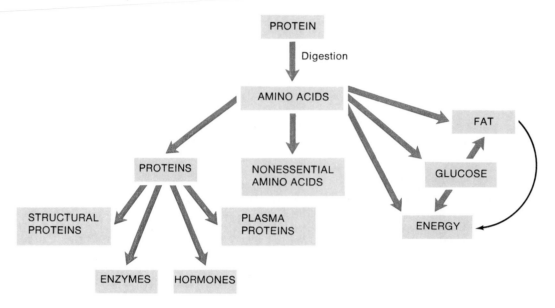

Amino acids also represent potential sources of energy. When they are present in excess or when the supplies of carbohydrates and fats are insufficient to provide needed energy, amino acids are oxidized using various metabolic pathways. (See fig. 13.27.)

Protein Requirements. In addition to supplying essential amino acids, proteins are needed to provide molecular parts and nitrogen for the synthesis of nonessential amino acids. Consequently, the amount of protein required by individuals varies according to body size, metabolic rate, and nitrogen needs.

For an average adult, nutritionists recommend a daily protein intake of about 0.8 grams per kilogram (0.4 grams per pound) of body weight. For a pregnant woman, the recommendation is increased by an additional 30 grams of protein per day. Similarly, a nursing mother requires an additional 20 grams of protein per day to maintain a high level of milk production.

The amount of energy contained in foods consisting of carbohydrates, lipids, or proteins can be expressed in units of heat energy called *calories.* Although a calorie is defined as the amount of heat needed to raise the temperature of a gram of water by one degree Celsius (°C.), the calorie used in the measurement of food energy is 1,000 times greater. This larger calorie (Cal) is called a *kilocalorie,* but in nutritional studies it usually is referred to simply as a Calorie or a food calorie.

1. What are the physiological functions of proteins?
2. How much protein is recommended for an adult diet?

Vitamins

Vitamins are organic compounds (other than carbohydrates, lipids, and proteins) that must be present in small amounts for normal metabolic processes, but cannot be synthesized in adequate amounts by body cells. They are, then, essential nutrients that must be supplied in foods.

For convenience, vitamins are often grouped on the basis of their solubilities, since some are soluble in fats (or fat solvents) and others are soluble in water. Those that are *fat-soluble* include vitamins A, D, E, and K; the *water-soluble* group includes the B vitamins and vitamin C.

Fat-Soluble Vitamins. Since the fat-soluble vitamins dissolve in fats, they occur in association with lipids and are influenced by the same factors that affect lipid absorption. For example, the presence of bile salts in the intestine promotes the absorption of these compounds. As a group, the fat-soluble vitamins are stored in moderate quantities within various tissues, and because they are fairly resistant to the effects of heat, they usually are not destroyed by cooking and food processing.

Chart 13.7 lists the characteristics, functions, and sources of these vitamins.

Chart 13.7 Fat-soluble vitamins

Vitamin	Characteristics	Functions	Sources
Vitamin A	Occurs in several forms; synthesized from carotenes; stored in liver; stable in heat, acids, and alkalis; unstable in light	Necessary for synthesis of visual pigments, mucoproteins, and mucopolysaccharides; for normal development of bones and teeth; and for maintenance of epithelial cells	Liver, whole milk, butter, eggs, leafy green vegetables, and yellow and orange vegetables and fruits
Vitamin D	A group of sterols; resistant to heat, oxidation, acids, and alkalis; stored in liver, skin, brain, spleen, and bones	Promotes absorption of calcium and phosphorus	Produced in skin exposed to ultraviolet light; in milk, egg yolk, fish liver oils, fortified foods
Vitamin E	A group of compounds; resistant to heat and visible light; unstable in presence of oxygen and ultraviolet light; stored in muscles and adipose tissue	An antioxidant; prevents oxidation of vitamin A and polyunsaturated fatty acids; may help maintain stability of cell membranes	Oils from cereal seeds, salad oils, margarine, shortenings, fruits, and vegetables
Vitamin K	Occurs in several forms; resistant to heat but destroyed by acids, alkalis, and light; stored in liver	Needed for synthesis of prothrombin	Leafy green vegetables, egg yolk, pork liver, soy oil, tomatoes, cauliflower

1. What are vitamins?
2. How do bile salts affect the absorption of fat-soluble vitamins?

Water-Soluble Vitamins. The water-soluble vitamins include the B vitamins and vitamin C. The **B vitamins** consist of several compounds that are essential for normal cellular metabolism and are particularly involved with the oxidation of carbohydrates, lipids, and proteins. Since the B vitamins often occur together in foods, they usually are referred to as a group called the *B complex.* Members of this group differ chemically, and they have unique functions. Some of them are easily destroyed by cooking and food processing.

Chart 13.8 lists the characteristics, functions, and sources of the water-soluble vitamins.

1. Name the water-soluble vitamins.
2. What is meant by the vitamin B complex?
3. Distinguish between fat-soluble vitamins and water-soluble vitamins.

Minerals

Dietary **minerals** are inorganic elements that play essential roles in human metabolism. They usually are extracted from the soil by plants. Humans, in turn, obtain them from plant foods or from animals that have eaten plants.

Characteristics of Minerals. Minerals are responsible for about 4% of the body weight and are most concentrated in the bones and teeth. In fact, the minerals *calcium* and *phosphorus,* which are very abundant in these tissues, account for nearly 75% of the body's minerals.

Minerals usually are incorporated into organic molecules. For example, phosphorus occurs in phospholipids, iron in hemoglobin, and iodine in thyroxine. However, some occur in inorganic compounds, such as the calcium phosphate of bone; others occur as free ions, such as the sodium, chloride, and calcium ions in blood plasma.

Minerals are present in all body cells, where they comprise parts of structural materials. They function as portions of enzyme molecules, help create the osmotic pressure of body fluids, and play vital roles in the conduction of nerve impulses, the contraction of muscle fibers, the coagulation of blood, and the maintenance of pH.

Chart 13.8 Water-soluble vitamins

Vitamin	Characteristics	Functions	Sources
Thiamine (Vitamin B_1)	Destroyed by heat and oxygen, especially in alkaline environment	Part of coenzyme needed for oxidation of carbohydrates, and coenzyme needed in synthesis of ribose	Lean meats, liver, eggs, whole grain cereals, leafy green vegetables, legumes
Riboflavin (Vitamin B_2)	Stable to heat, acids, and oxidation; destroyed by alkalis and light	Parts of enzymes and coenzymes needed for oxidation of glucose and fatty acids and for cellular growth	Meats, dairy products, leafy green vegetables, whole grain cereals
Niacin (Nicotinic acid)	Stable to heat, acids, and alkalis; converted to niacinamide by cells; synthesized from tryptophan	Part of coenzymes needed for oxidation of glucose and synthesis of proteins, fats, and nucleic acids	Liver, lean meats, poultry, peanuts, legumes
Vitamin B_6	Group of three compounds; stable to heat and acids; destroyed by oxidation, alkalis, and ultraviolet light	Coenzyme needed for synthesis of proteins and various amino acids, for conversion of tryptophan to niacin, for production of antibodies, and for synthesis of nucleic acids	Liver, meats, fish, poultry, bananas, avocados, beans, peanuts, whole grain cereals, egg yolk
Pantothenic acid	Destroyed by heat, acids, and alkalis	Part of coenzyme needed for oxidation of carbohydrates and fats	Meats, fish, whole grain cereals, legumes, milk, fruits, vegetables
Cyanocobalamin (Vitamin B_{12})	Complex, cobalt-containing compound; stable to heat; inactivated by light, strong acids, and strong alkalis; absorption regulated by intrinsic factor from gastric glands; stored in liver	Part of coenzyme needed for synthesis of nucleic acids and for metabolism of carbohydrates; plays role in synthesis of myelin	Liver, meats, poultry, fish, milk, cheese, eggs
Folacin (Folic acid)	Occurs in several forms; destroyed by oxidation in acid environment or by heat in alkaline environment; stored in liver where it is converted into folinic acid	Coenzyme needed for metabolism of certain amino acids and for synthesis of DNA; promotes production of normal red blood cells	Liver, leafy green vegetables, whole grain cereals, legumes
Biotin	Stable to heat, acids, and light; destroyed by oxidation and alkalis	Coenzyme needed for metabolism of amino acids and fatty acids and for synthesis of nucleic acids	Liver, egg yolk, nuts, legumes, mushrooms
Ascorbic acid (Vitamin C)	Closely related to monosaccharides; stable in acids, but destroyed by oxidation, heat, light, and alkalis	Needed for production of collagen, conversion of folacin to folinic acid, and metabolism of certain amino acids; promotes absorption of iron and synthesis of hormones from cholesterol	Citrus fruits, citrus juices, tomatoes, cabbage, potatoes, leafy green vegetables, fresh fruits

From Hole, John W., Jr., *Human Anatomy and Physiology 3d ed.* © 1978, 1981, 1984 Wm. C. Brown Publishers, Dubuque, Iowa. All Rights Reserved. Reprinted by permission.

The concentrations of various minerals in body fluids are regulated by homeostatic mechanisms, which ensure that the excretion of these substances will be balanced with the dietary intake. Thus toxic excesses are avoided, while minerals that are present in limited amounts are conserved.

1. How are minerals obtained?
2. What are the major functions of minerals?

Major Minerals. As mentioned, calcium and phosphorus account for nearly 75% of the mineral elements in the body; thus they are **major minerals.** Other major minerals, each of which accounts for 0.05% or more of the body weight, include potassium, sulfur, sodium, chlorine, and magnesium.

Chart 13.9 lists the distribution, functions, and sources of these substances.

Chart 13.9 Major minerals

Mineral	Distribution	Functions	Sources
Calcium (Ca)	Mostly in the inorganic salts of bones and teeth	Structure of bones and teeth; essential for nerve impulse conduction, muscle fiber contraction, and blood coagulation; increases permeability of cell membranes; activates certain enzymes	Milk, milk products, leafy green vegetables
Phosphorus (P)	Mostly in the inorganic salts of bones and teeth	Structure of bones and teeth; component in nearly all metabolic reactions; constituent of nucleic acids, many proteins, some enzymes, and some vitamins; occurs in cell membrane, ATP, and phosphates of body fluids	Meats, poultry, fish, cheese, nuts, whole grain cereals, milk, legumes
Potassium (K)	Widely distributed; tends to be concentrated inside cells	Helps maintain intracellular osmotic pressure and regulate pH; promotes metabolism; needed for nerve impulse conduction and muscle fiber contraction	Avocados, dried apricots, meats, nuts, potatoes, bananas
Sulfur (S)	Widely distributed	Essential part of various amino acids, thiamine, insulin, biotin, and mucopolysaccharides	Meats, milk, eggs, legumes
Sodium (Na)	Widely distributed; large proportion occurs in extracellular fluids and bonded to inorganic salts of bone	Helps maintain osmotic pressure of extracellular fluids and regulate water balance; needed for conduction of nerve impulses and contraction of muscle fibers; aids in regulation of pH and in transport of substances across cell membranes	Table salt, cured ham, sauerkraut, cheese, graham crackers
Chlorine (Cl)	Closely associated with sodium; most highly concentrated in cerebrospinal fluid and gastric juice	Helps maintain osmotic pressure of extracellular fluids, regulate pH, and maintain electrolyte balance; essential in formation of hydrochloric acid; aids transport of carbon dioxide by red blood cells	Same as for sodium
Magnesium (Mg)	Abundant in bones	Needed in metabolic reactions that occur in mitochondria and are associated with the production of ATP; plays role in conversion of ATP to ADP	Milk, dairy products, legumes, nuts, leafy green vegetables

Trace Elements. **Trace elements** are essential minerals that occur in minute amounts, each one making up less than 0.005% of the adult body weight. They include iron, manganese, copper, iodine, cobalt, and zinc.

Chart 13.10 lists the distribution, functions, and sources of the trace elements.

1. Distinguish between a major mineral and a trace element.
2. Name the major minerals and trace elements.

Adequate Diets

An adequate diet is one that provides sufficient *energy, essential fatty acids, essential amino acids, vitamins,* and *minerals* to support optimal growth and to maintain and repair body tissues. Since individual needs for nutrients vary greatly with age, sex, growth rate, and amount of physical activity, as well as with genetic and environmental factors, it is not possible to design a diet that will be adequate for everyone.

Chart 13.10   Trace elements

Trace Element	Distribution	Functions	Sources
Iron (Fe)	Primarily in blood; stored in liver, spleen, and bone marrow	Part of hemoglobin molecule; catalyzes formation of vitamin A; incorporated into a number of enzymes	Liver, lean meats, dried apricots, raisins, enriched whole grain cereals, legumes, molasses
Manganese (Mn)	Most concentrated in liver, kidneys, and pancreas	Occurs in enzymes needed for synthesis of fatty acids and cholesterol, formation of urea, and normal functioning of the nervous system	Nuts, legumes, whole grain cereals, leafy green vegetables, fruits
Copper (Cu)	Most highly concentrated in liver, heart, and brain	Essential for synthesis of hemoglobin, development of bone, production of melanin, and formation of myelin	Liver, oysters, crabmeat, nuts, whole grain cereals, legumes
Iodine (I)	Concentrated in thyroid gland	Essential component for synthesis of thyroid hormones	Food content varies with soil content in different geographic regions; iodized table salt
Cobalt (Co)	Widely distributed	Component of cyanocobalamin; needed for synthesis of several enzymes	Liver, lean meats, poultry, fish, milk
Zinc (Zn)	Most concentrated in liver, kidneys, and brain	Constituent of several enzymes involved in digestion, respiration, bone metabolism, liver metabolism; necessary for normal wound healing and maintaining integrity of the skin	Seafoods, meats, cereals, legumes, nuts, vegetables

From Hole, John W., Jr., *Human Anatomy and Physiology 3d ed.* © 1978, 1981, 1984 Wm. C. Brown Publishers, Dubuque, Iowa. All Rights Reserved. Reprinted by permission.

Chart 13.11   Recommended daily dietary allowances for various age groups in the United States

	Infants	Children	Adolescents 15–18 yrs		Adults 23–50 yrs	
	0–6 mos	4–6 yrs	Males	Females	Males	Females
Weight, kg (lb)	6 (13)	20 (44)	66 (145)	55 (120)	70 (154)	55 (120)
Height, cm (in)	60 (24)	112 (44)	176 (69)	163 (64)	178 (70)	163 (64)
Protein, g	kg × 2.2	30	56	46	56	44
Fat-soluble vitamins†						
Vitamin A, µg	420	500	1000	800	1000	800
Vitamin D, µg	10	10	10	10	5	5
Vitamin E activity, mg	3	6	10	8	10	8
Water-soluble vitamins						
Ascorbic acid, mg	35	45	60	60	60	60
Folacin, µg	30	200	400	400	400	400
Niacin, mg	6	11	18	14	18	13
Riboflavin, mg	0.4	1.0	1.7	1.3	1.6	1.2
Thiamine, mg	0.3	0.9	1.4	1.1	1.4	1.0
Vitamin B_6, mg	0.3	1.3	2.0	2.0	2.2	2.0
Vitamin B_{12}, µg	0.5	2.5	3.0	3.0	3.0	3.0
Minerals						
Calcium, mg	360	800	1200	1200	800	800
Phosphorus, mg	240	800	1200	1200	800	800
Iodine, µg	40	90	150	150	150	150
Iron, mg	10	10	18	18	10	18
Magnesium, mg	50	200	400	300	350	300
Zinc, mg	3	10	15	15	15	15

Source: Food and Nutrition Board, *Recommended Dietary Allowances,* 9th ed., National Academy of Sciences—National Research Council, Washington, D.C., 1980.
†Microgram

On the other hand, nutrients are so widely distributed in foods that satisfactory amounts and combinations of essential substances usually can be obtained in spite of individual food preferences that are related to cultural backgrounds, life-styles, and emotional attitudes.

Chart 13.11 lists the recommended daily allowance for each of the major nutrients.

Malnutrition is poor nutrition that results from a lack of essential nutrients or a failure to use available foods to best advantage. It may involve *undernutrition* and include the symptoms of deficiency diseases, or it may be due to *overnutrition* arising from an excessive intake of nutrients.

The factors leading to malnutrition are varied. A deficiency condition may stem, for example, from lack of availability or poor quality of food. On the other hand, malnutrition may result from excessive intake of vitamin supplements or from excessive caloric intake.

---

1. What is meant by an adequate diet?
2. What factors influence individual needs for nutrients?

Adults 51 yrs and over	
Males	*Females*
70 (154)	55 (120)
178 (70)	163 (64)
56	44
1000	800
5	5
10	8
60	60
400	400
16	13
1.4	1.2
1.2	1.0
2.2	2.0
3.0	3.0
800	800
800	800
150	150
10	10
350	300
15	15

Clinical Terms Related to the Digestive System

achalasia (ak″ah-la′ze-ah)—failure of the smooth muscle to relax at some junction in the digestive tube, such as that between the esophagus and stomach.

achlorhydria (ah″klor-hi′dre-ah)—a lack of hydrochloric acid in gastric secretions.

aphagia (ah-fa′je-ah)—an inability to swallow.

cholecystitis (ko″le-sis-ti′tis)—inflammation of the gallbladder.

cholecystolithiasis (ko″le-sis″to-li-thi′ah-sis)—the presence of stones in the gallbladder.

cirrhosis (si-ro′sis)—a liver condition in which the hepatic cells degenerate and the surrounding connective tissues thicken.

diverticulitis (di″ver-tik″u-li′tis)—inflammation of small pouches (diverticula) that sometimes form in the lining and wall of the colon.

dumping syndrome (dum′ping sin′drōm)—a set of symptoms, including diarrhea, that often occur following a gastrectomy.

dysentery (dis′en-ter″e)—an intestinal infection, caused by viruses, bacteria, or protozoans, that is accompanied by diarrhea and cramps.

dyspepsia (dis-pep′se-ah)—indigestion; difficulty in digesting a meal.

dysphagia (dis-fa′ze-ah)—difficulty in swallowing.

enteritis (en″tĕ-ri′tis)—inflammation of the intestine.

esophagitis (e-sof″ah-ji′tis)—inflammation of the esophagus.

gastrectomy (gas-trek′to-me)—partial or complete removal of the stomach.

gastritis (gas-tri′tis)—inflammation of the stomach lining.

gastrostomy (gas-tros′to-me)—the creation of an opening in the stomach wall to allow the administration of food and liquids when swallowing is not possible.

gingivitis (jin″ji-vi′tis)—inflammation of the gums.

glossitis (glŏ-si′tis)—inflammation of the tongue.

hemorrhoidectomy (hem″ō-roi-dek′to-me)—removal of hemorrhoids.

hemorrhoids (hem′ō-roids)—a condition in which veins associated with the lining of the anal canal become enlarged.

hepatitis (hep″ah-ti′tis)—inflammation of the liver.

ileitis (il″e-i′tis)—inflammation of the ileum.

pharyngitis (far″in-ji′tis)—inflammation of the pharynx.

proctitis (prok-ti′tis)—inflammation of the rectum.

pyloric stenosis (pi-lor′ik stĕ-no′sis)—a congenital obstruction at the pyloric sphincter due to an enlargement of the muscle.

pylorospasm (pi-lor′o-spazm)—a spasm of the pyloric portion of the stomach or of the pyloric sphincter.

pyorrhea (pi″o-re′ah)—an inflammation of the dental periosteum accompanied by the formation of pus.

stomatitis (sto″mah-ti′tis)—inflammation of the lining of the mouth.

vagotomy (va-got′o-me)—sectioning of the vagus nerve fibers.

Chapter Summary

Introduction

Digestion is the process of changing food substances into forms that can be absorbed.

The digestive system consists of an alimentary canal and several accessory organs.

General Characteristics of the Alimentary Canal

Various regions of the canal are specialized to perform specific functions.

1. Structure of the wall
   The wall consists of four layers.

2. Movements of the tube
   Motor functions include mixing and propelling movements.

The Mouth

The mouth is adapted to receive food and begin preparing it for digestion.

1. The cheeks and lips
   a. Cheeks consist of outer layers of skin, pads of fat, muscles associated with expression and chewing, and inner linings of epithelium.
   b. Lips are highly mobile and possess a variety of sensory receptors.

2. The tongue
   a. Its rough surface contains taste buds and aids in handling food.
   b. Lingual tonsils are located on the root of the tongue.

3. The palate
   a. The palate includes hard and soft portions.
   b. The soft palate closes the opening to the nasal cavity during swallowing.
   c. Palatine tonsils are located at either side of the tongue in the back of the mouth.

4. The teeth
   a. There are twenty deciduous and thirty-two permanent teeth.
   b. They function to break food into smaller pieces, increasing the surface area of food.
   c. Each tooth consists of a crown and root and is composed of enamel, dentine, pulp, nerves, and blood vessels.
   d. A tooth is attached to the alveolar bone by a periodontal ligament.

The Salivary Glands

Salivary glands secrete saliva, which moistens food, helps bind food particles together, begins digestion of carbohydrates, makes taste possible, and helps cleanse the mouth.

1. Salivary secretions
   A salivary gland includes serous cells that secrete digestive enzymes and mucous cells that secrete mucus.

2. Major salivary glands
   a. The parotid glands secrete saliva rich in amylase that begins the digestion of carbohydrates.
   b. The submandibular glands produce saliva that is more viscous than that of the parotid glands.
   c. The sublingual glands primarily secrete mucus.

The Pharynx and Esophagus

The pharynx and esophagus serve only as passageways.

1. Structure of the pharynx
   The pharynx is divided into a nasopharynx, oropharynx, and laryngopharynx.

2. The swallowing mechanism
   a. The act of swallowing occurs in three stages.
      (1) Food is mixed with saliva and forced into the pharynx.
      (2) Involuntary reflexes move the food into the esophagus.
      (3) Food is transported to the stomach.

3. The esophagus
   a. The esophagus passes through the diaphragm and joins the stomach.
   b. Some circular muscle fibers at the end of the esophagus help to prevent the regurgitation of food from the stomach.

The Stomach

The stomach receives food, mixes it with gastric juice, carries on a limited amount of absorption, and moves food into the small intestine.

1. Parts of the stomach
   a. The stomach is divided into cardiac, fundic, body, and pyloric regions.
   b. The pyloric sphincter serves as a valve between the stomach and the small intestine.

2. Gastric secretions
   a. Gastric glands secrete gastric juice.
   b. Gastric juice contains pepsin, hydrochloric acid, and intrinsic factor.
   c. Regulation of gastric secretions
      (1) Gastric secretions are enhanced by parasympathetic impulses and the hormone, gastrin.
      (2) Presence of food in the small intestine reflexly inhibits gastric secretions.

3. Gastric absorption
   A few substances such as water and other small molecules may be absorbed through the stomach wall.
4. Mixing and emptying actions
   a. Mixing movements aid in producing chyme; peristaltic waves move the chyme into the small intestine.
   b. The rate of emptying depends on the fluidity of the chyme and the type of food present.

The Pancreas

1. Structure of the pancreas
   a. The pancreas produces pancreatic juice that is secreted into a pancreatic duct.
   b. The pancreatic duct leads to the duodenum.
2. Pancreatic juice
   a. Pancreatic juice contains enzymes that can split carbohydrates, proteins, fats, and nucleic acids.
   b. Regulation of pancreatic secretion
      (1) Secretin stimulates the release of pancreatic juice that has a high bicarbonate ion concentration.
      (2) Cholecystokinin stimulates the release of pancreatic juice that has a high concentration of digestive enzymes.

The Liver

1. Functions of the liver
   a. The liver carries on a variety of important functions involving the metabolism of carbohydrates, lipids, and proteins; the storage of substances; the filtering of blood; the destruction of toxic chemicals; and the secretion of bile.
   b. Bile is the only liver secretion that directly affects digestion.
2. Structure of the liver
   a. Each lobe contains hepatic lobules, the functional units of the liver.
   b. Bile from the lobules is carried by bile canals to hepatic ducts.
3. Composition of bile
   a. Bile contains bile salts, bile pigments, cholesterol, and various electrolytes.
   b. Only the bile salts have digestive functions.
4. The gallbladder and its functions
   a. The gallbladder stores bile between meals.
   b. Release of bile from the common bile duct is controlled by a sphincter muscle.
   c. Release is stimulated by cholecystokinin from the small intestine.
   d. Sphincter muscle at the base of the common bile duct relaxes as a peristaltic wave in the duodenal wall passes by.
   e. Bile salts emulsify fats and aid in the absorption of fatty acids, cholesterol, and certain vitamins.

The Small Intestine

The small intestine receives secretions from the pancreas and liver, completes the digestion of nutrients, absorbs the products of digestion, and transports the residues to the large intestine.

1. Parts of the small intestine
   The small intestine consists of the duodenum, jejunum, and ileum.
2. Structure of the small intestinal wall
   a. The wall is lined with villi that aid in mixing and absorption.
   b. Intestinal glands are located between the villi.
3. Secretions of the small intestine
   a. Secretions include mucus and digestive enzymes.
   b. Digestive enzymes can split molecules of sugars, proteins, fats, and nucleic acids.
   c. Small intestinal secretions are enhanced by the presence of gastric juice and chyme and by the mechanical stimulation of distension.
4. Absorption in the small intestine
   a. Villi increase the surface area of the intestinal wall.
   b. Monosaccharides, amino acids, fatty acids, and glycerol are absorbed by the villi.
   c. Fat molecules with longer chains of carbon atoms enter the lacteals of the villi.
   d. Other digestive products enter the blood capillaries of the villi.
5. Movements of the small intestine
   a. Movements include mixing and peristalsis.
   b. The ileocecal valve controls movement from the small intestine into the large intestine.

The Large Intestine

The large intestine functions to reabsorb water and electrolytes and to form and store feces.

1. Parts of the large intestine
   a. The large intestine consists of the cecum, colon, rectum, and anal canal.
   b. The colon is divided into ascending, transverse, descending, and sigmoid portions.
2. Structure of the large intestinal wall
   a. Basically the large intestinal wall is like the wall in other parts of the alimentary canal.
   b. Unique features include a layer of longitudinal muscle fibers that are arranged in three distinct bands.
3. Functions of the large intestine
   a. The large intestine has little or no digestive function.
   b. The only significant secretion is mucus.
   c. Absorption in the large intestine is generally limited to water and electrolytes.
   d. Feces is formed and stored in the large intestine.

4. Movements of the large intestine
   a. Movements are similar to those in the small intestine.
   b. Mass movements occur two to three times each day.
   c. Defecation is stimulated by a defecation reflex.
5. The feces
   Feces consist largely of water, undigested material, mucus, and bacteria.

Nutrition and Nutrients

Nutrition is the process of ingestion and utilization of necessary food substances or nutrients.

1. Carbohydrates
   Carbohydrates are organic compounds that are used primarily to supply cellular energy.
   a. Sources of carbohydrates
      (1) Starch, glycogen, disaccharides, and monosaccharides are carbohydrates.
      (2) Cellulose is a polysaccharide that cannot be digested by human enzymes.
   b. Utilization of carbohydrates
      (1) Energy is released from glucose by oxidation.
      (2) Excessive glucose is stored as glycogen or converted to fat.
   c. Carbohydrate requirements
      (1) Most carbohydrates are utilized to supply energy.
      (2) Some cells depend on a continuous supply of glucose to survive.
      (3) Humans survive with a wide range of carbohydrate intakes.
2. Lipids
   Lipids are organic compounds that supply energy and are used to build cell structures.
   a. Sources of lipids
      (1) Triglycerides are obtained from foods of plant and animal origins.
      (2) Cholesterol is obtained in foods of animal origin only.
   b. Utilization of lipids
      (1) Metabolism of triglycerides is controlled mainly by the liver and adipose tissues.
      (2) Linoleic acid is an essential fatty acid.
   c. Lipid requirements
      (1) Humans survive with a wide range of lipid intakes.
      (2) Some fats contain fat-soluble vitamins.
3. Proteins
   Proteins are organic compounds that serve as structural materials, act as enzymes, and provide energy.
   a. Sources of proteins
      (1) Proteins are obtained mainly from meats, dairy products, cereals, and legumes.
      (2) Complete proteins contain adequate amounts of all essential amino acids.
      (3) Incomplete proteins lack one or more essential amino acids.
   b. Utilization of amino acids
      Amino acids are incorporated into various structural and functional proteins, including enzymes.
   c. Protein requirements
      Proteins and amino acids are needed to supply essential amino acids and nitrogen for the synthesis of various molecules.
4. Vitamins
   Vitamins are organic compounds (other than carbohydrates, lipids, and proteins) that are essential for normal metabolic processes, but cannot be synthesized by body cells in adequate amounts.
   a. Fat-soluble vitamins
      (1) These include vitamins A, D, E, and K.
      (2) They occur in association with lipids and are influenced by the same factors that affect lipid absorption.
      (3) They are fairly resistant to the effects of heat; thus they are not destroyed by cooking or food processing.
   b. Water-soluble vitamins
      (1) Group includes the B vitamins and vitamin C.
      (2) B vitamins make up a group called the B complex and are generally involved with the oxidation of carbohydrates, lipids, and proteins.
      (3) Some are destroyed by cooking and food processing.
5. Minerals
   a. Characteristics of minerals
      (1) Most of the minerals are found in bones and teeth.
      (2) Minerals usually are incorporated into organic molecules, although some occur in inorganic compounds or as free ions.
      (3) They comprise structural materials, function in enzymes, and play vital roles in various metabolic processes.
   b. Major minerals
      They include calcium, phosphorus, potassium, sulfur, sodium, chlorine, and magnesium.
   c. Trace elements
      They include iron, manganese, copper, iodine, cobalt, and zinc.

Adequate Diets

1. An adequate diet provides sufficient energy and essential nutrients to support optimal growth as well as maintenance and repair of tissues.
2. Individual needs vary so greatly that it is not possible to design a diet that is adequate for everyone.

Application of Knowledge

1. If a patient has 95% of the stomach removed (subtotal gastrectomy) as treatment for severe ulcers or cancer, how would the digestion and absorption of foods be affected? How would the patient's eating habits have to be altered? Why?

2. Why is it that a person with inflammation of the gallbladder (cholecystitis) also may develop an inflammation of the pancreas (pancreatitis)?

3. How would you explain the fact that the blood sugar concentration level of a person whose diet is relatively low in carbohydrates remains stable?

4. Examine the label information on the packages of a variety of dry breakfast cereals. Which types of cereals provide the best sources of vitamins and minerals?

Review Activities

1. List and describe the location of the major parts of the alimentary canal.
2. List and describe the location of the accessory organs of the digestive system.
3. Name the four layers of the wall of the alimentary canal.
4. Distinguish between mixing movements and propelling movements.
5. Define *peristalsis.*
6. Discuss the functions of the mouth and its parts.
7. Distinguish between lingual, palatine, and pharyngeal tonsils.
8. Compare the deciduous and permanent teeth.
9. Describe the structure of a tooth.
10. Explain how a tooth is anchored to its socket.
11. List and describe the location of the major salivary glands.
12. Explain how the secretions of the salivary glands differ.
13. Discuss the digestive functions of saliva.
14. Explain the function of the esophagus.
15. Describe the structure of the stomach.
16. List the enzymes in gastric juice and explain the function of each enzyme.
17. Explain how gastric secretions are regulated.

18. Describe the location of the pancreas and the pancreatic duct.
19. List the enzymes found in pancreatic juice and explain the function of each enzyme.
20. Explain how pancreatic secretions are regulated.
21. List the major functions of the liver.
22. Describe the structure of the liver.
23. Describe the composition of bile.
24. Define *cholecystokinin.*
25. Explain the functions of bile salts.
26. List and describe the location of the parts of the small intestine.
27. Name the enzymes of the intestinal mucosa and explain the function of each enzyme.
28. Explain how the secretions of the small intestine are regulated.
29. Describe the functions of the intestinal villi.
30. Summarize how the major types of digestive products are absorbed.
31. List and describe the location of the parts of the large intestine.
32. Explain the general functions of the large intestine.
33. Describe the defecation reflex.
34. List some common sources of carbohydrates.
35. Summarize the importance of cellulose in the diet.
36. Explain why a temporary drop in the glucose concentration may produce functional disorders of the nervous system.
37. List some common sources of lipids.
38. Describe the role of the liver in fat metabolism.
39. List some common sources of proteins.
40. Distinguish between essential and nonessential amino acids.
41. Distinguish between complete and incomplete proteins.
42. Discuss the general characteristics of fat-soluble vitamins.
43. List the fat-soluble vitamins and describe the major functions of each vitamin.
44. List the water-soluble vitamins and describe the major functions of each vitamin.
45. Discuss the general characteristics of the mineral nutrients.
46. List the major minerals and describe the major functions of each mineral.
47. List the trace elements and describe the major functions of each element.
48. Define *adequate diet.*

The Respiratory System

Before body cells can oxidize nutrients and release energy, they must be supplied with oxygen. Also, the carbon dioxide that results from oxidation must be excreted. These two general processes—obtaining oxygen and removing carbon dioxide—are the primary functions of the *respiratory system.*

In addition, the respiratory organs filter particles from incoming air, help to control the temperature and water content of the air, aid in producing the sounds used in speech, and play important roles in the sense of smell and the regulation of pH.

14

After you have studied this chapter, you should be able to

1. List the general functions of the respiratory system.

2. Name and describe the location of the organs of the respiratory system.

3. Describe the functions of each organ of the respiratory system.

4. Explain how inspiration and expiration are accomplished.

5. Name and define each of the respiratory air volumes.

6. Locate the respiratory center and explain how it controls normal breathing.

7. Discuss how various factors affect the respiratory center.

8. Describe the structure and function of the respiratory membrane.

9. Explain how oxygen and carbon dioxide are transported in the blood.

10. Complete the review activities at the end of this chapter. Note that the items are worded in the form of specific learning objectives. You may want to refer to them before reading the chapter.

alveolus   (al-ve′o-lus)

bronchial tree   (brong′ke-al tre)

carbonic anhydrase   (kar-bon′ik an-hi′drās)

cellular respiration   (sel′u-lar res″pĭ-ra′shun)

expiration   (ek″spĭ-ra′shun)

glottis   (glot′is)

hemoglobin   (he″mo-glo′bin)

hyperventilation   (hi″per-ven″tĭ-la′shun)

inspiration   (in″spĭ-ra′shun)

partial pressure   (par′shil presh′ur)

pleural cavity   (ploo′ral kav′ĭ-te)

respiratory center   (re-spi′rah-to″re sen′ter)

respiratory membrane
     (re-spi′rah-to″re mem′brān)

respiratory volume   (re-spi′rah-to″re vol′ūm)

surface tension   (ser′fas ten′shun)

surfactant   (ser-fak′tant)

alveol-, a small cavity: *alveol*us—a microscopic air sac within a lung.

bronch-, the windpipe: *bronch*us—a primary branch of the trachea.

cric-, a ring: *cric*oid cartilage—a ring-shaped mass of cartilage at the base of the larynx.

epi-, upon: *epi*glottis—a flaplike structure that partially covers the opening into the larynx during swallowing.

hem-, blood: *hem*oglobin—pigment in red blood cells that serves to transport oxygen and carbon dioxide.

Fig. 14.1 Major organs of the respiratory system.

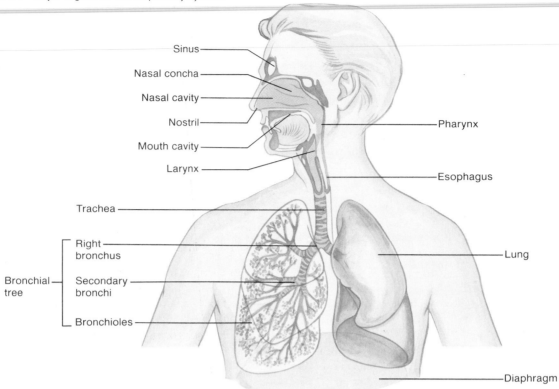

The **respiratory system** contains a group of passages that filter incoming air and transport it from the outside of the body into the lungs. It also includes numerous microscopic air sacs in which gas exchanges takes place.

Although the entire process of exchanging gases between the atmosphere and the body cells is called **respiration,** this process actually involves several events. These include the movement of air in and out of the lungs—commonly called breathing or pulmonary ventilation; the exchange of gases between the air in the lungs and the blood; the transport of gases by the blood between the lungs and body cells; and the exchange of gases between the blood and the cells. The utilization of oxygen and the production of carbon dioxide by the cells is called **cellular respiration.**

Organs of the Respiratory System

The organs of the respiratory system include the nose, nasal cavity, sinuses, pharynx, larynx, trachea, bronchial tree, and lungs. (See fig. 14.1.)

The Nose

The **nose** is supported internally by bone and cartilage. Its two *nostrils* provide openings through which air can enter and leave the nasal cavity. These openings are guarded by numerous internal hairs that help prevent the entrance of relatively coarse particles sometimes carried by the air.

The Nasal Cavity

The **nasal cavity,** which is a hollow space behind the nose, is divided medially into right and left portions by the **nasal septum.**

Nasal conchae (fig. 8.7) curl out from the lateral walls of the nasal cavity on each side, dividing the cavity into passageways. They support the mucous membrane that lines the nasal cavity and help to increase its surface area.

The mucous membrane contains pseudostratified ciliated epithelium that is rich in mucus-secreting goblet cells. (See chapter 5.) It also includes an extensive network of blood vessels, and as air passes over the membrane, heat leaves the blood and warms the air. In this way, the temperature of the incoming air quickly adjusts to that of the body. In addition, the incoming air tends to become moistened by the evaporation of water from the mucous lining. The sticky mucus secreted by the mucous membrane entraps dust and other small particles entering with the air.

As the cilia of the epithelial lining move, a thin layer of mucus and any entrapped particles are pushed toward the pharynx. When the mucus reaches the pharynx, it is swallowed. In the stomach, any microorganisms in the mucus are likely to be destroyed by the action of gastric juice. (See figs. 14.2 and 14.3.)

---

1. *What is meant by respiration?*
2. *What organs constitute the respiratory system?*
3. *What are the functions of the mucous membrane that lines the nasal cavity?*

The Sinuses

As discussed in chapter 8, the **sinuses** are air-filled spaces located within (and named from) the *maxillary, frontal, ethmoid,* and *sphenoid bones* of the skull. (See fig. 8.8.) These spaces open into the nasal cavity and are lined with mucous membranes that are continuous with the lining of the nasal cavity.

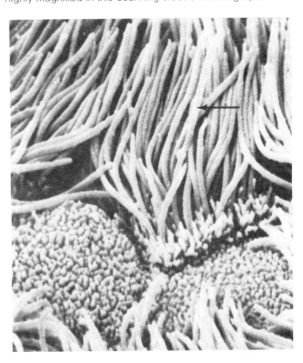

Fig. 14.2 Ciliated cells line the air passages of the respiratory tract. The arrow indicates the cilia, which are highly magnified in this scanning electron micrograph.

Fig. 14.3 Cilia move mucus and trapped particles from the nasal cavity to the pharynx.

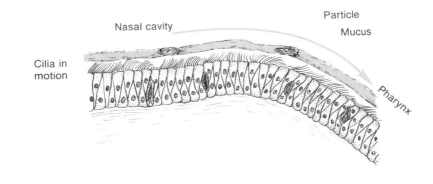

Fig. 14.7 The bronchial tree consists of the passageways that connect the trachea and the alveoli.

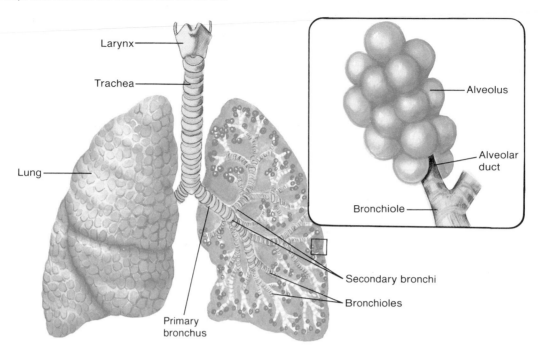

Larynx

Trachea

Lung

Primary bronchus

Secondary bronchi

Bronchioles

Alveolus

Alveolar duct

Bronchiole

Fig. 14.8 The tubes of the alveolar ducts end in tiny alveoli, and each alveolus is surrounded by a capillary net.

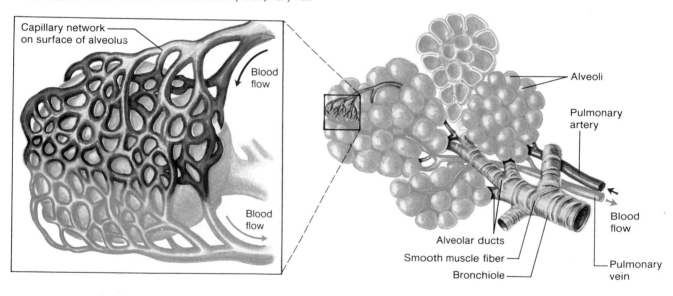

Capillary network on surface of alveolus

Blood flow

Blood flow

Alveoli

Pulmonary artery

Alveolar ducts

Smooth muscle fiber

Bronchiole

Blood flow

Pulmonary vein

Fig. 14.9 Oxygen diffuses from the air within the alveolus into the capillary, while carbon dioxide diffuses from the blood in the capillary into the alveolus.

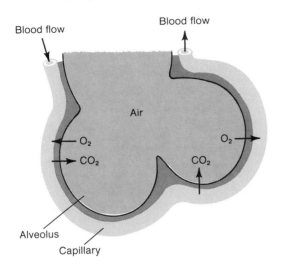

muscle surrounding the tube becomes more prominent. This muscular layer remains in the walls of the bronchioles, but only a few muscle fibers occur in the alveolar ducts.

The branches of the bronchial tree serve as air passages, distributing incoming air to the alveoli in all parts of the lungs. The alveoli, in turn, provide a large surface area through which gas exchanges can occur. During these exchanges, oxygen diffuses through the alveolar walls and enters the blood in nearby capillaries, while carbon dioxide diffuses from the blood through the walls and enters the alveoli. (See fig. 14.9.)

It is estimated that there are about 300 million alveoli in an adult lung and that these spaces have a total surface area of between 70 and 80 square meters. (This is roughly equivalent to the area of the floor in a room measuring 25 feet by 30 feet.)

---

1. *What is the function of the cartilaginous rings in the tracheal wall?*
2. *Describe the bronchial tree.*
3. *How are gases exchanged in the alveoli?*

The Lungs

The **lungs** (fig. 14.1) are soft, spongy, cone-shaped organs located in the thoracic cavity. The right and left lungs are separated medially by the heart and the mediastinum, and they are enclosed by the diaphragm and the thoracic cage.

Each lung occupies most of the thoracic space on its side and is suspended in this cavity by its attachments, which include a bronchus and some large blood vessels. These tubular parts enter the lung on its medial surface. A layer of serous membrane, the **visceral pleura,** is firmly attached to the surface of each lung, and this membrane folds back to become the **parietal pleura.** The parietal pleura, in turn, forms part of the mediastinum and lines the inner wall of the thoracic cavity. (See fig. 7.2.)

The potential space between the visceral and parietal pleurae is called the **pleural cavity,** and it contains a thin film of serous fluid. This fluid lubricates the adjacent pleural surfaces, reducing friction as they move against one another during breathing. It also helps to hold the pleural membranes together, as explained in the next section of this chapter.

The right lung is larger than the left one and is divided into three lobes, while the left lung consists of two lobes.

Each lobe is supplied by a major branch of the bronchial tree. It also has connections to blood and lymphatic vessels and is enclosed by connective tissues. Thus the substance of a lung includes air passages, alveoli, blood vessels, connective tissues, lymphatic vessels, and nerves.

Chart 14.1 summarizes the characteristics of the major parts of the respiratory system.

Lung cancer, is the most common form of cancer in males today and is becoming increasingly common among females. Although it may arise from epithelial cells, connective tissue cells, or various blood cells, the most common form arises from epithelium that lines the tubes of the bronchial tree. This type of cancer seems to occur in response to excessive irritation, such as that produced by prolonged exposure to tobacco smoke.

Once lung cancer cells have appeared, they are likely to produce masses that obstruct air passages and reduce the amount of alveolar surface available for gas exchanges. Furthermore, they are likely to spread to other tissues relatively quickly and establish secondary cancers.

---

1. *Where are the lungs located?*
2. *What is the function of the serous fluid within the pleural cavity?*
3. *What kinds of structures make up a lung?*

Chart 14.1   Parts of the respiratory system

Part	Description	Function
Nose	Part of face centered above the mouth and below the space between the eyes	Nostrils provide entrance to nasal cavity; internal hairs begin to filter incoming air
Nasal cavity	Hollow space behind nose	Conducts air to pharynx; mucous lining filters, warms, and moistens air
Sinuses	Hollow spaces in various bones of the skull	Reduce weight of the skull; serve as resonant chambers
Pharynx	Chamber behind mouth cavity and between nasal cavity and larynx	Passageway for air moving from nasal cavity to larynx and for food moving from mouth cavity to esophagus
Larynx	Enlargement at the top of the trachea	Passageway for air; prevents foreign objects from entering trachea; houses vocal cords
Trachea	Flexible tube that connects larynx with bronchial tree	Passageway for air; mucous lining continues to filter air
Bronchial tree	Branched tubes that lead from the trachea to the alveoli	Conducts air from the trachea to the alveoli; mucous lining continues to filter air
Lungs	Soft, cone-shaped organs that occupy a large portion of the thoracic cavity	Contain the air passages, alveoli, blood vessels, connective tissues, lymphatic vessels, and nerves of the lower respiratory tract

From Hole, John W., Jr., *Human Anatomy and Physiology 3d ed.* © 1978, 1981, 1984 Wm. C. Brown Publishers, Dubuque, Iowa. All Rights Reserved. Reprinted by permission.

Fig. 14.10   Atmospheric pressure is sufficient to support a column of mercury 760 mm high.

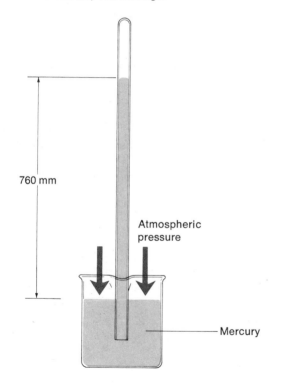

760 mm

Atmospheric pressure

Mercury

Mechanism of Breathing

Breathing, or *pulmonary ventilation,* entails the movement of air from outside the body into the bronchial tree and alveoli, followed by a reversal of this air movement. These actions are termed **inspiration** (inhalation) and **expiration** (exhalation), and they are accompanied by changes in the size of the thoracic cavity.

Inspiration

Atmospheric pressure, due to the weight of air, is the force that causes air to move into the lungs. At sea level, this pressure is sufficient to support a column of mercury about 760 millimeters (30 inches) high in a tube. (See fig. 14.10.) Thus normal air pressure is said to equal 760 millimeters of mercury (Hg).

Air pressure is exerted on all surfaces in contact with air, and since persons breathe air, the inside surfaces of their lungs also are subjected to pressure. In other words, the pressures on the inside of the lungs and alveoli and on the outside of the thoracic wall are about the same.

If the pressure on the inside of the lungs and alveoli decreases, air from the outside will be pushed into the airways by atmospheric pressure. This is what happens during inspiration. Muscle fibers in the dome-shaped *diaphragm* below the lungs are stimulated to contract by impulses carried on the phrenic nerves.

Fig. 14.11 The size of the thoracic cavity enlarges as the diaphragm and external intercostal muscles contract.

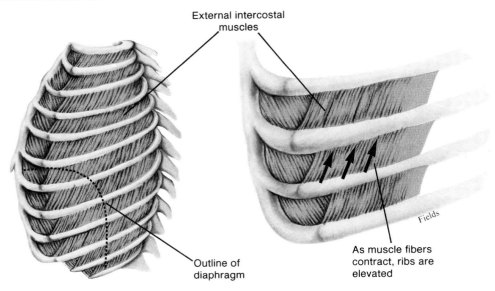

External intercostal muscles

Outline of diaphragm

As muscle fibers contract, ribs are elevated

As this happens, the diaphragm moves downward, the size of the thoracic cavity is enlarged, and the pressure within the alveoli is reduced. At the same time, air is forced into the airways by atmospheric pressure, and the lungs expand.

While the diaphragm is contracting and moving downward, the *external intercostal muscles* between the ribs may be stimulated to contract. This action raises the ribs and elevates the sternum, so that the size of the thoracic cavity increases even more. As a result, the pressure inside is further reduced and more air is forced into the airways by atmospheric pressure. (See fig. 14.11.)

The expansion of the lungs is aided by the fact that the parietal pleura, on the inner wall of the thoracic cavity, and the visceral pleura, attached to the surface of the lungs, are separated only by a thin film of serous fluid. The *water molecules* in this fluid have a great attraction for one another, and the resulting **surface tension** holds the moist surfaces of the pleural membranes tightly together. Consequently, when the thoracic wall is moved upward and outward by the action of the external intercostal muscles, the parietal pleura is moved too, and the visceral pleura follows it. This helps to expand the lungs in all directions.

The attraction between adjacent moist membranes is sufficient to cause the collapse of the tiny air sacs, which also have moist surfaces. Fortunately, the alveolar cells secrete a mixture of lipoproteins, called **surfactant** that acts to reduce surface tension and decreases the tendency to collapse.

Sometimes the lungs of a newborn fail to produce enough surfactant, and the newborn's breathing mechanism may be unable to overcome the force of surface tension. Consequently, the lungs cannot be ventilated and the newborn is likely to die of suffocation. This condition is called *respiratory distress syndrome,* and it is the primary cause of respiratory difficulty in immature newborns.

If a person needs to take a deeper than normal breath, the diaphragm and external intercostal muscles usually can be contracted to a greater degree. Additional muscles, such as the pectoralis minors and sternocleidomastoids, can be used to pull the thoracic cage further upward and outward. (See fig. 14.12.)

Expiration

The forces responsible for normal expiration come from the *elastic recoil* of tissues and from surface tension. The lungs and thoracic wall, for example, contain a considerable amount of elastic tissue, and as the lungs expand during inspiration, these tissues are stretched. As the diaphragm lowers, the abdominal organs beneath it are compressed. As the diaphragm and the external intercostal muscles relax following inspiration, these elastic tissues cause the lungs and thoracic cage to recoil, and they return to their original positions. Similarly, the abdominal organs spring back into their previous locations, pushing

Fig. 14.14 Respiratory air volumes. White line represents a recording of air volumes measured using an instrument called a spirometer.

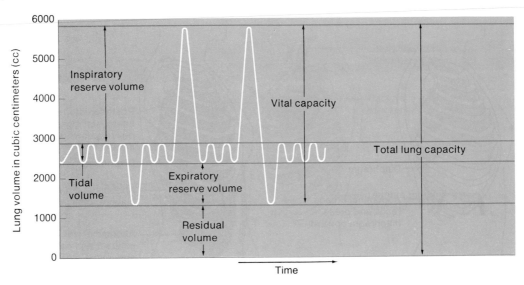

Chart 14.3 Nonrespiratory air movements

Air Movement	Mechanism	Function
Coughing	Deep breath is taken, glottis is closed, and air is forced against the closure; suddenly the glottis is opened and a blast of air passes upward	Clears lower respiratory passages
Sneezing	Same as coughing, except air moving upward is directed into the nasal cavity by depressing the uvula	Clears upper respiratory passages
Laughing	Deep breath is released in a series of short expirations	Expresses emotional happiness
Crying	Same as laughing	Expresses emotional sadness
Hiccuping	Diaphragm contracts spasmodically while the glottis is closed	No useful function
Yawning	Deep breath taken	Ventilates a large proportion of the alveoli and aids oxygenation of the blood
Speech	Air is forced through the larynx, causing vocal cords to vibrate; words are formed by lips, tongue, and soft palate	Communication

Nonrespiratory Air Movements

Air movements that occur in addition to breathing are called *nonrespiratory movements*. They are used to clear air passages, as in coughing and sneezing, or to express emotional feelings, as in laughing and crying, or to speak.

Chart 14.3 summarizes some of the characteristics of these movements.

Control of Breathing

Although respiratory muscles can be controlled voluntarily, normal breathing is a rhythmic, involuntary act that continues even when a person is unconscious.

The Respiratory Center

Breathing is controlled by a poorly defined region in the brain stem called the **respiratory center**. The neurons that make up this center are widely scattered throughout the pons and the upper 2/3 of the medulla

Fig. 14.15 In the process of inspiration, motor impulses travel from the respiratory center to the diaphragm and external intercostal muscles, which contract and cause the lungs to expand. This expansion stimulates stretch receptors to send inhibiting impulses to the respiratory center.

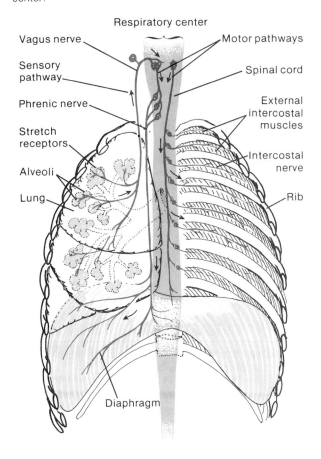

Respiratory center

Vagus nerve

Motor pathways

Sensory pathway

Spinal cord

Phrenic nerve

External intercostal muscles

Stretch receptors

Intercostal nerve

Alveoli

Lung

Rib

Diaphragm

oblongata. (See fig. 10.28.) They initiate impulses that travel on cranial and spinal nerves to various breathing muscles and are able to adjust the rate and depth of breathing. As a result, cellular needs for a supply of oxygen and the removal of carbon dioxide are met even during periods of strenuous physical exercise.

The mechanism responsible for maintaining the basic breathing pattern is not well understood. It seems to involve two groups of intermingled neurons. One group is concerned primarily with inspiration, and the other functions in expiration.

It is believed that when one of the inspiratory neurons becomes excited, it stimulates other neurons, which in turn stimulate still others. This action seems to establish a self-perpetuating circuit, and impulses travel around this circuit until the neurons become fatigued in about 2 seconds. Meanwhile, impulses also travel out from the inspiratory circuit to inspiratory muscles that contract and cause inspiration. At the

same time, other impulses from the inspiratory circuit are directed to an expiratory circuit, inhibiting it.

When the inspiratory circuit becomes fatigued, the expiratory neurons, which are no longer inhibited, become excited, and impulses begin to flow through the expiratory circuit. Impulses from this circuit cause expiration. At the same time, inhibiting impulses are sent to the neurons of the inspiratory circuit until the expiratory circuit becomes fatigued.

Thus normal breathing seems to result from alternating activity in the neurons of the inspiratory and the expiratory circuits.

---

1. *Where is the respiratory center located?*
2. *Describe how the respiratory center seems to function to maintain a normal breathing pattern.*

Factors Affecting Breathing

In addition to the controls exerted by the respiratory center, breathing rate and depth are influenced by a variety of other factors. These include reflexes and the presence of certain chemicals in body fluids.

For example, an **inflation reflex** helps to regulate the depth of breathing and to maintain the respiratory rhythm. This reflex occurs when *stretch receptors* in parts such as the visceral pleura, bronchioles, and alveoli are stimulated during inspiration. The resulting sensory impulses travel via the vagus nerves to the respiratory center, and further inspiration is inhibited (fig. 14.15). This action prevents overinflation of the lungs.

Chemical factors in the blood that affect breathing include carbon dioxide, hydrogen ions, and oxygen. More specifically, breathing rate increases as the blood concentrations of carbon dioxide or hydrogen ions increase, or when the concentration of oxygen decreases. Of these factors, however, carbon dioxide and hydrogen ions are the most important.

If a person decides to stop breathing, the blood concentrations of carbon dioxide and hydrogen ions begin to rise, and the concentration of oxygen falls. These changes stimulate the respiratory center, and soon the need to inhale overpowers the desire to hold the breath. On the other hand, a person can increase the breath-holding time by breathing rapidly and deeply in advance. This action, termed **hyperventilation,** causes a lowering of the blood carbon dioxide

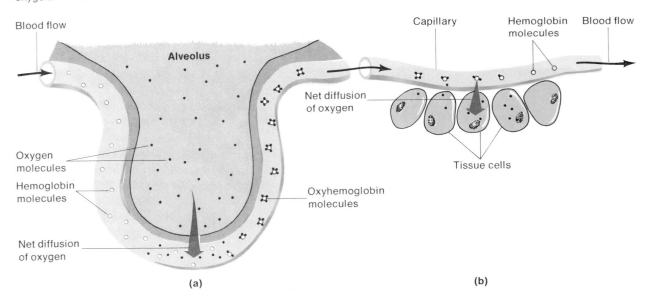

(a)

(b)

The chemical bonds that form between oxygen and hemoglobin molecules are relatively unstable, and as the P_{O_2} decreases, oxygen is released from oxyhemoglobin molecules. This happens in tissues where cells have used oxygen in their respiratory processes, and the free oxygen diffuses from the blood into nearby cells, as shown in figure 14.19. Therefore, more oxygen is released from the blood to skeletal muscle during periods of exercise, and less active cells receive relatively smaller amounts of oxygen.

Carbon monoxide (CO) is a toxic gas produced in gasoline engines as a result of incomplete combustion of fuel. It is also a component of tobacco smoke. Its toxic effect occurs because it combines with hemoglobin more effectively than oxygen does. Furthermore, carbon monoxide does not dissociate readily from the hemoglobin. Thus when a person breathes carbon monoxide, increasing quantities of hemoglobin become unavailable for oxygen transport, and the body cells soon begin to suffer.

A victim of carbon monoxide poisoning may be treated by administering oxygen in high concentration to replace some of the carbon monoxide bound to hemoglobin molecules. Carbon dioxide is usually administered simultaneously to stimulate the respiratory center, which in turn causes an increase in the breathing rate. Rapid breathing is desirable because it helps to reduce the concentration of carbon monoxide in the alveoli.

1. *How is oxygen transported from the lungs to the body cells?*
2. *What stimulates oxygen to be released from the blood to the various tissues?*

Carbon Dioxide Transport

Blood flowing through the capillaries of body tissues gains carbon dioxide because the tissues have a relatively high P_{CO_2}. This carbon dioxide is transported to the lungs in one of three forms: it may be carried as carbon dioxide dissolved in blood, as part of a compound formed by bonding to hemoglobin, or as part of a bicarbonate ion.

The amount of carbon dioxide that dissolves in blood is determined by its partial pressure. The higher the P_{CO_2} of the tissues, the more carbon dioxide will go into solution. However, only about 7% of the carbon dioxide is transported in this form.

Unlike oxygen, which combines with the iron atoms of hemoglobin molecules, carbon dioxide bonds with the amino groups ($-NH_2$) of these molecules. Consequently, oxygen and carbon dioxide do not compete for bonding sites, and both gases can be transported by a hemoglobin molecule at the same time.

When carbon dioxide combines with hemoglobin, a loosely bound compound called **carbaminohemoglobin** is formed. This substance decomposes

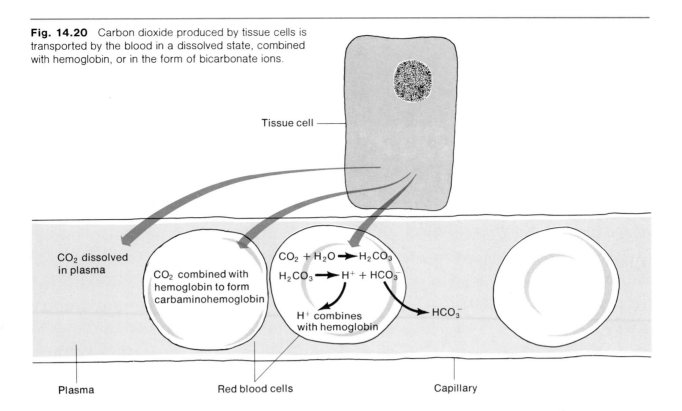

Fig. 14.20 Carbon dioxide produced by tissue cells is transported by the blood in a dissolved state, combined with hemoglobin, or in the form of bicarbonate ions.

Tissue cell

CO_2 dissolved in plasma

CO_2 combined with hemoglobin to form carbaminohemoglobin

$CO_2 + H_2O \rightarrow H_2CO_3$

$H_2CO_3 \rightarrow H^+ + HCO_3^-$

H^+ combines with hemoglobin

HCO_3^-

Plasma

Red blood cells

Capillary

readily in regions where the P_{CO_2} is low, and thus its carbon dioxide is released. Although this method of transporting carbon dioxide is theoretically quite effective, carbaminohemoglobin forms relatively slowly. It is believed that only about 23% of the total carbon dioxide is carried this way.

The most important carbon dioxide transport mechanism involves the formation of **bicarbonate ions** (HCO_3^-). Carbon dioxide reacts with water to form carbonic acid (H_2CO_3).

$$CO_2 + H_2O \rightarrow H_2CO_3$$

Although this reaction occurs slowly in blood plasma, much of the carbon dioxide diffuses into red blood cells, and these cells contain an enzyme called **carbonic anhydrase** that speeds the reaction between carbon dioxide and water.

The resulting carbonic acid then dissociates, releasing hydrogen ions (H^+) and bicarbonate ions (HCO_3^-).

$$H_2CO_3 \rightarrow H^+ + HCO_3^-$$

Most of the hydrogen ions combine quickly with hemoglobin molecules and thus are prevented from accumulating and causing a change in the blood pH.

The bicarbonate ions tend to diffuse out of the red blood cells and enter the blood plasma. It is estimated that nearly 70% of the carbon dioxide transported in blood is carried in this form. (See fig. 14.20.)

When blood passes through the capillaries of the lungs, it loses its dissolved carbon dioxide by diffusion into the alveoli. This occurs in response to the relatively low P_{CO_2} of the alveolar air. At the same time, hydrogen ions and bicarbonate ions in the red blood cells recombine to form carbonic acid molecules, and under the influence of carbonic anhydrase, the carbonic acid gives rise to carbon dioxide and water.

$$H^+ + HCO_3^- \rightarrow H_2CO_3 \rightarrow CO_2 + H_2O$$

Carbaminohemoglobin also releases its carbon dioxide, and as carbon dioxide continues to diffuse out of the blood, an equilibrium is established between the P_{CO_2} of the blood and the P_{CO_2} of the alveolar air. This process is summarized in figure 14.21.

1. Describe three ways carbon dioxide can be carried from body cells to the lungs.
2. How is it possible for hemoglobin to carry oxygen and carbon dioxide at the same time?
3. How is carbon dioxide released from the blood into the lungs?

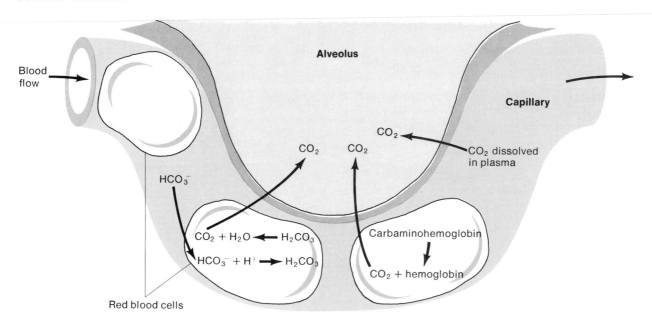

Fig. 14.21 In the lungs, carbon dioxide diffuses from the blood into the alveolus.

Clinical Terms Related to the Respiratory System

anoxia (ah-nok′se-ah)—an absence or deficiency of oxygen within tissues.

apnea (ap-ne′ah)—temporary absence of breathing.

asphyxia (as-fik′se-ah)—condition in which there is a deficiency of oxygen and an excess of carbon dioxide in the blood and tissues.

atelectasis (at″e-lek′tah-sis)—the collapse of a lung or some portion of it.

bradypnea (brad″e-ne′ah)—abnormally slow breathing.

bronchiolectasis (brong″ke-o-lek′tah-sis)—chronic dilation of the bronchioles.

bronchitis (brong-ki′tis)—inflammation of the bronchial lining.

Cheyne-Stokes respiration (chān stōks res″pi-ra′shun)—irregular breathing characterized by a series of shallow breaths that increase in depth and rate, followed by breaths that decrease in depth and rate.

dyspnea (disp′ne-ah)—difficulty in breathing.

eupnea (up-ne′ah)—normal breathing.

hemothorax (he″mo-tho′raks)—the presence of blood in the pleural cavity.

hypercapnia (hi″per-kap′ne-ah)—excessive carbon dioxide in the blood.

hyperpnea (hi″prep-ne′ah)—increased depth and rate of breathing.

hyperventilation (hi″per-ven″ti-la′-shun)—prolonged, rapid, and deep breathing.

hypoxemia (hi″pok-se′me-ah)—a deficiency in the oxygenation of the blood.

hypoxia (hi-pok′se-ah)—a diminished availability of oxygen in the tissues.

lobar pneumonia (lo′ber nu-mo′ne-ah)—pneumonia that affects an entire lobe of a lung.

pleurisy (ploo′ri-se)—inflammation of the pleural membranes.

pneumoconiosis (nu″mo-ko″ne-o′sis)—a condition characterized by the accumulation of particles from the environment in the lungs and the reaction of the tissues to their presence.

pneumothorax (nu″mo-tho′raks)—entrance of air into the space between the pleural membranes, followed by collapse of the lung.

rhinitis (ri-ni′tis)—inflammation of the nasal cavity lining.

sinusitis (si″nŭ-si′tis)—inflammation of the sinus cavity lining.

tachypnea (tak″ip-ne′ah)—rapid, shallow breathing.

tracheotomy (tra″ke-ot′o-me)—an incision in the trachea for exploration or the removal of a foreign object.

Chapter Summary

Introduction

The respiratory system includes passages that transport air to and from the lungs, and the air sacs of the lungs in which gas exchanges occur.

Respiration is the entire process by which gases are exchanged between the atmosphere and the body cells, and it involves several major events.

Organs of the Respiratory System

The respiratory system includes the nose, nasal cavity, sinuses, pharynx, larynx, trachea, bronchial tree, and lungs.

1. The nose
   a. The nose is supported by bone and cartilage.
   b. Nostrils provide entrances for air.

2. The nasal cavity
   a. Nasal conchae divide the cavity into passageways and help increase the surface area of mucous membrane.
   b. Mucous membrane filters, warms, and moistens incoming air.
   c. Particles trapped in the mucus are carried to the pharynx by ciliary action and swallowed.

3. The sinuses
   a. Sinuses are spaces in the bones of the skull that open into the nasal cavity.
   b. They are lined with mucous membrane.

4. The pharynx
   a. The pharynx is located behind the mouth and between the nasal cavity and the larynx.
   b. It functions as a common passage for air and food.

5. The larynx
   a. The larynx serves as a passageway for air and helps prevent foreign objects from entering the trachea.
   b. It is composed of muscles and cartilages.
   c. It contains the vocal cords, which produce sounds by vibrating as air passes over them.
   d. The glottis and epiglottis help prevent foods and liquids from entering the trachea.

6. The trachea
   a. The trachea extends into the thoracic cavity in front of the esophagus.
   b. It divides into right and left bronchi.

7. The bronchial tree
   a. The bronchial tree consists of branched air passages that lead from the trachea to the air sacs.
   b. Alveoli are located at the distal ends of the finest tubes.

8. The lungs
   a. The left and right lungs are separated by the mediastinum and enclosed by the diaphragm and the thoracic cage.
   b. The visceral pleura is attached to the surface of the lungs; parietal pleura lines the thoracic cavity.
   c. Each lobe is composed of alveoli, blood vessels, and supporting tissues.

Mechanism of Breathing

Inspiration and expiration movements are accompanied by changes in the size of the thoracic cavity.

1. Inspiration
   a. Air is forced into the lungs by atmospheric pressure.
   b. Inspiration occurs when the pressure inside the alveoli is reduced.
   c. Pressure within the alveoli is reduced when the diaphragm moves downward and the thoracic cage moves upward and outward.
   d. Expansion of the lungs is aided by surface tension.

2. Expiration
   a. The forces of expiration come from the elastic recoil of tissues and surface tension within the alveoli.
   b. Expiration can be aided by thoracic and abdominal wall muscles.

3. Respiratory air volumes
   a. The amount of air that normally moves in and out during quiet breathing is the tidal volume.
   b. Additional air that can be inhaled is the inspiratory reserve volume; additional air that can be exhaled is the expiratory reserve volume.
   c. Residual air remains in the lungs and is mixed with newly inhaled air.
   d. The vital capacity is the maximum amount of air a person can exhale after taking the deepest breath possible.
   e. The total lung capacity is equal to the vital capacity plus the residual air volume.

4. Nonrespiratory air movements
   Nonrespiratory air movements include coughing, sneezing, laughing, crying, hiccuping, yawning, and speaking.

Control of Breathing

Normal breathing is rhythmic and involuntary.

1. The respiratory center
   a. This center is located in the brain stem.
   b. It can adjust the rate and depth of breathing.
   c. Maintaining a normal breathing pattern involves alternating actions of the inspiratory and expiratory circuits, which can inhibit each other.

2. Factors affecting breathing
   a. Tissue stretch and the presence of certain chemicals affect breathing.
   b. An inflation reflex helps to regulate breathing and maintain rhythm.
   c. Increasing levels of carbon dioxide and hydrogen ions cause the rate to increase.

Alveolar Gas Exchanges

Alveoli carry on gas exchanges between the air and the blood.

1. The alveoli
   The alveoli are tiny air sacs clustered at the distal ends of the alveolar ducts.

2. The respiratory membrane
   a. This membrane consists of the alveolar and capillary walls.
   b. Gas exchanges take place through these walls.
   c. Diffusion through the respiratory membrane
      (1) The partial pressure of a gas is determined by the concentration of that gas in a mixture or the concentration dissolved in a liquid.
      (2) Gases diffuse from regions of higher partial pressure toward regions of lower partial pressure.
      (3) Oxygen diffuses from the alveolar air into the blood; carbon dioxide diffuses from the blood into the alveolar air.

Transport of Gases

Blood transports gases between the lungs and the body cells.

1. Oxygen transport
   a. Oxygen is transported in combination with hemoglobin molecules.
   b. The resulting oxyhemoglobin is relatively unstable and releases its oxygen in regions where the Po_2 is low.

2. Carbon dioxide transport
   a. Carbon dioxide may be carried in solution, bound to hemoglobin, or as a bicarbonate ion.
   b. Most carbon dioxide is transported in the form of bicarbonate ions.
   c. The enzyme, carbonic anhydrase, speeds the reaction between carbon dioxide and water.
   d. Carbonic acid dissociates to release hydrogen ions and bicarbonate ions.

Application of Knowledge

1. In certain respiratory disorders, such as emphysema, the capacity of the lungs to recoil elastically is reduced. Which respiratory air volumes will be affected by a condition of this type? Why?

2. What changes would you expect to occur in the relative concentrations of blood oxygen and carbon dioxide in a patient who breathes rapidly and deeply for a prolonged time? Why?

3. If a person has stopped breathing and is receiving pulmonary resuscitation, would it be better to administer pure oxygen or a mixture of oxygen and carbon dioxide? Why?

Review Activities

1. Describe the general functions of the respiratory system.

2. Explain how the nose and nasal cavity function in filtering incoming air.

3. Describe the locations of the major sinuses.

4. Distinguish between the pharynx and the larynx.

5. Name and describe the locations of the larger cartilages of the larynx.

6. Distinguish between the false vocal cords and the true vocal cords.

7. Compare the structure of the trachea and the branches of the bronchial tree.

8. Distinguish between visceral pleura and parietal pleura.

9. Explain how normal inspiration and forced inspiration are accomplished.

10. Define *surface tension* and explain how it aids the breathing mechanism.

11. Define *surfactant* and explain its function.

12. Explain how normal expiration and forced expiration are accomplished.

13. Distinguish between vital capacity and total lung capacity.

14. Describe the location of the respiratory center and explain how it functions in the control of breathing.

15. Describe the inflation reflex.

16. Define *hyperventilation* and explain how it affects the respiratory center.

17. Define *respiratory membrane* and explain its function.

18. Explain the relationship between the partial pressure of a gas and its rate of diffusion.

19. Summarize the gas exchanges that occur through the respiratory membrane.

20. Describe how oxygen is transported in blood.

21. List three ways that carbon dioxide can be transported in blood.

The Blood

The blood, heart, and blood vessels constitute the *circulatory system* and provide a link between the body's internal parts and its external environment. More specifically, the blood transports nutrients and oxygen to the body cells and wastes from these cells to the respiratory and excretory organs. It transports hormones from the endocrine glands to target tissues and bathes the body cells in a liquid of relatively stable composition. It also aids in temperature control by distributing heat from skeletal muscles and other active organs to all body parts. Thus the blood provides vital support for cellular activities and aids in maintaining a favorable cellular environment.

Introduction   364

Blood and Blood Cells   364
Volume and Composition of Blood
Characteristics of Red Blood Cells
Red Blood Cell Counts
Destruction of Red Blood Cells
Red Blood Cell Production and Its Control
Types of White Blood Cells
White Blood Cell Counts
Functions of White Blood Cells
Blood Platelets

Blood Plasma   372
Plasma Proteins
Nutrients and Gases
Nonprotein Nitrogenous Substances
Plasma Electrolytes

Hemostasis   374
Blood Vessel Spasm
Platelet Plug Formation
Blood Coagulation

Blood Groups and Transfusions   376
Agglutinogens and Agglutinins
The ABO Blood Group
The Rh Blood Group
Agglutination Reactions

Clinical Terms Related to the Blood   381

After you have studied this chapter, you should be able to

1. Describe the general characteristics of blood and discuss its major functions.

2. Distinguish between the various types of cells found in blood.

3. Explain how red cell production is controlled.

4. List the major components of blood plasma and describe the functions of each.

5. Define *hemostasis* and explain the mechanisms that help to achieve it.

6. Review the major steps in blood coagulation.

7. Explain the basis for blood typing and how it is used to avoid adverse reactions following blood transfusions.

8. Describe how blood reactions may occur between fetal and maternal tissues.

9. Complete the review activities at the end of this chapter. Note that the items are worded in the form of specific learning objectives. You may want to refer to them before reading the chapter.

agglutinin   (ah-gloo'ti-nin)

agglutinogen   (ag''loo-tin'o-jen)

albumin   (al-bu'min)

coagulation   (ko-ag''u-la'shun)

erythrocyte   (ĕ-rith'ro-sīt)

erythropoietin   (ĕ-rith''ro-poi'ĕ-tin)

fibrinogen   (fi-brin'o-jen)

globulin   (glob'u-lin)

hemostasis   (he''mo-sta'sis)

leukocyte   (lu'ko-sīt)

plasma   (plaz'mah)

platelet   (plat'let)

agglutin-, to glue together: *agglutin*ation—the clumping together of red blood cells.

bil-, bile: *bil*irubin—pigment excreted in bile.

embol-, a stopper: *embol*us—an obstruction in a blood vessel.

erythr-, red: *erythr*ocyte—a red blood cell.

hemo-, blood: *hemo*globin—red pigment responsible for the color of blood.

leuko-, white: *leuko*cyte—a white blood cell.

-osis, increased production: leukocyt*osis*— condition in which white blood cells are produced in abnormally large numbers.

-poiet, making: erythro*poiet*in—a hormone that stimulates the production of red blood cells.

-stas, a halting: hemo*stas*is—arrest of bleeding from damaged blood vessels.

thromb-, a clot: *thromb*ocyte—blood platelet that plays a role in the formation of a blood clot.

As is mentioned in chapter 5, blood is a type of connective tissue whose cells are suspended in a liquid intercellular material.

Blood and Blood Cells

Whole blood is slightly heavier and three to four times more viscous than water. Its cells, which are formed mostly in red bone marrow, include *red blood cells* and *white blood cells*. Blood also contains some cellular fragments called *platelets*. (See fig. 15.1.)

Volume and Composition of Blood

The volume of blood varies with body size. It also varies with changes in fluid and electrolyte concentrations and the amount of fat tissue present. An average-size male (70 kilograms or 154 pounds), however, will have a blood volume of about 5 liters (5.2 quarts).

A blood sample is usually about 45% red cells. This percentage is called the *hematocrit* (HCT). The remaining 55% of a blood sample consists of straw-colored liquid called **plasma.** Plasma is composed of a complex mixture that includes water, amino acids, proteins, carbohydrates, lipids, vitamins, hormones, electrolytes, and cellular wastes. The normal concentrations of various plasma components are listed in the appendix.

---

1. What factors affect blood volume?
2. What are the major components of blood?

Characteristics of Red Blood Cells

Red blood cells, or **erythrocytes,** are tiny, biconcave disks that are thin near their centers and thicker around their rims. This special shape is related to the red cell's function of transporting gases in that it provides an increased surface area through which gases can diffuse. The shape also places the cell membrane closer to various interior parts, where oxygen-carrying *hemoglobin* is found.

Each red blood cell is about ⅓ hemoglobin by volume, and this substance is responsible for the color of the blood. Specifically, when the hemoglobin is combined with oxygen, the resulting *oxyhemoglobin* is bright red, and when the oxygen is released, the remaining *deoxyhemoglobin* is darker.

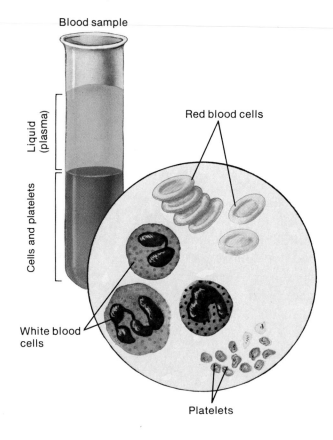

Fig. 15.1 Blood consists of a liquid portion called plasma and a cellular portion that includes red blood cells, white blood cells, and platelets.

Blood sample

Liquid (plasma)

Cells and platelets

Red blood cells

White blood cells

Platelets

A person experiencing a prolonged oxygen deficiency (hypoxia) may develop a symptom called *cyanosis*. In this condition, the skin and mucous membranes appear bluish due to an abnormally high concentration of deoxyhemoglobin in the blood. Cyanosis also may occur as a result of exposure to low temperature. In this case, superficial blood vessels become constricted, and more oxygen than usual is removed from the blood flowing through them.

Although red blood cells have nuclei during their early stages of development in the red bone marrow, these nuclei are lost as the cells mature. This characteristic, like the shape of a red blood cell, seems to be related to the function of transporting oxygen, since the space previously occupied by the nucleus is available to hold hemoglobin.

Fig. 15.2 Low P_{O_2} causes the release of erythropoietin from the kidneys and liver. This substance stimulates the production of red blood cells, and these cells carry additional oxygen to the tissues.

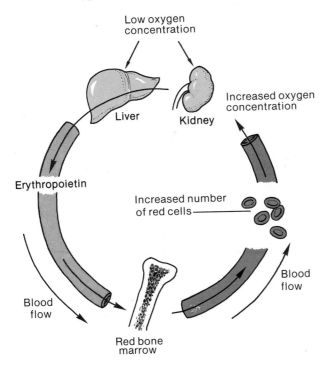

1. Describe a red blood cell.
2. What is the function of hemoglobin?
3. What changes occur in a red blood cell as it matures?

Red Blood Cell Counts

The number of red blood cells in a cubic millimeter (mm^3) of blood is called the *red cell count*. Although this number varies from time to time, the normal range for adult males is 4,600,000–6,200,000 cells per mm^3, while that for adult females is 4,200,000–5,400,000 cells per mm^3.

Since the number of circulating red blood cells is closely related to the blood's *oxygen-carrying capacity*, any changes in this number may be significant. For this reason, red cell counts are routinely made to help in diagnosing and evaluating the courses of various diseases.

Destruction of Red Blood Cells

Red blood cells are quite elastic and flexible, and they readily change shape as they pass through small blood vessels. With age, however, these cells become more fragile, and they often are damaged or ruptured when they squeeze through capillaries, particularly those in active muscles.

Damaged red cells are phagocytized and destroyed, primarily in the liver and spleen, by **macrophages.**

The hemoglobin molecules from red cells undergoing destruction are broken down into molecular subunits of *heme,* an iron-containing portion, and *globin,* a protein. The heme is further decomposed into iron and a greenish pigment called **biliverdin.** The iron, combined with a protein, may be carried by the blood to the blood-cell-forming tissue in the red bone marrow and reused in the synthesis of a new hemoglobin. Otherwise it may be stored in liver cells. In time, the biliverdin is converted to an orange pigment called **bilirubin.** These substances are excreted in bile as bile pigments. (See chapter 13.)

It is estimated that the average life span of a red blood cell is about 120 days, and that a large number of these cells are removed from circulation each day. Yet the number of cells in the circulating blood remains relatively stable. This observation suggests that a *homeostatic mechanism* regulates the rate of red cell production.

1. What is the normal red blood cell count for a male? For a female?
2. What happens to damaged red blood cells?
3. What are the end products of hemoglobin breakdown?

Red Blood Cell Production and Its Control

As is mentioned in chapter 8, red blood cell formation (hematopoiesis) occurs in the yolk sac, liver, and spleen of an embryo; but after an infant is born, the cells are produced almost exclusively by tissue that lines the vascular sinuses of the red bone marrow.

The rate of red blood cell formation seems to be controlled by a *negative feedback mechanism* involving a hormone called **erythropoietin.** In response to prolonged oxygen deficiency, this glycoprotein is released, primarily from the kidneys and to a lesser extent from the liver.

At high altitudes, for example, where the P_{O_2} is reduced, the amount of oxygen delivered to the tissues decreases. As figure 15.2 shows, this drop in oxygen concentration triggers the release of erythropoietin, which travels via the blood to the red bone marrow and stimulates increased red cell production.

Fig. 15.3 Stages in the development of blood cells and platelets.

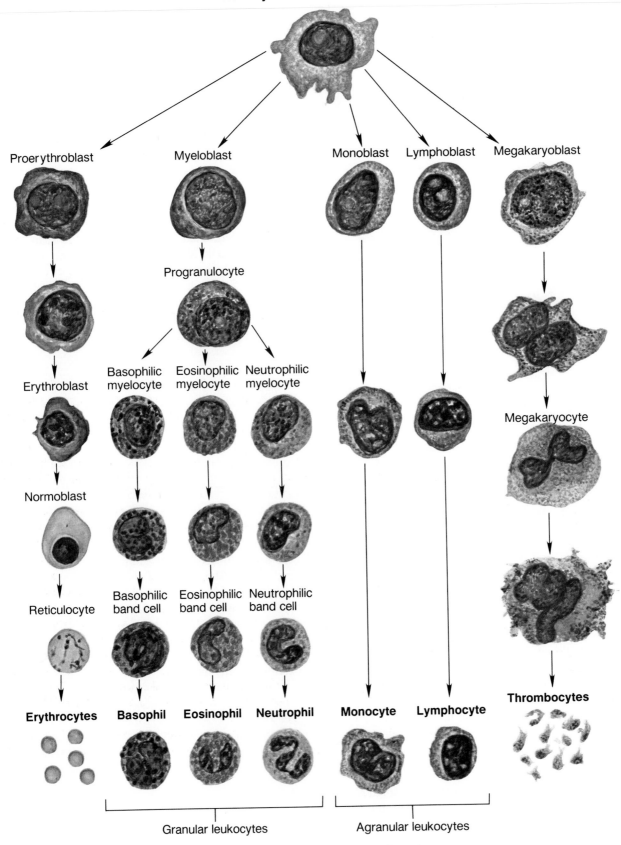

Hemocytoblast in Red Bone Marrow

Proerythroblast   Myeloblast   Monoblast   Lymphoblast   Megakaryoblast

Progranulocyte

Erythroblast   Basophilic myelocyte   Eosinophilic myelocyte   Neutrophilic myelocyte

Normoblast   Megakaryocyte

Reticulocyte   Basophilic band cell   Eosinophilic band cell   Neutrophilic band cell

Erythrocytes   **Basophil**   **Eosinophil**   **Neutrophil**   **Monocyte**   **Lymphocyte**   **Thrombocytes**

Granular leukocytes   Agranular leukocytes

After a few days, a large number of newly formed red cells begin to appear in the circulating blood, and the increased rate of production continues as long as the kidney and liver tissues experience an oxygen deficiency. Eventually, the number of red cells in circulation may be sufficient to supply these tissues with their oxygen needs. When this happens or when the P_{O_2} returns to normal, the release of erythropoietin ceases, and the rate of red cell production is reduced.

Figure 15.3 illustrates the stages in the development of red blood cells, white blood cells, and platelets.

Some individuals produce an abnormal type of hemoglobin molecules that tends to form long chains when exposed to low oxygen concentrations. This causes the red blood cells containing the molecules to become distorted or sickle-shaped, and the person is said to have *sickle-cell anemia.*

In this condition, the distorted cells may block capillaries and thus interrupt the flow of blood to tissues. The resulting symptoms may include severe joint and abdominal pain, skin ulceration, and chronic kidney disease.

---

1. *Where are red blood cells produced?*
2. *How is the production of red blood cells controlled?*

Dietary Factors Affecting Red Blood Cell Production. Red blood cell production is significantly influenced by the availability of two of the B-complex vitamins—*vitamin B_{12} and folic acid.* These substances are required for the synthesis of DNA molecules, and thus are needed by all cells for growth and reproduction. Since cellular reproduction occurs at a particularly high rate in red-blood-cell-forming tissue, this tissue is especially affected by a lack of either vitamin.

In addition, *iron* is needed for hemoglobin synthesis and normal red blood cell production. It can be absorbed from food passing through the small intestine; also, much of the iron released during the decomposition of hemoglobin from damaged red blood cells is made available for reuse. Because of this conservation mechanism, relatively small quantities of iron are needed in the daily diet.

Figure 15.4 illustrates the life cycle of a red blood cell.

Whenever a rapid loss or an inadequate production of red cells causes a deficiency of these cells, the resulting condition is termed *anemia.* In disorders of this type, the oxygen-carrying capacity of the blood is reduced, and the person may appear pale and lack energy. Among the more common types of anemia are: *hemorrhagic anemia,* caused by an abnormal loss of blood; *aplastic anemia,* due to malfunction in the red bone marrow that is characterized by decreased production of red blood cells; and *hemolytic anemia,* characterized by an abnormally high rate of red blood cell rupture (hemolysis).

In the absence of an adequate supply of iron, a person may develop a condition called *hypochromic anemia,* which is characterized by the presence of small, pale red blood cells as those shown in figure 15.5.

---

1. *What vitamins are necessary for red blood cell production?*
2. *Why is iron needed for the normal development of red blood cells?*

Types of White Blood Cells

White blood cells, or **leukocytes,** function primarily to defend the body against microorganisms. Although these cells do most of their work outside the circulatory system, they use the blood for transportation.

Normally, five types of white cells can be found in circulating blood. They are distinguished by size, the nature of the cytoplasm, the shape of the nucleus, and the staining characteristics. For example, some types have granular cytoplasm and make up a group called *granulocytes,* while others lack cytoplasmic granules and are called *agranulocytes.* (See fig. 15.3.)

A **granulocyte** is typically about twice the size of a red cell. Members of this group include three types of white cells—neutrophils, eosinophils, and basophils. These cells are produced in the red bone marrow in much the same manner as red cells but have a relatively short life span, averaging about 12 hours.

Neutrophils are characterized by the presence of fine cytoplasmic granules that stain pinkish in neutral stain. The nucleus of a neutrophil is lobed and consists of two to five parts connected by thin strands of chromatin. Neutrophils account for 54 to 62% of the white cells in a normal blood sample. (See fig. 15.6.)

Fig. 15.4 Life cycle of a red blood cell. (1) Essential nutrients are absorbed from the intestine; (2) nutrients are transported by blood to red bone marrow; (3) red blood cells are produced in red bone marrow by mitosis; (4) mature red blood cells are released into blood and circulate for about 120 days; (5) damaged red blood cells are destroyed in liver by macrophages; (6) hemoglobin from red blood cells is decomposed into heme and globin; (7) iron from heme is returned to red blood marrow to be reused, and biliverdin is excreted in the bile.

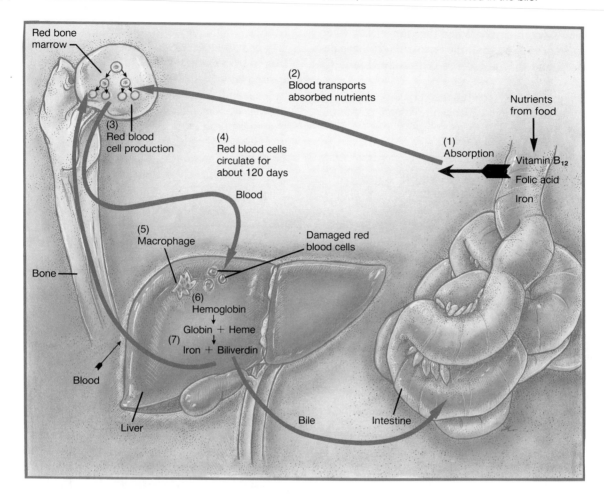

Fig. 15.5 (a) Normal human erythrocytes; (b) erythrocytes from a person with hypochromic anemia. (The larger cells with the stained nuclei are white blood cells.)

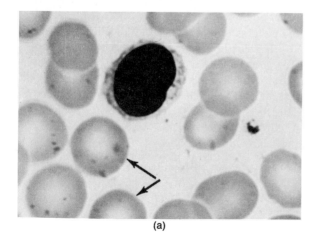

(a)

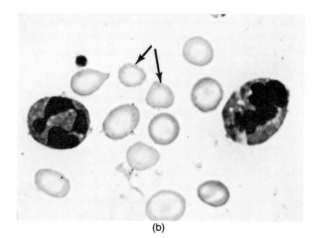

(b)

Fig. 15.6 A neutrophil has a lobed nucleus with two to five parts.

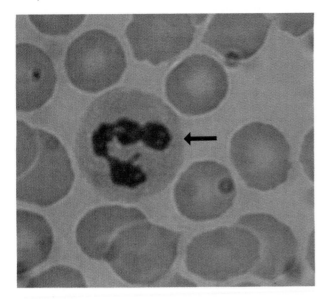

Fig. 15.8 A basophil has cytoplasmic granules that stain deep blue.

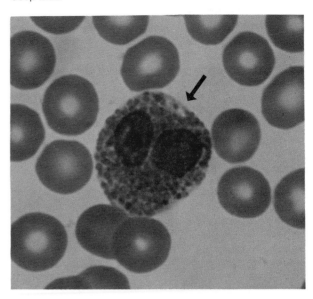

Fig. 15.7 An eosinophil is characterized by the presence of red-staining cytoplasmic granules.

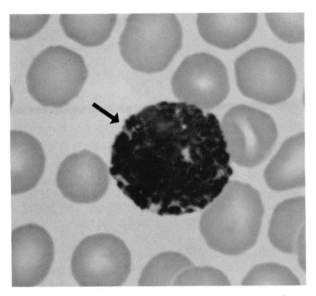

Fig. 15.9 A monocyte is the largest type of blood cell.

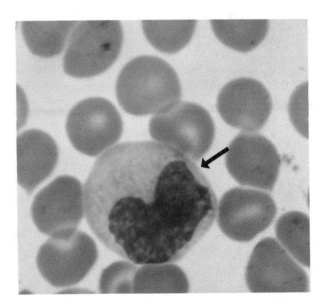

Eosinophils contain coarse, uniformly sized cytoplasmic granules that stain deep red in acid stain. The nucleus usually has only two lobes. These cells make up 1 to 3% of the total number of circulating leukocytes. (See fig. 15.7.)

Basophils are similar to eosinophils in size and in the shape of their nuclei, but they have fewer, more irregularly shaped cytoplasmic granules that stain deep blue in basic stain. This type of leukocyte usually accounts for less than 1% of the white cells. (See fig. 15.8.)

The leukocytes of the **agranulocyte** group include monocytes and lymphocytes. Both of these arise from red bone marrow; however lymphocytes also are formed in organs of the lymphatic system.

Monocytes are the largest cells found in the blood, having diameters two to three times greater than red cells. Their nuclei vary in shape and are described as round, kidney-shaped, oval, or lobed. They usually make up 3 to 9% of the leukocytes in a blood sample and seem to live for several weeks or even months. (See fig. 15.9.)

The Blood

Although large **lymphocytes** are sometimes found in the blood, usually they are only slightly larger than the red cells. Typically, a lymphocyte contains a relatively large, round nucleus with a thin rim of cytoplasm surrounding it. These cells account for 25 to 33% of the circulating white blood cells. They seem to have relatively long life spans that may extend for years. (See fig. 15.10.)

---

1. Distinguish between granulocytes and agranuloctyes.
2. List five types of white blood cells and explain how they differ from one another.

White Blood Cell Counts

Normally there are from 5000 to 10,000 white cells per mm³ of blood. Since the total number of white blood cells may change in response to abnormal conditions, however, white blood cell counts are of clinical interest. For example, some infectious diseases are accompanied by a rise in the number of circulating white cells, and if the total number of white cells exceeds 10,000 per mm³ of blood, the person is said to have **leukocytosis.** This condition occurs during certain acute infections, such as appendicitis.

If the total white cell count drops below 5000 per mm³ of blood, the condition is called **leukopenia.** Such a deficiency may accompany typhoid fever, influenza, measles, mumps, chicken pox, and poliomyelitis.

A *differential white blood cell count* is one in which the percentages of the various types of leukocytes in a blood sample are determined. This test is useful because the relative proportions of white cells may change in particular diseases. Neutrophils, for instance, usually increase during bacterial infections, while eosinophils may increase during certain parasitic infections and allergies.

---

1. What is the normal white blood cell count?
2. Distinguish between leukocytosis and leukopenia.
3. What is a differential white blood cell count?

Functions of White Blood Cells

As was mentioned, white blood cells generally function to protect the body against microorganisms. For example, some leukocytes phagocytize microorganisms that get into the body, and others produce antibodies that react with such foreign particles to destroy or disable them.

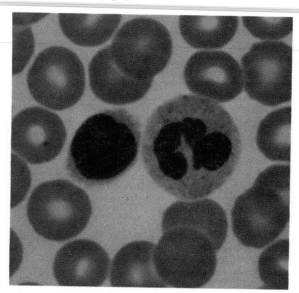

Fig. 15.10 The lymphocyte (left) contains a large, round nucleus; neutrophil (right).

Leukocytes can squeeze between the cells that make up blood vessel walls. This movement allows the white cells to leave the circulation. Once outside the blood, they move through interstitial spaces with an amebalike self-propulsion, called *ameboid motion.*

The most mobile and active phagocytic leukocytes are the *neutrophils* and the *monocytes.* Although the neutrophils are unable to ingest particles much larger than bacterial cells, monocytes can engulf relatively large objects. Both types of phagocytes contain numerous *lysosomes* that are filled with digestive enzymes capable of breaking down various proteins and lipids like those in bacterial cell membranes. (See chapter 3.) As a result of their activities, neutrophils and monocytes often become so engorged with digestive products and bacterial toxins that they also die.

When tissues are invaded by microorganisms, the body cells often respond by releasing substances that cause local blood vessels to dilate. One such substance, **histamine,** also causes an increase in the permeability of nearby capillaries and the walls of small veins (venules). As these changes occur, the tissues become reddened and large quantities of fluids leak into the interstitial spaces. The swelling produced by this *inflammatory reaction* tends to delay the spread of the invading microorganisms into other regions. (See chapter 17.) At the same time, leukocytes are somehow attracted by chemicals released from damaged cells, and they move toward these substances. This results in large numbers of white blood cells moving quickly into inflamed areas. (See fig. 15.11.)

Fig. 15.11 When bacteria invade the tissues, leukocytes migrate into the region and destroy bacteria by phagocytosis.

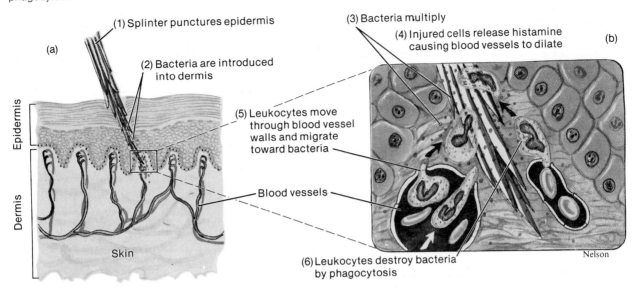

(1) Splinter punctures epidermis

(2) Bacteria are introduced into dermis

(3) Bacteria multiply

(4) Injured cells release histamine causing blood vessels to dilate

(5) Leukocytes move through blood vessel walls and migrate toward bacteria

Blood vessels

(6) Leukocytes destroy bacteria by phagocytosis

(a)

(b)

Epidermis

Dermis

Skin

Nelson

As bacteria, leukocytes, and damaged cells accumulate in an inflamed area, a thick fluid, called *pus* often forms and continues to do so while the invading microorganisms are active. If the pus cannot escape to the outside of the body or into a body cavity, it may remain walled in the tissues for some time. Eventually it may be absorbed by the surrounding cells.

The function of *eosinophils* is uncertain, although there is evidence that they act to detoxify foreign proteins that enter the body through the lungs or intestinal tract. They also may release enzymes that break down blood clots.

Some of the cytoplasmic granules of *basophils* contain a blood-clot-inhibiting substance, called *heparin,* while other granules contain *histamine.* It is thought that basophils may aid in preventing intravascular blood clot formation by releasing heparin, and that they may cause an increase in blood flow to injured tissues by releasing histamine.

Lymphocytes play an important role in the process of *immunity.* Some, for example, can form **antibodies** that act against various foreign substances that enter the body. This function is discussed in chapter 17.

1. What are the primary functions of white blood cells?
2. Which white blood cells are the most active phagocytes?
3. What are the functions of eosinophils and basophils?

Blood Platelets

Platelets, or **thrombocytes,** are not complete cells. They arise from very large cells in the red bone marrow, called **megakaryocytes.** These cells give off tiny cytoplasmic fragments, and as they detach and enter the circulation, they become platelets. (See fig. 15.3.)

Each platelet is a round disk that lacks a nucleus and is less than half the size of a red blood cell. It is capable of ameboid movement and may live for about 10 days. In normal blood, the platelet count will vary from 130,000 to 360,000 platelets per mm^3.

Platelets help close breaks in damaged blood vessels and function to initiate the formation of blood clots, as is explained in a later section of this chapter.

Chart 15.1 summarizes the characteristics of blood cells and platelets.

1. What is the normal blood platelet count?
2. What is the function of blood platelets?

Chart 15.1   Cellular components of blood

Component	Description	Number Present	Function
Red blood cell (erythrocyte)	Biconcave disk without nucleus, about ⅓ hemoglobin	4,000,000 to 6,000,000 per mm³	Transports oxygen and carbon dioxide
White blood cells (leukocytes)		5000 to 10,000 per mm³	Aids in defense against infections by microorganisms
Granulocytes	About twice the size of red cells, cytoplasmic granules present		
1. Neutrophil	Nucleus with two to five lobes, cytoplasmic granules stain pink in neutral stain	54 to 62% of white cells present	Destroys relatively small particles by phagocytosis
2. Eosinophil	Nucleus bilobed, cytoplasmic granules stain red in acid stain	1 to 3% of white cells present	Helps to detoxify foreign substances and secretes enzymes that break down clots
3. Basophil	Nucleus lobed, cytoplasmic granules stain blue in basic stain	Less than 1% of white cells present	Releases anticoagulant, heparin, and histamine
Agranulocytes	Cytoplasmic granules absent		
1. Monocyte	Two to three times larger than red cell, nuclear shape varies from round to lobed	3 to 9% of white cells present	Destroys relatively large particles by phagocytosis
2. Lymphocyte	Only slightly larger than red cell, nucleus nearly fills cell	25 to 33% of white cells present	Responsible for immunity
Platelet (thrombocyte)	Cytoplasmic fragment	130,000 to 360,000 per mm³	Helps control blood loss from broken vessels

From Hole, John W., Jr., *Human Anatomy and Physiology 3d ed.* © 1978, 1981, 1984 Wm. C. Brown Publishers, Dubuque, Iowa. All Rights Reserved. Reprinted by permission.

Blood Plasma

Plasma is the clear, straw-colored, liquid portion of the blood in which the cells and platelets are suspended. It is approximately 92% water and contains a complex mixture of organic and inorganic substances that function in a variety of ways. These functions include transporting nutrients, gases, and vitamins, regulating fluid and electrolyte balances, and maintaining a favorable pH.

Plasma Proteins

The most abundant dissolved substances (solutes) in plasma are **plasma proteins.** These proteins remain in the blood and interstitial fluids and ordinarily are not used as energy sources. There are three main groups—albumins, globulins, and fibrinogen. The members of each group differ in their chemical structures and in their physiological functions.

Albumins account for about 60% of the plasma proteins and have the smallest molecules of these proteins. They are synthesized in the liver, and because they are so plentiful, they are of particular significance in maintaining the *osmotic pressure* of the blood.

As is explained in chapter 3, whenever the concentration of dissolved substances changes on either side of a cell membrane, water is likely to move through the membrane toward the region where the dissolved molecules are in higher concentration. For this reason, it is important that the concentration of dissolved substances in plasma remain relatively stable. Otherwise, water will tend to leave the blood and enter the tissues, or leave the tissues and enter the blood by *osmosis.* The presence of albumins (and other plasma proteins) adds to the osmotic pressure of the plasma and aids in regulating the water balance between the blood and the tissues. It also helps to control the blood volume, which in turn is directly related to the blood pressure.

The concentration of plasma proteins may decrease significantly if a person is starving, and thus is forced to use body protein as an energy source, or has a protein-deficient diet. Similarly, the plasma protein level may drop if liver disease interferes with the synthesis of these proteins. In either case, as the blood protein concentration decreases, the osmotic pressure of the blood decreases, and water tends to accumulate in the intercellular spaces, causing the tissues to swell (edema).

Chart 15.2 Plasma proteins

Protein	Percentage of Total	Origin	Function
Albumin	60%	Liver	Helps in the maintenance of blood osmotic pressure
Globulin	36%		
Alpha globulins		Liver	Transport of lipids and fat-soluble vitamins
Beta globulins		Liver	Same as above
Gamma globulins		Lymphatic tissues	Constitute antibodies of immunity
Fibrinogen	4%	Liver	Plays key role in blood clot formation

From Hole, John W. Jr., *Human Anatomy and Physiology 3d ed.* © 1978, 1981, 1984 Wm. C. Brown Publishers, Dubuque, Iowa. All Rights Reserved. Reprinted by permission.

The **globulins,** which make up about 36% of the plasma proteins, can be separated further into fractions called *alpha globulins, beta globulins,* and *gamma globulins.* The alpha and beta globulins are synthesized in the liver, and they have a variety of functions including the transport of lipids and fat-soluble vitamins. The gamma globulins are produced in lymphatic tissues, and they include the proteins that function as *antibodies of immunity.* (See chapter 17.)

Fibrinogen plays a primary role in the blood clotting mechanism. It is synthesized in the liver and has the largest molecules of the plasma proteins. Its function is described in a subsequent section of this chapter.

Chart 15.2 summarizes the characteristics of the plasma proteins.

1. *List three types of plasma proteins.*
2. *How does albumin help to maintain a water balance between the blood and the tissues?*
3. *What are the functions of globulins?*

Nutrients and Gases

The *plasma nutrients* include amino acids, simple sugars, and various lipids that have been absorbed from the digestive tract and are being transported to various organs and tissues by the blood.

The most important blood gases are oxygen and carbon dioxide. While plasma also contains a considerable amount of dissolved nitrogen, this gas ordinarily has no physiological function.

Nonprotein Nitrogenous Substances

Molecules that contain nitrogen atoms but are not proteins comprise a group called **nonprotein nitrogenous substances.** Within the plasma this group includes amino acids, urea, and uric acid. The *amino acids* are present as a result of protein digestion and amino acid absorption. *Urea* and *uric acid* are the products of protein and nucleic acid catabolism, respectively, and are excreted in the urine.

Normally, the concentration of nonprotein nitrogenous (NPN) substances remains relatively stable because protein intake and utilization are balanced with the excretion of nitrogenous wastes. Because about half of the NPN is urea, which is ordinarily excreted by the kidneys, a rise in the plasma NPN level may suggest a kidney disorder. Such an increase may also occur as a result of excessive protein catabolism or the presence of an infection.

Plasma Electrolytes

Plasma contains a variety of *electrolytes* that have been absorbed from the intestine or have been released as by-products of cellular metabolism. They include sodium, potassium, calcium, magnesium, chloride, bicarbonate, phosphate, and sulfate ions. Of these, sodium and chloride ions are the most abundant.

Such ions are important in maintaining the osmotic pressure and the pH of the plasma, and like other plasma constituents, they are regulated so that their blood concentrations remain relatively stable. These electrolytes are discussed in chapter 19 in connection with water and electrolyte balance.

1. *What nutrients are found in blood plasma?*
2. *What gases occur in plasma?*
3. *What is meant by a nonprotein nitrogenous substance?*
4. *What are the sources of plasma electrolytes?*

The term **hemostasis** refers to the stoppage of bleeding, which is vitally important when blood vessels are ruptured. Following such an injury, several things may occur that help to prevent excessive blood loss. These include blood vessel spasm, platelet plug formation, and blood coagulation.

Blood Vessel Spasm

When a blood vessel is cut or broken, the smooth muscles in its wall are stimulated to contract, and blood loss is decreased almost immediately. In fact, the ends of a severed vessel may be closed completely by such a *spasm.*

Although this response may last only a few minutes, by then the platelet plug and blood coagulation mechanisms normally are operating. Also, as the platelet plug forms, the platelets release a substance called *serotonin,* which causes smooth muscles in the blood vessel wall to contract. This vasoconstricting action helps to maintain a prolonged vascular spasm.

Platelet Plug Formation

Platelets tend to stick to any rough surfaces and to the *collagen* in connective tissue. Consequently, when a blood vessel is broken, the platelets adhere to the collagen that underlies the lining of the blood vessel. At the same time, they tend to stick to each other creating a *platelet plug* in the vascular break. Such a plug may be able to control blood loss if the break is relatively small.

The steps in platelet plug formation are shown in figure 15.12.

---

1. *What is meant by hemostasis?*
2. *How does blood vessel spasm help to control bleeding?*
3. *Describe the formation of a platelet plug.*

Blood Coagulation

Coagulation, which is the most effective of the hemostatic mechanisms, results in the formation of a *blood clot.*

Although the mechanism by which blood coagulates is poorly understood, it is known that many substances are involved in the process. Some of these substances promote coagulation, and others inhibit it. Whether or not the blood coagulates depends on the balance that exists between these two groups of factors. Normally the anticoagulants prevail, and the

Fig. 15.12 Steps in platelet plug formation.

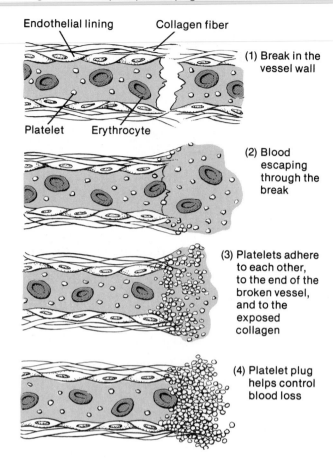

Endothelial lining   Collagen fiber

(1) Break in the vessel wall

Platelet   Erythrocyte

(2) Blood escaping through the break

(3) Platelets adhere to each other, to the end of the broken vessel, and to the exposed collagen

(4) Platelet plug helps control blood loss

blood does not clot. As a result of injury (trauma), however, substances that favor coagulation may increase in concentration, and the blood may coagulate.

The basic event in blood clot formation is the conversion of the soluble plasma protein *fibrinogen* into relatively insoluble threads of the protein **fibrin.**

When tissues are damaged, the clotting mechanism initiates a series of reactions resulting in the production of a substance called *prothrombin activator.* This series of changes depends upon the presence of *calcium ions* as well as certain proteins and phospholipids for its completion.

Prothrombin is an alpha globulin that is continually produced by the liver and thus is normally present in plasma. In the presence of calcium ions, prothrombin is converted into **thrombin** by the action of prothrombin activator. Thrombin, in turn, acts as an enzyme and causes a reaction in molecules of fibrinogen. As a result, fibrinogen molecules join, end to end, forming long threads of *fibrin.* The production of fibrin threads also is enhanced by the presence of calcium ions and other protein factors.

Once threads of fibrin have formed, they tend to stick to the exposed surfaces of damaged blood vessels and create a meshwork in which various blood cells and platelets become entangled. (See fig. 15.13.)

Fig. 15.13 When fibrin threads form, they entangle blood cells and help to create a clot. What factors serve to initiate this reaction?

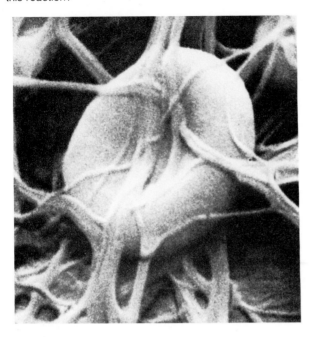

The resulting mass is the *blood clot,* which may effectively block a vascular break and prevent further loss of blood.

Since the liver plays an important role in the synthesis of various plasma proteins such as prothrombin, it is not surprising that liver diseases often are accompanied by a tendency to bleed. Also, bile salts from the liver are necessary for the efficient absorption of *vitamin K* from the intestine, and this vitamin is essential for the synthesis of prothrombin. If the liver fails to produce enough bile, or if the bile ducts become obstructed, a vitamin K deficiency is likely to develop, and the ability to form blood clots may be diminished. For this reason, vitamin K is often administered to patients with liver diseases or bile duct obstructions before they are treated surgically.

The amount of prothrombin activator that appears in the blood is directly proportional to the degree of tissue damage. Once a blood clot begins to form, it promotes still more clotting. This happens because thrombin also acts directly on blood clotting factors other than fibrinogen, and it can cause prothrombin to form still more thrombin. This type of self-initiating action is an example of a **positive feedback system** (see chapter 12).

Normally, the formation of a massive clot throughout the blood system is prevented by blood movement, which rapidly carries excessive thrombin away and thus keeps its concentration too low to enhance further clotting. As a result, blood clot formation usually is limited to blood that is standing still, and clotting stops where a clot comes in contact with circulating blood.

Blood clots that form in ruptured vessels are soon invaded by *fibroblasts.* (See chapter 5.) These cells produce fibrous connective tissue throughout the clots, which helps to strengthen and seal vascular breaks. Many clots, including those that form in tissues as a result of blood leakage (hematomas), disappear in time. This dissolution involves the activation of a plasma protein that can digest fibrin threads and other proteins associated with blood clots. Clots that fill large blood vessels, however, are seldom removed by natural processes.

If a blood clot forms in a vessel abnormally, it is termed a **thrombus.** If the clot becomes dislodged or if a fragment of it breaks loose and is carried away by the blood flow, it is called an **embolus.** Generally, emboli continue to move until they reach narrow places in vessels where they become lodged and interfere with the blood flow.

Such abnormal clot formations often are associated with conditions that cause changes in the endothelial linings of vessels. In the disease called *atherosclerosis,* for example, arterial linings are changed by accumulations of fatty deposits. These changes may initiate the clotting mechanism (fig. 15.14).

Coagulation also may occur in blood that is flowing too slowly. In this instance, the concentration of clot-promoting substances may increase to a critical level instead of being carried away by more rapidly moving blood, and a clot may form.

Figure 15.15 summarizes the three primary hemostatic mechanisms.

Hemophilia is a hereditary disease that appears almost exclusively in males. Hemophiliacs are deficient in a blood factor necessary for coagulation, and consequently, they usually experience repeated episodes of serious bleeding.

Treatment for hemophilia may involve pressure or packing of accessible bleeding sites in an effort to control blood loss. Transfusions are used to replace missing blood factors. For example, a common type of hemophilia caused by a deficiency of a substance, called blood factor VIII may be treated with fresh plasma, fresh-frozen plasma, or plasma concentrates (cryoprecipitates).

Fig. 15.14 (a) A normal artery; (b) the inner wall of an artery that has changed as a result of atherosclerosis. How might this condition promote the formation of blood clots?

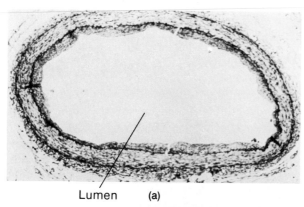

Lumen   (a)

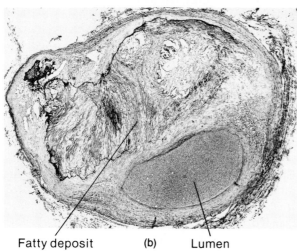

Fatty deposit   (b)   Lumen

Fig. 15.15 Hemostasis following tissue damage is likely to involve blood vessel spasm, platelet plug formation, and the blood-clotting mechanism.

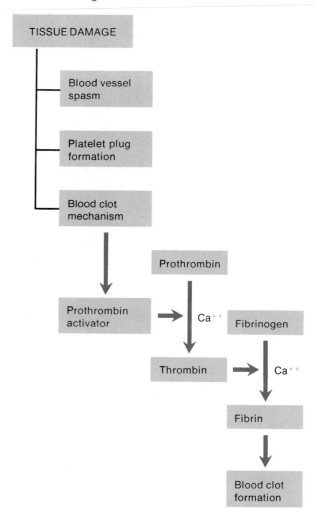

1. *Review the major steps in the formation of a blood clot.*
2. *What prevents the formation of massive clots throughout the blood system?*
3. *Distinguish between a thrombus and an embolus.*

Blood Groups and Transfusions

Early attempts to transfer blood from one person to another produced varied results. Sometimes the person receiving the transfusion was aided by the procedure. Other times the recipient suffered a blood reaction in which the red blood cells clumped together, obstructing vessels and producing other serious consequences.

Eventually, it was discovered that each individual has a particular combination of substances in his or her blood. Some of these substances reacted with those in another person's blood. These discoveries led to the development of procedures for typing blood. It is now known that safe transfusions of whole blood depend upon properly matching the blood types of the donors and recipients.

Agglutinogens and Agglutinins

The clumping of red cells following a transfusion reaction is called **agglutination.** This phenomenon is due to the presence of substances called **agglutinogens** (antigens) in the red cell membranes and substances called **agglutinins** (antibodies) dissolved in the plasma.

Fig. 15.16 Each blood type is characterized by a different combination of agglutinogens and agglutinins.

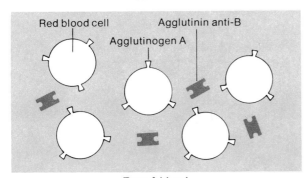

Type A blood

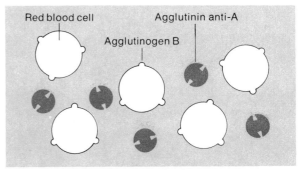

Type B blood

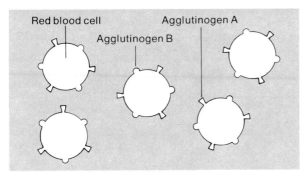

Type AB blood

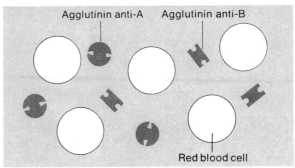

Type O blood

Blood typing involves identifying the agglutinogens that are present in a person's red cells. Although there are many different agglutinogens associated with human erythrocytes, only a few of them are likely to produce serious transfusion reactions. These include the agglutinogens of the ABO group and those of the Rh group.

Avoiding the mixture of certain kinds of agglutinogens and agglutinins prevents adverse transfusion reactions.

The ABO Blood Group

The *ABO blood group* is based upon the presence (or absence) of two major agglutinogens in red cell membranes—*agglutinogen A* and *agglutinogen B*. The erythrocytes of each person contain one of the four following combinations of agglutinogens: only A, only B, both A and B, or neither A nor B.

A person with only agglutinogen A is said to have *type A blood;* a person with only agglutinogen B has *type B blood;* one with both agglutinogen A and B has *type AB blood;* and one with neither agglutinogen A nor B has *type O blood.* Thus all humans have one of four possible blood types—A, B, AB, or O.

Chart 15.3 Agglutinogens and agglutinins of the ABO blood group

Blood Type	Agglutinogen	Agglutinin
A	A	anti-B
B	B	anti-A
AB	A and B	Neither anti-A nor anti-B
O	Neither A nor B	Both anti-A and anti-B

From Hole, John W. Jr., *Human Anatomy and Physiology 3d ed.* © 1978, 1981, 1984 Wm. C. Brown Publishers, Dubuque, Iowa. All Rights Reserved. Reprinted by permission.

Certain agglutinins in the plasma accompany the agglutinogens in the red cell membranes of each person's blood. Specifically, whenever agglutinogen A is absent, an agglutinin called *anti-A* is present; and whenever agglutinogen B is absent, an agglutinin called *anti-B* is present. Therefore, persons with type A blood also have agglutinin anti-B in their plasma; those with B blood have agglutinin anti-A; those with type AB blood have neither agglutinin; and those with type O blood have both agglutinin anti-A and anti-B. (See fig. 15.16.)

Chart 15.3 summarizes the agglutinogens and agglutinins of the ABO blood group.

Fig. 15.17 (a) If red blood cells with agglutinogen A are added to blood containing agglutinin anti-A, (b) the agglutinins will react with the agglutinogens of the red blood cells and cause them to clump together.

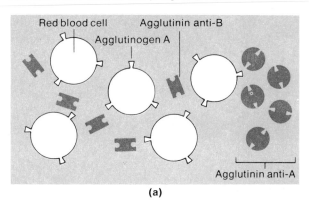

(a)

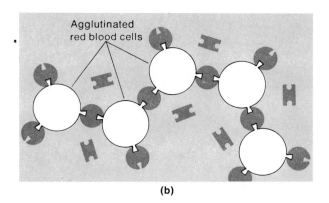

(b)

Since an agglutinin of one kind will react with an agglutinogen of the same kind and cause red blood cells to clump together, such combinations are avoided whenever possible. Actually, the major concern in blood transfusion procedures is that the cells in the *transfused blood* not be agglutinated by the agglutinins in the recipient's plasma. For this reason, a person with type A (anti-B) blood should never be given blood of type B or AB, because the red cells of both types would be agglutinated by the anti-B in the recipient's type A blood. Likewise, a person with type B (anti-A) blood should never be given type A or AB blood, and a person with type O (anti-A and anti-B) blood should never be given type A, B, or AB blood. (See fig. 15.17.)

Since type AB blood lacks both anti-A and anti-B agglutinins, it would appear that an AB person could receive a transfusion of blood of any other type. For this reason, type AB persons are sometimes called *universal recipients*. It should be noted, however, that type A (anti-B) blood, type B (anti-A) blood, and type O (anti-A and anti-B) blood still contain agglutinins (either anti-A or anti-B) that could cause agglutination of type AB cells. Consequently, it is always best to use donor blood (AB) of the same type as the recipient blood (AB). If the matching type is not available and type A, B, or O is used, it should be transfused slowly so that the donor blood is well diluted by the recipient's larger blood volume. This precaution usually avoids serious reactions between the donor's agglutinins and the recipient's agglutinogens.

Similarly, because type O blood lacks agglutinogens A and B, it would seem that this type could be transfused into persons with blood of any other type. Persons with type O blood, therefore, are sometimes called *universal donors*. Type O blood, however, does contain both anti-A and anti-B agglutinins,

Chart 15.4 Preferred and permissible blood types for transfusions

Blood Type	Preferred Transfusion	Permissible Transfusion
A	A	O
B	B	O
AB	AB	A, B, O
O	O	none

From Hole, John W. Jr., *Human Anatomy and Physiology* 3d ed. © 1978, 1981, 1984 Wm. C. Brown Publishers, Dubuque, Iowa. All Rights Reserved. Reprinted by permission.

and if it is given to a person with blood type A, B, or AB, it too should be transfused slowly to minimize the chance of an adverse reaction.

Chart 15.4 summarizes preferred blood type for normal transfusions and permissible blood type for emergency transfusions.

The agglutinogens of the ABO group are *inherited factors* that are present in the red cell membranes at the time of birth. The plasma agglutinins begin to appear spontaneously, for unknown reasons, about 2 to 8 months after birth, and they reach a maximum concentration between 8 and 10 years of age.

1. Distinguish between agglutinogens and agglutinins.
2. What is the main concern when blood is transfused from one individual to another?
3. Why is a type AB person called a universal recipient?

Fig. 15.18 (a) If an Rh-negative woman is pregnant with an Rh-positive fetus, (b) some of the fetal red blood cells with Rh agglutinogens may enter the maternal blood at the time of birth. (c) As a result, the woman's cells may produce anti-Rh agglutinins.

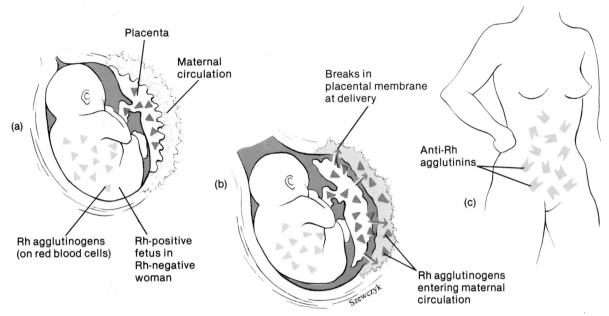

The Rh Blood Group

The *Rh blood group* was named after the *rhesus monkey,* in which it was first observed. In humans this group is based upon several different Rh agglutinogens (factors). The most important of these is *agglutinogen D;* however, if any of the Rh factors are present in the red cell membranes, the blood is said to be *Rh positive.* Conversely, if the red cells lack Rh agglutinogens, the blood is called *Rh negative.*

As in the case of agglutinogens A and B, the presence (or absence) of an Rh agglutinogen is an inherited trait. Unlike anti-A and anti-B, agglutinins for Rh (*anti-Rh*) do not appear spontaneously. Instead, they form only in Rh-negative persons in response to special stimulation.

If an Rh-negative person receives a transfusion of Rh-positive blood, the recipient's antibody-producing cells will be stimulated by the presence of the Rh agglutinogen and will begin producing an *anti-Rh agglutinin.* Generally there are no serious consequences from this initial transfusion, but if the Rh-negative person—now sensitized to Rh-positive blood—receives a subsequent transfusion of Rh-positive blood some months later, the donor's red cells are likely to agglutinate.

A related condition may occur when an Rh-negative woman is pregnant with an Rh-positive fetus for the first time. Such a pregnancy may be uneventful; however, at the time of this infant's birth, the placental membranes, which had separated the maternal blood from the fetal blood, may be broken, and some of the infant's Rh-positive blood cells may get into the maternal circulation. These Rh-positive cells may then stimulate the maternal tissues to begin producing anti-Rh agglutinins. (See fig. 15.18.)

If the mother, who has already developed anti-Rh agglutinin, becomes pregnant with a second Rh-positive fetus, these anti-Rh agglutinins can pass slowly through the placental membrane and react with the fetal red cells, causing them to agglutinate. The fetus then develops a disease called **erythroblastosis fetalis.** (See fig. 15.19.)

Erythroblastosis fetalis can be prevented in future offspring by treating Rh-negative mothers with a special blood serum within 72 hours following the birth of each Rh-positive child. This serum is obtained from the blood of another Rh-negative person who has formed anti-Rh agglutinin. Thus it is able to inactivate any Rh-positive cells that may have entered the maternal blood at the time of the infant's birth. The maternal tissues, consequently, are not stimulated to manufacture anti-Rh agglutinins, and the mother's blood should not be a hazard to her next Rh-positive child.

Fig. 15.19 (a) If a woman who has developed anti-Rh agglutinins is pregnant with an Rh-positive fetus, (b) agglutinins may pass through the placental membrane and cause the fetal red blood cells to agglutinate.

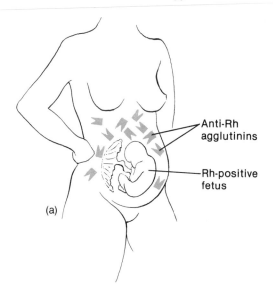

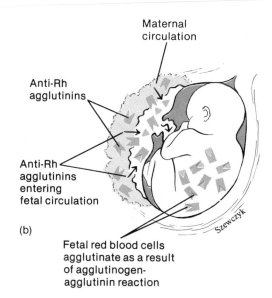

Anti-Rh agglutinins

Rh-positive fetus

(a)

Maternal circulation

Anti-Rh agglutinins

Anti-Rh agglutinins entering fetal circulation

(b)

Fetal red blood cells agglutinate as a result of agglutinogen-agglutinin reaction

Agglutination Reactions

In any blood reaction involving agglutinogens and agglutinins, the agglutinated red cells usually degenerate or are destroyed by phagocytic cells. At the same time, hemoglobin and other red cell contents are released, and the blood concentration of *free hemoglobin* increases greatly. Some of this hemoglobin may diffuse out of the vascular system and enter the body tissues, where it is gradually converted into bilirubin. As a result, the tissues may develop a yellowish stain, which is called *jaundice.*

Free hemoglobin also may pass into the kidneys and interfere with the vital functions of these organs, so that a person with a blood transfusion reaction may experience kidney failure later.

Infants with erythroblastosis fetalis usually are jaundiced and severely anemic. As their blood cell-forming tissues respond to the need for more red cells, various immature erythrocytes including *erythroblasts* are released into the blood. (See fig. 15.3.)

An affected infant may suffer permanent brain damage as a result of bilirubin precipitating in the brain tissues and injuring neurons. This condition is called *kernicterus,* and if the infant survives, it may have motor or sensory losses and exhibit mental deficiencies.

Treatment for erythroblastosis fetalis usually involves exposing the affected infant to bright blue or white *fluorescent light.* Bilirubin is a light-sensitive substance, and this exposure (phototherapy) causes a decrease in the blood bilirubin concentration.

In more severe cases, the infant's Rh-positive blood may be replaced slowly with Rh-negative blood. This procedure is called an *exchange transfusion,* and it reduces the concentration of bilirubin in the infant's tissues in addition to removing the agglutinating red cells, the anti-Rh agglutinins, the free hemoglobin, and the other products of erythrocyte destruction. It also provides a temporary supply of red cells that will not be agglutinated by any remaining anti-Rh agglutinins. In time, the infant's blood-cell-forming tissues will replace the donor's blood cells with Rh-positive cells, but by then the maternal agglutinins will have disappeared.

1. Under what conditions might a person with Rh-negative blood develop Rh agglutinins?
2. What happens to red blood cells that are agglutinated?

Clinical Terms Related to the Blood

anisocytosis (an-i″so-si-to′sis)—condition in which there is an abnormal variation in the size of erythrocytes.

antihemophilic plasma (an″ti-he″mo-fil′ik plaz′mah)—normal blood plasma that has been processed to preserve an antihemophilic factor.

Christmas disease (kris′mas di-zēz′)—a hereditary bleeding disease that is due to a deficiency in a clotting factor; also called *hemophilia B.*

citrated whole blood (sit′rāt-ed hōl blud)—normal blood in a solution of acid citrate solution to prevent coagulation.

dried plasma (drid plas′mah)—normal bood plasma that has been vacuum dried to prevent the growth of microorganisms.

hemochromatosis (he″mo-kro″mah-to′sis)—a disorder in iron metabolism in which excessive iron is deposited in the tissues.

hemorrhagic telangiectasia (hem″o-raj′ik tel-an″je-ek-ta′ze-ah)—a hereditary disorder in which there is a tendency to bleed from localized lesions of capillaries.

heparinized whole blood (hep′er-i-nizd″ hōl blud)—normal blood in a solution of heparin to prevent coagulation.

macrocytosis (mak″ro-si-to′sis)—condition characterized by the presence of abnormally large erythrocytes.

microcytosis (mi″kro-si-to′sis)—condition characterized by the presence of abnormally small erythrocytes.

neutrophilia (nu″tro-fil′e-ah)—condition in which there is an increase in the number of circulating neutrophils.

normal plasma (nor′mal plaz′mah)—plasma from which the blood cells have been removed by centrifugation or sedimentation.

packed red cells (pakd red selz)—a concentrated suspension of red blood cells from which the plasma has been removed.

pancytopenia (pan″si-to-pe′ne-ah)—condition characterized by an abnormal depression of all the cellular components of blood.

poikilocytosis (poi″ki-lo-si-to′sis)—condition in which the erythrocytes are irregularly shaped.

pseudoagglutination (su″do-ah-gloo″ti-na′shun)—clumping of erythrocytes due to some factor other than an agglutinogen-agglutinin reaction.

purpura (per′pu-rah)—a disease characterized by spontaneous bleeding into the tissues and through the mucous membrane.

spherocytosis (sfe″ro-si-to′sis)—a hereditary form of hemolytic anemia characterized by the presence of spherical erythrocytes (spherocytes).

thalassemia (thal″ah-se′me-ah)—a group of hereditary hemolytic anemias characterized by the presence of very thin, fragile erythrocytes.

von Willebrand's disease (fon vil′e-brandz di-zēz′)—a hereditary condition due to a deficiency of an antihemophilic blood factor and capillary defects, which is characterized by bleeding from the nose, gums, and genitalia.

Chapter Summary

Introduction

Blood is a type of connective tissue whose cells are suspended in liquid.

Blood and Blood Cells

1. Blood contains red blood cells, white blood cells, and platelets.

2. Volume and composition of blood
   a. Volume varies with body size, fluid and electrolyte balance, and fat content.
   b. Blood can be separated into cellular and liquid portions.
   c. Plasma includes water, nutrients, hormones, electrolytes, and cellular wastes.

3. Characteristics of red blood cells
   a. Red blood cells are biconcave disks whose shapes provide increased surface area.
   b. They contain hemoglobin that combines with oxygen.

4. Red blood cell counts
   a. The red blood cell count equals the number of cells per mm^3 of blood.
   b. The normal red cell count varies from 4,600,000–6,200,000 cells per mm^3 in males and from 4,200,000–5,400,000 cells per mm^3 in females.
   c. Red cell count is related to the oxygen-carrying capacity of the blood.

5. Destruction of red blood cells
   a. Damaged red cells are phagocytized by macrophages in the liver and spleen.
   b. Hemoglobin molecules are decomposed and the iron they contain is conserved.
   c. The number of red blood cells remains relatively stable.

6. Red blood cell production and its control
   a. Red cells are produced by the red bone marrow.
   b. The rate of red cell production is controlled by a negative feedback mechanism.
   c. Dietary factors affecting red blood cell production
      (1) Production is affected by the availability of vitamin B_{12} and folic acid.
      (2) Iron is needed for hemoglobin synthesis.

7. Types of white blood cells
   a. White blood cells function to defend the body against infections by microorganisms.
   b. Granulocytes include neutrophils, eosinophils, and basophils.
   c. Agranulocytes include monocytes and lymphocytes.
8. White blood cell counts
   a. Normal total white cell counts vary from 5000 to 10,000 cells per mm³.
   b. Number of white cells may change in abnormal conditions.
   c. A differential white cell count indicates the percentages of various types of leukocytes present.
9. Functions of white blood cells
   a. Some leukocytes phagocytize foreign particles; others produce antibodies of immunity.
   b. Leukocytes may be stimulated by the presence of chemicals released by damaged cells and move toward these chemicals.
   c. Basophils release heparin, which inhibits blood clotting.
10. Blood platelets
    a. Blood platelets are fragments of giant cells.
    b. The normal count varies from 130,000 to 360,000 platelets per mm³.
    c. They function to help close breaks in blood vessels.

Blood Plasma

Plasma functions to transport nutrients and gases, regulate fluid and electrolyte balance, and maintain pH.

1. Plasma proteins
   Three major groups exist.
   a. Albumins help maintain the osmotic pressure of blood.
   b. Globulins function to transport lipids and fat-soluble vitamins, and they include the antibodies of immunity.
   c. Fibrinogen functions in blood clotting.
2. Nutrients and gases
   a. Nutrients include amino acids, simple sugars, and lipids.
   b. Gases in plasma include oxygen, carbon dioxide, and nitrogen.

3. Nonprotein nitrogenous substances (NPN)
   a. These are composed of molecules that contain nitrogen atoms but are not proteins.
   b. They include amino acids, urea, and uric acid.
4. Plasma electrolytes
   a. They include ions of sodium, potassium, calcium, magnesium, chlorine, bicarbonate, phosphate, and sulfate.
   b. They are important in the maintenance of osmotic pressure and pH.

Hemostasis

Hemostasis refers to the stoppage of bleeding.

1. Blood vessel spasm
   a. Smooth muscles in blood vessel walls contract following injury.
   b. Platelets release serotonin that stimulates vasoconstriction.
2. Platelet plug formation
   a. Platelets adhere to rough surfaces and exposed collagen.
   b. Platelets stick together at the sites of injuries and form platelet plugs in broken vessels.
3. Blood coagulation
   a. Blood clotting is the most effective means of hemostasis.
   b. Clot formation depends on the balance between substances that promote clotting and those that inhibit clotting.
   c. The basic event of clotting is the conversion of soluble fibrinogen into insoluble fibrin.
   d. Factors that promote clotting include the presence of prothrombin activator, prothrombin, and calcium ions.
   e. A thrombus is a blood clot in a vessel; an embolus is a clot or fragment of a clot that has moved in a vessel.

Blood Groups and Transfusions

Blood can be typed on the basis of the substances it contains.
Blood substances of certain types will react adversely with other types.

1. Agglutinogens and agglutinins
   a. Red blood cell membranes may contain agglutinogens and blood plasma may contain agglutinins.
   b. Blood typing involves identifying the agglutinogens present in the red cell membranes.

2. The ABO blood group
   a. Blood can be grouped according to the presence or absence of agglutinogens A and B.
   b. Adverse transfusion reactions are avoided by preventing the mixing of red cells that contain an agglutinogen with plasma that contains the corresponding agglutinin.
3. The Rh blood group
   a. Rh agglutinogens are present in the red cell membranes of Rh-positive blood; they are absent in Rh-negative blood.
   b. Mixing Rh-positive red cells with plasma that contains anti-Rh agglutinins results in agglutination of the positive cells.
   c. Anti-Rh agglutinins in maternal blood may pass through the placental tissues and react with the red cells of an Rh-positive fetus, causing it to develop erythroblastosis fetalis.
4. Agglutination reactions
   a. Agglutinated cells are destroyed by phagocytic cells.
   b. Free hemoglobin is likely to cause jaundice and interfere with kidney functions.

Application of Knowledge

1. If a patient with an inoperable cancer is treated by using a drug that reduces the rate of cell division, what changes might occur in the patient's white blood cell count? How might the patient's environment be modified to compensate for the effects of these changes?

2. Hypochromic (iron-deficiency) anemia is relatively common among aging persons who are admitted to hospitals for other conditions. What environmental and sociological factors might promote this?

Review Activities

1. List the major components of blood.
2. Describe a red blood cell.
3. Distinguish between oxyhemoglobin and deoxyhemoglobin.
4. Describe the life cycle of a red blood cell.
5. Distinguish between biliverdin and bilirubin.
6. Define *erythropoietin* and explain its function.
7. Explain how vitamin B_{12} and folic acid deficiencies affect red blood cell production.
8. Distinguish between granulocytes and agranulocytes.
9. Name five types of leukocytes and list the major functions of each.
10. Explain the significance of white blood cell counts as aids to diagnosing diseases.
11. Describe a blood platelet and explain its functions.
12. Name three types of plasma proteins and list the major functions of each.
13. Define *nonprotein nitrogenous substances* and name those commonly present in plasma.
14. Name several plasma electrolytes.
15. Define *hemostasis*.
16. Explain how blood vessel spasms are stimulated following an injury.
17. Explain how a platelet plug forms.
18. List the major steps leading to the formation of a blood clot.
19. Distinguish between fibrinogen and fibrin.
20. Provide an example of a positive feedback system.
21. Distinguish between thrombus and embolus.
22. Distinguish between agglutinogen and agglutinin.
23. Explain the basis of ABO blood types.
24. Explain why a person with blood type AB is sometimes called a universal recipient.
25. Explain why a person with blood type O is sometimes called a universal donor.
26. Distinguish between Rh-positive and Rh-negative blood.
27. Describe how a person may become sensitized to Rh-positive blood.
28. Define *erythroblastosis fetalis* and explain how this condition may develop.
29. Describe the consequences of an agglutination reaction.

The Cardiovascular System

The *cardiovascular system* is the portion of the circulatory system that includes the heart and blood vessels. It functions to move blood between the body cells and organs of the integumentary, digestive, respiratory, and urinary systems, which communicate with the external environment.

In performing this function, the heart acts as a pump that forces blood through the blood vessels. The blood vessels, in turn, form a closed system of ducts that transports blood and allows exchanges of gases, nutrients, and wastes between the blood and the body cells.

16

Chapter Objectives

After you have studied this chapter, you should be able to

1. Name the organs of the cardiovascular system and discuss their functions.

2. Name and describe the location of the major parts of the heart and discuss the function of each part.

3. Trace the pathway of blood through the heart and the vessels of the coronary circulation.

4. Discuss the cardiac cycle and explain how it is controlled.

5. Identify the parts of a normal ECG pattern and discuss the significance of this pattern.

6. Compare the structures and functions of the major types of blood vessels.

7. Describe the mechanism that aids in the return of venous blood to the heart.

8. Explain how blood pressure is created and controlled.

9. Compare the pulmonary and systemic circuits of the cardiovascular system.

10. Identify and locate the major arteries and veins of the pulmonary and systemic circuits.

11. Complete the review activities at the end of this chapter. Note that the items are worded in the form of specific learning objectives. You may want to refer to them before reading the chapter.

atrium   (a′tre-um)

cardiac cycle   (kar′de-ak si′kl)

cardiac output   (kar′de-ak owt′poot)

diastolic pressure   (di″ah-stol′ik presh′ur)

electrocardiogram   (e-lek″tro-kar′de-o-gram″)

functional syncytium   (funk′shun-al sin-sish′e-um)

myocardium   (mi″o-kar′de-um)

peripheral resistance   (pĕ-rif′er-al re-zis′tans)

pulmonary circuit   (pul′mo-ner″e sur′kit)

systemic circuit   (sis-tem′ik sur′kit)

systolic pressure   (sis-tol′ik presh′ur)

vasoconstriction   (vas″o-kon-strik′shun)

vasodilation   (vas″o-di-la′shun)

ventricle   (ven′trĭ-kl)

brady-, slow: *brady*cardia—an abnormally slow heartbeat.

diastol-, a dilation: *diastol*ic pressure—the blood pressure that occurs when the ventricle is relaxed (thus dilated).

-gram, something written: electrocardio*gram*—a recording of the electrical changes that occur in the heart muscle during a cardiac cycle.

papill-, nipple: *papill*ary muscle—a small mound of muscle within a chamber of the heart.

syn-, together: *syn*cytium—a mass of merging cells that act together.

systol-, contraction: *systol*ic pressure—the blood pressure that occurs during a ventricular contraction.

tachy-, rapid: *tachy*cardia—an abnormally fast heartbeat.

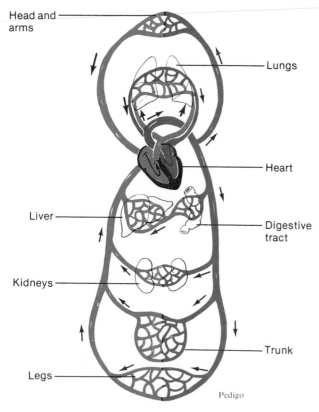

Head and arms

Lungs

Heart

Liver

Digestive tract

Kidneys

Trunk

Legs

Pedigo

A functional cardiovascular system is vital for survival because without circulation, tissues lack a supply of oxygen and nutrients, and waste substances begin to accumulate. Under such conditions cells soon begin to undergo irreversible changes that quickly lead to the death of the organism. The general pattern of the cardiovascular system is shown in figure 16.1.

Structure of the Heart

The heart is a hollow, cone-shaped, muscular pump located within the thorax and resting upon the diaphragm. (See fig. 16.2.)

Size and Location of the Heart

Although the size of the heart varies with body size, it is generally about 14 centimeters (5.5 inches) long and 9 centimeters (3.5 inches) wide in an average adult.

The heart is enclosed laterally by the lungs, posteriorly by the backbone, and anteriorly by the sternum. Its *base,* which is attached to several large

blood vessels, lies beneath the second rib. Its distal end extends downward and to the left, terminating as a bluntly pointed *apex* at the level of the fifth intercostal space.

Coverings of the Heart

The heart and the proximal ends of the large blood vessels to which it is attached are enclosed by a double-layered **pericardium.** The inner layer of this membrane is the *visceral pericardium* (epicardium). At the base of the heart, it turns back upon itself and becomes the serous part of a loose-fitting *parietal pericardium* (fig. 16.2).

The parietal pericardium is a tough, protective sac composed largely of white fibrous connective tissue. It is attached to the central portion of the diaphragm, the back of the sternum, the vertebral column, and the large blood vessels emerging from the heart. Between the parietal and visceral membranes is a potential space, the *pericardial cavity,* that contains a small amount of serous fluid. This fluid serves to reduce friction between the pericardial membranes as the heart moves within them.

If the pericardium becomes inflamed due to a bacterial or viral infection, the condition is called *pericarditis.* As a result of this inflammation, the layers of the pericardium sometimes become stuck together by adhesions, and this may interfere with heart movements. If this happens, surgery may be required to separate the surfaces and make unrestricted heart actions possible again.

1. Where is the heart located?
2. Distinguish between the visceral pericardium and the parietal pericardium.

Wall of the Heart

The wall of the heart is composed of three distinct layers—an outer epicardium, a middle myocardium, and an inner endocardium. (See fig. 16.3.)

The outer **epicardium** provides a protective layer and is composed of the visceral pericardium. This serous membrane consists of connective tissue covered by epithelium. The deeper portion often contains fat, particularly along the paths of larger blood vessels.

The **myocardium** is relatively thick and consists largely of the cardiac muscle tissue responsible for forcing blood out of the heart chambers. The muscle

Fig. 16.2 The pericardial cavity is a potential space between the visceral and parietal pericardial membranes, as shown in this anterior view of the heart.

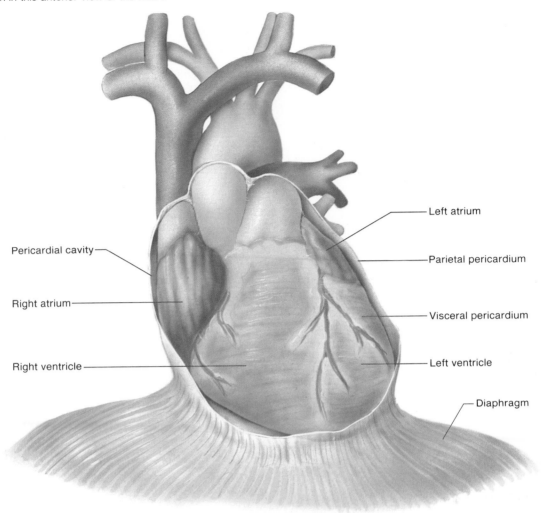

Pericardial cavity

Right atrium

Right ventricle

Left atrium

Parietal pericardium

Visceral pericardium

Left ventricle

Diaphragm

fibers are arranged in planes separated by connective tissues that are richly supplied with blood capillaries, lymph capillaries, and nerve fibers.

The **endocardium** consists of epithelium and connective tissue that contains many elastic and collagenous fibers. The connective tissue also contains some specialized cardiac muscle fibers called *Purkinje fibers,* whose function is described in a subsequent section of this chapter. This inner lining is continuous with the inner linings of the blood vessels attached to the heart.

Heart Chambers and Valves

Internally, the heart is divided into four chambers, two on the left and two on the right. The upper chambers, called **atria** (singular, *atrium*), have relatively thin walls and receive blood from veins. The lower chambers, the **ventricles,** force blood out of the heart into arteries. (Note: Veins are blood vessels that carry blood toward the heart; arteries carry blood away from the heart.)

The atrium and ventricle on the right side are separated from those on the left by a *septum*. The atrium on each side communicates with its corresponding ventricle through an opening, which is guarded by a *valve*. These parts of the heart are shown in figure 16.4.

The right atrium receives blood from two large veins—the *superior vena cava* and *inferior vena cava*. (See fig. 16.4.) A smaller vein, the *coronary sinus,* also drains blood into the right atrium from the wall of the heart.

Fig. 16.3 The wall of the heart consists of three layers: the epicardium, the myocardium, and the endocardium.

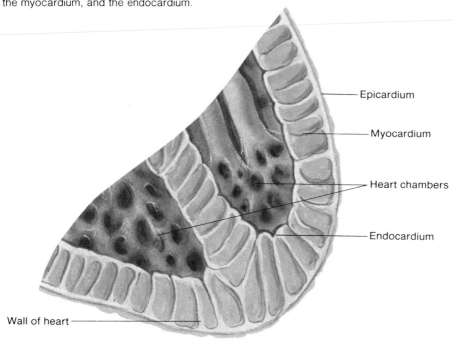

Epicardium

Myocardium

Heart chambers

Endocardium

Wall of heart

Fig. 16.4 A frontal section of the heart.

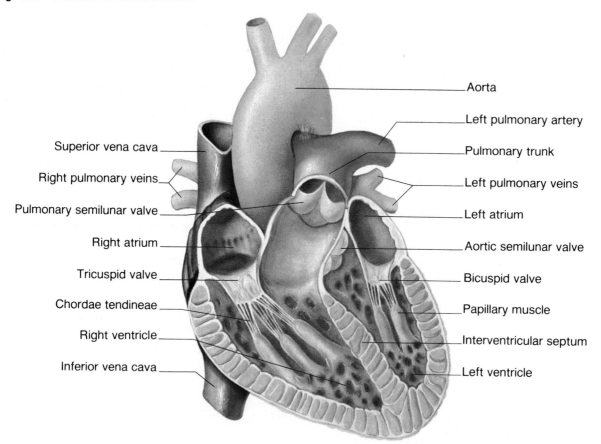

Superior vena cava

Right pulmonary veins

Pulmonary semilunar valve

Right atrium

Tricuspid valve

Chordae tendineae

Right ventricle

Inferior vena cava

Aorta

Left pulmonary artery

Pulmonary trunk

Left pulmonary veins

Left atrium

Aortic semilunar valve

Bicuspid valve

Papillary muscle

Interventricular septum

Left ventricle

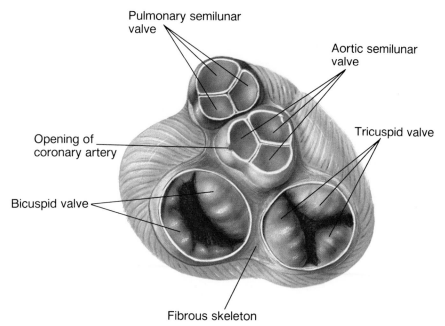

The opening between the right atrium and the right ventricle is guarded by a large **tricuspid valve,** which is composed of three leaflets or cusps. This valve permits blood to move from the right atrium into the right ventricle and prevents passage in the opposite direction.

Strong, fibrous strings, called *chordae tendineae,* are attached to the cusps on the ventricular side. These strings originate from small mounds of muscle tissue, the **papillary muscles,** that project inward from the walls of the ventricle. When the cusps close, the chordae tendineae prevent them from swinging back into the atrium. The papillary muscles help open the valve by pulling the cusps downward.

The right ventricle has a much thinner muscular wall than the left ventricle. This chamber pumps blood a fairly short distance to the lungs against a relatively low resistance to blood flow. The left ventricle, on the other hand, must force blood to all other parts of the body against a much greater resistance to flow.

An exit from the right ventricle is provided by the *pulmonary trunk,* which divides to form the left and right *pulmonary arteries* that lead to the lungs. At the base of this trunk is a **pulmonary semilunar valve** that consists of three cusps. This valve allows blood to leave the right ventricle and prevents a return flow into the ventricular chamber.

The left atrium receives blood from the lungs through four *pulmonary veins*—two from the right lung and two from the left lung. Blood passes from the left atrium into the left ventricle through the **bicuspid** (mitral) **valve,** which prevents blood from flowing back into the left atrium from the ventricle. Like the tricuspid valve, the bicuspid valve is aided by chordae tendineae and papillary muscles.

The only exit from the left ventricle is through a large artery called the *aorta.* At the base of the aorta, there is an **aortic semilunar valve** that consists of three cusps. It opens and allows blood to leave the left ventricle, but prevents blood from backing up into the ventricle when it closes.

---

1. *Describe the layers of the heart wall.*
2. *Name and locate the four chambers and valves of the heart.*

Skeleton of the Heart

At their proximal ends, the pulmonary trunk and aorta are surrounded by rings of dense fibrous connective tissue. The rings provide firm attachments for the heart valves and for various muscle fibers. In addition, they prevent the outlets of the atria and ventricles from dilating during myocardial contraction. The fibrous rings together with other masses of dense fibrous tissue in the upper portion of the interventricular septum constitute the skeleton of the heart (fig. 16.5).

Fig. 16.6 The right ventricle forces blood to the lungs, while the left ventricle forces blood to all other body parts. How does the composition of the blood in these two chambers differ?

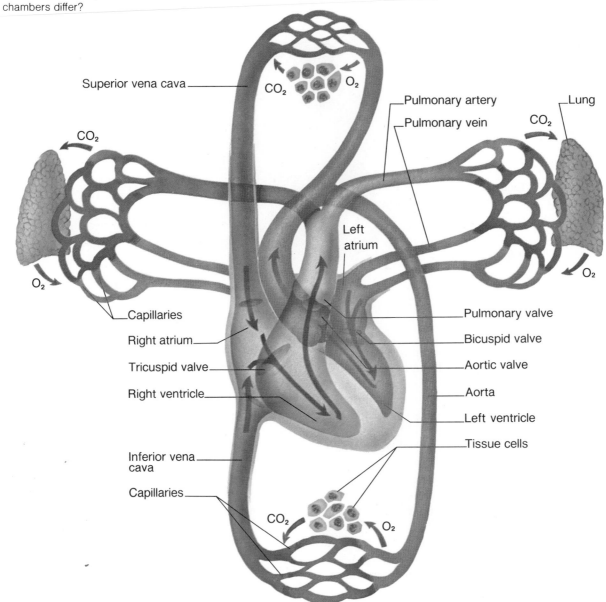

Superior vena cava

CO_2 O_2

CO_2

O_2

Pulmonary artery

Pulmonary vein

Lung

CO_2

O_2

Left atrium

Capillaries

Right atrium

Tricuspid valve

Right ventricle

Pulmonary valve

Bicuspid valve

Aortic valve

Aorta

Left ventricle

Tissue cells

Inferior vena cava

Capillaries

CO_2 O_2

Path of Blood through the Heart

Blood that is relatively low in oxygen and relatively high in carbon dioxide concentration enters the right atrium through the venae cavae and the coronary sinus. As the right atrial wall contracts, blood passes through the tricuspid valve and enters the chamber of the right ventricle. (See fig. 16.6.)

When the right ventricular wall contracts, the tricuspid valve closes, and the blood moves through the pulmonary semilunar valve and into the pulmonary trunk and its branches. From these vessels, the blood enters the capillaries associated with the alveoli of the lungs. Gas exchanges occur between the blood in the capillaries and the air in the alveoli. The freshly oxygenated blood, which is now relatively low in carbon dioxide concentration, returns to the heart through the pulmonary veins that lead to the left atrium.

Fig. 16.7 The coronary arteries provide blood to the tissues of the heart.

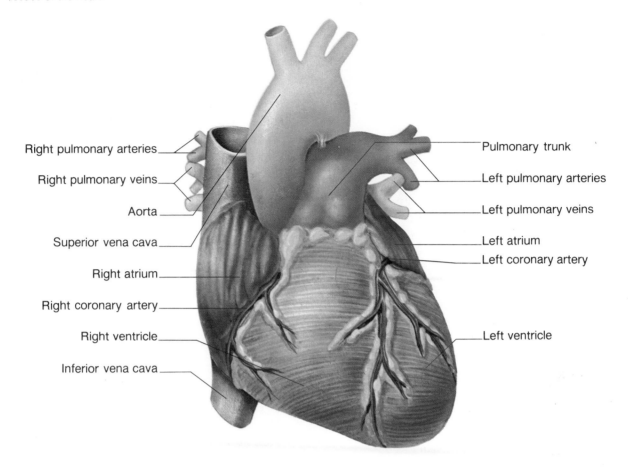

Right pulmonary arteries

Right pulmonary veins

Aorta

Superior vena cava

Right atrium

Right coronary artery

Right ventricle

Inferior vena cava

Pulmonary trunk

Left pulmonary arteries

Left pulmonary veins

Left atrium

Left coronary artery

Left ventricle

The left atrial wall contracts, and the blood moves through the bicuspid valve and into the chamber of the left ventricle. When the left ventricular wall contracts, the bicuspid valve closes and the blood passes through the aortic semilunar valve and into the aorta and its branches.

Blood Supply to the Heart

Blood is supplied to the tissues of the heart by the first two branches of the aorta, called the right and left **coronary arteries.** Their openings lie just beyond the aortic semilunar valve. (See fig. 16.7.)

Since the heart must beat continually to supply blood to the body tissues, the myocardial cells require a constant supply of freshly oxygenated blood. The myocardium contains many capillaries fed by branches of the coronary arteries. The larger branches of these arteries have interconnections between vessels that provide alternate pathways for blood.

If a branch of a coronary artery becomes abnormally constricted or obstructed by a thrombus or embolus, the myocardial cells it supplies may experience a blood deficiency, called *ischemia.* As a result of ischemia, the person may experience a painful condition called *angina pectoris.* The discomfort of angina pectoris usually occurs during physical activity or an emotional disturbance, and is relieved by rest. It may take the form of pain or a sensation of heavy pressure, tightening, or squeezing in the chest. Although it is usually felt in the region behind the sternum or in the anterior portion of the upper thorax, the pain may radiate to other parts, including the neck, jaw, throat, arm, shoulder, elbow, back, or upper abdomen.

Sometimes a portion of the heart dies because of ischemia, and this condition, a *myocardial infarction,* or more commonly called a heart attack, is one of the leading causes of death.

Fig. 16.8 The cardiac veins carry blood to the coronary sinus that empties into the right atrium.

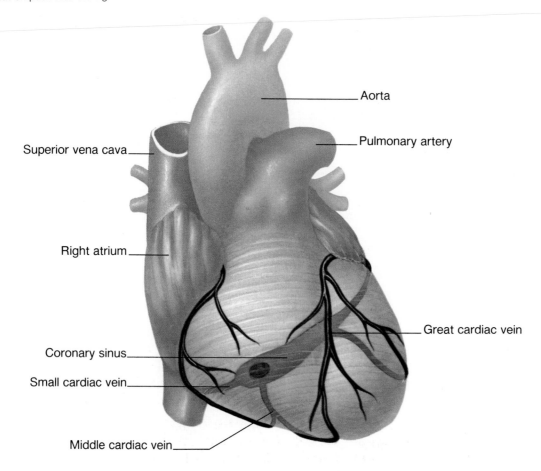

Aorta

Pulmonary artery

Superior vena cava

Right atrium

Great cardiac vein

Coronary sinus

Small cardiac vein

Middle cardiac vein

Blood that has passed through the capillaries of the myocardium is drained by branches of **cardiac veins,** whose paths roughly parallel those of the coronary arteries. As figure 16.8 shows, these veins join an enlarged vessel, the **coronary sinus,** which is on the posterior surface of the heart and empties into the right atrium.

---

1. *What structures make up the skeleton of the heart?*
2. *Review the path of blood through the heart.*
3. *What vessels supply blood to the heart?*
4. *What vessels drain blood from the heart?*

Actions of the Heart

Although the previous discussion described the actions of the heart chambers separately, they do not function independently. Instead, their actions are regulated so that the atrial walls contract while the ventricular walls are relaxed; and ventricular walls contract while the atrial walls are relaxed. Such a series of events, shown in figure 16.9, constitutes a complete heartbeat or **cardiac cycle.**

The Cardiac Cycle

During a cardiac cycle, the pressure within the chambers rises and falls. For example, when the atria are relaxed, blood flows into them from the large, attached veins. Then, the atrial walls contract (atrial systole), and the atrial pressure rises suddenly, forcing the atrial contents into the ventricles. This is followed by atrial relaxation (atrial diastole).

Fig. 16.9 (a) The ventricles fill with blood during ventricular diastole and (b) empty during ventricular systole.

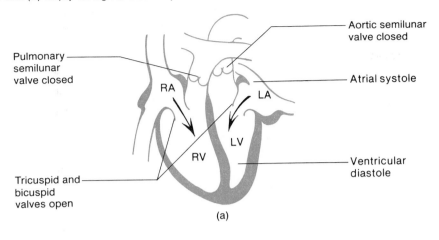

Pulmonary semilunar valve closed

Aortic semilunar valve closed

Atrial systole

Tricuspid and bicuspid valves open

Ventricular diastole

RA   LA   LV   RV

(a)

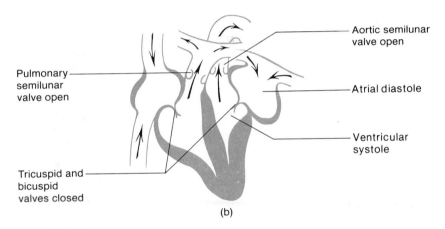

Pulmonary semilunar valve open

Aortic semilunar valve open

Atrial diastole

Tricuspid and bicuspid valves closed

Ventricular systole

(b)

As the ventricles contract (ventricular systole), the tricuspid and bicuspid valves close, and blood flows out of the ventricles into the arteries. When the ventricles relax (ventricular diastole), the tricuspid and bicuspid valves open, and blood flows through them from the atria into the ventricles.

Heart Sounds

The sounds associated with a heartbeat can be heard with a stethoscope and are described as *lub*-dup sounds. These sounds are due to vibrations in heart tissues that are created as blood flow is suddenly speeded or slowed with the contraction and relaxation of heart chambers and the opening and closing of valves.

The first part of a heart sound occurs during the ventricular contraction, when the tricuspid and bicuspid valves are closing. The second part occurs during ventricular relaxation, when the pulmonary and aortic semilunar valves are closing.

Heart sounds are of particular interest because they provide information concerning the condition of the heart valves. For example, inflammation of the endocardium (endocarditis) may cause changes in the shapes of the valvular cusps (valvular stenosis). Then, when the cusps close, the closure may be incomplete, and some blood may leak back through the valve. If this happens, an abnormal sound called a *murmur* may be heard. The seriousness of a heart murmur depends on the amount of valvular damage. Fortunately for those who have serious problems, it may be possible to repair the damaged valves or to replace them by open heart surgery.

Cardiac Muscle Fibers

As is mentioned in chapter 9, cardiac muscle fibers function much like those of skeletal muscles. In cardiac muscle, however, the fibers are interconnected in branching networks that spread in all directions through the heart. When any portion of this net is stimulated, an impulse travels to all of its parts, and the whole structure contracts as a unit.

A mass of merging cells that act together in this way is called a **functional syncytium.** There are two such structures in the heart—one in the atrial walls and another in the ventricular walls. These masses of muscle fibers are separated from each other by portions of the heart's fibrous skeleton, except for a small area in the right atrial floor. In this region, the *atrial syncytium* and the *ventricular syncytium* are connected by fibers of the cardiac conduction system.

1. Describe a cardiac cycle.
2. What causes heart sounds?
3. What is meant by a functional syncytium?

Cardiac Conduction System

Throughout the heart there are clumps and strands of specialized cardiac muscle tissue whose fibers contain only a few myofibrils. Instead of contracting, these parts function to initiate and distribute impulses (cardiac impulses) throughout the myocardium. They comprise the **cardiac conduction system,** which functions to coordinate the events occurring during the cardiac cycle.

A key portion of this conduction system is called the **sinoatrial (S-A) node.** It consists of a small mass of specialized muscle tissue just beneath the epicardium. It is located in the posterior wall of the right atrium, below the opening of the superior vena cava, and its fibers are continuous with those of the *atrial syncytium.*

The cells of the S-A node have a special ability to excite themselves. Without being stimulated by nerve fibers or any other outside agents, these cells initiate impulses that spread into the myocardium and stimulate cardiac muscle fibers to contract. Furthermore, this activity is rhythmic. The S-A node initiates one impulse after another, seventy to eighty times a minute. Thus it is responsible for the rhythmic contractions of the heart and is often called the **pacemaker.**

As a cardiac impulse travels from the S-A node into the atrial syncytium, the right and left atria contract almost simultaneously. Instead of passing directly into the ventricular syncytium, which is separated from the atrial syncytium by the fibrous skeleton of the heart, the cardiac impulse passes along fibers of the conduction system that are continuous with atrial muscle fibers. These conducting fibers lead to a mass of specialized muscle tissue called the **atrioventricular (A-V) node.** This node, located in the floor of the right atrium near the interatrial septum and just beneath the endocardium, provides the only normal conduction pathway between the atrial and ventricular syncytia.

The fibers that conduct the cardiac impulse into the A-V node (junctional fibers) have very small diameters, and because small fibers conduct impulses slowly, they cause the impulse to be delayed. The impulse is delayed still more as it travels through the A-V node, and this delay allows time for the atria to empty and the ventricles to fill with blood.

Once the cardiac impulse reaches the other side of the A-V node, it passes into a group of large fibers that make up the **A-V bundle** (bundle of His), and the impulse moves rapidly through them. The A-V bundle enters the upper part of the interventricular septum, and divides into right and left branches that lie just beneath the endocardium. About halfway down the septum, the branches give rise to enlarged **Purkinje fibers.**

The Purkinje fibers spread from the interventricular septum, into the papillary muscles that project inward from the ventricular walls, and continue downward to the apex of the heart. There they curve around the tips of the ventricles and pass upward over the lateral walls of these chambers. Along the way, the Purkinje fibers give off many small branches that become continuous with cardiac muscle fibers. These parts of the conduction system are shown in figure 16.10.

The muscle fibers in the ventricular walls are arranged in irregular whorls, so that when they are stimulated by impulses on Purkinje fibers, the ventricular walls contract with a twisting motion (fig. 16.11). This action squeezes or wrings the blood out of the ventricular chambers and forces it into the arteries.

Fig. 16.10 The cardiac conduction system. What is the function of this system?

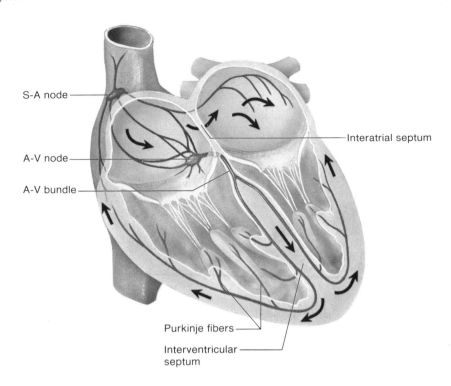

S-A node

A-V node

A-V bundle

Interatrial septum

Purkinje fibers

Interventricular septum

Fig. 16.11 The muscle fibers within the ventricular walls are arranged in patterns of whorls. The fibers of groups (a) and (b) surround both ventricles.

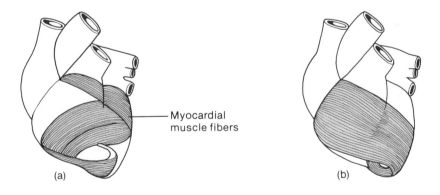

Myocardial muscle fibers

(a)

(b)

Fig. 16.12 A normal ECG.

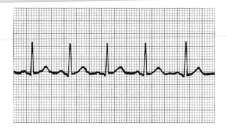

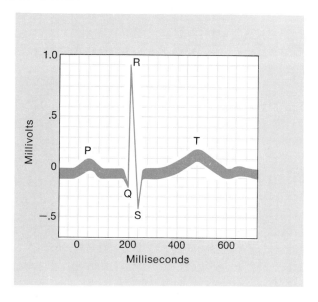

Sometimes a patient suffering from a disorder of the cardiac conduction system can be helped by an *artificial pacemaker.* Such a device contains a small, battery- or nuclear-powered electrical stimulator (generator pack). When the artificial pacemaker is installed, electrodes may be threaded through veins into the right ventricle, and the stimulator may be implanted beneath the skin in the shoulder or abdomen. It then transmits rhythmic electrical impulses to the heart, and the myocardium responds by contracting rhythmically.

1. *What kinds of tissues make up the cardiac conduction system?*
2. *How is a cardiac impulse initiated?*
3. *How is this impulse transmitted from the atrium to the ventricles?*

The Electrocardiogram

An **electrocardiogram** (ECG) is a recording of the electrical changes that occur in the *myocardium* during a cardiac cycle. These changes result from the depolarization and repolarization associated with the contraction of muscle fibers. Because body fluids can conduct electrical currents, such changes can be detected on the surface of the body.

To record an ECG, metal electrodes are placed in certain locations on the skin. These electrodes are connected by wires to an instrument that responds to very weak electrical changes by causing a pen or stylus to mark on a moving strip of paper. When the instrument is operating, up-and-down movements of the pen correspond to electrical changes occurring as a result of myocardial activity.

Since the paper moves past the pen at a known rate, the distance between pen deflections can be used to measure the time elapsing between various phases of the cardiac cycle.

As figure 16.12 illustrates, the ECG pattern includes several deflections, or *waves,* during each cardiac cycle. Between cycles, the muscle fibers remain polarized, and no detectable electrical changes occur. Consequently, the pen simply marks along the base line as the paper moves through the instrument. When the S-A node triggers a cardiac impulse, however, the atrial fibers are stimulated to depolarize, and an electrical change occurs. As a result, the pen is deflected, and when this electrical change is completed, the pen returns to the base position. This pen movement produces a *P wave* that is caused by the depolarization of the atrial fibers just before they contract. (See fig. 16.13.)

When the cardiac impulse reaches the ventricular fibers, they are stimulated to depolarize rapidly. Because the ventricular walls are much more extensive than those of the atria, the amount of electrical change is greater, and the pen is deflected to a greater degree than before. Once again, when the electrical change is completed, the pen returns to the base line, leaving a mark called the *QRS complex.* This wave appears just prior to the contraction of the ventricular walls.

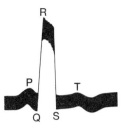

Near the end of the ECG pattern, the pen is deflected for a third time, producing a *T wave*. This wave is caused by electrical changes occurring as the ventricular muscle fibers become repolarized relatively slowly. The record of the atrial repolarization is missing from the pattern because the atrial fibers repolarize at the same time that the ventricular fibers depolarize. The recording of the atrial repolarization is thus obscured by the QRS complex.

ECG patterns are especially important because they allow a physician to assess the heart's ability to conduct impulses and thus to judge its condition. For example, the time period between the beginning of a P wave and the beginning of a QRS complex (*P-Q* or *P-R interval*) indicates how long it takes for the cardiac impulse to travel from the S-A node through the A-V node and into the ventricular walls. If ischemia or other problems involving the fibers of the A-V conduction pathways are present, this P-Q interval sometimes increases. Similarly, if the Purkinje fibers are injured, the duration of the QRS complex may increase, because it may take longer for an impulse to spread throughout the ventricular walls. (See fig. 16.14.)

1. *What is an electrocardiogram?*
2. *What cardiac event is represented by the P wave? By the QRS complex? By the T wave?*

Regulation of the Cardiac Cycle

The primary function of the heart is to pump blood to the body cells, and when the needs of these cells change, the quantity of blood pumped must change also. For example, during strenuous exercise, the amount of blood required by skeletal muscles increases greatly, and the rate of the heartbeat increases in response to this need. Since the S-A node normally controls the heart rate, changes in this rate often involve factors that affect the pacemaker. These include motor impulses on parasympathetic and sympathetic nerve fibers. (See chapter 10.)

The parasympathetic fibers that supply the heart arise from neurons in the medulla oblongata. Most of these fibers branch to the S-A node and the A-V node. When nerve impulses reach their endings, these fibers secrete acetylcholine that causes a decrease in S-A and A-V nodal activity. As a result, the rate of heartbeat decreases.

Parasympathetic fibers seem to carry impulses continually to the S-A and A-V nodes, and these impulses impose a braking action on the heart. Consequently, parasympathetic activity can cause the heart rate to change in either direction. An increase in impulses causes a slowing of the heart, and a decrease in impulses releases the parasympathetic brake and allows the heartbeat to increase.

Sympathetic fibers also reach the heart and join the S-A and A-V nodes as well as other areas of the atrial and ventricular myocardium. (See fig. 16.15.) The endings of these fibers secrete norepinephrine, and this substance causes an increase in the rate and the force of myocardial contractions.

A normal balance between the inhibitory effects of the parasympathetic fibers and the excitatory effects of the sympathetic fibers is maintained by the *cardiac center* of the medulla oblongata. This center receives sensory impulses from various parts of the circulatory system and relays motor impulses to the heart in response.

For example, there are *pressoreceptors* in certain regions of the aorta (aortic sinus and aortic arch) and in the carotid arteries (carotid sinuses). These receptors are sensitive to changes in blood pressure, and if the pressure rises, they signal the cardiac center in the medulla. In response, it sends *parasympathetic* motor impulses to the heart, causing the heart rate and force of contraction to decrease. This also causes blood pressure to drop toward the normal level. (See fig. 16.16.)

Two other factors that influence heart rate are temperature and various ions. Heart action is increased by a rising body temperature and is decreased by abnormally low body temperature. Consequently, a patient's body temperature is sometimes deliberately lowered (hypothermia) to slow the heart during surgery.

Of the ions that influence heart action, the most important are potassium (K^+) and calcium (Ca^{++}). An excess of *potassium ions* (hyperkalemia), for example, results in a decrease in the rate and force of

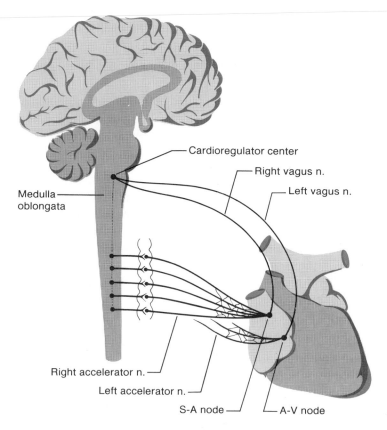

Cardioregulator center
Right vagus n.
Left vagus n.
Medulla oblongata
Right accelerator n.
Left accelerator n.
S-A node
A-V node

contractions. If the potassium concentration falls below normal (hypokalemia), the heart may develop a serious abnormal rhythm (arrhythmia).

Excessive *calcium ions* (hypercalcemia) cause increased heart actions, and there is a danger that the heart will undergo a prolonged contraction. Conversely, low calcium (hypocalcemia) depresses heart action.

---

1. *How do parasympathetic and sympathetic impulses help control heart rate?*
2. *How do changes in body temperature affect heart rate?*
3. *Describe the effects of abnormal concentrations of potassium and calcium on the heart.*

Although slight variations in heart actions sometimes occur normally, marked changes in the usual rate or rhythm may suggest cardiovascular disease. Such abnormal actions, termed cardiac *arrhythmias,* include the following:

1. *Tachycardia*—an abnormally fast heartbeat, usually over one-hundred beats per minute.

2. *Bradycardia*—a slow heart rate, usually less than sixty beats per minute.

3. *Flutter*—a very rapid heart rate, such as 250–350 contractions per minute.

4. *Fibrillation*—characterized by rapid heart actions but, unlike flutter, the accompanying contractions are *uncoordinated.* In fibrillation small regions of the myocardium contract and relax independently of all other areas. As a result, the myocardium fails to contract as a whole, and the walls of the fibrillating chambers are completely ineffective in pumping blood. Although a person may survive atrial fibrillation because venous blood pressure may continue to force blood into the ventricles, ventricular fibrillation is very likely to cause death.

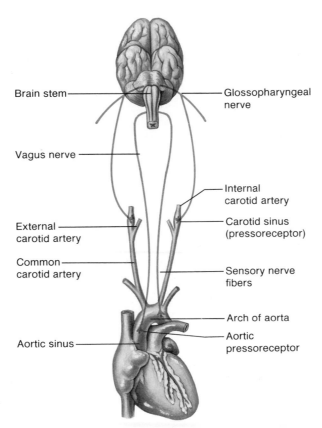

Blood Vessels

The vessels of the cardiovascular system form a closed circuit of tubes that carry blood from the heart to the body cells and back again. These tubes include arteries, arterioles, capillaries, venules, and veins.

Arteries and Arterioles

Arteries are strong, elastic vessels that are adapted for carrying blood away from the heart under relatively high pressure. These vessels subdivide into progressively thinner tubes and eventually give rise to fine branches called **arterioles.**

The wall of an artery consists of three distinct layers, shown in figure 16.17. The innermost layer is composed of epithelium, called *endothelium,* resting on a connective tissue membrane that is rich in elastic and collagenous fibers.

The middle layer makes up the bulk of the arterial wall. It includes smooth muscle fibers that encircle the tube and a thick layer of elastic connective tissue.

The outer layer is relatively thin and consists chiefly of connective tissue with irregularly arranged elastic and collagenous fibers. This layer attaches the artery to the surrounding tissues.

The smooth muscles in the walls of arteries and arterioles are innervated by sympathetic branches of the autonomic nervous system. Impulses on these *vasomotor* fibers cause the smooth muscles to contract, reducing the diameter of the vessel. This action is called **vasoconstriction.** If such vasomotor impulses are inhibited, the muscle fibers relax and the diameter of the vessel increases. In this case, the artery is said to undergo **vasodilation.** Changes in the diameters of arteries greatly influence the flow and pressure of the blood.

Arterioles, which are microscopic continuations of arteries, join capillaries. Although the walls of the larger arterioles have three layers, similar to those of arteries, these walls become thinner and thinner as

Fig. 16.17 The wall of an artery.

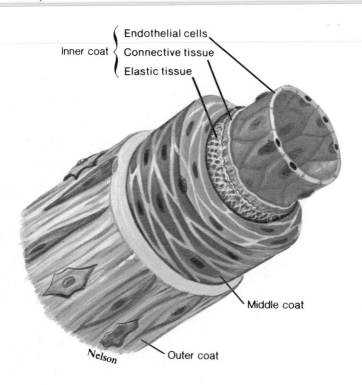

Inner coat { Endothelial cells
Connective tissue
Elastic tissue

Middle coat

Nelson — Outer coat

Fig. 16.18 Small arterioles have some smooth muscle fibers in their walls; capillaries lack these fibers.

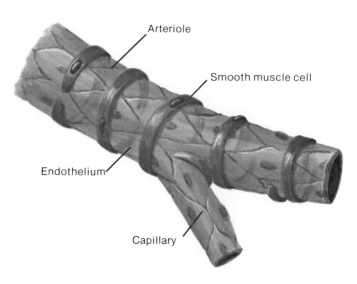

Arteriole

Smooth muscle cell

Endothelium

Capillary

Fig. 16.19 Scanning electron micrograph of an arteriole cross section (3900 ×).

From *Tissues and Organs: A Text-Atlas of Scanning Electron Microscopy* by Richard G. Kessel and Randy H. Kardon. W. H. Freeman and Company. Copyright © 1979.

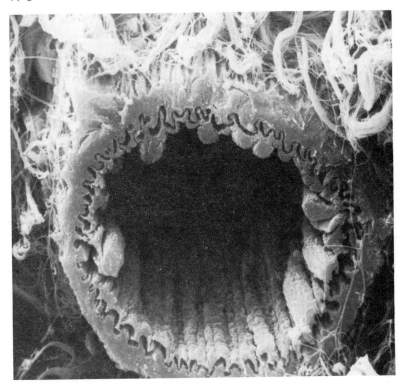

the arterioles approach the capillaries. The wall of a very small arteriole consists only of an endothelial lining and some smooth muscle, surrounded by a small amount of connective tissue (figs. 16.18 and 16.19).

---

1. Describe the wall of an artery.
2. What is the function of the smooth muscle in the arterial wall?
3. How is the structure of an arteriole different from that of an artery?

Capillaries

Capillaries are the smallest blood vessels. They form the connections between the smallest arterioles and the smallest venules. Capillaries are essentially extensions of the inner linings of these larger vessels in that their walls consist of endothelium—a single layer of squamous epithelial cells. (See fig. 16.18.) These thin walls form the semipermeable membranes through which substances in the blood are exchanged for substances in the tissue fluid surrounding body cells.

The density of capillaries within tissues varies directly with the tissues' rates of metabolism. Thus muscle and nerve tissues, which utilize relatively large quantities of oxygen and nutrients, are richly supplied with capillaries, while hyaline cartilage, the epidermis, and the cornea, whose metabolic rates are very slow, lack capillaries.

The patterns of capillary arrangement also differ in various body parts. For example, some capillaries pass directly from arterioles to venules, while others lead to highly branched networks. Such arrangements make it possible for blood to follow different pathways through a tissue and to meet the varying demands of its cells. During periods of exercise, for example, blood can be directed into the capillary networks of the skeletal muscles, where cells are experiencing an increasing need for oxygen and nutrients. At the same time, blood can bypass some of the capillary nets in the tissues of the digestive tract, where the demand for blood is less critical.

The distribution of blood in the various capillary pathways is regulated mainly by smooth muscles that encircle the capillary entrances. These muscles form *precapillary sphincters* that may close a capillary by contracting or open it by relaxing. How the sphincters are controlled is not clear, but they seem to respond to the demands of the cells supplied by their individual capillaries. When the cells are low in oxygen and nutrients, the sphincter relaxes; when the cellular needs are met, the sphincter may contract again.

---

1. Describe the wall of a capillary.
2. What is the function of a capillary?
3. How is blood flow into capillaries controlled?

Exchanges in the Capillaries.   The vital function of exchanging gases, nutrients, and metabolic by-products between the blood and the tissue fluid surrounding body cells occurs in the capillaries. The substances exchanged move through the capillary walls primarily by the processes of diffusion, filtration, and osmosis, described in chapter 3. Of these processes, diffusion provides the most important means of transfer.

It is by diffusion that molecules and ions move from regions where they are more highly concentrated toward regions where they are in lower concentration. Since blood entering capillaries of tissues outside the lungs generally carries relatively high concentrations of oxygen and nutrients, these substances diffuse through the capillary walls and enter the tissue fluid. Conversely, the concentrations of carbon dioxide and various wastes are generally more highly concentrated in these tissues, and they tend to diffuse into the capillary blood.

In the brain, the endothelial cells of the capillary walls are more tightly fused than those in other body regions. Consequently, some substances that readily leave capillaries in other tissues enter brain tissues only slightly or not at all. This resistance to movement is called the *blood-brain barrier,* and it is of particular interest because it prevents certain drugs from entering the brain tissues or cerebrospinal fluid in sufficient concentrations to effectively treat certain diseases.

Plasma proteins generally remain in the blood because their molecular size is too great to permit diffusion through the membrane pores or slitlike openings between the endothelial cells of most capillaries.

Filtration involves the forcing of molecules through a membrane by *hydrostatic pressure*. In capillaries, the force is provided by blood pressure generated by contractions of the ventricular walls.

Blood pressure also is responsible for moving blood through the arteries and arterioles. Pressure tends to decrease, however, as the distance from the heart increases because friction (peripheral resistance) between the blood and the vessel walls slows the flow. For this reason, blood pressure is greater in arteries than in arterioles and greater in arterioles than in capillaries. It is similarly greater at the arteriole end of a capillary than at the venule end. Therefore, the filtration effect occurs primarily at the arteriole ends of capillaries.

The plasma proteins, which remain in the capillaries, help to make the *osmotic pressure* of the blood greater (hypertonic) than that of the tissue fluid. Although the capillary blood has a greater osmotic attraction for water than does the tissue fluid, this attraction is overcome by the greater force of the blood pressure. As a result, the net movement of water and dissolved substances is outward at the arteriole end of the capillary by filtration.

Since the blood pressure decreases as the blood moves through the capillary, however, the outward filtration force is less than the osmotic pressure of the blood at the venule end. Consequently, there is a net movement of water and dissolved materials into the venule end of the capillary by osmosis. This process is shown in figure 16.20.

Normally, more fluid leaves the capillaries than returns to them, and the excess is collected and returned to the venous circulation by *lymphatic vessels*. This mechanism is discussed in chapter 17.

Sometimes unusual events cause an increase in the permeability of the capillaries, and an excessive amount of fluid is likely to enter the interstitial spaces. This may occur, for instance, following a traumatic injury to the tissues or in response to certain chemicals such as *histamine*, which increase membrane permeability. In any case, so much fluid may leak out of the capillaries that the lymphatic drainage is overwhelmed, and the affected tissues become swollen (edematous) and painful.

Fig. 16.20 Substances leave the capillaries because of a net outward filtration pressure; substances enter the capillaries because of a net inward force of osmotic pressure.

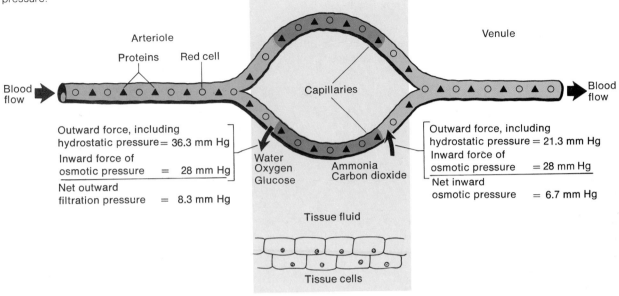

Arteriole

Proteins   Red cell

Blood flow

Capillaries

Venule

Blood flow

Outward force, including
hydrostatic pressure = 36.3 mm Hg

Inward force of
osmotic pressure   = 28 mm Hg

Net outward
filtration pressure   = 8.3 mm Hg

Water
Oxygen
Glucose

Ammonia
Carbon dioxide

Outward force, including
hydrostatic pressure = 21.3 mm Hg

Inward force of
osmotic pressure   = 28 mm Hg

Net inward
osmotic pressure   = 6.7 mm Hg

Tissue fluid

Tissue cells

---

1. *What forces are responsible for the exchange of substances between the blood and tissue fluid?*
2. *Why is the fluid movement out of a capillary greater at its arteriole end than at its venule end?*

Fig. 16.21 Note the structural differences in the cross sections of (a) this artery and (b) this vein.

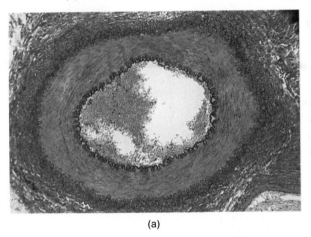

(a)

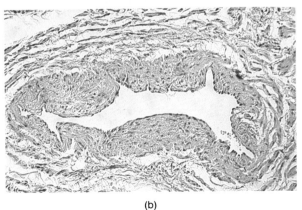

(b)

Venules and Veins

Venules are the microscopic vessels that continue from the capillaries and merge to form **veins.** The veins, which carry blood back to the atria, follow pathways that roughly parallel those of the arteries.

The walls of veins are similar to those of arteries in that they are composed of three distinct layers. Because the middle layer of the venous wall is poorly developed, however, veins have thinner walls and contain less smooth muscle and less elastic tissue than arteries. (See fig. 16.21.)

Many veins, particularly those in the arms and legs, contain flaplike *valves* that project inward from their linings. These valves, shown in figure 16.22, are usually composed of two leaflets that close if the blood begins to back up in a vein. In other words, the valves aid in returning blood to the heart, since the valves open as long as the flow is toward the heart, but close if it is in the opposite direction.

In addition to providing pathways for blood returning to the heart, the veins function as blood reservoirs that can be drawn on in times of need. For example, if a hemorrhage accompanied by a drop in

Chart 16.1 Characteristics of blood vessels

Vessel	Type of Wall	Function
Artery	Thick, strong wall with three layers—endothelial lining, middle layer of smooth muscle and elastic tissue, and outer layer of connective tissue	Carries high pressure blood from heart to arterioles
Arteriole	Thinner than artery but with three layers; smaller arterioles have endothelial lining, some smooth muscle tissue, and small amount of connective tissue	Connects artery to capillary; helps to control blood flow into capillary by undergoing vasoconstriction or vasodilation
Capillary	Single layer of squamous epithelium	Provides semipermeable membrane through which nutrients, gases, and wastes are exchanged between blood and tissue cells; connects arteriole to venule
Venule	Thinner wall, less smooth muscle and elastic tissue than arteriole	Connects capillary to vein
Vein	Thinner wall than artery but with similar layers; middle layer more poorly developed; some with flaplike valves	Carries low pressure blood from venule to heart; valves prevent backflow of blood; serves as blood reservoir

From Hole, John W., Jr., *Human Anatomy and Physiology 3d ed.* © 1978, 1981, 1984 Wm. C. Brown Publishers, Dubuque, Iowa. All Rights Reserved. Reprinted by permission.

Fig. 16.22 (a) Venous valves allow blood to move toward the heart, but (b) prevent blood from moving away from the heart.

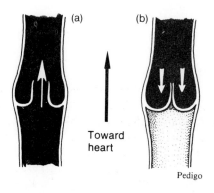

Toward heart

Pedigo

arterial blood pressure occurs, the muscular walls of veins are stimulated reflexly by sympathetic nerve impulses. The resulting venous constrictions help to raise the blood pressure. This mechanism ensures a nearly normal blood flow even when as much as 25% of the blood volume has been lost.

The characteristics of the blood vessels are summarized in chart 16.1.

---

1. *How does the structure of a vein differ from that of an artery?*
2. *How does the venous circulation help to maintain blood pressure when blood is lost by hemorrhage?*

Blood Pressure

Blood pressure is the force exerted by the blood against the inner walls of blood vessels. Although such a force occurs throughout the vascular system, the term *blood pressure* is most commonly used to refer to arterial pressure in various branches of the aorta.

Arterial Blood Pressure

Arterial blood pressure rises and falls in a pattern corresponding to the phases of the cardiac cycle. That is, when the ventricles contract (ventricular systole), their walls squeeze the blood inside their chambers and force it into the pulmonary trunk and aorta. As a result, the pressures in these arteries and their branches rise sharply. The maximum pressure achieved during ventricular contraction is called the **systolic pressure.** When the ventricles relax (ventricular diastole), the arterial pressure drops, and the lowest pressure that remains in the arteries before the next ventricular contraction is termed the **diastolic pressure.** (See fig. 16.23.)

The surge of blood entering the arterial system during a ventricular contraction causes the elastic walls of the arteries to swell, but the pressure drops almost immediately as the contraction is completed, and the arterial walls recoil. This alternate expanding and recoiling of an arterial wall can be felt as a *pulse* in an artery that runs close to the surface.

---

1. *What is meant by blood pressure?*
2. *Distinguish between systolic and diastolic blood pressure.*
3. *What causes a pulse in an artery?*

Fig. 16.23 Blood pressure decreases as the distance from the left ventricle increases.

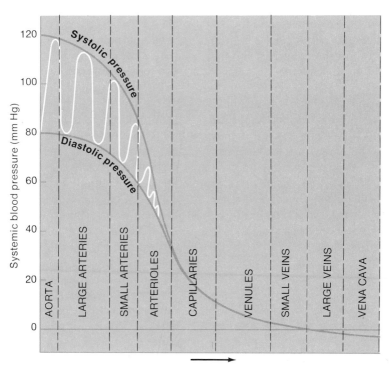

Distance from left ventricle

Factors That Influence Arterial Blood Pressure

The arterial pressure depends on a variety of factors, including heart action, blood volume, resistance to flow, and the viscosity of the blood.

Heart Action. In addition to creating blood pressure by forcing blood into the arteries, the heart action determines how much blood will enter the arterial system with each ventricular contraction. The heart also determines the rate of this fluid output.

The volume of blood discharged from the ventricle with each contraction is called the **stroke volume** and equals about 70 milliliters. The volume discharged from the ventricle per minute is called the **cardiac output.** It is calculated by multiplying the stroke volume by the heart rate in beats per minute. (Cardiac output = stroke volume × heart rate.) Thus if the stroke volume is 70 milliliters, and the heart rate is 72 beats per minute, the cardiac output is 5040 milliliters per minute.

Blood pressure varies with the cardiac output. If either the stroke volume or the heart rate increases, so does the cardiac output, and as a result, the blood pressure rises. Conversely, if the stroke volume or the heart rate decreases, so do the cardiac output and the blood pressure.

Blood Volume. The **blood volume** is equal to the sum of the blood cell and plasma volumes in the cardiovascular system. Although the blood volume varies somewhat with age, body size, and sex, for adults it usually remains about 5 liters.

Blood pressure is directly proportional to the volume of blood. Thus, any changes in blood volume are accompanied by changes in blood pressure. For example, if blood volume is reduced by a hemorrhage, the blood pressure drops. If the normal blood volume is restored by a blood transfusion, normal pressure may be reestablished.

Peripheral Resistance. Friction between the blood and the walls of the blood vessels creates a force called **peripheral resistance,** which hinders blood flow. This force must be overcome by blood pressure if the blood is to continue flowing. Consequently, factors that alter peripheral resistance cause changes in blood pressure.

For example, if the smooth muscles in the walls of arterioles contract, the peripheral resistance of these constricted vessels increases. Blood tends to back up into the arteries supplying the arterioles, and the arterial pressure rises. Dilation of the arterioles has the opposite effect—peripheral resistance lessens, and the arterial blood pressure drops in response.

Viscosity.   **Viscosity** is a physical property of a fluid and is related to the ease with which its molecules slide past one another. A fluid with a high viscosity tends to be sticky like syrup, and one with a low viscosity flows easily like water.

The presence of blood cells and plasma proteins increases the viscosity of the blood. Since the greater the blood's resistance to flowing, the greater the force needed to move it through the vascular system, it is not surprising that blood pressure rises as blood viscosity increases and drops as viscosity decreases.

Although the viscosity of blood normally remains relatively stable, any condition that alters the concentrations of blood cells or plasma proteins may cause changes in viscosity. For example, anemia and hemorrhage may be accompanied by a decreasing viscosity and a consequent drop in blood pressure. If red blood cells are present in abnormally high numbers (polycythemia), the viscosity increases, and there is likely to be a rise in blood pressure.

1. What is the relationship between cardiac output and blood pressure?
2. How does blood volume affect blood pressure?
3. What is the relationship between peripheral resistance and blood pressure?

Control of Blood Pressure

Two important mechanisms for maintaining normal arterial pressure involve the regulation of cardiac output and peripheral resistance.

As was mentioned, cardiac output depends on the volume of blood discharged from the ventricle with each contraction (stroke volume) and the rate of heartbeat.

For example, the volume of blood entering the ventricle affects the stroke volume. As the blood enters, myocardial fibers in the ventricular wall are mechanically stretched. Within limits, the greater the length of these fibers, the greater the force with which they contract. This relationship between fiber length and force of contraction is called *Starling's law of the heart.* Because of it, the heart can respond to the demands placed on it by the varying quantities of blood that return from the venous system.

In other words, the more blood that enters the heart from the veins, the stronger the ventricular contraction, the greater the stroke volume, and the greater the cardiac output. This mechanism ensures that the volume of blood discharged from the heart is equal to the volume entering its chambers.

The neural regulation of heart rate was described previously in this chapter. To review, pressoreceptors located in the walls of the aorta and carotid arteries are sensitive to changes in blood pressure. If the arterial pressure rises, nerve impulses travel from the receptors to the cardiac center of the medulla oblongata. This center relays parasympathetic impulses to the S-A node in the heart, and its rate decreases in response. As a result of this reflex, the cardiac output is reduced, and blood pressure falls toward the normal level.

Conversely, if the arterial blood pressure drops, a reflex involving sympathetic impulses to the S-A node is initiated, and the heart beats faster. This response increases the cardiac output, and the arterial pressure rises.

Other factors that cause an increase in heart rate and a rise in blood pressure include the presence of chemicals such as epinephrine, emotional responses such as fear and anger, physical exercise, and an increase in body temperature.

Peripheral resistance is regulated primarily by changes in the diameters of arterioles. Since blood vessels with smaller diameters offer greater resistance to blood flow, factors that cause arteriole vasoconstriction bring about increasing peripheral resistance, while those causing vasodilation produce a decrease in resistance.

The *vasomotor center* of the medulla oblongata continually sends sympathetic impulses to the smooth muscles in the arteriole walls. As a result, these muscles are kept in a state of tonic contraction, which helps to maintain the peripheral resistance associated with normal blood pressure. Since the vasomotor center is responsive to changes in blood pressure, it can cause an increase in peripheral resistance by increasing its outflow of sympathetic impulses, or it can cause a decrease in such resistance by decreasing its sympathetic outflow. In the latter case, the vessels undergo vasodilation as the sympathetic stimulation is decreased.

For instance, whenever the arterial blood pressure suddenly rises, the pressoreceptors in the aorta and carotid arteries signal the vasomotor center, and the sympathetic outflow to the arteriole walls is inhibited. The resulting vasodilation causes a decrease in peripheral resistance, and the blood pressure falls toward the normal level.

Chemical substances, including carbon dioxide, oxygen, and hydrogen ions, also influence peripheral resistance by affecting the smooth muscles in the walls of arterioles and the actions of precapillary sphincters. For example, an increasing P_{CO_2}, a decreasing P_{O_2}, and a decreasing pH cause relaxation of these muscles and a consequent drop in blood pressure.

High blood pressure, or *hypertension,* is characterized by a persistently elevated arterial pressure and is one of the more common diseases of the cardiovascular system.

Since the cause of high blood pressure is often unknown, it may be called *essential* (or idiopathic) *hypertension.* Sometimes the elevated pressure is related to another problem, such as arteriosclerosis or kidney disease.

Arteriosclerosis, for example, is accompanied by decreasing elasticity of the arterial walls and narrowing of the lumens of these vessels. Both of these effects promote an increase in blood pressure.

---

1. What factors affect cardiac output?
2. What is the function of the pressoreceptors in the walls of the aorta and carotid arteries?
3. How does the vasomotor center control the diameter of arterioles?

Venous Blood Flow

Blood pressure decreases as the blood moves through the arterial system and into the capillary networks. In fact, little pressure remains at the venule ends of the capillaries. (See fig. 16.23.) Therefore, blood flow through the venous system is not the direct result of heart action, but depends on other factors, such as skeletal muscle contraction and breathing movements.

For example, when skeletal muscles contract, they thicken and press on nearby vessels, squeezing the blood inside. As was mentioned, many veins, particularly those in the arms and legs, contain flaplike valves. These valves offer little resistance to blood flowing toward the heart, but they close if blood moves in the opposite direction. (See fig. 16.22.) Consequently, as *skeletal muscles* exert pressure on veins with valves, some blood is moved from one valve section to another. This massaging action of contracting skeletal muscles helps to push blood through the venous system toward the heart.

Respiratory movements provide another means of moving venous blood. During inspiration, the pressure within the thoracic cavity is reduced as the diaphragm contracts and the rib cage moves upward and outward. (See chapter 14.) At the same time, the pressure within the abdominal cavity is increased as the diaphragm presses downward on the abdominal viscera. Consequently, blood tends to be squeezed out of the abdominal veins and into the thoracic veins. Backflow into the legs is prevented by the valves in the veins of the legs.

During exercise, respiratory movements act together with skeletal muscle contractions to increase the return of venous blood.

Another mechanism that promotes the return of venous blood involves constriction of veins. When venous pressure is low, sympathetic reflexes stimulate smooth muscles in the walls of veins to contract. Such *venoconstriction* serves to increase the venous pressure and forces more blood toward the heart.

Also, as was mentioned, the veins provide a blood reservoir that can adapt its capacity to changes in the blood volume. If blood is lost and the blood pressure decreases, venoconstriction can help to return the blood pressure to normal by forcing blood out of this reservoir.

---

1. What is the function of the venous valves?
2. How do skeletal muscles and respiratory movements affect venous blood flow?
3. What factors stimulate venoconstriction?

The blood vessels of the cardiovascular system can be divided into two major pathways—a pulmonary circuit and a systemic circuit. The **pulmonary circuit** consists of those vessels that carry blood from the heart to the lungs and back to the heart, while the **systemic circuit** is responsible for carrying blood from the heart to all other parts of the body and back again. (See fig. 16.1.)

The circulatory pathways described in the following sections are those of an adult. The fetal pathways are somewhat different and are described in chapter 20.

The Pulmonary Circuit

Blood enters the pulmonary circuit as it leaves the right ventricle through the *pulmonary trunk*. The pulmonary trunk extends upward and posteriorly from the heart and, about 2 inches above its origin, divides into right and left *pulmonary arteries*. These branches penetrate the right and left lungs respectively. After repeated divisions, they give rise to arterioles that continue into the capillary networks associated with the walls of the alveoli, where gas exchanges occur between the blood and the air. (See chapter 14.)

From the pulmonary capillaries, blood enters venules, which merge to form small veins. They in turn converge to form still larger ones. Four *pulmonary veins*, two from each lung, return blood to the left atrium, and this completes the vascular loop of the pulmonary circuit.

The Systemic Circuit

The freshly oxygenated blood received by the left atrium is forced into the systemic circuit by the contraction of the left ventricle. This circuit includes the aorta and its branches that lead to all body tissues, as well as the companion system of veins that returns blood to the right atrium.

---

1. *Distinguish between the pulmonary and systemic circuits.*
2. *Trace a drop of blood through the pulmonary circuit from the right ventricle.*

The **aorta** is the largest artery in the body. It extends upward from the left ventricle, arches over the heart to the left, and descends just in front of the vertebral column. Figure 16.24 shows the aorta and its main branches.

Principal Branches of the Aorta

The first portion of the aorta is called the *ascending aorta*. At its base are located the three cusps of the aortic semilunar valve, and opposite each one is a swelling in the aortic wall called an **aortic sinus.** The right and left *coronary arteries* spring from two of these sinuses.

Three major arteries originate from the *arch of the aorta* (aortic arch). They are the **brachiocephalic artery,** the left **common carotid artery,** and the left **subclavian artery.**

Although the upper part of the *descending aorta* is positioned to the left of the midline, it gradually moves medially and finally lies directly in front of the vertebral column at the level of the twelfth thoracic vertebra.

The portion of the descending aorta above the diaphragm is known as the **thoracic aorta,** and it gives off numerous small branches to the thoracic wall and the thoracic visceral organs.

Below the diaphragm, the descending aorta becomes the **abdominal aorta,** and it gives off branches to the abdominal wall and various abdominal visceral organs.

The descending aorta terminates near the brim of the pelvis, where it divides into the right and left *common iliac arteries*. These vessels supply blood to the lower regions of the abdominal wall, the pelvic organs, and the lower extremities.

The names of the major branches of the aorta and the general regions or organs they supply with blood are listed in chart 16.2.

If the aortic wall becomes weakened as a result of injury or disease, a blood-filled dilation or sac may appear in the vessel. This condition is called an *aortic aneurysm,* and it occurs most frequently in the abdominal aorta below the level of the renal arteries. An aneurysm usually increases in size with time, and the swelling is likely to rupture unless the affected portion of the artery is removed and replaced with a vessel graft.

Fig. 16.24 The principal branches of the aorta. (a. stands for *artery.*)

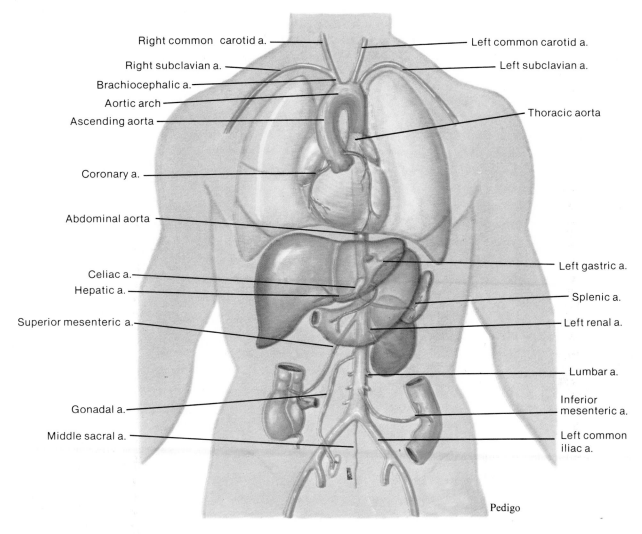

Right common carotid a.

Right subclavian a.

Brachiocephalic a.

Aortic arch

Ascending aorta

Coronary a.

Abdominal aorta

Celiac a.

Hepatic a.

Superior mesenteric a.

Gonadal a.

Middle sacral a.

Left common carotid a.

Left subclavian a.

Thoracic aorta

Left gastric a.

Splenic a.

Left renal a.

Lumbar a.

Inferior mesenteric a.

Left common iliac a.

Pedigo

---

1. *Name the parts of the aorta.*
2. *List the major branches of the aorta.*

Arteries to the Neck, Head, and Brain

Blood is supplied to parts within the neck, head, and brain through branches of the subclavian and common carotid arteries. (See fig. 16.25.) The main divisions of the subclavian artery to these regions are the vertebral and thyrocervical arteries. The common carotid communicates to parts of the head and neck by means of the internal and external carotid arteries.

The **vertebral arteries** pass upward through the foramina of the transverse processes of the cervical vertebrae and enter the skull by way of the foramen magnum. Along their paths, these vessels supply blood to the vertebrae and to the ligaments and muscles associated with them.

Within the cranial cavity, the vertebral arteries unite to form a single *basilar artery*. This vessel passes along the ventral brain stem and gives rise to branches leading to the pons, midbrain, and cerebellum.

The **thyrocervical arteries** are short vessels that give off branches to the thyroid glands, parathyroid glands, larynx, trachea, esophagus, and pharynx, as well as to various muscles in the neck, shoulder, and back.

The left and right **common carotid arteries** divide to form the internal and external carotid arteries.

Fig. 16.25 The main arteries of the head, neck, and brain. (a. stands for *artery*.)

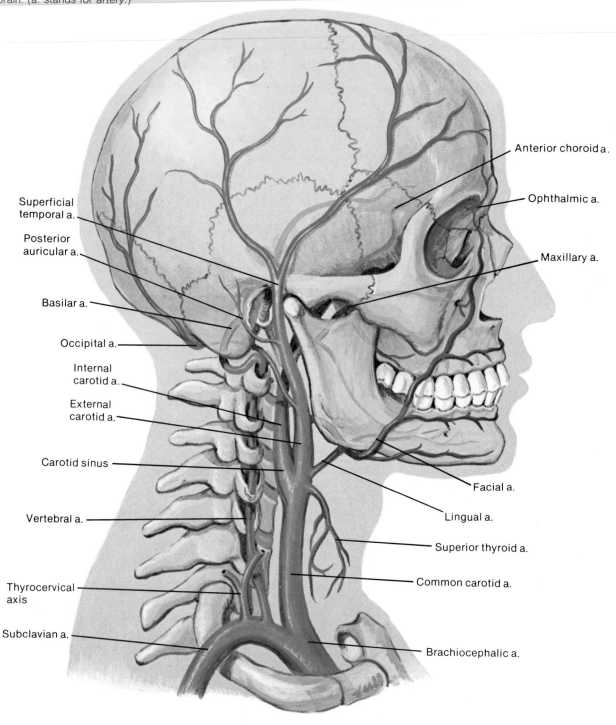

Superficial temporal a.

Posterior auricular a.

Basilar a.

Occipital a.

Internal carotid a.

External carotid a.

Carotid sinus

Vertebral a.

Thyrocervical axis

Subclavian a.

Anterior choroid a.

Ophthalmic a.

Maxillary a.

Facial a.

Lingual a.

Superior thyroid a.

Common carotid a.

Brachiocephalic a.

The *external carotid artery* courses upward on the side of the head, giving off branches to various structures in the neck, face, jaw, scalp, and base of the skull.

The *internal carotid artery* follows a deep course upward along the pharynx to the base of the skull. Entering the cranial cavity, it provides the major blood supply to the brain.

Near the base of each internal carotid artery is an enlargement called a **carotid sinus.** Like the aortic sinuses, these structures contain pressoreceptors that function in the reflex control of blood pressure.

The major branches of the external and internal carotid arteries and the general regions and organs they supply with blood are listed in chart 16.3.

Chart 16.2 Aorta and its principal branches

Portion of Aorta	Major Branch	General Regions or Organs Supplied
Ascending aorta	Right and left coronary arteries	Heart
Arch of aorta	Brachiocephalic artery	Right arm, right side of head
	Left common carotid artery	Left side of head
	Left subclavian artery	Left arm
Descending aorta		
Thoracic aorta	Bronchial artery	Bronchi
	Pericardial artery	Pericardium
	Esophageal artery	Esophagus
	Mediastinal artery	Mediastinum
	Intercostal artery	Thoracic wall
Abdominal aorta	Celiac artery	Organs of upper digestive system
	Phrenic artery	Diaphragm
	Superior mesenteric artery	Portions of small and large intestines
	Suprarenal artery	Adrenal gland
	Renal artery	Kidney
	Gonadal artery	Ovary or testis
	Inferior mesenteric artery	Lower portions of large intestine
	Lumbar artery	Posterior abdominal wall
	Middle sacral artery	Sacrum and coccyx
	Common iliac artery	Lower abdominal wall, pelvic organs, and leg

From Hole, John W. Jr., *Human Anatomy and Physiology 3d ed.* © 1978, 1981, 1984 Wm. C. Brown Publishers, Dubuque, Iowa. All Rights Reserved. Reprinted by permission.

Arteries to the Thoracic and Abdominal Walls

Blood reaches the thoracic wall through several vessels. (See fig. 16.26.) They include:

1. **Internal thoracic artery,** a branch of the subclavian artery, which gives off two *anterior intercostal arteries.* They, in turn, supply the intercostal muscles and the mammary glands.

2. **Posterior intercostal arteries,** which arise from the thoracic aorta and enter the intercostal spaces. They supply the intercostal muscles, the vertebrae, the spinal cord, and various deep muscles of the back.

The blood supply to the anterior abdominal wall is provided primarily by branches of the *internal thoracic* and *external iliac arteries.* Structures in the

Chart 16.3 Major branches of the external and internal carotid arteries

Artery	Major Branch	General Region or Organs Supplied
External carotid artery	Superior thyroid artery	Larynx and thyroid gland
	Lingual artery	Tongue and salivary glands
	Facial artery	Pharynx, chin, lips, and nose
	Occipital artery	Posterior scalp, meninges, and neck muscles
	Posterior auricular artery	Ear and lateral scalp
	Maxillary artery	Teeth, jaw, cheek, and eyelids
	Superficial temporal artery	Parotid salivary gland and surface of face and scalp
Internal carotid artery	Ophthalmic artery	Eye and eye muscles
	Anterior choroid artery	Choroid plexus and brain

posterior and lateral abdominal wall are supplied by paired vessels originating from the abdominal aorta, including the *phrenic* and *lumbar arteries.*

Arteries to the Shoulder and Arm

The subclavian artery, after giving off branches to the neck, continues into the upper arm. (See fig. 16.27.) It passes between the clavicle and the first rib and becomes the axillary artery.

The **axillary artery** supplies branches to various muscles and other structures in the axilla and the chest wall and becomes the brachial artery.

The **brachial artery,** which courses along the humerus to the elbow, gives rise to a *deep brachial artery.* Within the elbow, the brachial artery divides into an ulnar and a radial artery.

The **ulnar artery** leads downward on the ulnar side to the forearm and wrist. Some of its branches supply the elbow joint, while others supply blood to muscles in the lower arm.

The **radial artery** travels along the radial side of the forearm to the wrist, supplying muscles of the forearm.

At the wrist, branches of the ulnar and radial arteries join to form an interconnecting network of vessels. Arteries arising from this network supply blood to structures in the wrist, hand, and fingers.

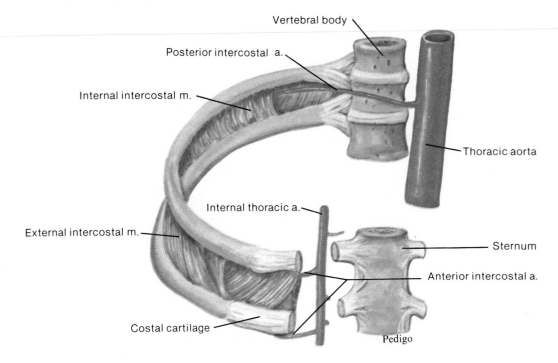

Vertebral body

Posterior intercostal a.

Internal intercostal m.

Thoracic aorta

Internal thoracic a.

External intercostal m.

Sternum

Anterior intercostal a.

Costal cartilage

Pedigo

Arteries to the Pelvis and Leg

The abdominal aorta divides to form the **common iliac arteries** at the level of the pelvic brim, and these vessels provide blood to the pelvic organs, gluteal region, and legs. (See fig. 16.27.) Each common iliac artery divides into an internal and an external branch.

The **internal iliac artery** gives off numerous branches to various pelvic muscles and visceral structures, as well as to the gluteal muscles and the external genitalia.

The **external iliac artery** provides the main blood supply to the legs. It passes downward along the brim of the pelvis and gives off branches that supply muscles and skin in the lower abdominal wall.

Midway between the pubic symphysis and the anterior superior iliac spine of the ilium, the external iliac artery passes beneath the inguinal ligament and becomes the femoral artery.

The **femoral artery,** which passes fairly close to the anterior surface of the upper thigh, gives off many branches to the muscles and superficial tissues of the thigh. These branches also supply the skin of the groin and lower abdominal wall.

As the femoral artery reaches the proximal border of the space behind the knee, it becomes the **popliteal artery.** Branches of this artery supply blood to the knee joint and to certain muscles in the thigh and calf. The popliteal artery divides into the anterior and posterior tibial arteries.

The **anterior tibial artery** passes downward between the tibia and the fibula, giving off branches to the skin and muscles in the anterior and lateral regions of the lower leg. This vessel continues into the foot as the *dorsalis pedis artery,* which supplies blood to the foot and toes.

The **posterior tibial artery,** the larger of the two popliteal branches, descends beneath the calf muscles, giving off branches to skin, muscles, and other tissues of the lower leg along the way.

---

1. *Name the branches of the thoracic and abdominal aorta.*
2. *What vessels supply blood to the head? The arm? The abdominal wall? The leg?*

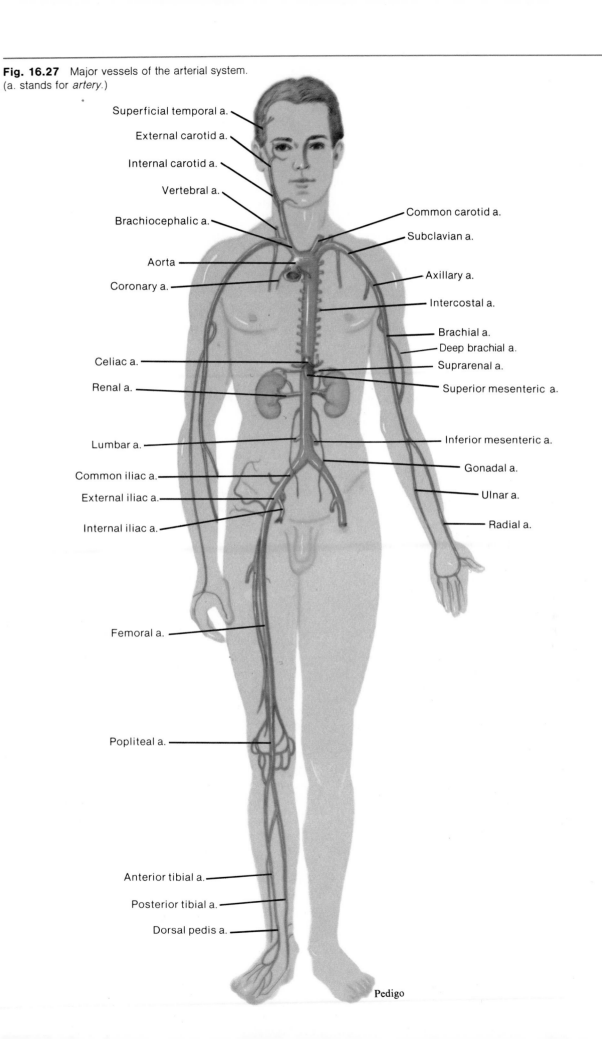

Fig. 16.27 Major vessels of the arterial system.
(a. stands for *artery*.)

Superficial temporal a.

External carotid a.

Internal carotid a.

Vertebral a.

Brachiocephalic a.

Aorta

Coronary a.

Celiac a.

Renal a.

Lumbar a.

Common iliac a.

External iliac a.

Internal iliac a.

Femoral a.

Popliteal a.

Anterior tibial a.

Posterior tibial a.

Dorsal pedis a.

Common carotid a.

Subclavian a.

Axillary a.

Intercostal a.

Brachial a.

Deep brachial a.

Suprarenal a.

Superior mesenteric a.

Inferior mesenteric a.

Gonadal a.

Ulnar a.

Radial a.

Pedigo

Fig. 16.28 The main veins of the brain, head, and neck. (v. stands for *vein.*)

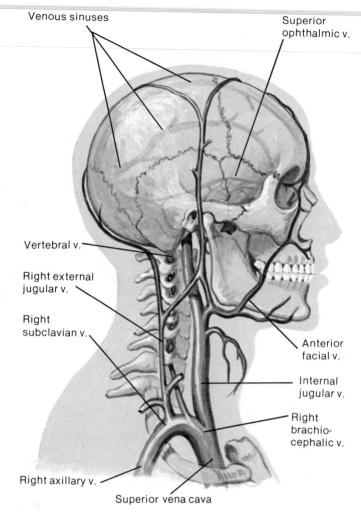

Venous sinuses

Superior ophthalmic v.

Vertebral v.

Right external jugular v.

Right subclavian v.

Anterior facial v.

Internal jugular v.

Right brachio-cephalic v.

Right axillary v.

Superior vena cava

The Venous System

Venous circulation is responsible for returning blood to the heart after exchanges of gases, nutrients, and wastes have been made between the blood and the body cells.

Characteristics of Venous Pathways

The vessels of the venous system begin with the merging of capillaries into venules, venules into small veins, and small veins into larger ones. Unlike arterial pathways, however, those of the venous system are difficult to follow. This is because the vessels are commonly interconnected in irregular networks, so that many unnamed tributaries may join to form a relatively large vein.

On the other hand, larger veins typically parallel the courses taken by named arteries, and these veins often have the same names as their companions in the arterial system. Thus, with some exceptions, the name of a major artery also provides the name of the vein next to it. For example, the renal vein parallels the renal artery, the common iliac vein accompanies the common iliac artery and so forth.

The veins that carry blood from the lungs and myocardium back to the heart have already been described. The veins from all other parts of the body converge into two major pathways that lead to the right atrium. They are the *superior* and *inferior venae cavae.*

Veins from the Brain, Head, and Neck

Blood from the face, scalp, and superficial regions of the neck is drained by the **external jugular veins.** These vessels descend on either side of the neck and empty into the *subclavian veins.* (See fig. 16.28.)

The **internal jugular veins,** which are somewhat larger than the externals, arise from numerous veins and venous sinuses of the brain and from deep veins

Fig. 16.29 Veins that drain the thoracic wall. (v. stands for *vein.*)

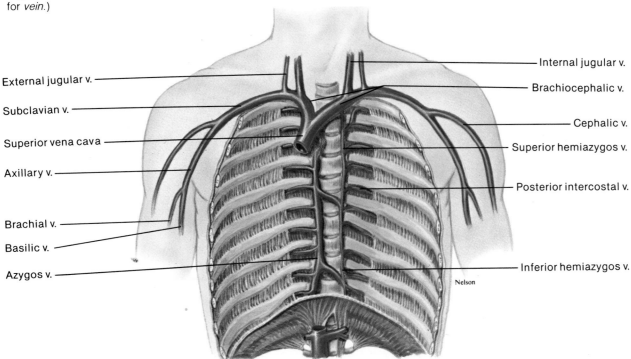

External jugular v.

Subclavian v.

Superior vena cava

Axillary v.

Brachial v.

Basilic v.

Azygos v.

Internal jugular v.

Brachiocephalic v.

Cephalic v.

Superior hemiazygos v.

Posterior intercostal v.

Inferior hemiazygos v.

Nelson

in various parts of the face and neck. They pass downward through the neck and join the subclavian veins. These unions of the internal jugular and subclavian veins form large **brachiocephalic veins** on each side. These vessels then merge and give rise to the **superior vena cava,** which enters the right atrium.

Veins from the Abdominal and Thoracic Walls

The abdominal and thoracic walls are drained mainly by tributaries of the brachiocephalic and azygos veins. For example, the *brachiocephalic vein* receives blood from the *internal thoracic vein,* which generally drains the tissues supplied by the internal thoracic artery. Some *intercostal veins* also empty into the brachiocephalic vein. (See fig. 16.29.)

The **azygos vein** originates in the dorsal abdominal wall and ascends through the mediastinum on the right side of the vertebral column to join the superior vena cava. It drains most of the muscular tissue in the abdominal and thoracic walls.

Tributaries of the azygos vein include the *posterior intercostal veins,* which drain the intercostal spaces on the right side, and the *superior* and *inferior hemiazygos veins,* which connect to the posterior intercostal veins on the left. Right and left *ascending lumbar veins,* whose tributaries include vessels from the lumbar and sacral regions, also connect to the azygos system.

Veins from the Abdominal Viscera

Although veins usually carry blood directly to the atria of the heart, those that drain the abdominal viscera are exceptions. (See fig. 16.30.) They originate in the capillary networks of the stomach, intestines, pancreas, and spleen and carry blood from these organs through a **portal vein** to the liver. This unique venous pathway is called the **hepatic portal system.**

The tributaries of the portal vein include the following vessels.

1. Right and left *gastric veins* from the stomach.

2. *Superior mesenteric vein* from the small intestine, ascending colon, and transverse colon.

3. *Splenic vein* from a convergence of several veins draining the spleen, pancreas, and a portion of the stomach. Its largest tributary, the *inferior mesenteric vein,* brings blood upward from the descending colon, sigmoid colon, and rectum.

About 80% of the blood flowing to the liver in the hepatic portal system comes from capillaries in the stomach and intestines.

After passing through the hepatic sinusoids of the liver, blood in the hepatic portal system is carried through a series of merging vessels into **hepatic veins.** These veins empty into the *inferior vena cava;* thus the blood is returned to the general circulation.

Fig. 16.30 Veins that drain the abdominal viscera.

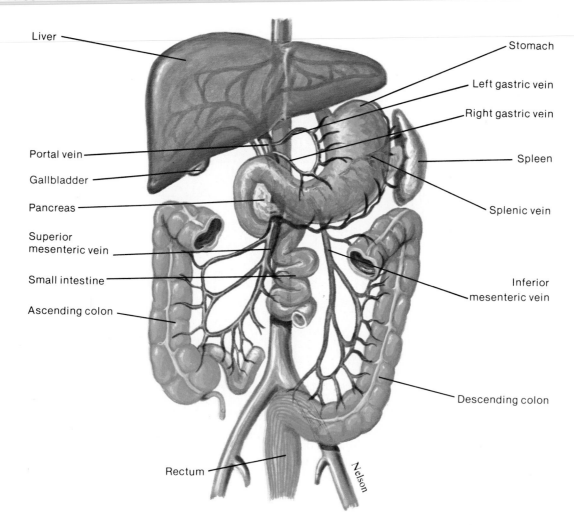

Liver

Portal vein

Gallbladder

Pancreas

Superior
mesenteric vein

Small intestine

Ascending colon

Rectum

Stomach

Left gastric vein

Right gastric vein

Spleen

Splenic vein

Inferior
mesenteric vein

Descending colon

Nelson

Veins from the Arm and Shoulder

The arm is drained by a set of deep veins and a set of superficial ones. The deep veins generally parallel the arteries in each region and are given similar names, such as *radial vein, ulnar vein, brachial vein,* and *axillary vein.* The superficial veins are interconnected in complex networks just beneath the skin. They also communicate with the deep vessels of the arm, providing many alternate pathways through which blood can leave the tissues. (See fig. 16.31.) The main vessels of the superficial network are the basilic and cephalic veins.

The **basilic vein** ascends from the forearm to the middle of the upper arm. There it penetrates the tissues deeply and joins the *brachial vein.* As the basilic and brachial veins merge, they form the *axillary vein.*

The **cephalic vein** courses upward from the hand to the shoulder. In the shoulder it pierces the tissues and empties into the axillary vein. Beyond the axilla, the axillary vein becomes the *subclavian vein.*

In the bend of the elbow, a *median cubital vein* ascends from the cephalic vein on the lateral side of the arm to the basilic vein on the medial side. This vein is often used as a site for *venipuncture* when it is necessary to remove a sample of blood for examination, or to add fluids to the blood.

Veins from the Leg and Pelvis

As in the arm, veins that drain blood from the leg can be divided into deep and superficial groups. (See fig. 16.31.)

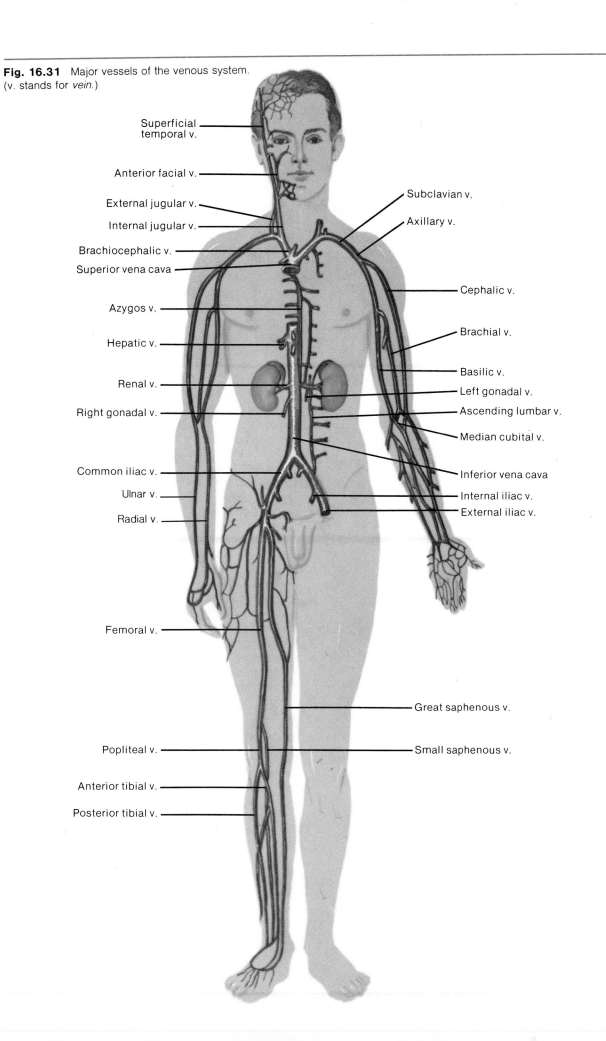

Fig. 16.31 Major vessels of the venous system. (v. stands for *vein.*)

Superficial temporal v.

Anterior facial v.

External jugular v.

Internal jugular v.

Brachiocephalic v.

Superior vena cava

Azygos v.

Hepatic v.

Renal v.

Right gonadal v.

Common iliac v.

Ulnar v.

Radial v.

Femoral v.

Popliteal v.

Anterior tibial v.

Posterior tibial v.

Subclavian v.

Axillary v.

Cephalic v.

Brachial v.

Basilic v.

Left gonadal v.

Ascending lumbar v.

Median cubital v.

Inferior vena cava

Internal iliac v.

External iliac v.

Great saphenous v.

Small saphenous v.

The deep veins of the lower leg, such as the *anterior* and *posterior tibial veins,* have names that correspond with the arteries they accompany. At the level of the knee, these vessels form a single trunk, the **popliteal vein.** This vein continues upward through the thigh as the **femoral vein,** which in turn becomes the **external iliac vein** just behind the inguinal ligament.

The superficial veins of the foot and leg interconnect to form a complex network beneath the skin. These vessels drain into two major trunks—the small and great saphenous veins.

The **small saphenous vein** ascends along the back of the calf, enters the popliteal fossa, and joins the *popliteal vein.*

The **great saphenous vein,** which is the longest vein in the body, ascends in front of the medial malleolus and extends upward along the medial side of the leg. In the thigh, it penetrates deeply and joins the femoral vein. Near its termination, the great saphenous vein receives tributaries from a number of vessels that drain the upper thigh, the groin, and the lower abdominal wall.

In addition to communicating freely with each other, the *saphenous veins* communicate extensively with the deep veins of the leg. As a result, there are many pathways by which blood can be returned to the heart from the lower extremities.

Whenever a person stands, the *saphenous veins* are subjected to increased blood pressure because of gravitational forces. When such pressure is prolonged, these veins are especially prone to developing abnormal dilations characteristic of varicose veins.

In the pelvic region, blood is carried away from organs of the reproductive, urinary, and digestive systems by vessels leading to the **internal iliac vein.**

The internal iliac veins unite with the right and left *external iliac veins* to form the **common iliac veins.** These vessels in turn merge to produce the *inferior vena cava.*

---

1. *Name the veins that return blood to the right atrium.*
2. *What major veins drain blood from the head? The abdominal viscera? The arm? The leg?*

Clinical Terms Related to the Cardiovascular System

aneurysm (an′u-rizm)—a saclike swelling in the wall of a blood vessel, usually an artery.

angiocardiography (an″je-o-kar″de-og′rah-fe)—injection of radiopaque solution into the vascular system for X-ray examination of the heart and pulmonary circuit.

angiospasm (an′je-o-spazm″)—a muscular spasm in the wall of a blood vessel.

arteriography (ar″te-re-og′rah-fe)—injection of radiopaque solution into the vascular system for X-ray examination of arteries.

asystole (a-sis′to-le)—condition in which the myocardium fails to contract.

cardiac tamponade (kar′de-ak tam″pon-ād)—compression of the heart by an accumulation of fluid within the pericardial cavity.

congestive heart failure (kon-jes′tiv hart fāl′yer)—condition in which the heart is unable to pump an adequate amount of blood to the body cells.

cor pulmonale (kor pul-mo-na′le)—a heart-lung disorder characterized by pulmonary hypertension and hypertrophy of the right ventricle.

embolectomy (em″bo-lek′to-me)—removal of an embolus through an incision in a blood vessel.

endarterectomy (en″dar-ter-ek′to-me)—removal of the inner wall of an artery to reduce an arterial occlusion.

palpitation (pal″pĭ-ta′shun)—an awareness of a heartbeat that is unusually rapid, strong, or irregular.

pericardiectomy (per″ĭ-kar′de-ek′to-me)—an excision of the pericardium.

phlebitis (flĕ-bi′tis)—inflammation of a vein, usually in the legs.

phlebosclerosis (fleb″o-sklĕ-ro′sis)—abnormal thickening or hardening of the walls of veins.

phlebotomy (flĕ-bot′o-me)—incision of a vein for the purpose of withdrawing blood.

sinus rhythm (si′nus rithm)—normal cardiac rhythm regulated by the S-A node.

thrombophlebitis (throm″bo-flĕ-bi′tis)—formation of a blood clot in a vein in response to inflammation of the venous wall.

valvotomy (val-vot′o-me)—an incision of a valve.

venography (ve-nog′rah-fe)—injection of radiopaque solution into the vascular system for X-ray examination of veins.

Chapter Summary

Introduction

The cardiovascular system is vital for providing oxygen and nutrients to tissues and for removing wastes.

Structure of the Heart

1. Size and location of the heart
   a. The heart is about 14 cm long and 9 cm wide.
   b. It is located within the mediastinum and rests on the diaphragm.
2. Coverings of the heart
   a. The heart is enclosed in a pericardium.
   b. The pericardial cavity is a potential space.
3. Wall of the heart
   The wall of the heart is composed of an epicardium, myocardium, and endocardium.
4. Heart chambers and valves
   a. The heart is divided into two atria and two ventricles.
   b. Right chambers and valves
      (1) The right atrium receives blood from the venae cavae and coronary sinus.
      (2) The right chambers are separated by the tricuspid valve.
      (3) The base of the pulmonary trunk is guarded by a pulmonary semilunar valve.
   c. Left chambers and valves
      (1) The left atrium receives blood from the pulmonary veins.
      (2) The left chambers are separated by the bicuspid valve.
      (3) The base of the aorta is guarded by an aorta semilunar valve.
5. Skeleton of the heart
   The skeleton of the heart consists of fibrous rings that enclose bases of the pulmonary artery and aorta.
6. Path of blood through the heart
   a. Blood that is relatively low in oxygen and high in carbon dioxide concentration enters the right side of the heart and is pumped into the pulmonary circulation.
   b. After blood is oxygenated in the lungs and some of its carbon dioxide is removed, it returns to the left side of the heart.
7. Blood supply to the heart
   a. Blood is supplied through the coronary arteries.
   b. It is returned to the right atrium through the cardiac veins and coronary sinus.

Actions of the Heart

1. The cardiac cycle
   a. Atria contract while ventricles relax; ventricles contract while atria relax.
   b. This series of events constitutes a cardiac cycle.
2. Heart sounds
   Heart sounds are due to vibrations produced as a result of blood and valve movements.
3. Cardiac muscle fibers
   a. Fibers are interconnected to form a functional syncytium.
   b. If any part of the syncytium is stimulated, the whole structure contracts as a unit.
4. Cardiac conduction system
   a. This system functions to initiate and conduct impulses through the myocardium.
   b. Impulses from the S-A node pass slowly to the A-V node; impulses travel rapidly along the A-V bundle and Purkinje fibers.
5. The electrocardiogram (ECG)
   a. The ECG is a recording of the electrical changes occurring in the myocardium during a cardiac cycle.
   b. The pattern contains several waves.
      (1) The P wave represents atrial depolarization.
      (2) The QRS complex represents ventricular depolarization.
      (3) The T wave represents ventricular repolarization.
6. Regulation of the cardiac cycle
   a. Heartbeat is affected by physical exercise, body temperature, and the concentration of various ions.
   b. S-A and A-V nodes are innervated by branches of sympathetic and parasympathetic nerve fibers.
   c. Autonomic impulses are regulated by the cardiac center in the medulla oblongata.

Blood Vessels

The blood vessels form a closed circuit of tubes.

1. Arteries and arterioles
   a. Arteries are adapted to carry high pressure blood away from the heart.
   b. Walls of these vessels consist of endothelium, smooth muscle, and connective tissue.
   c. The smooth muscles are innervated by autonomic fibers.

2. Capillaries
   a. The capillary wall consists of a single layer of cells.
   b. Blood flow into the capillary is controlled by the precapillary sphincter.
   c. Exchanges in capillaries
      (1) Gases, nutrients, and metabolic by-products are exchanged between the capillary blood and tissue fluid.
      (2) Diffusion provides the most important means of transport.
      (3) Filtration causes a net outward movement of fluid at the arterial end of a capillary.
      (4) Osmosis causes a net inward movement of fluid at the venule end of the capillary.

3. Venules and veins
   a. Venules continue from capillaries and merge to form veins.
   b. Veins carry blood to the heart.
   c. Venous walls are similar to arterial walls, but are thinner and contain less muscle and elastic tissue.

Blood Pressure

Blood pressure is the force exerted by the blood against the inside of the blood vessels.

1. Arterial blood pressure
   a. Arterial blood pressure is created primarily by heart action.
   b. Systolic pressure occurs when the ventricle contracts; diastolic pressure occurs when the ventricle relaxes.

2. Factors that influence arterial blood pressure
   Arterial pressure increases as the cardiac output, blood volume, peripheral resistance, or blood viscosity increases.

3. Control of blood pressure
   a. Blood pressure is controlled in part by the mechanisms that regulate cardiac output and peripheral resistance.
   b. The more blood that enters the heart, the stronger the ventricular contraction, the greater the stroke volume, and the greater the cardiac output.
   c. Regulation of heart rate and peripheral resistance involves the cardiac and vasomotor center of the medulla oblongata.

4. Venous blood flow
   a. Venous blood flow depends on skeletal muscle contraction and breathing movements.
   b. Many veins contain flaplike valves that prevent blood from backing up.
   c. Venoconstriction can increase venous pressure and blood flow.

Paths of Circulation

1. The pulmonary circuit
   The pulmonary circuit is composed of vessels that carry blood from the right ventricle to the lungs and back to the left atrium.

2. The systemic circuit
   a. The systemic circuit is composed of vessels that lead from the heart to the body cells and back to the heart.
   b. It includes the aorta and its branches.

The Arterial System

1. Principal branches of the aorta
   a. The aorta is the largest artery.
   b. Major branches include the coronary, brachiocephalic, left common carotid, and left subclavian arteries.
   c. Branches of the descending aorta include thoracic and abdominal groups.
   d. The descending aorta terminates by dividing into right and left common iliac arteries.

2. Arteries to the neck, head, and brain
   These include branches of the subclavian and common carotid arteries.

3. Arteries to the thoracic and abdominal walls
   a. The thoracic wall is supplied by branches of the subclavian artery and thoracic aorta.
   b. The abdominal wall is supplied by branches of the abdominal aorta and other arteries.

4. Arteries to the shoulder and arm
   a. The subclavian artery passes into the upper arm and in various regions is called the axillary and the brachial artery.
   b. Branches of the brachial artery include the ulnar and radial arteries.
5. Arteries to the pelvis and leg
   The common iliac artery supplies the pelvic organs, gluteal region, and leg.

The Venous System

1. Characteristics of venous pathways
   a. Veins are responsible for returning blood to the heart.
   b. Larger veins usually parallel the courses of major arteries.
2. Veins from the brain, head, and neck
   a. These regions are drained by the jugular veins.
   b. Jugular veins unite with subclavian veins to form the brachiocephalic veins.
3. Veins from the abdominal and thoracic walls
   These are drained by tributaries of the brachiocephalic and azygos veins.
4. Veins from the abdominal viscera
   a. Blood from the abdominal viscera generally enters the hepatic portal system and is carried to the liver.
   b. From the liver, the blood is carried by hepatic veins to the inferior vena cava.
5. Veins from the arm and shoulder
   a. The arm is drained by sets of superficial and deep veins.
   b. Deep veins parallel arteries with similar names.
6. Veins from the leg and pelvis
   a. These are drained by sets of deep and superficial veins.
   b. Deep veins include the tibial veins, and the superficial veins include the saphenous veins.

Application of Knowledge

1. Based upon your understanding of the way capillary blood flow is regulated, do you think it is wiser to rest or to exercise following a heavy meal? Give a reason for your answer.
2. If a patient develops a blood clot in the femoral vein of the left leg and a portion of the clot breaks loose, where is the blood flow likely to carry the embolus? What symptoms is this condition likely to produce?
3. In the case of a patient with a heart weakened by damage to the myocardium, would it be better to keep the legs raised or lowered below the level of the heart? Why?

Review Activities

1. Describe the general structure, function, and location of the heart.
2. Distinguish between the visceral pericardium and the parietal pericardium.
3. Compare the layers of the cardiac wall.
4. Identify and describe the location of the chambers and the valves of the heart.
5. Describe the skeleton of the heart and explain its function.
6. Trace the path of blood through the heart.
7. Trace the path of blood through the coronary circulation.
8. Describe a cardiac cycle.
9. Explain the origin of heart sounds.
10. Distinguish between the S-A node and A-V node.
11. Explain how the cardiac conduction system functions in the control of the cardiac cycle.
12. Describe a normal ECG pattern and explain the significance of its various waves.
13. Discuss how the nervous system functions in the regulation of the cardiac cycle.
14. Distinguish between an artery and an arteriole.
15. Explain how vasoconstriction and vasodilation are controlled.
16. Describe the structure and function of a capillary.
17. Explain how the blood flow through a capillary is controlled.
18. Explain how diffusion functions in the exchange of substances between plasma and tissue fluid.
19. Explain why water and dissolved substances leave the arteriole end of a capillary and enter the venule end.
20. Distinguish between a venule and a vein.
21. Explain how veins function as blood reservoirs.
22. Distinguish between systolic and diastolic blood pressures.
23. Name several factors that influence blood pressure and explain how each produces its effect.

24. Describe how blood pressure is controlled.

25. List the major factors that promote the flow of venous blood.

26. Distinguish between the pulmonary and systemic circuits of the cardiovascular system.

27. Trace the path of blood through the pulmonary circuit.

28. Describe the aorta and name its principal branches.

29. On a diagram, locate and identify the major arteries that supply the abdominal visceral organs.

30. On a diagram, locate and identify the major arteries that supply parts in the head, neck, and brain.

31. On a diagram, locate and identify the major arteries that supply parts in the thoracic and abdominal walls.

32. On a diagram, locate and identify the major arteries that supply parts in the shoulder and arm.

33. On a diagram, locate and identify the major arteries that supply parts in the pelvis and leg.

34. Describe the relationship between the major venous pathways and the major arterial pathways.

35. On a diagram, locate and identify the major veins that drain parts in the brain, head, and neck.

36. On a diagram, locate and identify the major veins that drain parts in the abdominal and thoracic walls.

37. On a diagram, locate and identify the major veins that drain parts of the abdominal viscera.

38. On a diagram, locate and identify the major veins that drain parts in the arm and shoulder.

39. On a diagram, locate and identify the major veins that drain parts of the leg and pelvis.

The Lymphatic System

When substances are exchanged between the blood and tissue fluid, more fluid leaves the blood capillaries than returns to them. If the fluid remaining in the interstitial spaces were allowed to accumulate, the hydrostatic pressure in tissues would increase. The *lymphatic system* helps to prevent such an imbalance by providing pathways through which tissue fluid can be transported as lymph from the interstitial spaces to veins, where it becomes part of the blood.

The lymphatic system also helps to defend the tissues against infections by filtering particles from the lymph and by supporting the activities of lymphocytes that furnish immunity against specific disease-causing agents.

17

Chapter Objectives

After you have studied this chapter, you should be able to

1. Describe the general functions of the lymphatic system.

2. Describe the location of the major lymphatic pathways.

3. Describe how tissue fluid and lymph are formed and explain the function of lymph.

4. Explain how lymphatic circulation is maintained.

5. Describe a lymph node and its major functions.

6. Discuss the functions of the thymus and spleen.

7. Distinguish between specific and nonspecific body defenses and provide examples of each defense.

8. Explain how two major types of lymphocytes are formed and how they function in immune mechanisms.

9. Name the major types of immunoglobulins and discuss their origins and actions.

10. Distinguish between primary and secondary immune responses.

11. Distinguish between active and passive immunity.

12. Explain how allergic reactions and tissue rejection reactions are related to immune mechanisms.

13. Complete the review activities at the end of this chapter. Note that the items are worded in the form of specific learning objectives. You may want to refer to them before reading the chapter.

allergen   (al'-er-jen)

antibody   (an'ti-bod''e)

antigen   (an'ti-jen)

clone   (klōn)

immunity   (i-mu'ni-te)

immunoglobulin   (im''u-no-glob'u-lin)

lymph   (limf)

lymphatic pathway   (lim-fat'ik path'wa)

lymph node   (limf nōd)

lymphocyte   (lim'fo-sīt)

pathogen   (path'o-jen)

spleen   (splēn)

thymus   (thi'mus)

gen-, to be produced: aller*gen*—substance that stimulates an allergic response.

humor-, fluid: *humor*al immunity—immunity resulting from soluble antibodies in body fluids.

immun-, free: *immun*ity—resistance to (freedom from) a specific disease.

inflamm-, setting on fire: *inflamm*ation—a condition characterized by localized redness, heat, swelling, and pain in the tissues.

nod-, knot: *nod*ule—a small mass of lymphocytes surrounded by connective tissue.

patho-, disease: *patho*gen—a disease-causing agent.

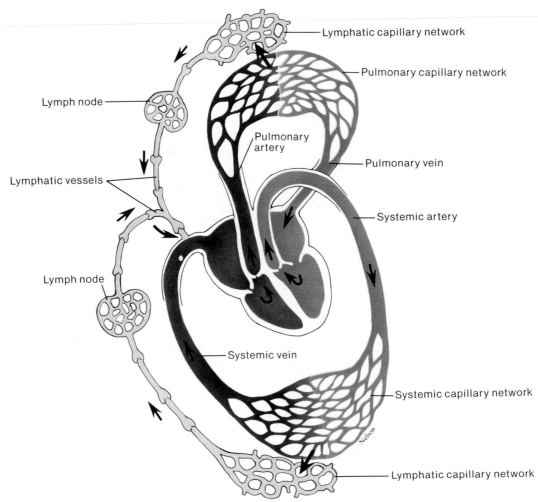

- Lymphatic capillary network
- Pulmonary capillary network
- Lymph node
- Pulmonary artery
- Pulmonary vein
- Lymphatic vessels
- Systemic artery
- Lymph node
- Systemic vein
- Systemic capillary network
- Lymphatic capillary network

The lymphatic system is closely associated with the *cardiovascular system* since it includes a network of vessels that assist in circulating body fluids. These vessels transport excess fluid away from the interstitial spaces and return it to the bloodstream. The organs of the lymphatic system also help to defend the body against invasion by disease-causing agents. (See fig. 17.1.)

Lymphatic Pathways

The lymphatic pathways begin as lymphatic capillaries. These tiny tubes merge to form larger lymphatic vessels, and they in turn lead to collecting ducts that unite with veins in the thorax.

Lymphatic Capillaries

Lymphatic capillaries are microscopic, closed-ended tubes. They extend into the interstitial spaces of most tissues, forming complex networks that parallel the networks of blood capillaries. (See fig. 17.2) The walls of lymphatic capillaries, like those of blood capillaries, consist of a single layer of squamous epithelial cells. This thin wall makes it possible for tissue fluid from the interstitial space to enter the lymphatic capillary. Once the fluid is inside the capillary, it is called **lymph.**

Lymphatic Vessels

Lymphatic vessels, which are formed by the merging of lymphatic capillaries, have walls similar to those of veins. Also like veins, the lymphatic vessels have flap-like *valves* that help to prevent the backflow of lymph. Figure 17.3 shows one of these valves.

Fig. 17.2 Lymph capillaries are microscopic closed-ended tubes that begin in the interstitial spaces of most tissues.

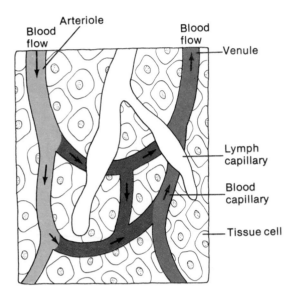

Fig. 17.3 A micrograph of the flaplike valve within a lymphatic vessel. What is the function of this valve?

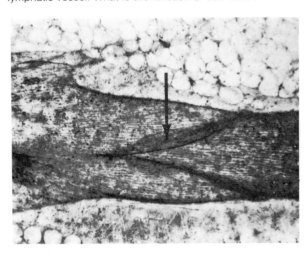

The skin is richly supplied with lymphatic capillaries. Consequently, if the skin is broken or something is injected into the skin, such as venom from a stinging insect, foreign particles are likely to enter the lymphatic system relatively rapidly.

Typically, lymphatic vessels lead to specialized organs called **lymph nodes,** and after leaving these structures, the vessels merge to form still larger lymphatic trunks.

1. What is the general function of the lymphatic system?

2. Distinguish between the thoracic duct and the right lymphatic duct.

Lymphatic Trunks and Collecting Ducts

The **lymphatic trunks** drain lymph from relatively large regions of the body, and they are named for the regions they serve. These lymphatic trunks then join one of two **collecting ducts**—the thoracic duct or the right lymphatic duct. The major lymphatic trunks and collecting ducts are shown and named in figure 17.4.

The **thoracic duct** is the larger and longer of the collecting ducts. It receives lymph from the lower body regions, left arm, and left side of the head and neck, and empties into the left subclavian vein near the junction of the left jugular vein.

The **right lymphatic duct** receives lymph from the right side of the head and neck, right arm, and right thorax. It empties into the right subclavian vein near the junction of the right jugular vein.

After leaving the collecting ducts, lymph enters the venous system and becomes part of the plasma just before the blood returns to the right atrium.

Chart 17.1 summarizes a typical lymphatic pathway.

Tissue Fluid and Lymph

Lymph is essentially tissue fluid that has entered a lymphatic capillary. The formation of lymph is, then, closely associated with the formation of tissue fluid.

Tissue Fluid Formation

As is explained in chapter 15, *tissue fluid* originates from blood plasma. This fluid is composed of water and dissolved substances that leave the blood capillaries as a result of diffusion and filtration.

Fig. 17.4 Lymphatic vessels merge into larger lymphatic trunks, which in turn, drain into collecting ducts.

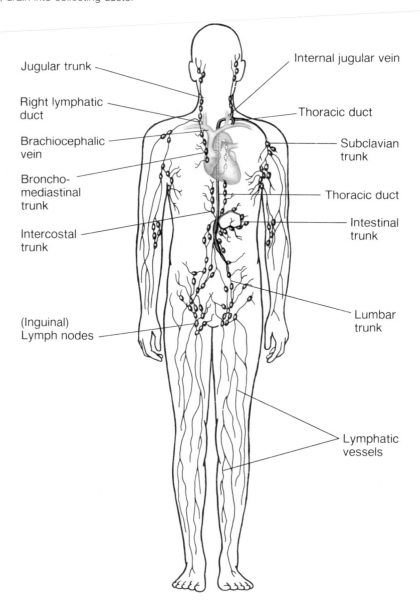

Jugular trunk

Right lymphatic duct

Brachiocephalic vein

Broncho-mediastinal trunk

Intercostal trunk

(Inguinal) Lymph nodes

Internal jugular vein

Thoracic duct

Subclavian trunk

Thoracic duct

Intestinal trunk

Lumbar trunk

Lymphatic vessels

Chart 17.1 Typical lymphatic pathway

Tissue fluid
  leaves the interstitial space and becomes

Lymph
  as it enters the

Lymphatic capillary
  that merges with other capillaries to form a

Lymphatic vessel
  that enters the

Lymph node
  where lymph is filtered and leaves via a

Lymphatic vessel
  that merges with other vessels to form the

Lymphatic trunk
  that merges with other trunks and joins the

Collecting duct
  that empties into the

Subclavian vein
  where lymph is added to the blood

Fig. 17.5 As tissue fluid enters a lymphatic capillary, it becomes lymph. What factors are responsible for this movement of tissue fluid?

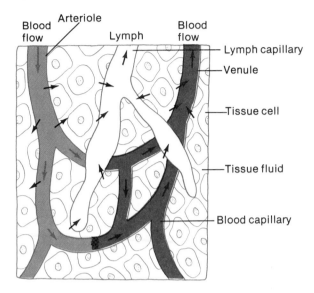

Blood flow
Arteriole
Lymph
Blood flow

- Lymph capillary
- Venule
- Tissue cell
- Tissue fluid
- Blood capillary

1. *Explain the relationship between tissue fluid and lymph.*
2. *How does the presence of protein in tissue fluid affect the formation of lymph?*
3. *What are the major functions of lymph?*

Although tissue fluid contains various nutrients and gases found in plasma, it generally lacks proteins of large molecular size. Some proteins with smaller molecules do leak out of the blood capillaries and enter the interstitial space. Usually, these proteins are not reabsorbed when water and other dissolved substances move back into the venule ends of these capillaries by diffusion and osmosis. As a result, the protein concentration of the tissue fluid tends to rise, causing the *osmotic pressure* of the fluid to rise also.

Lymph Formation

As the osmotic pressure of the tissue fluid rises, it interferes with the osmotic reabsorption of water by the blood capillaries. The volume of fluid in the interstitial spaces then tends to increase, as does the pressure within the spaces.

This increasing interstitial pressure is responsible for forcing some of the tissue fluid into the lymphatic capillaries, where it becomes lymph. (See fig. 17.5.)

Function of Lymph

Most of the protein molecules that leak out of the blood capillaries are carried away by lymph and are returned to the bloodstream. At the same time, lymph transports various foreign particles, such as bacterial cells or viruses that may have entered the tissue fluids, to lymph nodes.

Movement of Lymph

Although the entrance of lymph into the lymphatic capillaries is influenced by the *osmotic pressure* of tissue fluid, the movement of lymph through the lymphatic vessels is controlled largely by *muscular activity.*

Flow of Lymph

Lymph, like venous blood, is under relatively low pressure and may not flow readily through the lymphatic vessels without the aid of outside forces. These forces include contraction of skeletal muscles, the action of breathing muscles, and contraction of smooth muscles in the walls of larger lymphatic vessels.

As skeletal muscles contract, they compress lymphatic vessels. This squeezing action causes the lymph inside a vessel to move, but since lymphatic vessels contain valves that prevent backflow, the lymph only can move toward a collecting duct. Similarly, the smooth muscles in the walls of larger lymphatic vessels may contract and compress the lymph inside. This action also helps to force the fluid onward.

Breathing muscles aid the circulation of lymph (as they do that of venous blood) by creating relatively low pressure in the thorax during inhalation. At the same time, the pressure in the abdominal cavity is increased by the contracting diaphragm. Consequently, lymph (and venous blood as well) is squeezed out of the abdominal vessels and into the thoracic vessels. Once again, backflow of lymph (and blood) is prevented by valves within the lymphatic (and blood) vessels.

Conditions sometimes occur that interfere with lymph movement, and tissue fluids accumulate in the spaces, causing a form of *edema.*

For example, lymphatic vessels may be obstructed as a result of surgical procedures in which portions of the lymphatic system are removed. Since the affected pathways no longer can drain lymph from the tissues, proteins tend to accumulate in the interstitial spaces. This causes an increase in the osmotic pressure of the tissue fluid, which in turn promotes the accumulation of water within the tissues.

---

1. How do skeletal muscles promote the flow of lymph?
2. How does breathing aid the movement of lymph?

Lymph Nodes

Lymph nodes (lymph glands) are structures located along lymphatic pathways. They contain large numbers of *lymphocytes* that are vital in the defense against invasion by microorganisms.

Structure of a Lymph Node

Lymph nodes vary in size and shape; however, they are usually less than 2.5 centimeters (1 inch) in length and are somewhat bean shaped. A section of a typical lymph node is illustrated in figure 17.6.

The indented region of a bean-shaped node is called the **hilum,** and it is the portion through which blood vessels and nerves enter the structure. The lymphatic vessels leading to a node (afferent vessels) enter separately at various points on its convex surface, while the lymphatic vessels leaving the node (efferent vessels) exit from the hilum.

Each lymph node is divided into compartments that contain dense masses of lymphocytes and macrophages. These masses, called **nodules,** represent the structural units of the node.

Spaces within the node, called **lymph sinuses,** provide a complex network of chambers and channels through which lymph circulates as it passes through the node.

Fig. 17.6 A section of a lymph node. What factors promote the flow of lymph through a node?

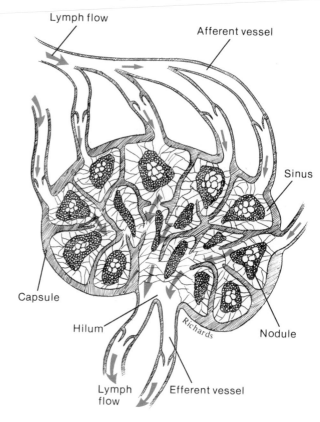

Locations of Lymph Nodes

Lymph nodes generally occur in groups or chains along the larger lymphatic vessels. Although they are widely distributed throughout the body, they are lacking in the tissues of the central nervous system.

The major lymph nodes are located and named in figure 17.7.

Sometimes lymphatic vessels become inflamed due to a bacterial infection. When this happens in superficial vessels, painful reddish streaks may appear beneath the skin. This condition is called *lymphangitis,* and it is usually followed by *lymphadenitis,* an inflammation of the lymph nodes. Affected nodes may become greatly enlarged and quite painful.

Fig. 17.7 Major locations of lymph nodes.

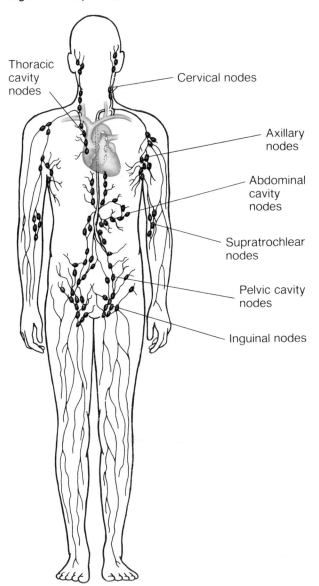

- Thoracic cavity nodes
- Cervical nodes
- Axillary nodes
- Abdominal cavity nodes
- Supratrochlear nodes
- Pelvic cavity nodes
- Inguinal nodes

Functions of Lymph Nodes

As was mentioned, lymph nodes contain large numbers of lymphocytes. These nodes, in fact, are centers for lymphocyte production, although such cells are produced in other tissues as well. Lymphocytes act against foreign particles, such as bacterial cells and viruses, that are carried to the lymph nodes by the lymphatic vessels.

The nodes also contain macrophages that engulf and destroy foreign substances, damaged cells, and cellular debris.

Fig. 17.8 The thymus gland is a bilobed organ located between the lungs and above the heart.

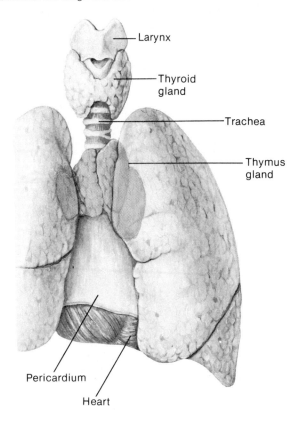

- Larynx
- Thyroid gland
- Trachea
- Thymus gland
- Pericardium
- Heart

---

1. *How would you distinguish between a lymph node and a lymph nodule?*
2. *What are the major functions of lymph nodes?*

Thymus and Spleen

Two other lymphatic organs whose functions are closely related to those of the lymph nodes are the thymus and the spleen.

The Thymus

The **thymus,** shown in figure 17.8, is a soft, bilobed structure whose lobes are surrounded by connective tissue. It is located in front of the aorta and behind the upper part of the sternum.

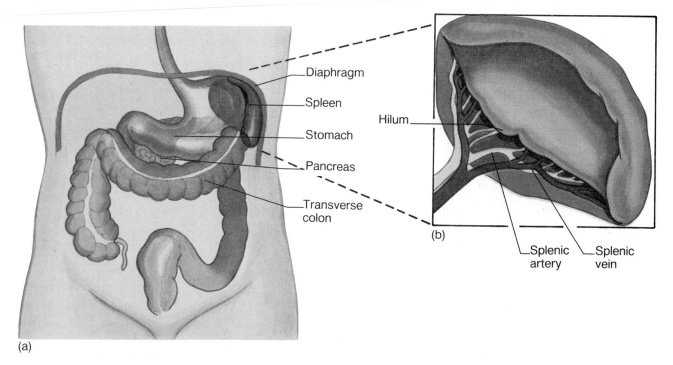

This organ is composed of lymphatic tissue that is subdivided into *lobules* by connective tissues. The lobules contain large numbers of lymphocytes. The majority of these cells (thymocytes) remain inactive; however, some of them develop into a group (T-lymphocytes) that leaves the thymus and functions in immunity.

The thymus also may secrete a hormone called *thymosin,* which is thought to stimulate the activity of these lymphocytes after they leave the thymus and migrate to other lymphatic tissues.

The Spleen

The **spleen** is the largest of the lymphatic organs. It is located in the upper left portion of the abdominal cavity, just beneath the diaphragm and behind the stomach. (See fig. 17.9.)

The spleen resembles a large lymph node and is divided into chambers or lobules. Unlike a lymph node, however, the spaces (venous sinuses) within the chambers of the spleen are filled with *blood* instead of lymph. (See fig. 17.10.)

In addition to being very vascular, the spleen is soft and elastic, and can be distended by blood filling its venous sinuses. These characteristics are related to the spleen's function as a *blood reservoir.* During times of rest when circulation of blood is decreased,

some blood can be stored within the spleen by vasodilation of its blood vessels. Then, during periods of exercise or in response to low oxygen concentration or hemorrhage, these blood vessels are stimulated to undergo vasoconstriction, and some of the stored blood is expelled into the circulation.

The spleen contains numerous phagocytic macrophages in the linings of its venous sinuses. These cells engulf and destroy foreign particles, such as bacteria, that may be carried in the blood as it flows through the sinuses. In addition, the macrophages help to destroy damaged red blood cells and the remains of ruptured cells carried in the blood.

---

1. Why are the thymus and spleen considered to be organs of the lymphatic system?
2. What are the major functions of the thymus and the spleen?

Body Defenses against Infection

An infection is a condition caused by the presence of some kind of disease-causing agent within the body. Such agents are termed **pathogens,** and they include viruses and microorganisms such as bacteria, fungi, and protozoans, as well as various other parasitic forms of life.

Fig. 17.10 A cross section of the spleen.

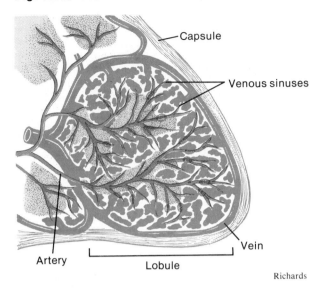

Capsule

Venous sinuses

Vein

Artery

Lobule

Richards

The human body is equipped with a variety of *defense mechanisms* that help to prevent the entrance of pathogens or act to destroy them if they do get into the tissues. Some of these mechanisms are quite general, or *nonspecific,* in that they protect against many types of pathogens. These mechanisms include species resistance, mechanical barriers, actions of enzymes, interferon, inflammation, and phagocytosis.

Other defense mechanisms are very *specific* in their actions, providing protection against particular disease-causing agents; they are responsible for the type of resistance called *immunity.*

Nonspecific Resistance

Species Resistance

Species resistance refers to the fact that a given kind of organism or *species* (such as humans, *Homo sapiens*) develops diseases that are unique to it. At the same time, a species may be resistant to diseases that affect other species. For example, humans are subject to infections by the microorganisms that cause gonorrhea and syphilis, but other animal species generally are resistant to these diseases.

Mechanical Barriers

The *skin* and the *mucous membranes* lining the tubes of the respiratory, digestive, urinary, and reproductive systems create **mechanical barriers** against the entrance of infectious agents. As long as these barriers remain unbroken, many pathogens are unable to penetrate them.

Enzymatic Actions

Enzymatic actions against disease-causing agents are due to the presence of enzymes in various body fluids. Gastric juice, for example, contains the protein-splitting enzyme *pepsin.* It also has a low pH due to the presence of hydrochloric acid, and the combined effect of these substances is lethal to many pathogens that reach the stomach. Similarly, tears contain an enzyme called *lysozyme* that has an antibacterial action against certain pathogens that get onto the surfaces of the eyes.

Interferon

Interferon is the name given to a group of proteins produced by cells in response to the presence of *viruses.* Although the effect of interferon is nonspecific, it somehow interferes with the reproduction of viruses and thus helps to control the development of the diseases they cause.

Inflammation

Inflammation is a tissue response to injury from mechanical forces, chemical irritants, exposure to extreme temperatures, or exposure to excessive radiation. It also may accompany infections.

The major symptoms of inflammation include localized redness, swelling, heat, and pain. The *redness* is a result of blood vessel dilation and the consequent increase of blood volume within the affected tissues. This effect, coupled with an increase in the permeability of the capillaries, is responsible for tissue *swelling* (edema). The *heat* is due to the presence of blood from deeper body parts, which is generally warmer than that near the surface, while the *pain* results from the stimulation of pain receptors in the affected tissues.

White blood cells tend to accumulate at the sites of inflammation. Some of these cells help to control pathogens by *phagocytosis.*

Body fluids also tend to collect in inflamed tissues. These fluids contain *fibrinogen* and other clotting elements. As a result of clotting, a network of fibrin threads may develop within a region of damaged tissue. Later, *fibroblasts* may appear and form fibers around the affected area until it is enclosed in a sac of fibrous connective tissue. This action serves to confine an infection and to prevent the spread of pathogens to adjacent tissues.

A *pimple* or *boil* is an example of a confined infection. An abscess of this type should never be squeezed because the pressure acting upon the inflamed tissues may force pathogenic microorganisms into adjacent tissues and cause the infection to spread.

Once an infection has been controlled, phagocytic cells remove dead cells and other debris from the site of inflammation, and the damaged cells are replaced by cellular reproduction.

The process of inflammation is summarized in chart 17.2.

Phagocytosis

As is mentioned in chapter 15, the most active phagocytic cells in the blood are *neutrophils* and *monocytes*. These wandering cells can leave the bloodstream by squeezing between the cells of blood vessel walls (diapedesis). They are attracted toward sites of inflammation by chemicals released from injured tissues.

Monocytes give rise to macrophages (histiocytes), which become fixed in various tissues or to the inner walls of blood vessels and lymphatic vessels. These relatively nonmotile, phagocytic cells can divide and produce new macrophages and are found in such organs as lymph nodes, spleen, liver, and lungs. This diffuse group of macrophages constitutes the **reticuloendothelial tissue** or **system** (tissue macrophage system).

The name *reticuloendothelium* describes two characteristics of cells that make up this tissue. They are capable of forming fibrous networks or *reticula*, and they are often associated with the linings of blood vessels, the *endothelium*.

As a result of reticuloendothelial activities, foreign particles are removed from lymph as it moves from the interstitial spaces to the bloodstream. Any such particles that reach the blood are likely to be removed by phagocytes located in the vessels and tissues of the spleen, liver, or bone marrow.

---

1. *What is meant by an infection?*
2. *Explain six nonspecific defense mechanisms.*
3. *Define reticuloendothelial tissue.*

Chart 17.2 Major actions that may occur during an inflammation response

Blood vessels dilate

Capillary permeability increases
  Tissues become red, swollen, warm, and painful

White blood cells invade the region
  Pus may form as white blood cells, bacterial cells, and cellular debris accumulate

Body fluids seep into the area
  A clot containing threads of fibrin may form

Fibroblasts appear
  A connective tissue sac may be formed around the injured tissues

Phagocytes are active
  Dead cells and other debris are removed

Cells reproduce
  Newly formed cells replace injured ones

From Hole, John W. Jr., *Human Anatomy and Physiology 3d ed.* © 1978, 1981, 1984 Wm. C. Brown Publishers, Dubuque, Iowa. All Rights Reserved. Reprinted by permission.

Immunity

Immunity is resistance to specific foreign agents, such as pathogens, or to the toxins they release. It involves a number of *immune mechanisms,* whereby certain cells recognize the presence of particular foreign substances and act against them. The cells that function in immune mechanisms include lymphocytes and macrophages.

Origin of Lymphocytes

Lymphocytes originate from bone marrow cells, although later they proliferate in various lymphatic tissues.

Before they become specialized or *differentiated,* developing lymphocytes are released from the marrow and are carried away by the blood. About half of them reach the thymus gland, where they remain for a time. Within the thymus, these undifferentiated cells undergo special processing, and thereafter they are called *T-lymphocytes* (thymus-derived lymphocytes). Later, these T-cells are transported away from the thymus by the blood, where they comprise 70–80% of circulating lymphocytes. They tend to reside in various organs of the lymphatic system and are particularly abundant in lymph nodes, the thoracic duct, and the spleen.

Lymphocytes that have been released from the bone marrow and do not reach the thymus gland, differentiate into *B-lymphocytes* (bone marrow-derived lymphocytes).

B-lymphocytes are distributed by the blood and constitute 20–30% of the circulating lymphocytes. They settle in various lymphatic organs with the T-lymphocytes and are abundant in lymph nodes, spleen, bone marrow, secretory glands, intestinal lining, and reticuloendothelial tissue (figs. 17.11 and 17.12).

Fig. 17.11 Bone marrow releases undifferentiated lymphocytes, which, after processing, become T-lymphocytes and B-lymphocytes.

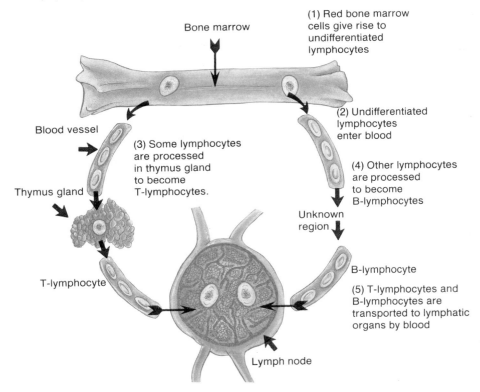

Bone marrow

(1) Red bone marrow cells give rise to undifferentiated lymphocytes

(2) Undifferentiated lymphocytes enter blood

Blood vessel

(3) Some lymphocytes are processed in thymus gland to become T-lymphocytes.

(4) Other lymphocytes are processed to become B-lymphocytes

Thymus gland

Unknown region

T-lymphocyte

B-lymphocyte

(5) T-lymphocytes and B-lymphocytes are transported to lymphatic organs by blood

Lymph node

Antigens

Prior to birth, body cells somehow make an inventory of the proteins and various other large molecules that are present in the body. Subsequently, the substances that are present (self-substances) can be distinguished from foreign (nonself) substances; and when foreign substances are recognized, lymphocytes can produce immune reactions against them.

Foreign substances, such as proteins, polysaccharides, and some glycolipids to which lymphocytes respond, are called **antigens.**

Functions of Lymphocytes

T-lymphocytes and B-lymphocytes respond to antigens in different ways. For example, some T-lymphocytes attach themselves to antigen-bearing cells, such as certain types of bacterial cells, and interact with these foreign cells directly—that is, with cell-to-cell contact. This type of response is called **cell-mediated immunity** (CMI).

T-cells also provide an important defense against many viral infections. Such infections are caused by viruses that grow inside body cells where they are

Fig. 17.12 Scanning electron micrograph of a human circulating lymphocyte.

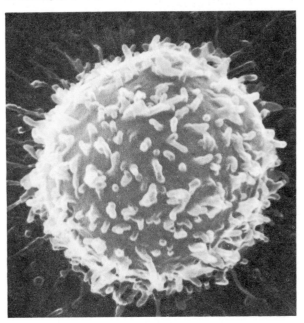

Chart 17.3 A comparison of T-lymphocytes and B-lymphocytes

Characteristic	T-lymphocytes	B-lymphocytes
Origin of undifferentiated cell	Bone marrow	Bone marrow
Site of differentiation	Thymus gland	Region outside thymus gland
Primary locations	Lymphatic tissues, 70–80% of circulating lymphocytes	Lymphatic tissues, 20–30% of circulating lymphocytes
Primary functions	Responsible for cell-mediated immunity, secrete lymphokines, trigger and regulate actions of B-lymphocytes	Responsible for antibody-mediated immunity, secrete antibodies into body fluids

somewhat protected. Most viruses however, cause antigens to be produced on the membranes of infected cells. T-lymphocytes can detect the antigen-bearing cells and destroy them along with the viruses they contain.

T-lymphocytes also can release a group of substances (*lymphokines*) that in turn, can cause a variety of effects. For example, they may attract macrophages and leukocytes into inflamed tissues and retain them there, or stimulate the activities of B-cells.

B-lymphocytes act indirectly against the antigens by producing and secreting proteins called **antibodies.** Antibodies are carried by body fluids and react in various ways with specific antigens or antigen-bearing particles to destroy them. This type of response is called **antibody-mediated immunity** (AMI) or humoral immunity.

T-lymphocytes and B-lymphocytes also interact with each other in complex ways. For example, one group of T-lymphocytes (T-helper cells) sometimes must interact with B-cells, either directly or indirectly by means of released substances, before the B-cells can synthesize antibodies. Another group of T-lymphocytes (T-suppressor cells) function to regulate the synthesis of antibodies by producing a substance (suppressor factor) that inhibits B-cell activity. Thus, both types of lymphocytes are required for normal immune responses. Chart 17.3 compares T- and B-lymphocytes.

The virus that causes the disease called AIDS (acquired immunodeficiency syndrome) infects T-helper cells. As a result, these cells lose their ability to recognize antigens and to interact with B-cells. Since immune responses are greatly weakened, the person is likely to experience a series of severe infections caused by a variety of common pathogens, and to develop certain unusual forms of cancer.

1. Define immunity.
2. Explain the difference in origin between a T-lymphocyte and a B-lymphocyte.
3. What is an antigen?
4. What are the functions of a T-lymphocyte? A B-lymphocyte?

Types of Antibodies

The antibodies produced and secreted by B-lymphocytes are soluble globular proteins. These proteins are called **immunoglobulins,** and they constitute a portion of the *gamma globulin* fraction of plasma proteins. (See chapter 15.)

The three most abundant types of immunoglobulins are immunoglobulin G, immunoglobulin A, and immunoglobulin M.

Immunoglobulin G (IgG) occurs in plasma and tissue fluids and is particularly effective against bacterial cells, viruses, and various toxins. It functions in a variety of ways. For example, it may cause antigen-bearing particles to form insoluble precipitates, digest the membranes of foreign cells, or alter cell membranes so the cells become more susceptible to phagocytosis.

Immunoglobulin A (IgA) is commonly found in the secretions of various *exocrine glands*. It occurs in milk, tears, nasal fluid, gastric juice, intestinal juice, bile, and urine. IgA helps to control certain respiratory viruses and various pathogens responsible for digestive disturbances. It seems to act by causing antigens to form insoluble precipitates and by causing antigen-bearing cells to clump together.

Immunoglobulin A can pass from a nursing mother to her baby via milk and other breast secretions. These antibodies provide the infant with some protection against digestive disturbances that might otherwise cause serious problems.

Fig. 17.13 Sensitized B-lymphocytes proliferate and give rise to antibody-secreting plasma cells and dormant memory cells. The antibodies combine with specific antigens.

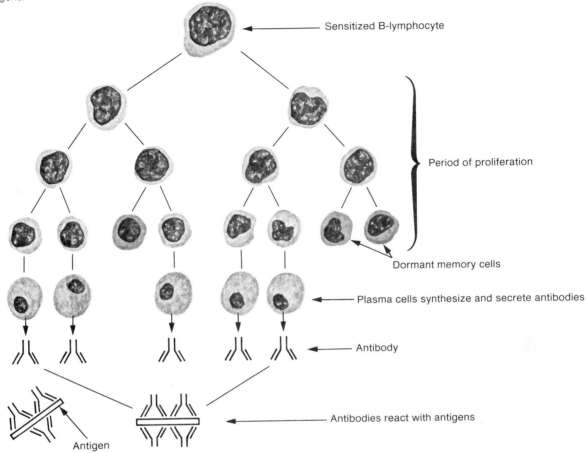

Immunoglobulin M (IgM) is the type of antibody that occurs naturally in blood plasma (agglutinins anti-A and anti-B), as described in chapter 15.

---

1. What is an immunoglobulin?
2. What immunoglobulins are most abundant and how do they differ?

Clones of Lymphocytes

Within the large lymphocyte population of each person, there are millions of varieties. The members of each variety originated from a single early cell, so that the members are all alike. Also the members of each variety are capable of responding only to the same specific antigen. Such a group is called a **clone.**

Before a lymphocyte can actually respond to its particular antigen, it must be *sensitized.* The process by which this occurs involves an activity of macrophages. An antigen-bearing agent that enters the body is phagocytized by a macrophage. This cell passes the antigen directly to lymphocytes, thus sensitizing those specialized to respond to that particular antigen.

Immune Responses

When an antigen first is encountered, the immune reaction that follows is called a **primary immune response.** During this response, members of a specialized clone of lymphocytes become sensitized. If, for example, B-lymphocytes are sensitized, they enlarge and undergo a period of rapid reproduction. Some of the newly formed offspring cells become **plasma cells** and make use of their DNA and protein-synthesis mechanism (see chapter 4) to produce the appropriate antibody molecules. The antibodies are released into the lymph and are transported to the blood. Such antibody production and release may continue for several weeks. (See fig. 17.13.)

If the antigen that stimulates the primary response is present on certain types of bacterial cells or cells infected with viruses, members of a clone of T-lymphocytes may be sensitized. As a result, the sensitized cells proliferate and some of them begin to function as **effector cells** that attack the invaders directly.

Following a primary immune response some of the B-cells and T-cells that are produced during proliferation remain dormant, serving as *memory cells.* In this way, if such antigens are encountered in the future, the original clones of lymphocytes are increased in sizes and can respond to the antigens to which they were sensitized more rapidly and effectively. A subsequent reaction of this type is called a **secondary immune response.**

As a result of a primary immune response, detectable concentrations of antibodies usually appear in the body fluids within five to ten days following the exposure to antigens. If the same antigen is encountered some time later, a secondary immune response may produce additional antibodies within a day or two. Although such newly formed antibodies may persist in the body for only a few months, or perhaps a few years, the memory cells live much longer. Consequently, the ability to produce a secondary immune response may be long-lasting.

---

1. What is meant by a clone of lymphocytes?
2. Distinguish between a primary and a secondary immune response.

Types of Immunity

One type of immunity is called *naturally acquired active immunity.* It occurs when a person who has been exposed to a live pathogen develops a disease and becomes resistant to that pathogen as a result of a primary immune response.

Another type of active immunity can be produced in response to a **vaccine.** Such a substance contains an antigen that can stimulate a primary immune response against a particular disease-causing agent.

A vaccine, for example, might contain bacteria or viruses that have been killed or weakened sufficiently so that they cannot cause a serious infection or produce the severe symptoms of an infection; or it may contain a toxin of an infectious organism that has been chemically altered to destroy its toxic effects. In any case, the antigens present still retain the characteristics needed to stimulate a primary immune response. Thus, a person who has been vaccinated is said to develop *artificially acquired active immunity.*

Sometimes a person needs protection against a disease-causing microorganism, but lacks the time needed to develop active immunity. In such a case, it may be possible to provide the person with an injection of ready-made antibodies. These antibodies may be obtained from *gamma globulin* separated from the blood of persons who have already developed immunity against the disease in question.

A person who receives an injection of gamma globulin is said to have *artificially acquired passive immunity.* This type of immunity is called passive because the antibodies involved are not produced by the recipient's cells. Such immunity is relatively short-term, seldom lasting more than a few weeks. Furthermore, since the recipient's lymphocytes have not been sensitized to respond against the pathogens for which the protection was needed, the person continues to be susceptible to those pathogens in the future.

During pregnancy, certain antibodies (IgG) are able to pass from the maternal blood through the placental membrane and into the fetal bloodstream. As a result, the fetus acquires some immunity against the pathogens for which the mother has developed active immunities. In this case, the fetus is said to have *naturally acquired passive immunity,* which may remain effective for six months to a year after birth.

The types of immunity are summarized in chart 17.4.

Injections of gamma globulin sometimes are used to provide passive immunity for pregnant women who have been exposed to rubella (German or three-day measles) viruses during the early stages of their pregnancies. Injections of antibodies also are used to provide temporary protection against the effects of poisonous snakebites and certain other toxins.

Allergic Reactions

Allergic reactions are closely related to immune responses in that both may involve the sensitizing of lymphocytes or the combining of antigens with antibodies. Allergic reactions, however, are likely to be excessive and cause damage to tissues.

Some forms of allergic reactions can occur in almost anyone, while another form affects only those people who have inherited from their parents an ability to produce exaggerated immune responses.

Chart 17.4 Types of immunity

Type	Stimulus	Result
Naturally acquired active immunity	Exposure to live pathogens	Symptoms of a disease and stimulation of an immune response
Artificially acquired active immunity	Exposure to a vaccine containing weakened or dead pathogens	Stimulation of an immune response without the severe symptoms of a disease
Artificially acquired passive immunity	Injection of gamma globulin containing antibodies	Immunity for a short time without stimulating an immune response
Naturally acquired passive immunity	Antibodies passed to fetus from mother with active immunity	Short-term immunity for infant, without stimulating an immune response

Delayed-reaction allergy is an example of a type of allergic response that may occur in anyone. It results from repeated exposure of the skin to certain chemical substances—commonly, household or industrial chemicals or some cosmetics. As a consequence of such repeated contacts, lymphocytes eventually become sensitized to this substance and a large number of these lymphocytes enter the skin. As a result of their cellular responses and the activities of the macrophages they attract, toxic factors are released, which in turn cause eruptions and inflammation of the skin.

People who have an inherited ability to produce exaggerated immune responses synthesize abnormally large quantities of an immunoglobulin following exposure to certain antigens. In this instance, the antigen that stimulates the production of the antibodies is called an **allergen.**

Typically, the person's lymphocytes become sensitized to the allergen when it is first encountered, and subsequent exposures trigger allergen-antibody reactions. The immunoglobulin involved (IgE), is attached to the membranes of certain widely distributed cells (mast cells) and when an allergen-antibody reaction occurs, the cells release substances such as *histamine.* These substances, in turn, cause a variety of physiological effects, including the dilation of blood vessels, swelling of tissues, and contraction of smooth muscles. The result is a severe inflammation reaction that is responsible for the symptoms of the allergy—hives, hay fever, asthma, eczema, or gastric disturbances.

Transplantation and Tissue Rejection

It is desirable occasionally to transplant some tissue or an organ, such as skin, kidney, heart, or liver, from one person to another to replace a nonfunctional, damaged, or lost body part. In such cases, there is a danger that the recipient's cells may recognize the donor's tissues as being foreign. This triggers the recipient's immune mechanisms, which may act to destroy the donor tissue. Such a response is called a **tissue rejection reaction.**

Tissue rejection involves the activities of lymphocytes and both cell- and antibody-mediated responses—responses similar to those that occur when any foreign substances are present. The greater the antigenic difference between the proteins of the recipient and the donor tissues, the more rapid and severe the rejection reaction will be. Thus, the reaction can sometimes be minimized by matching recipient and donor tissues. This means locating a donor whose tissues are antigenically similar to those of the person needing a transplant—a procedure much like matching the blood of a donor with that of a recipient before giving a blood transfusion.

Another approach to reducing the rejection of transplanted tissue involves the use of *immunosuppressive drugs.* These substances act by interfering with the recipient's immune mechanisms. A drug may, for example, suppress the formation of antibodies, or it may cause the destruction of lymphocytes and so prevent them from producing antibodies.

Unfortunately, the use of immunosuppressive drugs leaves the recipient relatively unprotected against infections. Although the drug may prevent a tissue rejection reaction, the recipient may develop a serious infectious disease that is difficult to control.

1. *Explain the difference between active and passive immunities.*
2. *In what ways is an allergic reaction related to an immune reaction?*
3. *In what ways is a tissue rejection reaction related to an immune response?*

Clinical Terms Related to the Lymphatic System and Immunity

anaphylaxis (an″ah-fi-lak′sis)—hypersensitivity to the presence of a foreign substance.

asplenia (ah-sple′ne-ah)—the absence of a spleen.

autograft (aw′to-graft)—transplantation of tissue from one part of a body to another part of the same body.

histocompatibility (his″to-kom-pati-bil′ĭ-te)—compatibility between the tissues of a donor and a recipient based on antigenic similarities.

homograft (ho′mo-graft)—transplantation of tissue from one person to another person.

immunocompetence (im″u-no-kom′pe-tens)—the ability to produce an immune response to the presence of antigens.

immunodeficiency (im″u-no-de-fish′en-se)—a lack of ability to produce an immune response.

lymphadenectomy (lim-fad″ĕ-nek′to-me)—surgical removal of lymph nodes.

lymphadenopathy (lim-fad″ĕ-nop′ah-the)—enlargement of the lymph nodes.

lymphadenotomy (lim-fad″ĕ-not′o-me)—an incision of a lymph node.

lymphocytopenia (lim″fo-si″to-pe′ne-ah)—an abnormally low concentration of lymphocytes in the blood.

lymphocytosis (lim″fo-si′to′sis)—an abnormally high concentration of lymphocytes in the blood.

lymphoma (lim-fo′mah)—a tumor composed of lymphatic tissue.

lymphosarcoma (lim″fo-sar-ko′mah)—a cancer within the lymphatic tissue.

splenectomy (sple-nek′to-me)—surgical removal of the spleen.

splenitis (sple-ni′tis)—an inflammation of the spleen.

splenomegaly (sple″no-meg′ah-le)—an abnormal enlargement of the spleen.

splenotomy (sple-not′o-me)—incision of the spleen.

thymectomy (thi-mek′to-me)—surgical removal of the thymus gland.

thymitis (thi-mi′tis)—an inflammation of the thymus gland.

Chapter Summary

Introduction

The lymphatic system is closely associated with the cardiovascular system and transports excess tissue fluid to the bloodstream.

Lymphatic Pathways

1. Lymphatic capillaries
   a. Lymphatic capillaries are microscopic closed-ended tubes.
   b. They receive lymph through their thin walls.
2. Lymphatic vessels
   a. Lymphatic vessels have walls similar to veins and possess valves.
   b. They lead to lymph nodes and then merge into lymphatic trunks.
3. Lymphatic trunks and collecting ducts
   a. Lymphatic trunks lead to two collecting ducts.
   b. Collecting ducts join the subclavian veins.

Tissue Fluid and Lymph

1. Tissue fluid formation
   a. Tissue fluid originates from blood plasma.
   b. It generally lacks proteins, but some smaller protein molecules leak into interstitial spaces.
   c. As protein concentration of tissue fluid increases, osmotic pressure increases also.
2. Lymph formation
   a. Rising osmotic pressure in tissue fluid interferes with the return of water to the blood.
   b. Increasing pressure within interstitial spaces forces some tissue fluid into lymphatic capillaries.
3. Function of lymph
   a. Lymph returns protein molecules to the bloodstream.
   b. It transports foreign particles to the lymph nodes.

Movement of Lymph

Forces that aid movement of lymph include squeezing action of skeletal muscles and breathing movements.

Lymph Nodes

1. Structure of a lymph node
   a. Lymph nodes are divided into nodules.
   b. Nodules contain masses of lymphocytes and macrophages.

2. Locations of lymph nodes
   Lymph nodes generally occur in groups along the paths of larger lymphatic vessels.

3. Functions of lymph nodes
   a. Lymph nodes are centers for production of lymphocytes.
   b. They also contain phagocytic cells.

Thymus and Spleen

1. The thymus
   a. The thymus is composed of lymphatic tissue that is subdivided into lobules.
   b. Some lymphocytes leave the thymus and function in immunity.

2. The spleen
   a. The spleen resembles a large lymph node.
   b. It acts as a blood reservoir.
   c. It contains numerous macrophages that filter foreign particles and damaged red blood cells from the blood.

Body Defenses against Infection

The body is equipped with specific and nonspecific defenses against infection.

Nonspecific Resistance

1. Species resistance
   Each species of organism is resistant to certain diseases that may affect other species.

2. Mechanical barriers
   Mechanical barriers include skin and mucous membranes which prevent the entrance of pathogens.

3. Enzymatic actions
   Enzymes of gastric juice and tears are lethal to many pathogens.

4. Interferon
   Interferon is a group of proteins produced by cells in response to the presence of viruses; it can interfere with the reproduction and spread of viruses.

5. Inflammation
   a. Inflammation response includes localized redness, swelling, heat, and pain.
   b. Chemicals released by damaged tissues attract various white blood cells to the site of inflammation.
   c. Fibrous connective tissue may form a sac around the injured tissue.

6. Phagocytosis
   a. The most active phagocytes in blood are neutrophils and monocytes; monocytes give rise to macrophages that remain fixed in tissues.
   b. Macrophages associated with the linings of blood vessels in bone marrow, liver, spleen, and lymph nodes constitute the reticuloendothelial tissue.

Immunity

1. Origin of lymphocytes
   a. Lymphocytes originate in bone marrow and are released into the blood before they become differentiated.
   b. Some reach the thymus where they become T-lymphocytes.
   c. Others become B-lymphocytes.
   d. Both T- and B-lymphocytes tend to reside in organs of the lymphatic system.

2. Antigens
   a. Before birth, cells make an inventory of the proteins and other large molecules present in the body.
   b. After the inventory, lymphocytes can recognize foreign substances.
   c. Antigens are foreign substances that stimulate lymphocytes to respond.

3. Functions of lymphocytes
   a. T-lymphocytes attack antigens or antigen-bearing agents directly, providing cell-mediated immunity.
   b. B-lymphocytes produce antibodies that act against specific antigens, providing antibody-mediated immunity.
   c. Normal immune responses require the interaction of T- and B-lymphocytes.

4. Types of antibodies
   a. Antibodies are composed of soluble proteins called immunoglobulins.
   b. The three most abundant types of immunoglobulins are IgG, IgA, and IgM.
   c. Immunoglobulins act in various ways, causing antigens or antigen-bearing particles to be destroyed.

5. Clones of lymphocytes
   a. The lymphocyte population includes a large number of clones, each of whose members are capable of responding to a specific antigen.
   b. Before lymphocytes can respond to antigens, they must be sensitized by macrophages.

6. Immune responses
   a. A primary immune response is stimulated by the first encounter with an antigen.
   b. During the primary response, a clone of lymphocytes is sensitized against the antigen.
   c. After being sensitized, some of the members of the clone proliferate and become plasma cells.
   d. Plasma cells synthesize antibodies and release them into the body fluids.
   e. Some of the sensitized T-cells become effector cells that attack antigen-bearing agents.
   f. Some of the sensitized B-cells and T-cells remain dormant, enlarge the original clone, and serve as memory cells.
   g. A secondary immune response occurs rapidly if the same antigen is encountered later.

7. Types of immunity
   a. A person who encounters a pathogen and has a primary immune response develops naturally acquired active immunity.
   b. A person who receives vaccine containing a dead or weakened pathogen develops artificially acquired active immunity.
   c. A person who receives an injection of gamma globulin that contains ready-made antibodies has artificially acquired passive immunity.
   d. When antibodies pass through a placental membrane from a pregnant woman to her fetus, the fetus develops naturally acquired passive immunity.
   e. Active immunity lasts much longer than passive immunity.

8. Allergic reactions
   a. Allergic reactions involve combining antigens with antibodies; reactions are likely to be excessive or violent, and may cause tissue damage.
   b. Delayed-reaction allergy results from repeated exposure to antigenic substances.
   c. A person with an inherited allergic response has an ability to produce an abnormally large amount of immunoglobulin.
   d. Allergic reactions may damage certain cells, which in turn release histamine and other chemicals.
   e. Released chemicals are responsible for the symptoms of the allergic reaction: hives, hayfever, asthma, eczema, or gastric disturbances.

9. Transplantation and tissue rejection
   a. If tissue is transplanted from one person to another, the recipient's cells may recognize the donor's tissue as foreign and act against it.
   b. Tissue rejection reaction may be reduced by matching the donor and recipient tissues.

Application of Knowledge

1. Based on your understanding of the functions of lymph nodes, how would you explain the fact that enlarged nodes are often removed for microscopic examination as an aid to diagnosing certain disease conditions?

2. Why is it true that an injection into the skin is, to a large extent, an injection into the lymphatic system.

3. Explain why vaccination provides long-lasting protection against a disease, while gamma globulin provides only short-term protection.

Review Activities

1. Explain how the lymphatic system is related to the cardiovascular system.

2. Trace the general pathway of lymph from the interstitial spaces to the bloodstream.

3. Describe the primary functions of lymph.

4. Explain how exercise promotes lymphatic circulation.

5. Describe the structure of a lymph node and list its major functions.

6. Describe the structure and functions of the thymus gland.

7. Describe the structure and functions of the spleen.

8. Distinguish between specific and nonspecific body defenses against infection.

9. List the major symptoms of inflammation and explain why each occurs.

10. Identify the major phagocytic cells in the blood and other tissues.

11. Define the *reticuloendothelial tissue* and explain its importance.

12. Review the origins of T-lymphocytes and B-lymphocytes.

13. Distinguish between an antigen and an antibody.

14. Explain what is meant by cell-mediated immunity.

15. Explain what is meant by antibody-mediated immunity.

16. Explain two ways in which T-cells and B-cells interact.

17. List three types of immunoglobulin and describe their differences.

18. Define a *clone of lymphocytes.*

19. Explain how lymphocytes become sensitized to antigens.

20. Distinguish between a primary and a secondary immune response.

21. Define *plasma cell.*

22. Distinguish between active and passive immunity.

23. Define a *vaccine* and explain its action.

24. Explain the relationship between an allergic reaction and an immune response.

25. Distinguish between an antigen and an allergen.

26. List the major events leading to a delayed-reaction allergic response.

27. Describe how an inherited allergic response may occur.

28. Explain the relationship between a tissue rejection and an immune response.

29. Describe a method used to reduce the severity of a tissue rejection reaction.

The Urinary System

Cells form a variety of wastes, and if these substances are allowed to accumulate, their effects are likely to be toxic.

Body fluids, such as blood and lymph, serve to carry wastes away from the tissues that produce them. Other parts remove these wastes from the blood and transport them to the outside. The respiratory system, for example, removes carbon dioxide from the blood, and the *urinary system* removes various salts and nitrogenous wastes.

The urinary system also helps to maintain the normal concentrations of water and electrolytes within body fluids, helps to regulate the pH and volume of body fluids, and aids in the control of red blood cell production and blood pressure.

18

After you have studied this chapter, you should be able to

1. Name the organs of the urinary system and list their general functions.

2. Describe the location of the kidneys and the structure of a kidney.

3. List the functions of the kidneys.

4. Trace the pathway of blood through the major vessels within a kidney.

5. Describe a nephron and explain the functions of its major parts.

6. Explain how glomerular filtrate is produced and describe its composition.

7. Explain how various factors affect the rate of glomerular filtration and how this rate is regulated.

8. Discuss the role of tubular reabsorption in the formation of urine.

9. Define *tubular secretion* and explain its role in urine formation.

10. Describe the structure of the ureters, urinary bladder, and urethra.

11. Discuss the process of micturition and explain how it is controlled.

12. Complete the review activities at the end of this chapter. Note that the items are worded in the form of specific learning objectives. You may want to refer to them before reading the chapter.

afferent arteriole   (af′er-ent ar-te′re-ōl)

Bowman's capsule   (bo′manz kap′sūl)

detrusor muscle   (de-truz′or mus′l)

efferent arteriole   (ef′er-ent ar-te′re-ōl)

glomerulus   (glo-mer′u-lus)

juxtaglomerular apparatus   (juks″tah-glo-mer′u-lar ap″ah-ra′tus)

micturition   (mik″tu-rish′un)

nephron   (nef′ron)

peritubular capillary   (per″i-tu′bu-lar kap′i-ler″e)

renal corpuscle   (re′nal kor′pusl)

renal cortex   (re′nal kor′teks)

renal medulla   (re′nal mĕ-dul′ah)

renal tubule   (re′nal tu′būl)

calyc-, a small cup: major *calyc*es—cuplike divisions of the renal pelvis.

detrus-, to force away: *detrus*or muscle—muscle within bladder wall that causes urine to be expelled.

glom-, little ball: *glom*erulus—cluster of capillaries within a renal corpuscle.

nephr-, pertaining to the kidney: *nephr*on—functional unit of a kidney.

mict-, to pass urine: *mict*urition—process of expelling urine from the bladder.

papill-, nipple: renal *papill*ae—small elevations that project into a renal calyx.

trigon-, a triangular shape: *trigon*e—triangular area on the internal floor of the bladder.

Fig. 18.1 The urinary system includes the kidneys, ureters, urinary bladder, and urethra. What are the general functions of this system?

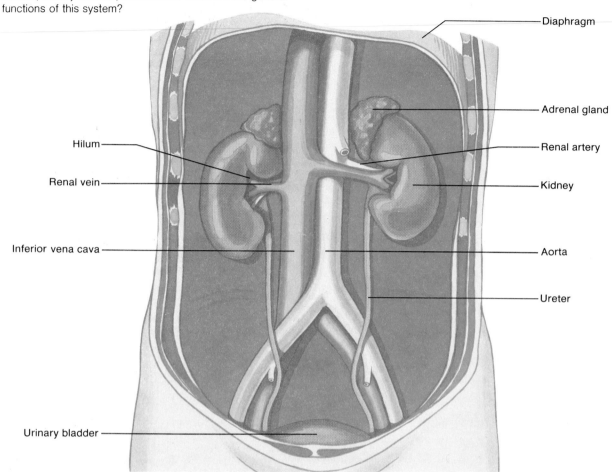

Diaphragm

Adrenal gland

Renal artery

Kidney

Aorta

Ureter

Hilum

Renal vein

Inferior vena cava

Urinary bladder

The urinary system consists of a pair of *kidneys,* which remove substances from the blood, form urine, and help to regulate various metabolic processes; a pair of tubular *ureters,* which transport urine away from the kidneys; a saclike *urinary bladder,* which serves as a urine reservoir; and a tubular *urethra,* which conveys urine to the outside of the body. These organs are shown in figure 18.1.

The Kidney

A **kidney** is a reddish brown, bean-shaped organ with a smooth surface. It is about 12 centimeters long, 6 centimeters wide, and 3 centimeters thick (4.7 × 2.3 × 1.2 inches) in an adult and is enclosed in a tough, fibrous capsule. (See fig. 18.1.)

Location of the Kidneys

The kidneys lie on either side of the vertebral column in a depression high on the posterior wall of the abdominal cavity.

The upper and lower borders of the kidneys are generally at the levels of the twelfth thoracic and third lumbar vertebrae, respectively. The left kidney is usually about 1.5 to 2 centimeters higher than the right one.

The kidneys are positioned *retroperitoneally,* which means they are behind the parietal peritoneum and against the deep muscles of the back. They are held in position by connective tissue and the masses of adipose tissue that surround them.

Structure of a Kidney

The lateral surface of each kidney is convex, while its medial side is deeply concave. The resulting medial depression leads into a hollow chamber called the **renal sinus.** The entrance to this sinus is termed the *hilum,* and through it pass various blood vessels, nerves, lymphatic vessels, and the ureter. (See fig. 18.2.)

Fig. 18.2 (a) Longitudinal section of a kidney; (b) a renal pyramid containing nephrons; (c) a single nephron.

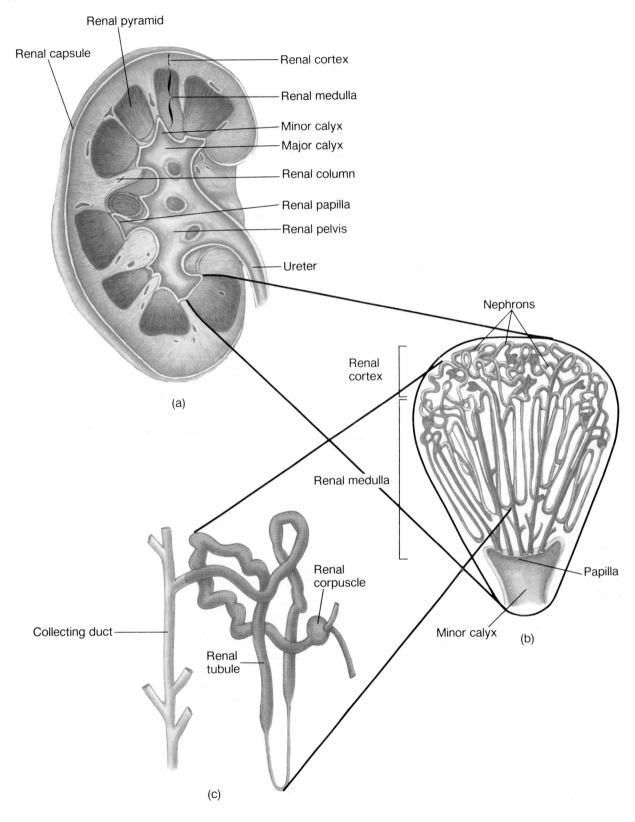

Renal pyramid

Renal capsule

Renal cortex

Renal medulla

Minor calyx

Major calyx

Renal column

Renal papilla

Renal pelvis

Ureter

(a)

Nephrons

Renal cortex

Renal medulla

Papilla

Minor calyx

(b)

Renal corpuscle

Collecting duct

Renal tubule

(c)

Fig. 18.3 Main branches of the renal artery and renal vein.

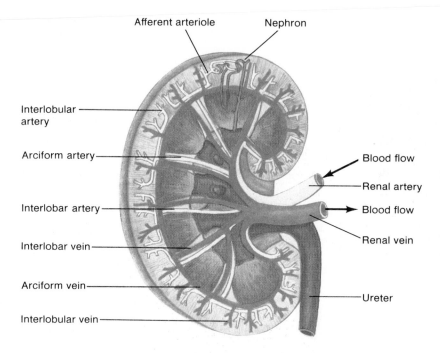

The superior end of the ureter is expanded to form a funnel-shaped sac called the **renal pelvis,** which is located inside the renal sinus. The pelvis is divided into two or three tubes called *major calyces* (singular, *calyx*), and they in turn are divided into several *minor calyces.*

A series of small elevations project into the renal sinus from its wall. These projections are called *renal papillae,* and each is pierced by tiny openings that lead into a minor calyx.

The substance of the kidney is divided into two distinct regions—an inner medulla and an outer cortex. The **renal medulla** is composed of conical masses of tissue called *renal pyramids.*

The **renal cortex,** which appears somewhat granular, forms a shell around the medulla. Its tissue dips into the medulla between adjacent renal pyramids to form *renal columns.* The granular appearance of the cortex is due to the random arrangement of tiny tubules associated with **nephrons,** the functional units of the kidney.

---

1. *Where are the kidneys located?*
2. *Describe the structure of a kidney.*
3. *Name the functional unit of the kidney.*

Functions of the Kidneys

The kidneys function to remove metabolic wastes from the blood and excrete them to the outside. They also carry on a variety of equally important regulatory activities including helping to control the rate of red blood cell formation by secreting the hormone *erythropoietin* (see chapter 15) and helping to regulate blood pressure by secreting the enzyme *renin.*

The kidneys also help to regulate the volume, composition, and pH of body fluids. These functions involve complex mechanisms that lead to the formation of urine. They are discussed in a subsequent section of this chapter and are explored in still more detail in chapter 19.

Renal Blood Vessels

Blood is supplied to the kidneys by means of **renal arteries** that arise from the abdominal aorta. These arteries transport a relatively large volume of blood; in fact, when a person is at rest they usually carry from 15–30% of the total cardiac output into the kidneys.

A renal artery enters a kidney through the hilum and gives off several branches, the *interlobar arteries,* that pass between the renal pyramids. At the junction between the medulla and cortex, the interlobar arteries branch to form a series of incomplete arches, the *arciform arteries* (arcuate arteries), which in turn

Fig. 18.4 A scanning electron micrograph of a cast of the renal blood vessels associated with the glomeruli (260 ×).
From *Tissues and Organs: A Text-Atlas of Scanning Electron Microscopy* by Richard G. Kessel and Randy H. Kardon. W. H. Freeman and Company. Copyright © 1979.

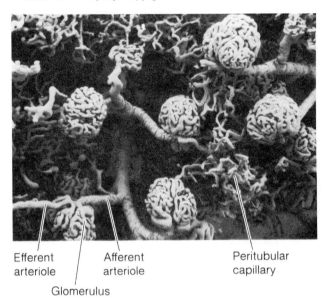

Efferent arteriole   Afferent arteriole   Peritubular capillary

Glomerulus

Fig. 18.5 The parts of the renal tubule connect the Bowman's capsule with a collecting duct.

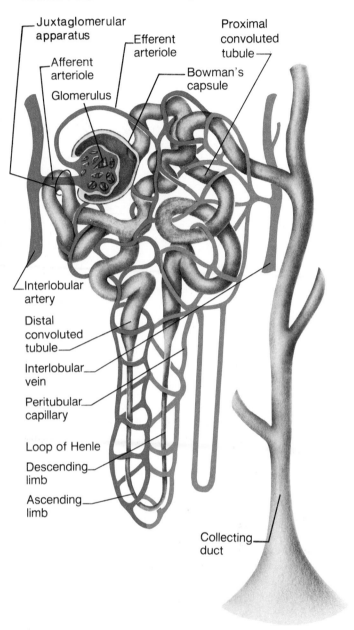

Juxtaglomerular apparatus

Afferent arteriole

Efferent arteriole

Proximal convoluted tubule

Glomerulus

Bowman's capsule

Interlobular artery

Distal convoluted tubule

Interlobular vein

Peritubular capillary

Loop of Henle

Descending limb

Ascending limb

Collecting duct

give rise to *interlobular arteries.* Lateral branches of the interlobular arteries, called **afferent arterioles,** lead to the nephrons.

Venous blood is returned through a series of vessels that correspond generally to the arterial pathways. The **renal vein** then joins the inferior vena cava as it courses through the abdominal cavity. (See figs. 18.3 and 18.4.)

The Nephrons

Structure of a Nephron. A kidney contains about one million nephrons, each consisting of a **renal corpuscle** and a **renal tubule.** (See fig. 18.2.)

A renal corpuscle is composed of a tangled cluster of blood capillaries, called a **glomerulus,** and a thin-walled, saclike structure, called **Bowman's capsule,** that surrounds the glomerulus.

The Bowman's capsule is an expansion at the closed end of a renal tubule. The renal tubule leads away from the Bowman's capsule and becomes highly coiled; this coiled portion of the tubule is appropriately named the *proximal convoluted tubule.*

The proximal tubule dips toward the renal pelvis, following a straight path into the deeper layers of the cortex to become the *descending limb of the loop of Henle.* The tubule then curves back toward its renal corpuscle and forms the *ascending limb of the loop of Henle.*

The ascending limb returns in a straight path to the region of the renal corpuscle, where it becomes highly coiled again, and is called the *distal convoluted tubule.*

Several distal convoluted tubules merge in the renal cortex to form a *collecting duct,* which in turn passes into the renal medulla, becoming larger and larger as it is joined by other collecting ducts. The resulting tube empties into a minor calyx through an opening in a renal papilla. The parts of a nephron are shown in figure 18.5.

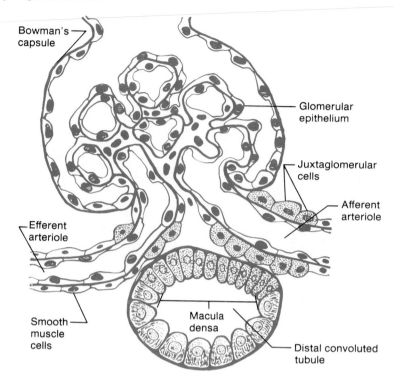

Bowman's capsule

Glomerular epithelium

Juxtaglomerular cells

Afferent arteriole

Efferent arteriole

Smooth muscle cells

Macula densa

Distal convoluted tubule

---

1. *List the general functions of the kidneys.*
2. *Trace the blood supply to the kidney.*
3. *Name the parts of a nephron.*

Blood Supply of a Nephron. The capillary cluster that forms a glomerulus arises from an **afferent arteriole.** After passing through the capillary of the glomerulus, blood enters an **efferent arteriole** (rather than a venule) whose diameter is somewhat less than that of the afferent vessel.

Because of its small diameter, the efferent arteriole creates some resistance to blood flow. This causes blood to back up into the glomerulus, producing a relatively high pressure in the glomerular capillary.

The efferent arteriole branches into a complex, freely interconnecting network of capillaries that surrounds the various portions of the renal tubule. This network is called the **peritubular capillary system,** and the blood it contains is under relatively low pressure (fig. 18.5).

After flowing through the capillary network, blood is returned to the renal cortex, where it joins blood from other branches of the peritubular capillary system and enters the venous system of the kidney.

Juxtaglomerular Apparatus. Near its beginning, the distal convoluted tubule passes between the afferent and efferent arterioles and contacts them. At the point of contact, the epithelial cells of the distal tubule are quite narrow and densely packed. These cells comprise a structure called the *macula densa.*

Close by, in the walls of the arterioles near their attachments to the glomerulus, some of the smooth muscle cells are enlarged. They are called *juxtaglomerular cells,* and together with the cells of the macula densa, they constitute the **juxtaglomerular apparatus** (complex). This structure plays an important role in regulating the flow of blood through various renal vessels (fig. 18.6).

---

1. *Describe the blood supply to a nephron.*
2. *What parts comprise the juxtaglomerular apparatus?*

Fig. 18.7 The first step in urine formation is the filtration of substances through the glomerular membrane into Bowman's capsule.

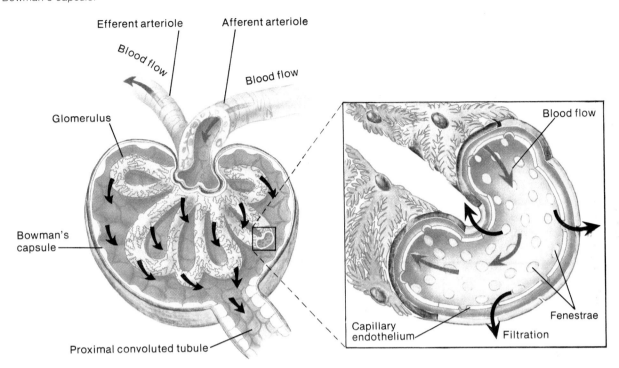

Efferent arteriole Afferent arteriole

Blood flow

Blood flow

Glomerulus

Bowman's capsule

Proximal convoluted tubule

Blood flow

Capillary endothelium

Fenestrae

Filtration

Urine Formation

The primary functions of the nephrons include the removal of waste substances from the blood and the regulation of water and electrolyte concentrations within the body fluids. The end product of these functions is **urine,** which is excreted to the outside of the body carrying with it wastes, excess water, and excess electrolytes.

Urine formation involves glomerular filtration, tubular reabsorption, and tubular secretion.

Glomerular Filtration

Urine formation begins when water and various dissolved substances are filtered out of the glomerular capillaries and into the Bowman's capsules. The filtration of these materials through the capillary walls is much like the filtration that occurs at the arteriole ends of other capillaries throughout the body. The glomerular capillaries, however, are many times more permeable than the capillaries in other tissues due to the presence of numerous tiny openings (fenestrae) in their walls. (See fig. 18.7.)

Filtration Pressure. As in the case of other capillaries, the force mainly responsible for the movement of substances through the glomerular capillary wall is the pressure of the blood inside (glomerular hydrostatic pressure). This movement also is influenced by the osmotic pressure of the plasma in the glomerulus and by the hydrostatic pressure inside the Bowman's capsule. An increase in either of these pressures will oppose movement out of the capillary and thus reduce filtration. The net pressure acting to force substances out of the glomerulus is called **filtration pressure.**

The **glomerular filtrate** consists of the substances received by the Bowman's capsule and has about the same composition as the filtrate that becomes tissue fluid elsewhere. That is, glomerular filtrate is largely water and contains essentially the same substances as blood plasma, except for the larger protein molecules, which the filtrate lacks. The relative concentrations of some substances, in plasma, glomerular filtrate, and urine are shown in chart 18.1.

Filtration Rate. The rate of glomerular filtration is directly proportional to the filtration pressure. Consequently, factors that affect the glomerular hydrostatic pressure, the glomerular plasma osmotic pressure, or the hydrostatic pressure in Bowman's capsule will affect the rate of filtration also.

For example, since the glomerular capillary is located between two arterioles—the *afferent* and *efferent arterioles*—any change in the diameters of these vessels is likely to cause a change in the glomerular hydrostatic pressure, accompanied by a change in the glomerular filtration rate. The afferent arteriole, through which blood enters the glomerulus, may become constricted as a result of mild stimulation by sympathetic nerve impulses. If this occurs, blood flow is diminished, the glomerular hydrostatic pressure is decreased, and the filtration rate drops. If on the other hand, the efferent arteriole (through which blood leaves the glomerulus) becomes constricted, blood backs up into the glomerulus, the glomerular hydrostatic pressure is increased, and the filtration rate rises. Converse effects are produced by vasodilation of these vessels.

In capillaries, the blood pressure, acting to force water and dissolved substances outward, is opposed by the effect of the plasma osmotic pressure that attracts water inward. (See chapter 15.) As filtration occurs through the capillary wall, proteins remaining in the plasma cause the osmotic pressure within the glomerular capillary to rise. When this pressure reaches a certain high level, filtration ceases. Conversely, conditions that tend to decrease plasma osmotic pressure, such as a decrease in plasma protein concentration, cause an increase in the filtration rate.

In the condition called *glomerulonephritis,* the glomerular capillaries are inflamed and become more permeable to proteins. Consequently, proteins appear in the glomerular filtrate and are excreted in the urine (proteinuria). At the same time, the protein concentration in the blood plasma decreases (hypoproteinemia), and this causes a drop in the osmotic pressure of the blood. As a result, the movement of tissue fluid into the capillaries is decreased, and edema develops.

Chart 18.1 Relative concentrations of certain substances in plasma, glomerular filtrate, and urine

| | Concentrations (mEq/l) | | |
Substance	Plasma	Glomerular Filtrate	Urine
Sodium (Na$^+$)	142	142	128
Potassium (K$^+$)	5	5	60
Calcium (Ca^{+2})	4	4	5
Magnesium (Mg^{+2})	3	3	15
Chlorine (Cl$^-$)	103	103	134
Bicarbonate (HCO$_3$$^-$)	27	27	14
Sulfate (SO$_4$$^{-2}$)	1	1	33
Phosphate (PO$_4$$^{-3}$)	2	2	40

| | Concentrations (mg/100 ml) | | |
Substance	Plasma	Glomerular Filtrate	Urine
Glucose	100	100	0
Urea	26	26	1820
Uric acid	4	4	53

From Hole, John W. Jr., *Human Anatomy and Physiology 3d ed.* © 1978, 1981, 1984 Wm. C. Brown Publishers, Dubuque, Iowa. All Rights Reserved. Reprinted by permission.

The hydrostatic pressure in Bowman's capsule sometimes changes as a result of an obstruction, such as may be caused by a stone in a ureter or an enlarged prostate gland pressing on the urethra. If this occurs, fluids tend to back up into the renal tubules and cause the hydrostatic pressure in Bowman's capsules to rise. Since any increase in capsular pressure opposes glomerular filtration, the rate of filtration may decrease significantly.

In an average adult, the glomerular filtration rate for the nephrons of both kidneys is about 125 milliliters per minute, or 180,000 milliliters (180 liters) in 24 hours. Since this 24-hour volume is nearly 45 gallons, it is obvious that not all of it is excreted as urine. Instead, most of the fluid that passes through the renal tubules is reabsorbed and reenters the plasma.

---

1. *What general processes are involved with urine formation?*
2. *What forces affect filtration pressure?*
3. *What factors influence the rate of glomerular filtration?*

Fig. 18.8 Two mechanisms involving the macula densa ensure a constant blood flow through the glomerulus.

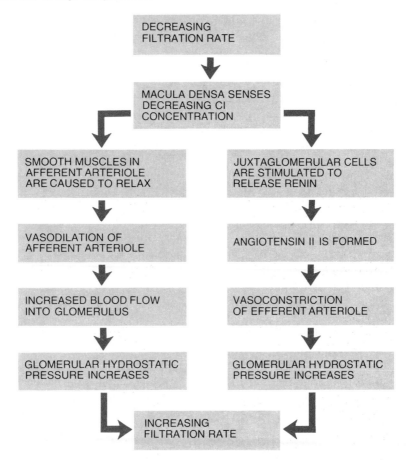

Regulation of Filtration Rate. The regulation of the filtration rate involves the *juxtaglomerular apparatus,* which was described previously (fig. 18.6), and two negative feedback mechanisms. These mechanisms are triggered whenever the filtration rate is decreasing. For example, as the rate decreases, the concentration of chloride ions reaching the *macula densa* in the distal convoluted tubule also decreases. In response, the macula densa signals the smooth muscles in the wall of the afferent arteriole to relax, and the vessel becomes dilated. This action allows more blood to flow into the glomerulus, increasing the glomerular pressure, and as a consequence, the filtration rate rises toward its previous level.

At the same time that the macula densa signals the afferent arteriole to dilate, it stimulates the *juxtaglomerular cells* to release *renin.* This enzyme causes a plasma globulin (angiotensinogen) to form a substance called *angiotensin I,* which is converted quickly to *angiotensin II* by an enzyme (converting enzyme) present in the lungs and plasma.

Angiotensin II is a vasoconstrictor, and it stimulates the smooth muscle cells in the wall of the efferent arteriole to contract, constricting the vessel. As a result of this action, blood tends to back up into the glomerulus, and as the glomerular hydrostatic pressure increases, the filtration rate increases also. (See fig. 18.8.)

These two mechanisms operate together to ensure a constant blood flow through the glomerulus and a relatively stable glomerular filtration rate, in spite of marked changes occurring in the arterial blood pressure.

---

1. *What is the function of the macula densa?*
2. *How does renin help to regulate the filtration rate?*

Tubular Reabsorption

If the composition of the glomerular filtrate entering the renal tubule is compared with that of the urine leaving the tubule, it is clear that changes have occurred as the fluid passed through the tubule (chart 18.1). For example, glucose is present in the filtrate, but is absent in the urine. Also, urea and uric acid are considerably more concentrated in urine than they are in the glomerular filtrate. Such changes in fluid composition are largely the result of **tubular reabsorption**, a process by which substances are transported out of the glomerular filtrate, through the epithelium of the renal tubule, and into the blood of the peritubular capillary (fig. 18.9).

Since the efferent arteriole is narrower than the peritubular capillary, blood flowing from the former into the latter is under relatively low pressure. Also, the wall of this capillary is more permeable than that of other capillaries. Both of these factors enhance the rate of fluid reabsorption from the renal tubule.

Although tubular reabsorption occurs throughout the renal tubule, most of it occurs in the proximal convoluted portion. The epithelial cells in this portion have numerous microscopic projections called *microvilli* that form a "brush border" on their free surfaces. These tiny extensions greatly increase the surface area exposed to glomerular filtrate and enhance the reabsorption process.

Various segments of the renal tubule are adapted to reabsorb specific substances, using particular modes of transport. Glucose reabsorption, for example, occurs primarily through the walls of the proximal tubule by active transport, while water is reabsorbed throughout the length of the renal tubule by osmosis.

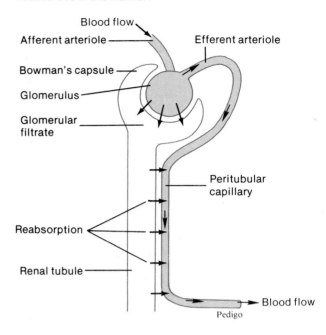

Fig. 18.9 Reabsorption is the process by which substances are transported from the glomerular filtrate into the blood of the peritubular capillary. What substances are reabsorbed in this manner?

Blood flow
Afferent arteriole
Efferent arteriole
Bowman's capsule
Glomerulus
Glomerular filtrate
Peritubular capillary
Reabsorption
Renal tubule
Blood flow
Pedigo

As is described in chapter 3, an active transport mechanism depends on the presence of carrier molecules in a cell membrane. These carriers transport passenger molecules through the membrane, release them, and return to the other side to transport more passenger molecules. Such a mechanism has a *limited transport capacity;* that is, it can only transport a certain number of molecules in a given amount of time, because the number of carriers is limited.

Usually all of the glucose in the glomerular filtrate is reabsorbed, since there are enough carrier molecules to transport them. Sometimes, however, the plasma glucose concentration increases, and if it reaches a critical level, called the *renal plasma threshold,* there will be more glucose molecules in the filtrate than the active transport mechanism can handle. As a result, some glucose will remain in the filtrate and be excreted in the urine.

The appearance of glucose in the urine is called *glucosuria* (or *glycosuria*). This condition may occur following the administration of glucose intravenously or in a patient with diabetes mellitus. If the cause is diabetes mellitus, the blood glucose concentration rises because of insufficient insulin from the pancreas.

Amino acids also enter the glomerular filtrate and are reabsorbed in the proximal convoluted tubule, apparently by three different active transport mechanisms. Each mechanism is thought to reabsorb a different group of amino acids, whose members have molecular similarities. As a result of their actions, only a trace of amino acids usually remains in urine.

Although the glomerular filtrate is nearly free of protein, some *albumin* may be present. These proteins have relatively small molecules, and they are reabsorbed by *pinocytosis* through the brush border of epithelial cells lining the proximal convoluted tubule. Once they are inside an epithelial cell, the proteins probably are converted to amino acids and moved into the blood of the peritubular capillary.

Other substances reabsorbed by the epithelium of the proximal convoluted tubule include creatine, lactic acid, citric acid, uric acid, ascorbic acid (vitamin C), phosphate ions, sulfate ions, calcium ions, potassium ions, and sodium ions. As a group, these substances are reabsorbed by active transport mechanisms with limited transport capacities (like that of glucose). Such a substance usually does not appear in the urine until its concentration in the glomerular filtrate exceeds its particular threshold.

Sodium and Water Reabsorption. Substances that remain in the renal tubule tend to become more and more concentrated as water is reabsorbed from the filtrate. Most of this water reabsorption occurs *passively* by *osmosis* in the proximal convoluted tubule and is closely associated with the active reabsorption of sodium ions. In fact, if sodium reabsorption increases, water reabsorption increases; if sodium reabsorption decreases, water reabsorption decreases also.

About 70% of the *sodium ion reabsorption* occurs in the proximal segment of the renal tubule by active transport (sodium pump mechanism). As these positively charged ions (Na^+) are moved through the tubular wall, negatively charged ions including chloride (Cl^-), phosphate (PO_4^{-3}), and bicarbonate (HCO_3^-) accompany them. This movement of negatively charged ions is due to the electrochemical attraction between particles of opposite charge. It is termed **passive transport** because it does not require a direct expenditure of cellular energy.

As more and more sodium ions are actively transported into the peritubular capillary, along with various negatively charged ions, the concentration of solutes within the peritubular blood is increased. Furthermore, since water moves through cell membranes

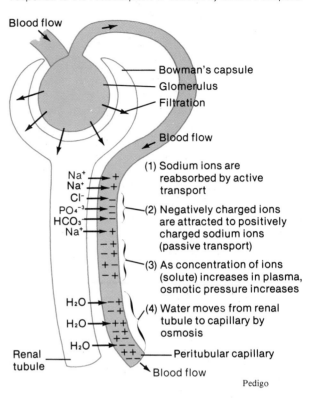

Fig. 18.10 Water reabsorption by osmosis occurs in response to the reabsorption of sodium by active transport.

Blood flow

Bowman's capsule
Glomerulus
Filtration

Blood flow

(1) Sodium ions are reabsorbed by active transport

(2) Negatively charged ions are attracted to positively charged sodium ions (passive transport)

(3) As concentration of ions (solute) increases in plasma, osmotic pressure increases

(4) Water moves from renal tubule to capillary by osmosis

Renal tubule

Peritubular capillary

Blood flow

Pedigo

from regions of lesser solute concentration (hypotonic) toward regions of greater solute concentration (hypertonic), water is transported by osmosis from the renal tubule into the peritubular capillary. Because of the movement of solutes and water into the peritubular capillary, the volume of fluid within the renal tubule is greatly reduced (fig. 18.10).

1. What substances present in glomerular filtrate are not normally present in urine?
2. What mechanisms are responsible for reabsorption of solutes from the glomerular filtrate?
3. Describe the role of passive transport in urine formation.

Sodium ions continue to be reabsorbed by active transport as the tubular fluid moves through the loop of Henle, the distal convoluted segment, and the collecting duct. As a result, almost all the sodium that enters the renal tubule as glomerular filtrate may be reabsorbed before the urine is excreted, and consequently, water also continues to be reabsorbed passively by osmosis in various segments of the renal tubule.

Additional water may be reabsorbed due to the action of antidiuretic hormone (ADH). As is discussed in chapter 12, ADH is produced by neurons in the *hypothalamus.* It is released from the posterior lobe of the pituitary gland in response to a decreasing concentration of water in the blood.

When it reaches the kidney, ADH causes an increase in the permeability of the epithelial linings of the distal convoluted tubule and the collecting duct, and water moves rapidly out of these segments by osmosis. Consequently, the urine volume is reduced, and it becomes more concentrated. (See chart 18.2.)

Thus ADH stimulates the production of concentrated urine, which contains soluble wastes and other substances in a minimum of water; it also inhibits the loss of body fluids whenever there is a danger of dehydration. If the water concentration of the body fluids is excessive, ADH secretion is decreased. In the absence of ADH, the epithelial linings of the distal segment and the collecting duct become less permeable to water, less water is reabsorbed, and the urine tends to be more dilute.

Urea and Uric Acid Excretion

Urea is a by-product of amino acid metabolism. Consequently, its plasma concentration is directly related to the amount of protein in the diet. Urea enters the renal tubule by filtration, and about 50% of it is reabsorbed (passively) by diffusion, while the remainder is excreted in the urine.

Uric acid, which results from the metabolism of certain organic bases in nucleic acids, is reabsorbed by active transport. Although this mechanism seems able to reabsorb all the uric acid normally present in glomerular filtrate, about 10% of the amount filtered is excreted in the urine. This amount is apparently *secreted* into the renal tubule.

Chart 18.2 Major events in the regulation of urine concentration and volume

1. Concentration of water in the blood decreases
2. Osmoreceptors in hypothalamus of brain are stimulated by increase in osmotic pressure of body fluid
3. Hypothalamus signals the posterior pituitary gland to release ADH
4. Blood carries ADH to kidneys
5. ADH causes distal convoluted tubules and collecting ducts to increase water reabsorption by osmosis
6. Urine becomes more concentrated and urine volume decreases

Gout is a disorder in which the plasma concentration of uric acid becomes abnormally high. Since uric acid is a relatively insoluble substance, it tends to precipitate when it is present in excess. As a result, crystals of uric acid may be deposited in joints and other tissues, where they produce inflammation and extreme pain. The joints of the great toes are affected most commonly, but other joints in the hands and feet often are involved.

This condition, which often seems to be inherited, is sometimes treated by administering drugs that inhibit the reabsorption of uric acid and thus increase its excretion.

1. What role does the hypothalamus play in regulating urine concentration and volume?
2. Explain how urea and uric acid are excreted.

Tubular Secretion

Tubular secretion is the process by which certain substances are transported from the plasma of the peritubular capillary into the fluid of the renal tubule. As a result, the amount of a particular substance excreted in the urine may be greater than the amount filtered from the plasma in the glomerulus. (See fig. 18.11.)

Some substances are secreted by active transport mechanisms similar to those that function in reabsorption. *Secretory mechanisms,* however, transport substances in the opposite direction. For example, certain organic compounds, including penicillin, creatinine, and histamine, are actively secreted into the tubular fluid by the epithelium of the proximal convoluted segment.

Fig. 18.11 Secretory mechanisms act to move substances from the plasma of the peritubular capillary into the fluid of the renal tubule.

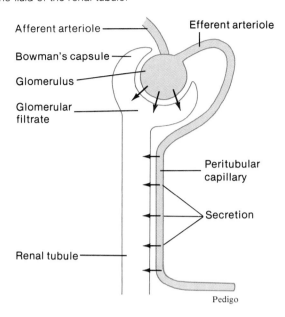

Afferent arteriole
Efferent arteriole
Bowman's capsule
Glomerulus
Glomerular filtrate
Peritubular capillary
Secretion
Renal tubule

Pedigo

Hydrogen ions also are actively secreted. In this case, the proximal segment of the renal tubule is specialized to secrete large quantities of hydrogen ions between plasma and tubular fluid, where hydrogen ion concentrations are similar. This secretion of hydrogen ions plays an important role in the regulation of the pH of body fluids, as is explained in chapter 19.

Although most of the *potassium ions* in the glomerular filtrate are actively reabsorbed in the proximal convoluted tubule, some may be secreted passively in the distal segment and collecting duct. During this process, the active reabsorption of sodium ions out of the tubular fluid creates a negative electrical charge within the tube. Since positively charged potassium ions (K$^+$) are attracted to regions that are negatively charged, these ions move through the tubular epithelium and enter the tubular fluid.

Chart 18.3 summarizes the functions of various parts of the nephron.

---

1. *Define tubular secretion.*
2. *What substances are actively secreted?*
3. *How does the reabsorption of sodium affect the secretion of potassium?*

Composition of Urine

The composition of urine varies considerably from time to time because of differences in dietary intake and physical activity. In addition to containing about

Chart 18.3 Functions of nephron parts

Part	Function
Renal Corpuscle	
Glomerulus	Filtration of water and dissolved substances from plasma
Bowman's capsule	Receives glomerular filtrate
Renal Tubule	
Proximal convoluted tubule	Reabsorption of glucose, amino acids, lactic acid, uric acid, ascorbic acid, phosphate ions, sulfate ions, calcium ions, potassium ions, and sodium ions by active transport
	Reabsorption of water by osmosis
	Reabsorption of chloride ions and other negatively charged ions by electrochemical attraction
	Active secretion of substances such as penicillin, histamine, and hydrogen ions
Descending limb of loop of Henle	Reabsorption of water by osmosis
Ascending limb of loop of Henle	Reabsorption of chloride ions by active transport and passive reabsorption of sodium ions
Distal convoluted tubule	Reabsorption of sodium ions by active transport
	Reabsorption of water by osmosis
	Active secretion of hydrogen ions
	Passive secretion of potassium ions by electrochemical attraction

From Hole, John W. Jr., *Human Anatomy and Physiology* 3d ed. © 1978, 1981, 1984 Wm. C. Brown Publishers, Dubuque, Iowa. All Rights Reserved. Reprinted by permission.

95% water, it usually contains *urea* and *uric acid.* It also may contain a trace of *amino acids,* as well as a variety of *electrolytes* whose concentrations tend to vary directly with the amounts included in the diet. (See chart 18.1.)

The volume of urine produced usually varies between 0.6 and 2.5 liters per day. The exact volume is influenced by such factors as the fluid intake, the environmental temperature, the relative humidity of the surrounding air, and the person's emotional condition, respiratory rate, and body temperature. An output of 50–60 cubic centimeters of urine per hour is considered normal, and an output of less than 30 cubic centimeters per hour may be an indication of kidney failure.

---

1. *List the normal constituents of urine.*
2. *What factors affect the volume of urine produced?*

After being formed by the nephrons, urine passes from the collecting ducts through openings in the renal papillae and enters the calyces of the kidney. From there it passes through the renal pelvis and is conveyed by a ureter to the urinary bladder. It is excreted to the outside of the body by means of the urethra.

The Ureter

The **ureter** is a tubular organ about 25 centimeters (10 inches) long, which begins as the funnel-shaped renal pelvis (fig. 18.1). It extends downward behind the parietal peritoneum and parallel to the vertebral column. Within the pelvic cavity, it courses forward and medially to join the urinary bladder from underneath.

The wall of the ureter is composed of three layers of tissues. The inner layer *(mucous coat)* is continuous with the linings of the renal tubules and the urinary bladder. The middle layer *(muscular coat)* consists largely of smooth muscle fibers. The outer layer *(fibrous coat)* is composed of connective tissue.

Although the ureter is simply a tube leading from the kidney to the urinary bladder, its muscular wall helps to move urine. Muscular peristaltic waves that originate in the renal pelvis force urine along the length of the ureter.

When such a peristaltic wave reaches the urinary bladder, it causes a jet of urine to spurt into the bladder. The opening through which the urine enters is covered by a flaplike fold of mucous membrane. This fold acts as a valve, allowing urine to move inward from the ureter but preventing it from backing up.

Kidney stones, which usually are composed of uric acid, calcium oxalate, calcium phosphate, or magnesium phosphate, sometimes form in the renal pelvis. If such a stone passes into a ureter it may stimulate severe pain. This pain commonly begins in the region of the kidney and tends to radiate into the abdomen, pelvis, and legs. It also may be accompanied by nausea and vomiting.

---

1. Describe the structure of a ureter.
2. How is urine moved from the renal pelvis?
3. What prevents urine from backing up from the urinary bladder into the ureters?

The Urinary Bladder

The **urinary bladder** is a hollow, distensible, muscular organ. It is located within the pelvic cavity, behind the symphysis pubis, and below the parietal peritoneum. (See fig. 18.12.)

Although the bladder is somewhat spherical, its shape is altered by the pressures of surrounding organs. When it is empty, the inner wall of the bladder is thrown into many folds, but as it fills with urine, the wall becomes smoother. At the same time, the superior surface of the bladder expands upward into a dome.

The internal floor of the bladder consists of a triangular area called the *trigone,* which has an opening at each of its three angles. (See fig. 18.13.) Posteriorly, at the base of the trigone, the openings are those of the ureters. The *internal urethral orifice,* which opens into the urethra, is located anteriorly at the apex of the trigone.

The wall of the urinary bladder consists of four layers. The inner layer, or *mucous coat,* includes several thicknesses of transitional epithelial cells. This tissue is adapted to changes in tension. More precisely, its thickness changes as the bladder expands and contracts, so that during distension it may be only two or three cells thick, while during contraction it may be five or six cells thick. (See chapter 5.)

The second layer of the wall is the *submucous coat.* It consists of connective tissue and contains many elastic fibers.

The third layer, or *muscular coat,* is composed primarily of coarse bundles of smooth muscle fibers. These bundles are interlaced in all directions and depths, and together they comprise the **detrusor muscle.** This muscle is supplied with parasympathetic nerve fibers that function in the micturition reflex, which is discussed in the next section of this chapter.

The outer layer, or *serous coat,* consists of the parietal peritoneum. This layer occurs only on the upper surface of the bladder. Elsewhere, the outer coat is composed of fibrous connective tissue.

Since the linings of the ureters and the urinary bladder are continuous, infectious agents such as bacteria may ascend from the bladder into the ureters. An inflammation of the bladder, which is called *cystitis,* occurs more commonly in women than men because the female urethral pathway is shorter. An inflammation of the ureter is called *ureteritis.*

Fig. 18.12 The urinary bladder is located within the pelvic cavity and behind the symphysis pubis.

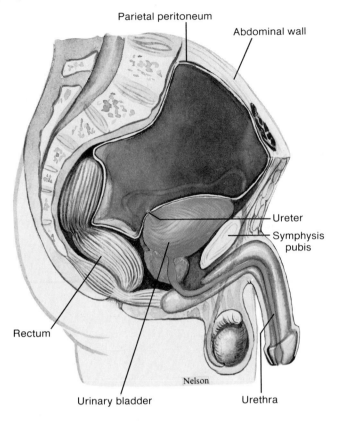

Parietal peritoneum

Abdominal wall

Ureter

Symphysis pubis

Rectum

Nelson

Urinary bladder

Urethra

---

1. *Describe the trigone of the urinary bladder.*
2. *Describe the structure of the bladder wall.*
3. *What kind of nerve fibers supply the detrusor muscle?*

Micturition

Micturition, or urination, is the process by which urine is expelled from the urinary bladder. It involves the contraction of the detrusor muscle and may be aided by contractions of muscles in the abdominal wall and pelvic floor and fixation of the thoracic wall and diaphragm. Micturition also involves the relaxation of the *external urethral sphincter.* This muscle surrounds the urethra about 3 centimeters (1.2 inches) from the bladder and is composed of voluntary muscle tissue.

The need to urinate usually is stimulated by distension of the bladder wall as it fills with urine. As the wall expands, stretch receptors are stimulated, and the micturition reflex is triggered.

Fig. 18.13 The trigone of the urinary bladder is marked by the openings of the ureters and urethra.

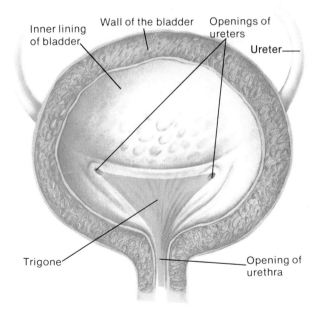

Inner lining of bladder

Wall of the bladder

Openings of ureters

Ureter

Trigone

Opening of urethra

The *micturition reflex center* is located in the sacral segments of the spinal cord. When it is signalled by sensory impulses from the stretch receptors, *parasympathetic* motor impulses travel out to the detrusor muscle, and it undergoes rhythmic contractions in response. This action is accompanied by a sensation of urgency.

Although the urinary bladder may hold as much as 600 milliliters of urine, the desire to urinate usually is experienced when it contains about 150 milliliters. Then, as the volume of urine increases to 300 milliliters or more, the sensation of fullness becomes increasingly uncomfortable.

Since the external urethral sphincter can be consciously controlled, it ordinarily remains contracted until a decision is made to urinate. This control is aided by nerve centers in the *midbrain* and *cerebral cortex* that are able to inhibit the micturition reflex. When a person decides to urinate, the external urethral sphincter is allowed to relax, and the micturition reflex is no longer inhibited. Nerve centers within the *pons* and *hypothalamus* of the brain function to make the micturition reflex more effective. Consequently, the detrusor muscle contracts, and urine is excreted to the outside through the urethra. Within a few moments, the neurons of the micturition reflex seem to fatigue, the detrusor muscle relaxes, and the bladder begins to fill with urine again.

Damage to the spinal cord above the sacral region may result in loss of voluntary control of urination. If the micturition reflex center and its sensory and motor fibers are uninjured, however, micturition may continue to occur reflexly. In this case, the bladder collects urine until its walls are stretched enough to trigger a micturition reflex, and the detrusor muscle contracts in response. This condition is called an *automatic bladder.*

The Urethra

The **urethra** (fig. 18.12) is a tube that conveys urine from the urinary bladder to the outside of the body. Its wall is lined with mucous membrane and contains a relatively thick layer of smooth muscle tissue, whose fibers are generally directed longitudinally. It also contains numerous mucous glands, called *urethral glands,* that secrete mucus into the urethral canal.

---

1. Describe the events of micturition.
2. How is it possible to inhibit the micturition reflex?
3. Describe the structure of the urethra.

Clinical Terms Related to the Urinary System

anuria (ah-nu're-ah)—an absence of urine due to failure of kidney function or to an obstruction in a urinary pathway.

bacteriuria (bak-te″re-u're-ah)—bacteria in the urine.

cystectomy (sis-tek'to-me)—surgical removal of the urinary bladder.

cystitis (sis-ti'tis)—inflammation of the urinary bladder.

cystoscope (sis'to-skōp)—instrument used for visual examination of the interior of the urinary bladder.

cystotomy (sis-tot'o-me)—an incision of the wall of the urinary bladder.

diuresis (di″u-re'sis)—an increased production of urine.

dysuria (dis-u're-ah)—painful or difficult urination.

enuresis (en″u-re'sis)—uncontrolled urination.

hematuria (hem″ah-tu're-ah)—blood in the urine.

nephrectomy (ne̅-frek'to-me)—surgical removal of a kidney.

nephrolithiasis (nef″ro-lĭ-thi'ah-sis)—presence of a stone(s) in the kidney.

nephroptosis (nef″rop-to'sis)—a movable or displaced kidney.

oliguria (ol″ĭ-gu're-ah)—a scanty output of urine.

polyuria (pol″e-u're-ah)—an excessive output of urine.

pyelolithotomy (pi″e̅-lo-lĭ-thot'o-me)—removal of a stone from the renal pelvis.

pyelonephritis (pi″e-lon-ne-fri'tis)—inflammation of the renal pelvis.

pyelotomy (pi″e̅-lot'o-me)—incision into the renal pelvis.

pyuria (pi-u're-ah)—pus in the urine.

uremia (u-re'me-ah)—condition in which substances ordinarily excreted in the urine accumulate in the blood.

ureteritis (u-re″ter-i'tis)—inflammation of the ureter.

urethritis (u″re-thri'tis)—inflammation of the urethra.

Introduction

The urinary system consists of kidneys, ureters, urinary bladder, and urethra.

The Kidney

1. Location of the kidneys
   a. The kidneys are high on the posterior wall of the abdominal cavity.
   b. They are positioned behind the parietal peritoneum.
2. Structure of a kidney
   a. A kidney contains a hollow renal sinus.
   b. The ureter expands into the renal pelvis.
   c. Renal papillae project into renal sinus.
   d. Kidney tissue is divided into medulla and cortex.
3. Functions of the kidneys
   a. Kidneys remove metabolic wastes from the blood and excrete them to the outside.
   b. They also help to regulate red blood cell production, blood pressure, and the volume, composition, and pH of the blood.
4. Renal blood vessels
   a. Arterial blood flows through the renal artery, interlobar arteries, arciform arteries, interlobular arteries, and afferent arterioles.
   b. Venous blood returns through a series of vessels that correspond to those of the arterial pathways.
5. The nephrons
   a. Structure of the nephron
      (1) The nephron is the functional unit of the kidney.
      (2) It consists of a renal corpuscle and a renal tubule.
         (a) The corpuscle consists of a glomerulus and Bowman's capsule.
         (b) Portions of the renal tubule include the proximal convoluted tubule, the loop of Henle (ascending and descending limbs), the distal convoluted tubule, and the collecting duct.
      (3) The collecting duct empties into the minor calyx of the renal pelvis.
   b. Blood supply of a nephron
      (1) The glomerular capillary receives blood from the afferent arteriole and passes it to the efferent arteriole.
      (2) The efferent arteriole gives rise to the peritubular capillary system that surrounds the renal tubule.
   c. Juxtaglomerular apparatus
      (1) The juxtaglomerular apparatus is located at the point of contact between the distal convoluted tubule and the afferent and efferent arterioles.
      (2) It consists of the macula densa and the juxtaglomerular cells.

Urine Formation

Nephrons function to remove wastes from blood and to regulate water and electrolyte concentrations. Urine is the end product of these functions.

1. Glomerular filtration
   a. Urine formation begins when water and dissolved materials are filtered out of the glomerular capillary.
   b. The glomerular capillaries are much more permeable than the capillaries in other tissues.
2. Filtration pressure
   a. Filtration is due mainly to hydrostatic pressure inside the glomerular capillaries.
   b. Osmotic pressure of the plasma and hydrostatic pressure in the Bowman's capsule also affect filtration.
   c. Filtration pressure is the net force acting to move material out of the glomerulus and into the Bowman's capsule.
   d. The composition of the filtrate is similar to tissue fluid.
3. Filtration rate
   a. The rate of filtration varies with the filtration pressure.
   b. Filtration pressure changes with the diameters of the afferent and efferent arterioles.
   c. As osmotic pressure in the glomerulus increases, filtration decreases.
   d. As the hydrostatic pressure in Bowman's capsule increases, the filtration rate decreases.
   e. The kidneys produce about 125 ml of glomerular fluid per minute, most of which is reabsorbed.
4. Regulation of filtration rate
   a. When the filtration rate decreases, the macula densa causes the afferent arteriole to dilate, increasing blood flow through the glomerulus and increasing the filtration rate.
   b. The macula densa also causes the juxtaglomerular cells to release renin, which triggers a series of changes leading to constriction of the efferent arteriole, increasing the glomerular hydrostatic pressure and increasing the filtration rate.

5. Tubular reabsorption
   a. Substances are selectively reabsorbed.
   b. The peritubular capillary is adapted for reabsorption by being very permeable.
   c. Most reabsorption occurs in the proximal tubule, where the epithelial cells possess microvilli.
   d. Various substances are reabsorbed in particular segments of the renal tubule by different modes of transport.
      (1) Glucose and amino acids are reabsorbed by active transport.
      (2) Water is reabsorbed by osmosis.
   e. Active transport mechanisms have limited transport capacities.
   f. Substances that remain in the filtrate are concentrated as water is reabsorbed.
   g. Sodium ions are reabsorbed by active transport.
      (1) As positively charged sodium ions are transported out of the filtrate, negatively charged ions accompany them.
      (2) Water is passively reabsorbed by osmosis.

6. Regulation of urine concentration and volume
   a. Most sodium is reabsorbed before urine is excreted.
   b. ADH causes the permeability of the distal tubule and collecting duct to increase, and thus promotes the reabsorption of water.

7. Urea and uric acid excretion
   a. Urea is reabsorbed passively by diffusion.
   b. Uric acid is reabsorbed by active transport and secreted into the renal tubule.

8. Tubular secretion
   a. This is the process by which certain substances are transported from the plasma to the tubular fluid.
   b. Various organic compounds and hydrogen ions are secreted actively.
   c. Potassium ions may be secreted passively.

9. Composition of urine
   a. Urine is about 95% water, and it usually contains urea and uric acid.
   b. It may contain a trace of amino acids and varying amounts of electrolytes.
   c. The volume of urine varies with the fluid intake and with certain environmental factors.

Elimination of Urine

1. The ureter
   a. The ureter extends from the kidney to the urinary bladder.
   b. Peristaltic waves in the ureter force urine to the bladder.

2. The urinary bladder
   a. The urinary bladder stores urine and forces it into the urethra.
   b. The openings for the ureters and urethra are located at the three angles of the trigone.

3. Micturition
   a. Micturition is the process by which urine is expelled.
   b. It involves contraction of the detrusor muscle and relaxation of the external urethral sphincter.
   c. Micturition reflex
      (1) Stretch receptors in the bladder wall are stimulated by distension.
      (2) The micturition reflex center in the sacral spinal cord sends parasympathetic motor impulses to the detrusor muscle.

4. The urethra
   The urethra conveys urine from the bladder to the outside.

Application of Knowledge

1. If an infant is born with a narrowing of the renal arteries, what effect would this condition have on the volume of urine produced?

2. If a patient who has had major abdominal surgery receives intravenous fluids equal to the volume of blood lost during surgery, would you expect the volume of urine produced to be greater or less than normal? Why?

Review Activities

1. Name the organs of the urinary system and list their general functions.

2. Describe the external and internal structure of a kidney.

3. List the functions of the kidneys.

4. Name the vessels through which blood passes as it travels from the renal artery to the renal vein.

5. Distinguish between a renal corpuscle and a renal tubule.

6. Name the parts through which fluid passes as it travels from the glomerulus to the collecting duct.

7. Describe the location and structure of the juxtaglomerular apparatus.

8. Define filtration pressure.

9. Compare the composition of the glomerular filtrate with that of blood plasma.

10. Explain how the diameters of the afferent and efferent arterioles affect the rate of glomerular filtration.

11. Explain how changes in the osmotic pressure of the blood plasma may affect the rate of glomerular filtration.

12. Explain how the hydrostatic pressure of Bowman's capsule affects the rate of glomerular filtration.

13. Describe two mechanisms by which the juxtaglomerular apparatus helps to regulate filtration rate.

14. Discuss the reason tubular reabsorption is said to be a selective process.

15. Explain how the peritubular capillary is adapted for reabsorption.

16. Explain how the epithelial cells of the proximal convoluted tubule are adapted for reabsorption.

17. Explain why active transport mechanisms have limited transport capacities.

18. Define renal plasma threshold.

19. Explain how amino acids and proteins are reabsorbed.

20. Describe the effect of sodium reabsorption on the reabsorption of negatively charged ions.

21. Explain how sodium reabsorption affects water reabsorption.

22. Describe the function of ADH.

23. Compare the processes by which urea and uric acid are reabsorbed.

24. Explain how potassium ions may be secreted passively.

25. List the more common wastes found in urine and their sources.

26. List some factors that affect the volume of urine produced each day.

27. Describe the structure and function of a ureter.

28. Explain how the muscular wall of the ureter aids in moving urine.

29. Describe the structure and location of the urinary bladder.

30. Define detrusor muscle.

31. Describe the micturition reflex.

32. Explain how the micturition reflex can be voluntarily controlled.

Water and Electrolyte Balance

Cell functions and, indeed, cell survival depend upon homeostasis—the existence of a stable cellular environment. In such an environment, body cells are supplied continually with oxygen and nutrients, and the waste products resulting from their metabolic activities are carried away continually. At the same time, the concentrations of water and dissolved electrolytes in the cellular fluids and their surroundings remain constant. This condition requires the maintenance of a *water* and *electrolyte balance.*

After you have studied this chapter, you should be able to

1. Explain what is meant by water and electrolyte balance and discuss the importance of this balance.

2. Describe how body fluids are distributed within compartments, how the fluid composition differs between compartments, and how fluids move from one compartment to another.

3. List the routes by which water enters and leaves the body and explain how water input and output are regulated.

4. Explain how electrolytes enter and leave the body and how the input and output of electrolytes are regulated.

5. List the major sources of hydrogen ions in the body.

6. Distinguish between strong and weak acids and bases.

7. Explain how changing pH values of body fluids are minimized by chemical buffer systems, the respiratory center, and the kidneys.

8. Complete the review activities at the end of this chapter. Note that the items are worded in the form of specific learning objectives. You may want to refer to them before reading the chapter.

acid   (as′id)

base   (bās)

buffer system   (buf′er sis′tem)

electrolyte balance   (e-lek′tro-līt bal′ans)

extracellular   (ek″strah-sel′u-lar)

intracellular   (in″trah-sel′u-lar)

transcellular   (trans-sel′u-lar)

water balance   (wot′er bal′ans)

de-, separation from: *de*hydration—removal of water from cells or body fluids.

extra-, outside: *extra*cellular fluid—fluid outside of body cells.

im- (or **in-**), not: *im*balance—condition in which factors are not in equilibrium.

intra-, within: *intra*cellular fluid—fluid within body cells.

neutr-, neither one nor the other: *neutr*al—solution that is neither acidic nor basic.

Fig. 19.1 Fluid within the intracellular compartment is separated from fluid in the extracellular compartment by cell membranes. What membranes separate the various components of the extracellular fluid compartment?

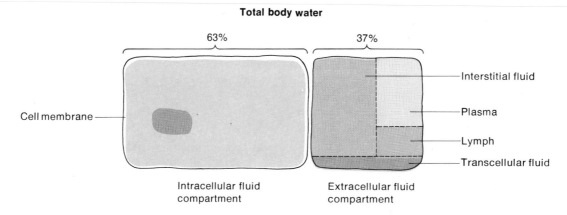

The term *balance* suggests a state of equilibrium, and in the case of water and electrolytes, it means the quantities entering the body are equal to the quantities leaving it. Maintaining such a balance requires mechanisms that ensure lost water and electrolytes will be replaced and any excesses will be expelled. As a result, the quantities within the body are relatively stable at all times.

It is important to remember that water balance and electrolyte balance are interdependent, since the electrolytes are dissolved in the water of body fluids. Consequently, anything that alters the concentrations of electrolytes will necessarily alter the concentration of water by adding solutes to it or by removing solutes from it. Likewise, anything that changes the concentration of water will change the concentrations of electrolytes by making them more concentrated or more diluted.

Distribution of Body Fluids

Water and electrolytes are not evenly distributed throughout the tissues. Instead, they occur in regions or *compartments* that contain fluids of varying compositions. Movement of water and electrolytes between these compartments is regulated, so that their distribution remains stable.

Fluid Compartments

The body of an average adult male is about 63% water by weight. This water (about 40 liters), together with its dissolved electrolytes, is distributed into two major compartments—an intracellular compartment and an extracellular compartment.

The **intracellular fluid compartment** includes all the water and electrolytes enclosed by cell membranes. In other words, intracellular fluid is the fluid within cells, and in an adult it represents about 63% of the total body water.

The **extracellular fluid compartment** includes all the fluid outside cells—within the tissue spaces (interstitial fluid), the blood vessels (plasma), and the lymphatic vessels (lymph). A specialized fraction of the extracellular fluid, called **transcellular fluid,** includes *cerebrospinal fluid* of the central nervous system, *aqueous and vitreous humors* of the eyes, *synovial fluid* of the joints, *serous fluid* within the various body cavities, and fluid *secretions* of glands. All together, the fluids of the extracellular compartment constitute about 37% of the total body water (fig. 19.1).

Composition of Body Fluids

Extracellular fluids generally have similar compositions. They are characterized by relatively high concentrations of sodium, chloride, and bicarbonate ions, and lesser concentrations of potassium, calcium, magnesium, phosphate, and sulfate ions. The plasma fraction of extracellular fluid contains considerably more protein than either interstitial fluid or lymph.

Intracellular fluid contains relatively high concentrations of potassium, phosphate, and magnesium ions. It includes a somewhat greater concentration of sulfate ions, and lesser concentrations of sodium, chloride, and bicarbonate ions than extracellular fluid. Intracellular fluid also has a greater concentration of protein than plasma. These relative concentrations are shown in figure 19.2.

Fig. 19.2 Extracellular fluid is relatively high in concentrations of sodium, chloride, and bicarbonate ions; intracellular fluid is relatively high in concentrations of potassium, magnesium, and phosphate ions.

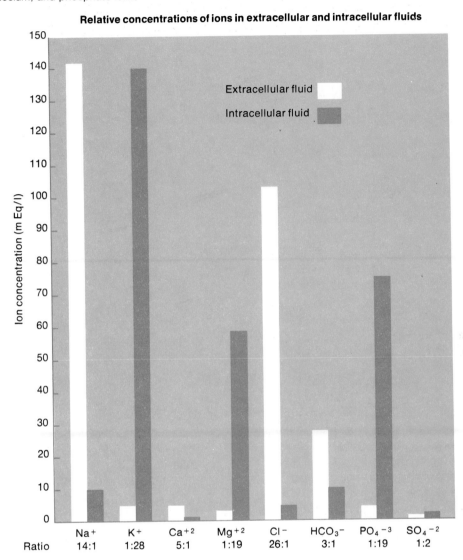

Relative concentrations of ions in extracellular and intracellular fluids

1. Describe the normal distribution of water within the body.
2. What electrolytes are in higher concentrations in extracellular fluid? In intracellular fluid?
3. How does the concentration of protein vary in various body fluids?

Movement of Fluid between Compartments

The movement of water and electrolytes from one compartment to another is largely regulated by two factors—hydrostatic pressure and osmotic pressure. For example, as is explained in chapter 16, fluid leaves the plasma at the arteriole ends of capillaries and enters the interstitial spaces because of the filtration effect of *hydrostatic pressure* (blood pressure). Fluid returns to the plasma from the interstitial spaces at the venule ends of capillaries because of *osmotic pressure*. Likewise, as is mentioned in chapter 17, fluid leaves the interstitial spaces and enters the lymph capillaries because of osmotic pressure that develops within these spaces. As a result of the circulation of lymph, interstitial fluid is returned to the plasma.

Fig. 19.3 Net movements of fluids between compartments result from differences in hydrostatic and osmotic pressures.

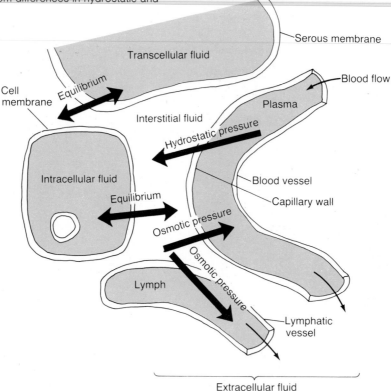

Water Balance

Water balance exists when the total intake of water is equal to the total loss of water.

Water Intake

Although the volume of water gained each day varies from individual to individual, an average adult living in a moderate environment takes in about 2500 milliliters.

Of this amount, probably 60% (1500 milliliters) will be obtained in drinking water or beverages, while another 30% (750 milliliters) will be gained from moist foods. The remaining 10% (250 milliliters) will be a by-product of oxidative metabolism of various nutrients, which is called **water of metabolism.** (See fig. 19.4.)

Regulation of Water Intake

The primary regulator of water intake is thirst. Although the thirst mechanism is poorly understood, it seems to involve the osmotic pressure of extracellular fluid and a *thirst center* in the hypothalamus of the brain.

The movement of fluid between the intracellular and extracellular compartments similarly is controlled by such pressures. Since the hydrostatic pressure within the cells and the surrounding interstitial fluid ordinarily is equal and remains stable, however, any fluid movement that occurs is likely to be the result of changes in osmotic pressure. (See fig. 19.3.)

For example, because the sodium concentration in the interstitial fluid is especially high, a decrease in this concentration will cause a net movement of water from the extracellular compartment into the intracellular compartment by osmosis. As a consequence, cells will tend to swell. Conversely, if the concentration of sodium in the interstitial fluid increases, the net movement of water will be outward from the intracellular compartment, and the cells will shrink as they lose water.

---

1. *What factors control movement of water and electrolytes from one fluid compartment to another?*
2. *How does the sodium concentration within body fluids affect the net movement of water between compartments?*

Fig. 19.4 Major sources of body water.

Fig. 19.5 Routes by which body water is lost. What factors influence water loss by each of these routes?

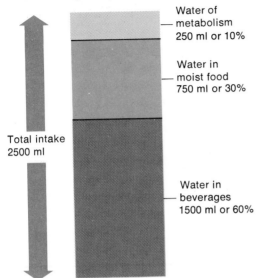

Average daily intake of water

Water of metabolism 250 ml or 10%

Water in moist food 750 ml or 30%

Total intake 2500 ml

Water in beverages 1500 ml or 60%

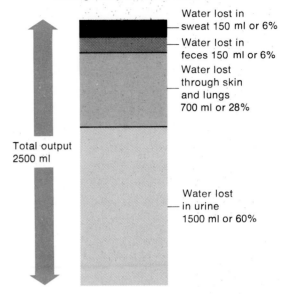

Average daily output of water

Water lost in sweat 150 ml or 6%

Water lost in feces 150 ml or 6%

Water lost through skin and lungs 700 ml or 28%

Total output 2500 ml

Water lost in urine 1500 ml or 60%

As water is lost from the body, the osmotic pressure of the extracellular fluid increases. *Osmoreceptors* in the thirst center are thought to be stimulated by such a change, and as a result, the hypothalamus causes the person to feel thirsty and to seek water.

The thirst mechanism usually is triggered whenever the total body water is decreased by 1 or 2%. As a person drinks water in response to thirst, the act of drinking and the resulting distension of the stomach wall seems to trigger nerve impulses that inhibit the mechanism. Thus the person usually stops drinking long before the swallowed water has been absorbed.

---

1. What is meant by water balance?
2. Where is the thirst center located?
3. What mechanism stimulates fluid intake? What inhibits it?

Water Output

Water normally enters the body through the mouth, but it can be lost by a variety of routes. These include losses in urine, feces, and sweat (sensible perspiration), as well as less obvious losses that occur by diffusion of water through the skin (insensible perspiration) and evaporation of water from the lungs during breathing.

If an average adult takes in 2500 milliliters of water each day, then 2500 milliliters must be eliminated if water balance is to be maintained. Of this volume, perhaps 60% (1500 milliliters) will be lost in urine, 6% (150 milliliters) in feces, and 6% (150 milliliters) in sweat. About 28% (700 milliliters) will be lost by diffusion through the skin and evaporation from the lungs. (See fig. 19.5.) These percentages will, of course, vary with such environmental factors as temperature and relative humidity, and with the amount of physical exercise.

Dehydration is a condition that occurs when the output of water exceeds the intake. This condition may develop following excessive sweating, or as a result of prolonged water deprivation accompanied by continued water output. In either case, as water is lost, the extracellular fluid becomes increasingly more concentrated, and water tends to leave cells by osmosis. Dehydration also may accompany illnesses in which excessive fluids are lost as a result of prolonged vomiting or diarrhea. During dehydration, the skin and mucous membranes of the mouth feel dry, body weight is lost, and a high fever may develop as the body temperature-regulating mechanism becomes less effective due to a lack of water. In severe cases, as waste products accumulate in the extracellular fluid, symptoms of cerebral disturbances including mental confusion, delirium, and coma may develop.

Regulation of Water Balance

The primary regulator of water output involves urine producton. As is discussed in chapter 18, the volume of water excreted in the urine is regulated mainly by activity in the distal convoluted tubule and collecting duct of the nephron. The epithelial linings of these segments of the renal tubule remain relatively impermeable to water unless antidiuretic hormone (ADH) is present. The action of ADH causes water to be reabsorbed in these segments and thus be conserved. In the absence of ADH, less water is reabsorbed and the urine volume increases. (See chart 18.2.)

Diuretics are substances that promote the production of urine. A number of common substances, such as caffeine in coffee and tea, have diuretic effects, as do a variety of drugs used to reduce the volume of body fluids.

Although diuretics produce their effects in different ways, some, such as alcohol and various narcotic drugs, promote urine formation by inhibiting the release of ADH. Others, including a group called *mercurial diuretics,* inhibit the reabsorption of sodium and other solutes in certain portions of the renal tubules. As a consequence, the osmotic pressure of the tubular fluid increases, and the reabsorption of water by osmosis is reduced.

---

1. By what routes is water lost from the body?
2. What role do the renal tubules play in the regulation of water balance?

Electrolyte Balance

An **electrolyte balance** exists when the quantities of the various electrolytes gained by the body are equal to those lost.

Electrolyte Intake

The electrolytes of greatest importance to cellular functions are those that release ions of sodium, potassium, calcium, magnesium, chloride, sulfate, phosphate, and bicarbonate. These substances are obtained primarily from foods, but also may occur in drinking water and other beverages. In addition, some electrolytes occur as by-products of various metabolic reactions.

Regulation of Electrolyte Intake

Ordinarily, a person obtains sufficient electrolytes by responding to hunger and thirst. When there is a severe electrolyte deficiency, however, a person may experience a *salt craving,* which is a strong desire to eat salty foods.

Electrolyte Output

Some electrolytes are lost from the body by perspiration. The quantities leaving by this route will vary with the amount of perspiration excreted. Greater quantities are lost on warmer days and during times of strenuous exercise. Also, varying amounts of electrolytes are lost with the feces. The greatest electrolyte output occurs in the urine as a result of kidney functions.

---

1. What electrolytes are most important to cellular functions?
2. What mechanisms ordinarily regulate electrolyte intake?
3. By what routes are electrolytes lost from the body?

Regulation of Electrolyte Balance

The concentrations of postively charged ions, such as sodium (Na^+), potassium (K^+), and calcium (Ca^{+2}), are particularly important. Certain levels of these ions, for example, are necessary for the conduction of nerve impulses, the contraction of muscle fibers, and for maintaining the normal permeability of cell membranes. Thus it is vital that their concentrations be regulated.

Sodium ions account for nearly 90% of the positively charged ions in extracellular fluid. The primary mechanism regulating these ions involves the kidneys and the hormone, *aldosterone.* As is explained in chapter 12, this hormone is secreted by the adrenal cortex. Its presence causes an increase in sodium reabsorption in various segments of the renal tubule.

Aldosterone also functions in regulating potassium. In fact, the most important stimulus for its secretion is a rising potassium concentration, which seems to stimulate the cells of the adrenal cortex directly. This hormone enhances the reabsorption of sodium and causes the secretion of potassium at the same time. (See fig. 19.6.)

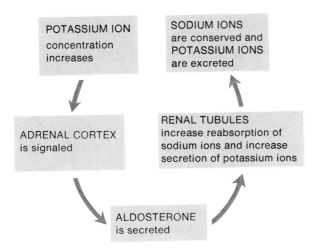

Acid-Base Balance

As is discussed in chapter 2, electrolytes that ionize in water and release hydrogen ions are called *acids*. Substances that combine with hydrogen ions are called *bases*. The concentrations of acids and bases within body fluids must be controlled if homeostasis is to be maintained.

Sources of Hydrogen Ions

Most of the hydrogen ions in the body fluids originate as by-products of metabolic processes, although small quantities may enter in foods.

The major metabolic sources of hydrogen ions include the following:

1. **Aerobic respiration of glucose.** This process results in the production of carbon dioxide and water. Carbon dioxide diffuses out of cells and reacts with water in extracellular fluid to form *carbonic acid*. The resulting carbonic acid then ionizes to release hydrogen ions and bicarbonate ions.

$$H_2CO_3 \rightarrow H^+ + HCO_3^-$$

2. **Anaerobic respiration of glucose.** When glucose is utilized anaerobically, *lactic acid* is produced and hydrogen ions are added to body fluids.

3. **Incomplete oxidation of fatty acids.** The incomplete oxidation of fatty acids results in the production of *acidic ketone bodies* that cause the hydrogen ion concentration to increase.

4. **Oxidation of amino acids containing sulfur.** The oxidation of sulfur-containing amino acids results in the production of *sulfuric acid* (H_2SO_4) that ionizes to release hydrogen ions and sulfate ions.

5. **Breakdown of phosphoproteins and nucleoproteins.** Phosphoproteins and nucleoproteins contain phosphorus, and their oxidation results in the production of *phosphoric acid* (H_3PO_4).

The acids produced by various metabolic processes vary in strength, and so their effects on the hydrogen ion concentration of body fluids vary also. (See fig. 19.7.)

As was mentioned in chapter 12, the concentration of *calcium ions* in extracellular fluid is regulated mainly by the parathyroid glands. Whenever the calcium concentration drops below the normal concentration, these glands are stimulated directly, and they secrete *parathyroid hormone* in response.

Parathyroid hormone causes the concentrations of calcium and phosphate in the extracellular fluid to increase.

Generally, the concentrations of negatively charged ions are controlled secondarily by the regulatory mechanisms that control positively charged ions. For example, chloride ions (Cl^-), the most abundant negatively charged ions in extracellular fluid, are passively reabsorbed from the renal tubule as a result of the active reabsorption of sodium ions. That is, the chloride ions are electrically attracted to the positively charged sodium ions and accompany them as they are reabsorbed. (See chapter 18.)

Some negatively charged ions, such as phosphate (PO_4^{-3}) and sulfate (SO_4^{-2}), are also partially regulated by active transport mechanisms that have limited transport capacities. Thus if the extracellular phosphate concentration is low, the phosphate ions in the renal tubules are conserved. On the other hand, if the renal plasma threshold is exceeded, the excess phosphate will be excreted in the urine.

---

1. *How are the levels of sodium and potassium ions controlled?*
2. *How is calcium regulated?*
3. *What mechanism functions to regulate the concentrations of most negatively charged ions?*

---

1. *Distinguish between an acid and a base.*
2. *What are the major sources of hydrogen ions in the body?*

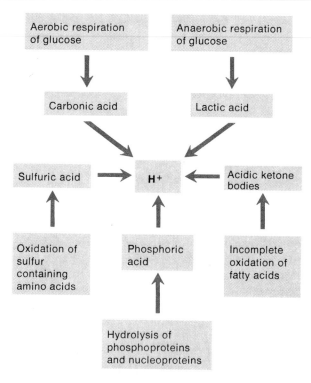

Strengths of Acids and Bases

Acids that ionize more completely are termed *strong acids,* while those that ionize less completely are termed *weak acids.* For example, the hydrochloric acid (HCl) of gastric juice is a strong acid, while the carbonic acid (H_2CO_3) produced when carbon dioxide reacts with water is weak.

Bases are substances, like hydroxyl ions (OH^-), that will combine with hydrogen ions. *Chloride ions* (Cl^-) and *bicarbonate ions* (HCO_3^-) are also bases. Furthermore, since chloride ions combine less readily with hydrogen ions, they are *weak* bases, while the bicarbonate ions, which combine more readily with hydrogen ions, are *strong bases.*

Regulation of Hydrogen Ion Concentration

The concentration of hydrogen ions in body fluids or pH is regulated primarily by acid-base buffer systems, the activity of the respiratory center in the brain stem, and the functions of nephrons in the kidneys.

Acid-Base Buffer Systems. **Acid-base buffer systems** occur in all body fluids and usually are composed of sets of two or more chemical substances. Such chemicals can combine with acids or bases when they occur in excess. More specifically, the substances of a buffer system can convert strong acids, which tend to release large quantities of hydrogen ions, into weak acids, which release fewer hydrogen ions. Likewise, they can combine with strong bases and change them into weak bases. Such actions help to minimize pH changes in body fluids.

The three most important acid-base buffer systems in body fluids are these:

1. **Bicarbonate buffer system.** The bicarbonate buffer system, which is present in intracellular and extracellular body fluids, consists of carbonic acid (H_2CO_3) and sodium bicarbonate ($NaHCO_3$). If a strong acid, like hydrochloric acid, is present, it reacts with the sodium bicarbonate. The products are carbonic acid, which is a weaker acid, and sodium chloride. Consequently, an increase in the hydrogen ion concentration in the body fluid is minimized.

$$HCl + NaHCO_3 \longrightarrow H_2CO_3 + NaCl$$
(strong   (weaker
acid)   acid)

If, on the other hand, a strong base like sodium hydroxide (NaOH) is present, it reacts with the carbonic acid. The products are sodium bicarbonate ($NaHCO_3$), which is a weaker base, and water. Thus a shift toward a more basic (alkaline) state is minimized.

$$NaOH + H_2CO_3 \longrightarrow NaHCO_3 + H_2O$$
(strong   (weaker
base)   base)

2. **Phosphate buffer system.** The phosphate acid-base buffer system also is present in intracellular and extracellular body fluids. It is particularly important as a regulator of the hydrogen ion concentration in the tubular fluid of the nephrons and the urine. This buffer system consists of two phosphate compounds—sodium monohydrogen phosphate (Na_2HPO_4) and sodium dihydrogen phosphate (NaH_2PO_4).

Sodium monohydrogen phosphate can react with strong acids to produce weaker ones, while sodium dihydrogen phosphate can react with strong bases to produce weaker ones.

3. **Protein buffer system.** The protein acid-base buffer system is the most important buffer system in plasma and intracellular fluid. It consists of the plasma proteins and various proteins within cells, including the hemoglobin of red blood cells.

As is discussed in chapter 2, proteins are composed of amino acids bound together in complex chains. Some of these amino acids have freely exposed groups of atoms, called *carboxyl groups*, (—COOH). Under certain conditions, a carboxyl group can become ionized, and a hydrogen ion is released.

$$—COOH \rightarrow —COO^- + H^+$$

Some of the amino acids within a protein molecule also contain freely exposed *amino groups* (—NH_2). Under certain conditions, these amino groups can accept hydrogen ions.

$$—NH_2 + H^+ \rightarrow —NH_3^+$$

Thus protein molecules can function as acids by releasing hydrogen ions from their carboxyl groups or as bases by accepting hydrogen ions into their amino groups. This special property allows protein molecules to operate as an acid-base buffer system. In the presence of excess hydrogen ions, the —COO^- portions of the protein molecules accept hydrogen ions and become —COOH groups again. This action decreases the number of free hydrogen ions in body fluid and minimizes the amount of pH change that occurs.

In the presence of excess hydroxyl ions (OH^-), the —NH_3^+ groups of protein molecules give up hydrogen ions and become —NH_2 groups again. These hydrogen ions then combine with the hydroxyl ions to form water molecules. Once again, the pH change is reduced to a minimum since water is a neutral substance.

$$H^+ + OH^- \rightarrow H_2O$$

Chart 19.1 summarizes the actions of the three major buffer systems.

Neurons are particularly sensitive to changes in the pH of body fluids. For example, if the interstitial fluid becomes more alkaline than normal (alkalosis), neurons tend to become more excitable and the person may experience seizures. Conversely, if conditions become more acidic (acidosis), neuron activity tends to be depressed, and the person's level of consciousness may decrease.

---

1. What is the difference between a strong acid or base and a weak acid or base?
2. How does a chemical buffer system help to regulate pH changes?
3. List the major buffer systems of the body.

Chart 19.1   Chemical acid-base buffer systems

Buffer System	Constituents	Actions
Bicarbonate system	Sodium bicarbonate ($NaHCO_3$)	Converts strong acid into weak acid
	Carbonic acid (H_2CO_3)	Converts strong base into weak base
Phosphate system	Sodium monohydrogen phosphate (Na_2HPO_4)	Converts strong acid into weak acid
	Sodium dihydrogen phosphate (NaH_2PO_4)	Converts strong base into weak base
Protein system (and amino acids)	—COO^- group of a molecule	Accepts hydrogen ions in presence of excess acid
	—NH_3^+ group of a molecule	Releases hydrogen ions in presence of excess base

From Hole, John W. Jr., *Human Anatomy and Physiology 3d ed.* © 1978, 1981, 1984 Wm. C. Brown Publishers, Dubuque, Iowa. All Rights Reserved. Reprinted by permission.

The Respiratory Center.   The **respiratory center** in the brain stem helps to regulate hydrogen ion concentrations in body fluids by controlling the rate and depth of breathing. (See chapter 10.) Specifically, if the cells increase their production of carbon dioxide, as occurs during periods of physical exercise, the production of carbonic acid increases. As the carbonic acid dissociates, the concentration of hydrogen ions increases, and the pH of the fluids tends to drop.

Such an increasing concentration of carbon dioxide and the consequent increase in hydrogen ion concentration in the plasma directly stimulate the cells of the respiratory center. In response, the center causes the depth and rate of breathing to increase, so that a greater amount of carbon dioxide is excreted through the lungs. This loss of carbon dioxide is accompanied by a drop in the hydrogen ion concentration in the body fluids, since the released carbon dioxide comes from carbonic acid. (See fig. 19.8.)

$$H_2CO_3 \rightarrow CO_2 + H_2O$$

Conversely, if the body cells are less active, the concentrations of carbon dioxide and hydrogen ions in the plasma remain relatively low. As a consequence, the breathing rate and depth are decreased.

The Kidneys.   The nephrons of the kidneys help to regulate hydrogen ion concentrations of the body fluids by secreting hydrogen ions. As is discussed in chapter 18, these ions are secreted into the urine of the renal tubules by the epithelial cells that line certain segments.

Fig. 19.8 An increase in carbon dioxide production is followed by an increase in carbon dioxide elimination.

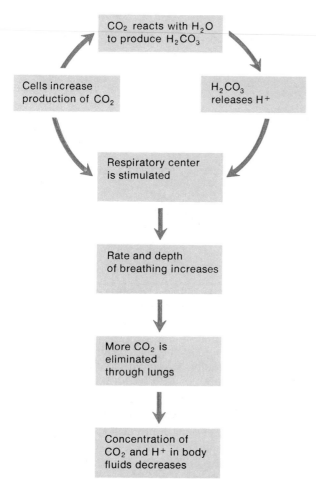

Fig. 19.8 An increase in carbon dioxide production is followed by an increase in carbon dioxide elimination.

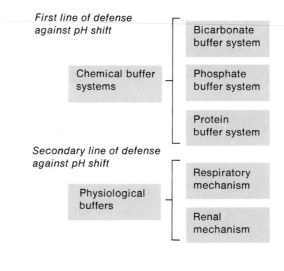

Fig. 19.9 Chemical buffers act rapidly, while physiological buffers may require several minutes to begin resisting a change in pH.

1. How does the respiratory system help to regulate acid-base balance?
2. How do the kidneys handle excessive concentrations of hydrogen ions?
3. What are the differences between the rates at which chemical and physiological buffer systems act?

Clinical Terms Related to Water and Electrolyte Balance

acetonemia (as″ĕ-to-ne′me-ah)—the presence of abnormal amounts of acetone in the blood.

acetonuria (as″ĕ-to-nu′re-ah)—the presence of abnormal amounts of acetone in the urine.

albuminuria (al-bu″mĭ-nu′re-ah)—the presence of albumin in the urine.

antacid (ant-as′id)—a substance that neutralizes an acid.

anuria (ah-nu′re-ah)—the absence of urine excretion.

azotemia (az″o-te′me-ah)—an accumulation of nitrogenous wastes in the blood.

diuresis (di″u-re′sis)—increased production of urine.

glycosuria (gli″ko-su′re-ah)—the presence of excessive sugar in the urine.

hyperkalemia (hi″per-kah-le′me-ah)—the presence of excessive potassium in the blood.

hypernatremia (hi″per-na-tre′me-ah)—the presence of excessive sodium in the blood.

hyperuricemia (hi″per-u″rĭ-se′me-ah)—the presence of excessive uric acid in the blood.

The various regulators of hydrogen ion concentration operate at different rates. Acid-base buffers, for example, function rapidly and can convert strong acids or bases into weak acids or bases almost immediately. For this reason, chemical buffer systems sometimes are called the body's *first line of defense* against shifts in pH.

Physiological buffer systems, such as the respiratory and renal mechanisms, function more slowly, and constitute *secondary defenses*. The respiratory mechanism may require several minutes to begin resisting a change in pH, while the renal mechanism may require one to three days to control a changing hydrogen ion concentration. (See fig. 19.9.)

hypoglycemia (hi″po-gli-se′me-ah)—an abnormally low level of blood sugar.

ketonuria (ke″to-nu′re-ah)—the presence of ketone bodies in the urine.

ketosis (ke″to′sis)—acidosis due to the presence of excessive ketone bodies in the body fluids.

proteinuria (pro″te-ĭ-nu′re-ah)—the presence of protein in the urine.

uremia (u-re′me-ah)—toxic condition resulting from the presence of excessive amounts of nitrogenous wastes in the blood.

Chapter Summary

Introduction

The maintenance of water and electrolyte balance requires that the quantities of these substances entering the body equal the quantities leaving it. Altering the water balance necessarily affects the electrolyte balance.

Distribution of Body Fluids

1. Fluid compartments
   a. The intracellular fluid compartment includes the fluids and electrolytes enclosed by cell membranes.
   b. The extracellular fluid compartment includes all fluids and electrolytes outside cell membranes.

2. Composition of body fluids
   a. Extracellular fluid
      (1) Extracellular fluid is characterized by high concentrations of sodium, chloride, and bicarbonate ions with lesser amounts of potassium, calcium, magnesium, phosphate, and sulfate ions.
      (2) Plasma contains more protein than either interstitial fluid or lymph.
   b. Intracellular fluid contains relatively high concentrations of potassium, phosphate, and magnesium ions; it also contains a greater concentration of sulfate ions and lesser concentrations of sodium, chloride, and bicarbonate ions than extracellular fluid.

3. Movement of fluid between compartments
   a. Movements are regulated by hydrostatic pressure and osmotic pressure.
      (1) Fluid leaves plasma because of hydrostatic pressure and returns to plasma because of osmotic pressure.
      (2) Fluid enters lymph vessels because of osmotic pressure.
      (3) Movement in and out of cells is regulated primarily by osmotic pressure.
   b. Sodium concentrations are especially important in the regulation of fluid movements.

Water Balance

1. Water intake
   a. Most water is gained in liquid or in moist foods.
   b. Some water is produced by oxidative metabolism.

2. Regulation of water intake
   a. The thirst mechanism is the primary regulator of water intake.
   b. The act of drinking and the resulting distension of the stomach inhibit the thirst mechanism.

3. Water output
   Water is lost in urine, feces, and sweat and by evaporation from skin and lungs.

4. Regulation of water balance
   Water balance is regulated mainly by the distal convoluted tubule and collecting duct of the nephron.

Electrolyte Balance

1. Electrolyte intake
   a. The most important electrolytes in body fluids are ions of sodium, potassium, calcium, magnesium, chloride, sulfate, phosphate, and bicarbonate.
   b. These are obtained in foods and beverages or as by-products of metabolic processes.

2. Regulation of electrolyte intake
   a. Electrolytes usually are obtained in sufficient quantities in response to hunger and thirst mechanisms.
   b. In severe deficiency, a person may experience a salt craving.

3. Electrolyte output
   a. Electrolytes are lost through perspiration, feces, and urine.
   b. Quantities lost vary with temperature and physical exercise.
   c. The greatest loss is usually in urine.

4. Regulation of electrolyte balance
   a. Concentrations of sodium, potassium, and calcium ions in body fluids are particularly important.
   b. Regulation of sodium and potassium ions involves the secretion of aldosterone from the adrenal glands.
   c. Regulation of calcium ions involves parathyroid hormone.
   d. In general, negatively charged ions are regulated secondarily by the mechanisms that control positively charged ions.

Acid-Base Balance

Acids are electrolytes that release hydrogen ions. Bases are substances that combine with hydrogen ions.

1. Sources of hydrogen ions
   a. Aerobic respiration of glucose results in carbonic acid.
   b. Anaerobic respiration of glucose gives rise to lactic acid.
   c. Incomplete oxidation of fatty acids gives rise to acidic ketone bodies.
   d. Oxidation of certain amino acids gives rise to sulfuric acid.
   e. Hydrolysis of phosphoproteins and nucleoproteins gives rise to phosphoric acid.

2. Strengths of acids and bases.
   a. Acids vary in the degrees to which they ionize.
      (1) Strong acids, such as hydrochloric acid, ionize more completely.
      (2) Weak acids, such as carbonic acid, ionize less completely.
   b. Bases also vary in strength.

3. Regulation of hydrogen ion concentration
   a. Acid-base buffer systems
      (1) They function to convert strong acids into weaker ones, or strong bases into weaker ones.
      (2) They include the bicarbonate buffer system, phosphate buffer system, and protein buffer system.
      (3) Buffer systems cannot prevent pH changes, but they can minimize them.
   b. Respiratory center helps regulate pH by controlling the rate and depth of breathing.
   c. Kidneys
      Nephrons regulate pH by secreting hydrogen ions.
   d. Chemical buffers act rapidly; physiological buffers act less rapidly.

Application of Knowledge

1. Some time ago a news story reported the death of several newborn infants due to an error in which sodium chloride was substituted for sugar in their formula. What symptoms would this produce? Why are infants more prone to the hazard of excess salt intake than adults?

2. Explain the threat to fluid and electrolyte balance in the following situation: a patient is being nutritionally maintained on concentrated solutions of hydrolyzed protein that are administered through a gastrostomy tube.

Review Activities

1. Explain how water balance and electrolyte balance are interdependent.

2. Name the body fluid compartments and describe their locations.

3. Explain how the fluids within these compartments differ in composition.

4. Describe how fluid movements between compartments are controlled.

5. Prepare a water budget to illustrate how the input of water equals the output of water.

6. Define *water of metabolism*.

7. Explain how water intake is regulated.

8. Explain how the nephrons function in the regulation of water output.

9. List the electrolytes of greatest importance in body fluids.

10. Explain how electrolyte intake is regulated.

11. List the routes by which electrolytes leave the body.

12. Explain how the adrenal cortex functions in the regulation of electrolyte output.

13. Describe the role of the parathyroid glands in regulating electrolyte balance.

14. Distinguish between an acid and a base.

15. List five sources of hydrogen ions in body fluids and name an acid that originates from each.

16. Distinguish between a strong acid and a weak acid and name an example of each.

17. Distinguish between a strong base and a weak base and name an example of each.

18. Explain how an acid-base buffer system functions.

19. Describe how the bicarbonate buffer system resists shifts in pH.

20. Explain why a protein has acidic as well as basic properties.

21. Explain how the respiratory center functions in the regulation of acid-base balance.

22. Explain how the kidneys function in the regulation of acid-base balance.

23. Distinguish between a chemical buffer system and a physiological buffer system.

Reproduction

U nit 5 is concerned with reproduction of the human organism. This chapter describes how the organs of the male and female reproductive systems function to produce an embryo and how this offspring grows and develops before birth.

This unit includes

The Reproductive Systems and Pregnancy

20

The male and female *reproductive systems* are specialized to produce offspring. These systems are unique in that their functions are not necessary for the survival of each individual. Instead, their functions are vital to the continuation of the human species. Without reproduction by at least some of its members, the species would soon become extinct.

Organs of each of these systems are adapted to produce sex cells, to sustain these cells, and to transport them to a location where fertilization may occur. In addition, some of the reproductive organs secrete hormones that play vital roles in the development and maintenance of sexual characteristics and the regulation of reproductive physiology. Also, specialized organs of the female system are adapted to support the life of an offspring that may develop from a fertilized egg cell and transport this offspring to the outside.

After you have studied this chapter, you should be able to

1. Name the parts of the male reproductive system and describe the general functions of each part.

2. Describe the structure of a testis and explain how sperm cells are formed.

3. Trace the path followed by sperm cells from the site of their formation to the outside.

4. Explain how hormones control the activities of the male reproductive organs.

5. Name the parts of the female reproductive system and describe the general functions of each part.

6. Describe the structure of an ovary and explain how egg cells and follicles are formed.

7. Trace the path followed by an egg cell after ovulation.

8. Describe how hormones control the activities of the female reproductive system.

9. Review the structure of the mammary glands.

10. Describe the major events that occur during a menstrual cycle.

11. Define *pregnancy* and describe the process of fertilization.

12. Describe the hormonal changes that occur in the maternal body during pregnancy.

13. Describe the major events that occur during the embryonic and fetal stages of development.

14. Trace the path of blood through the fetal circulatory system and describe the circulatory adjustments that occur in a newborn.

15. Describe the birth process and explain the role of hormones in this process.

16. Complete the review activities at the end of this chapter. Note that the items are worded in the form of specific learning objectives. You may want to refer to them before reading the chapter.

androgen   (an′dro-jen)

ejaculation   (e-jak″u-la′shun)

embryo   (em′bre-o)

emission   (e-mish′un)

estrogen   (es′tro-jen)

fertilization   (fer″tĭ-lĭ-za′shun)

fetus   (fe′tus)

follicle   (fol′i-kl)

menstrual cycle   (men′stroo-al si′kl)

oogenesis   (ō″o-jen′ĕ-sis)

ovulation   (o″vu-la′shun)

placenta   (plah-sen′tah)

postnatal   (pōst-na′tal)

pregnancy   (preg′nan-se)

progesterone   (pro-jes′tĕ-rōn)

puberty   (pu′ber-te)

seminal fluid   (se′men-al floo′id)

spermatogenesis   (sper″mah-to-jen′ĕ-sis)

chorio-, skin: *chorio*n—outermost membrane that surrounds the fetus and other fetal membranes.

cleav-, to divide: *cleav*age—period of development characterized by a division of the zygote into smaller and smaller cells.

ejacul-, to shoot forth: *ejacul*ation—process by which seminal fluid is expelled from the male reproductive tract.

fimb-, a fringe: *fimb*riae—irregular extensions on the margin of the infundibulum of the uterine tube.

follic-, small bag: *follic*le—ovarian structure that contains an egg.

genesis-, origin: spermato*genesis*—process by which sperm cells are formed.

labi-, lip: *labi*a minora—flattened, longitudinal folds that extend along the margins of the vestibule.

mens-, month: *mens*trual cycle—monthly female reproductive cycle.

puber-, adult: *puber*ty—time when a person becomes able to function reproductively.

umbil-, the naval: *umbil*ical cord—structure attached to the fetal navel (umbilicus) that connects the fetus to the placenta.

The organs of the male and female reproductive systems are concerned with the general process of reproduction, and each is adapted for specialized tasks. Some, for example, are concerned with the production of sex cells, and others function to sustain the lives of these cells or to transport them from one location to another. Still other parts produce and secrete hormones, which regulate the formation of sex cells and play roles in the development and maintenance of male and female sexual characteristics.

Organs of the Male Reproductive System

The primary function of the male reproductive system is to produce and maintain male sex cells called **sperm cells** (spermatozoa). Parts of this system are specialized to transport these cells and various fluids to the outside, and to produce and secrete sex hormones. The *primary organs* of the male system are the *testes* in which sperm cells and male sex hormones are formed. The other structures of the male reproductive system are termed *accessory organs,* and they include two groups—the internal and external reproductive organs. (See fig. 20.1.)

The Testes

The **testes** are ovoid structures, about 5 centimeters (2 inches) in length and 3 centimeters (1.2 inches) in diameter. Each is suspended by a spermatic cord within the cavity of the saclike *scrotum.*

Structure of the Testes. Each testis is enclosed by a tough, white fibrous capsule. Along its posterior border, the connective tissue thickens and extends into the organ forming thin septa, which divide the testis into about 250 *lobules.*

Each lobule contains one to four highly coiled, convoluted **seminiferous tubules,** each of which is up to 70 centimeters (28 inches) long when uncoiled. These tubules course posteriorly and become united into a complex network of channels. These, in turn, give rise to several ducts that join a tube called the **epididymis.** The epididymis is coiled on the outer surface of the testis.

The seminiferous tubules are lined with a specialized tissue called **germinal epithelium.** The cells of this tissue produce the male sex cells. Other specialized cells, called **interstitial cells,** are located in the spaces between the seminiferous tubules. They function in the production and secretion of male sex hormones. (See fig. 20.2.)

1. Describe the structure of a testis.
2. Where are the sperm cells produced within the testes?
3. What cells produce male sex hormones?

Germinal Epithelium. The germinal epithelium consists of two types of cells—supporting cells and spermatogenic cells. The *supporting cells* support and nourish the *spermatogenic cells,* which in turn give rise to sperm cells.

Sperm cell production occurs continually throughout the reproductive life of a male. The resulting sperm cells collect in the lumens of the seminiferous tubules. Then they pass to the epididymis, where they remain for a time and mature.

A mature sperm cell is a tiny, tadpole-shaped structure about 0.06 millimeter long. It consists of a flattened head, a cylindrical body, and an elongated tail. (See fig. 20.3.)

The *head* of a sperm cell is composed primarily of a nucleus. It has a small part at its anterior end called the *acrosome,* which contains enzymes that aid the sperm cell in penetrating an egg cell at the time of fertilization.

Spermatogenesis. In a young male, all the spermatogenic cells are undifferentiated and are called *spermatogonia.* Each of these cells contains 46 chromosomes in its nucleus, which is the usual number for human body cells. During early adolescence, hormones stimulate these cells to change. The spermatogonia begin to undergo mitosis (see chapter 3), and some of them enlarge to become *primary spermatocytes.* When these primary spermatocytes divide, however, they do so by a special type of cell division called **meiosis.**

In the course of meiosis, the primary spermatocytes each divide to form two *secondary spermatocytes.* Each of these cells, in turn, divides to form two *spermatids,* which mature into sperm cells. Also during meiosis, the number of chromosomes is reduced by one-half. Consequently, for each primary spermatocyte that undergoes meiosis, four sperm cells with 23 chromosomes in their nuclei are formed. This process by which sperm cells are produced is called **spermatogenesis** (fig. 20.4).

1. Explain the function of supporting cells in the germinal epithelium.
2. Describe the structure of a sperm cell.
3. Describe the process of spermatogenesis.

Fig. 20.1 Sagittal view of the male reproductive organs.

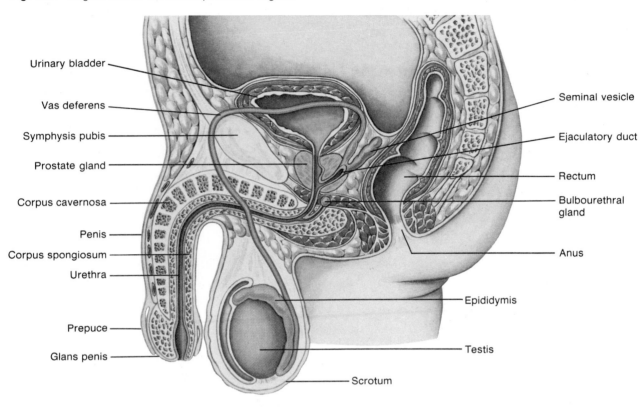

Urinary bladder

Vas deferens

Symphysis pubis

Prostate gland

Corpus cavernosa

Penis

Corpus spongiosum

Urethra

Prepuce

Glans penis

Seminal vesicle

Ejaculatory duct

Rectum

Bulbourethral gland

Anus

Epididymis

Testis

Scrotum

Fig. 20.2 (a) Sagittal section of a testis; (b) cross section of a seminiferous tubule.

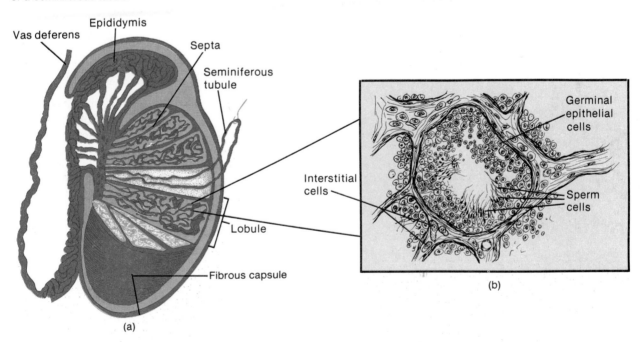

Vas deferens

Epididymis

Septa

Seminiferous tubule

Interstitial cells

Germinal epithelial cells

Sperm cells

Lobule

Fibrous capsule

(a)

(b)

Fig. 20.3 (a) Structure of a human sperm cell;
(b) scanning electron micrograph of a human sperm cell.
*Micrograph from Tissues and Organs: A Text-Atlas of Scanning
Electron Microscopy* by Richard G. Kessel and Randy H. Kardon.
W. H. Freeman and Company. Copyright © 1979.

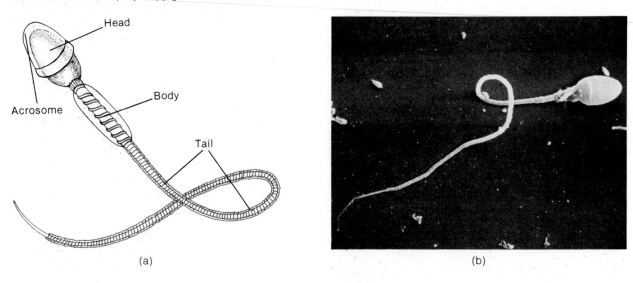

(a) (b)

Fig. 20.4 Spermatogonia give rise to primary
spermatocytes by mitosis; the spermatocytes, in turn, give
rise to sperm cells by meiosis.

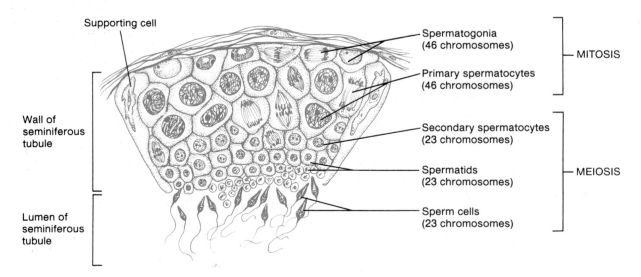

Male Internal Accessory Organs

The *internal accessory organs* of the male reproductive system include the epididymides, vasa deferentia, ejaculatory ducts, and urethra, as well as the seminal vesicles, prostate gland, and bulbourethral glands.

The Epididymis. The **epididymis** (plural, epididymides) is a tightly coiled, threadlike tube that is about 6 meters (20 feet) long. (See fig. 20.1.) This tube is connected to ducts within the testis. It emerges from the top of the testis, descends along its posterior surface, and then courses upward to become the *vas deferens.*

When immature sperm cells are moved from the ducts of the testis into the epididymis, they are completely nonmobile and are unable to fertilize an egg cell; however, as they travel slowly through the epididymis, they undergo *maturation* and become capable of moving independently.

The Vas Deferens. The **vas deferens** (plural, vasa deferentia) is a muscular tube about 45 centimeters (18 inches) long. (See fig. 20.1.) It passes upward along the medial side of the testis, through the inguinal canal, enters the abdominal cavity, and ends behind the urinary bladder.

Just outside the prostate gland, the vas deferens unites with the duct of a seminal vesicle. The fusion of these two ducts forms an **ejaculatory duct,** which passes through the substance of the prostate gland and empties into the urethra.

The Seminal Vesicle. A **seminal vesicle** (fig. 20.1) is a convoluted, saclike structure about 5 centimeters (2 inches) long that is attached to the vas deferens near the base of the urinary bladder.

The glandular tissue lining the inner wall of a seminal vesicle secretes a slightly alkaline fluid. This fluid is thought to help regulate the pH of the tubular contents as sperm cells are conveyed to the outside. The secretion of the seminal vesicle contains a variety of nutrients. For instance, it is rich in *fructose,* a monosaccharide that is thought to provide sperm cells with an energy source.

---

1. *Describe the structure of the epididymis.*
2. *Trace the path of the vas deferens.*
3. *What is the function of the seminal vesicle?*

The Prostate Gland. The **prostate gland** (fig. 20.1) is a chestnut-shaped structure about 4 centimeters (1.6 inches) across and 3 centimeters (1.2 inches) thick that surrounds the beginning of the urethra, just below the urinary bladder. It is enclosed by connective tissue and is composed of many branched tubular glands whose ducts open into the urethra.

The prostate gland secretes a thin, milky fluid with an alkaline pH. It functions to neutralize the seminal fluid, which is acidic due to an accumulation of metabolic wastes produced by stored sperm cells. It also enhances the motility of the sperm cells, which remain relatively immobile in the acidic contents of the epididymis.

Although the prostate gland is relatively small in male children, it begins to grow in early adolescence and reaches its adult size a few years later. As a rule, its size remains unchanged between the ages of 20 and 50 years. In older males the prostate gland usually enlarges. When this happens it tends to squeeze the urethra and interfere with urine excretion.

The treatment for an enlarged prostate gland is usually surgical. If the obstruction is slight, the procedure may be performed through the urethral canal (*transurethral prostatic resection*).

The Bulbourethral Glands. The **bulbourethral glands** (fig. 20.1) are two small structures about the size of peas, which are located below the prostate gland among the fibers of the external urethral sphincter muscle.

These glands are composed of numerous tubes whose epithelial linings secrete a mucuslike fluid. This fluid is released in response to sexual stimulation and provides some lubrication to the end of the penis in preparation for sexual intercourse. Most of the lubricating fluid for intercourse is secreted by female reproductive organs, however.

Seminal Fluid. The fluid conveyed by the urethra to the outside as a result of sexual stimulation is called **seminal fluid** (semen). It consists of sperm cells from the testes and secretions of the seminal vesicles, prostate gland, and bulbourethral glands. It has a slightly alkaline pH (about 7.5) and a milky appearance. It contains a variety of nutrients, as well as prostaglandins (see chapter 12) that function to enhance sperm cell survival and movement through the female reproductive tract.

The volume of seminal fluid released at one time varies from 2 to 6 milliliters, while the average number of sperm cells present in the fluid is about 120 million per milliliter.

1. Where is the prostate gland located?
2. What is the function of its secretion?
3. What are the characteristics of seminal fluid?

Male External Reproductive Organs

The male external reproductive organs are the scrotum, which encloses the testis, and the penis, through which the urethra passes.

The Scrotum. The **scrotum** (fig. 20.1) is a pouch of skin and subcutaneous tissue that hangs from the lower abdominal region behind the penis. It is divided into chambers by a medial septum, and each chamber is occupied by a testis. Each chamber also contains a serous membrane that provides a covering and helps ensure that the testis will move smoothly within the scrotum.

The Penis. The **penis** is a cylindrical organ that functions to convey urine and seminal fluid through the urethra to the outside. It also is specialized to become enlarged and stiffened by a process called *erection,* so that it can be inserted into the female vagina during sexual intercourse.

The *body,* or shaft, of the penis is composed of three columns of erectile tissue. These include a pair of dorsally located *corpora cavernosa* and a single *corpus spongiosum* below. These columns are enclosed by skin, a thin layer of subcutaneous tissue, and a layer of elastic tissue. In addition, each column is surrounded by a tough capsule of white fibrous connective tissue. (See fig. 20.1.)

The corpus spongiosum, through which the urethra extends, is enlarged at its distal end to form a sensitive, cone-shaped **glans penis.** The glans covers the ends of the corpora cavernosa and bears the urethral opening (*external urethral meatus*). The skin of the glans is very thin and hairless. A loose fold of skin called the *prepuce* (foreskin) begins just behind the glans and extends forward to cover it as a sheath. The prepuce sometimes is removed during infancy by a surgical procedure called *circumcision.*

1. Describe the structure of the penis.
2. What is circumcision?

Erection, Orgasm, and Ejaculation

The erectile tissue within the body of the penis contains a network of vascular spaces (venous sinusoids). These spaces are lined with endothelium and are separated from each other by cross bars of smooth muscle and connective tissue.

Ordinarily, the vascular spaces remain small as a result of partial contractions in the smooth muscle fibers that surround them. During sexual stimulation, however, the smooth muscles become relaxed. At the same time, *parasympathetic* nerve impulses pass from the sacral portion of the spinal cord to arteries leading into the penis, causing them to dilate. These impulses also stimulate veins leading away from the penis to constrict. As a result, arterial blood under relatively high pressure enters the vascular spaces of the erectile tissue, and the flow of venous blood away from the penis is reduced. Consequently, blood accumulates in the erectile tissues, and the penis swells, elongates, and becomes erect.

The culmination of sexual stimulation is called **orgasm** and involves a pleasurable feeling of physiological and psychological release. Also, orgasm in the male is accompanied by emission and ejaculation.

Emission is the movement of sperm cells from the testes and secretions from the prostate gland and seminal vesicles into the urethra, where they are mixed to form seminal fluid. Emission occurs in response to *sympathetic* impulses traveling from the spinal cord to smooth muscles in the walls of the testicular ducts, epididymides, vasa deferentia, and ejaculatory ducts, causing peristaltic contractions. At the same time, other sympathetic impulses stimulate rhythmic contractions of the seminal vesicles and prostate gland.

As the urethra fills with seminal fluid, sensory impulses are stimulated and pass into the sacral portion of the spinal cord. In response, motor impulses are transmitted from the cord to certain skeletal muscles at the base of the erectile columns of the penis, causing them to contract rhythmically. This increases the pressure within the erectile tissues and aids in forcing the seminal fluid through the urethra to the outside—a process called *ejaculation.*

Chart 20.1 Functions of male reproductive organs

Organ	Function
Testes	
Seminiferous tubules	Production of sperm cells
Interstitial cells	Production and secretion of male sex hormones
Epididymis	Storage and maturation of sperm cells; conveys sperm cells to vas deferens
Vas deferens	Conveys sperm cells to ejaculatory duct
Seminal vesicle	Secretes alkaline fluid containing nutrients and prostaglandins; fluid helps neutralize acidic seminal fluid
Prostate gland	Secretes alkaline fluid that helps neutralize acidic seminal fluid and enhances motility of sperm cells
Bulbourethral gland	Secretes fluid that lubricates end of penis
Scrotum	Encloses and protects testes
Penis	Conveys urine and seminal fluid to outside of the body; inserted into vagina during sexual intercourse; glans penis is richly supplied with sensory nerve endings associated with feelings of pleasure during sexual stimulation

From Hole, John W. Jr., *Human Anatomy and Physiology* 3d ed. © 1978, 1981, 1984 Wm. C. Brown Publishers, Dubuque, Iowa. All Rights Reserved. Reprinted by permission.

The sequence of events during emission and ejaculation is regulated so that the fluid from the bulbourethral glands is expelled first, followed by the release of fluid from the prostate gland, the passage of sperm cells, and finally the ejection of fluid from the seminal vesicles.

Immediately after ejaculation, sympathetic impulses cause vasoconstriction of the arteries that supply the erectile tissue with blood, reducing the inflow of blood. The smooth muscles within the walls of the vascular spaces partially contract again, and the veins of the penis carry the excess blood out of these spaces. Thus, the penis gradually returns to its former flaccid condition.

The functions of the male reproductive organs are summarized in chart 20.1.

---

1. *How is blood flow into the erectile tissues of the penis controlled?*
2. *Distinguish between orgasm, emission, and ejaculation.*
3. *Review the events associated with emission and ejaculation.*

Hormonal Control of Male Reproductive Functions

Male reproductive functions are largely controlled by hormones secreted from the hypothalamus of the brain, the *anterior pituitary gland,* and the *testes.*

Hypothalamic and Pituitary Hormones

For about the first 10 years, the male body is reproductively immature. It remains childlike, and the spermatogenic cells of the testes remain undifferentiated. Then a series of changes is triggered that leads to the development of a reproductively functional adult. Although the mechanism that initiates such changes is not well understood, it involves an action of the hypothalamus.

As explained in chapter 12, the hypothalamus secretes a gonadotropin-releasing hormone (GnRH), which enters blood vessels leading to the anterior pituitary gland. In response, the anterior pituitary gland secretes the **gonadotropins** called *luteinizing hormone* (LH) and *follicle-stimulating hormone* (FSH). LH, which also is called interstitial cell-stimulating hormone (ICSH), promotes the development of the interstitial cells of the testes, and they, in turn, secrete male sex hormones. FSH causes the supporting cells of the germinal epithelium to become responsive to the effects of the male sex hormone, testosterone. Then, in the presence of FSH and testosterone, these supporting cells somehow stimulate the spermatogenic cells to undergo spermatogenesis, giving rise to sperm cells.

Male Sex Hormones

As a group, the male sex hormones are termed **androgens,** and although most of them are produced by the interstitial cells of the testes, small amounts are synthesized in the adrenal cortex (see chapter 12).

The hormone called **testosterone** is the most abundant of the androgens, and when it is secreted, it is transported in the blood loosely attached to plasma proteins.

Although the secretion of testosterone begins during fetal development and continues for a few weeks following birth, its secretion nearly ceases during childhood. Sometime between the ages of 13 and 15 however, androgen production usually increases rapidly. This phase in development, during which a male becomes reproductively functional, is called **puberty.** After puberty, testosterone secretion continues throughout the life of the male.

Actions of Testosterone. During puberty, testosterone stimulates enlargement of the testes and various accessory organs of the reproductive system, and it causes the development of the male *secondary sexual characteristics*. These secondary sexual characteristics are special features associated with the adult male body, and they include the following:

1. Increased growth of body hair, particularly on the face, chest, axillary region, and pubic region, but sometimes accompanied by decreased growth of hair on the scalp.
2. Enlargement of the larynx and thickening of the vocal folds, accompanied by the development of a lower-pitched voice.
3. Thickening of the skin.
4. Increased muscular growth accompanied by the development of broader shoulders and a relatively narrow waist.
5. Thickening and strengthening of the bones.

Other actions of testosterone include increasing the rate of cellular metabolism and increasing the production of red blood cells, so the average number of red blood cells in a cubic millimeter of blood usually is greater in males than in females. Testosterone also stimulates sexual activity by influencing certain portions of the brain.

Regulation of Male Sex Hormones. The degree to which male secondary sexual characteristics develop is directly related to the amount of testosterone secreted by the interstitial cells. This quantity is regulated by a *negative feedback system* involving the hypothalamus. (See fig. 20.5.)

As the concentration of testosterone in the blood increases, the hypothalamus becomes inhibited, and its stimulation of the anterior pituitary gland by GnRH is decreased. As the pituitary's secretion of LH (ICSH) is reduced, the amount of testosterone released by the interstitial cells is reduced also.

But as the blood level of testosterone drops, the hypothalamus becomes less inhibited, and it once again stimulates the pituitary gland to release LH. The increasing secretion of LH causes the interstitial cells to release more testosterone, and the blood concentration of testosterone rises.

Thus the concentration of testosterone in the male body is regulated so that it remains relatively constant.

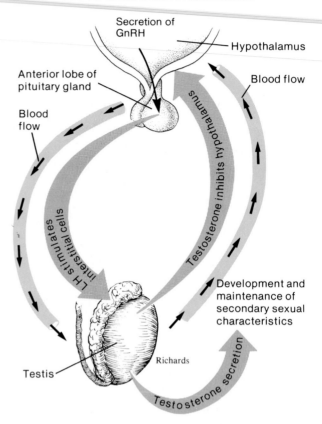

Fig. 20.5 A negative feedback mechanism operating between the anterior lobe of the pituitary gland and the testes controls the concentration of testosterone.

The amount of testosterone secreted by the testes usually declines gradually after about 40 years of age. Consequently, even though sexual activity may be continued into old age, males typically experience a decrease in sexual functions as they grow older. This decrease is sometimes called the *male climacteric.*

1. What initiates the changes associated with male sexual maturity?
2. Describe several male secondary sexual characteristics.
3. List the functions of testosterone.
4. Explain how the secretion of male sex hormones is regulated.

Fig. 20.6 Organs of the female reproductive system.

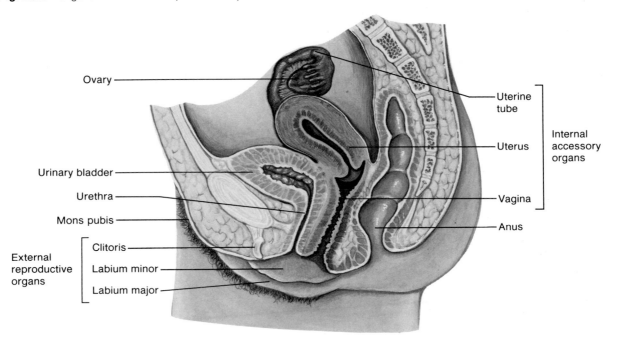

Organs of the Female Reproductive System

The organs of the female reproductive system are specialized to produce and maintain female sex cells, to transport these cells to the site of fertilization, to provide a favorable environment for a developing offspring, to move the offspring to the outside, and to produce female sex hormones.

The *primary organs* of this system are the ovaries, which produce the female sex cells, or **egg cells,** and female sex hormones. The other parts of the system comprise the internal and external *accessory organs.*

The Ovaries

The **ovaries** are solid, ovoid structures measuring about 3.5 centimeters (1.4 inches) in length, 2 centimeters (0.8 inch) in width, and 1 centimeter (0.4 inch) in thickness. They are located, one on each side, in a shallow depression of the lateral wall of the pelvic cavity. (See fig. 20.6.)

Structure of the Ovaries. The tissues of an ovary can be divided into two regions, an inner *medulla* and an outer *cortex.* These regions are not distinctly separated, however.

The ovarian medulla is composed largely of loose connective tissue and contains numerous blood vessels, lymphatic vessels, and nerve fibers. The ovarian cortex is composed of more compact tissue and has a somewhat granular appearance due to the presence of tiny masses of cells called *ovarian follicles.*

The free surface of the ovary is covered by a layer of cuboidal cells, called the **germinal epithelium.** Just beneath the epithelium, there is a dense layer of connective tissue.

---

1. What are the primary organs of the female reproductive system?
2. Describe the structure of an ovary.

Primordial Follicles. During prenatal development (before birth), small groups of cells in the outer region of the ovarian cortex form several million **primordial follicles.** Each of these structures consists of a single, large cell called a *primary oocyte* and several epithelial cells called *follicular cells* that closely surround the oocyte.

Also early in development, the primary oocytes begin to undergo *meiosis,* but the process is soon halted and is not continued until after puberty. After the primordial follicles appear, no new ones are formed. Instead, the number of oocytes in the ovary steadily decreases throughout the life of the female as many of these cells degenerate. Of the several million oocytes formed originally, only a million or so remain at the time of birth, and perhaps 400,000 are present at puberty. Of these, probably fewer than 1000 will be released from the ovary during the reproductive life of the person.

Oogenesis. Beginning at puberty, some of the primary oocytes are stimulated to continue meiosis, and as in the case of sperm cells, the resulting cells have one-half as many chromosomes (23) in their nuclei as the parent cells.

When a primary oocyte divides, the division of the cellular cytoplasm is grossly unequal. One of the resulting cells, called a *secondary oocyte,* is quite large, and the other, called the *first polar body,* is very small.

The large secondary oocyte represents a future *egg cell* (ovum) in that it can be fertilized by uniting with a sperm cell. If this happens, the oocyte divides unequally to produce a tiny *second polar body* and a relatively large fertilized egg cell or **zygote.** Meanwhile, the first polar body also may divide to form two more polar bodies (fig. 20.7).

Thus, the result of this process, which is called **oogenesis,** is one secondary oocyte or egg cell and a polar body. After being fertilized, the egg cell divides to produce a second polar body and a zygote, which can give rise to an embryo. Although the first polar body may divide again, the polar bodies have no further function and they soon degenerate.

---

1. *How does the timing of the production of egg cells differ from that of sperm cells?*
2. *Describe the major events of oogenesis.*

Maturation of a Follicle. The primordial follicles remain relatively unchanged throughout childhood. At puberty, however, the anterior pituitary gland secretes a greatly increased amount of FSH, and the ovaries begin to enlarge in response. At the same time, some of the primordial follicles begin to undergo maturation, becoming *primary follicles* (fig. 20.8).

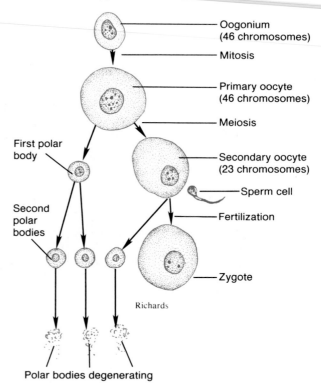

Fig. 20.7 During the process of oogenesis, a single egg cell (secondary oocyte) results from meiosis of a primary oocyte. If the egg is fertilized, it forms a second polar body and becomes a zygote.

Oogonium (46 chromosomes)

Mitosis

Primary oocyte (46 chromosomes)

Meiosis

First polar body

Secondary oocyte (23 chromosomes)

Sperm cell

Fertilization

Second polar bodies

Zygote

Richards

Polar bodies degenerating

During maturation, the oocyte of a primary follicle grows larger, and the follicular cells surrounding it divide actively by mitosis. These follicular cells soon organize themselves into layers, and in time a cavity appears in the cellular mass. As the cavity forms, it becomes filled with a clear *follicular fluid* that bathes the oocyte.

The fluid-filled follicular cavity continues to enlarge, and the oocyte is pressed to one side within the follicle. In time, the follicle reaches a diameter of 10 millimeters or more and bulges outward on the surface of the ovary like a blister.

The oocyte within such a mature follicle is a large, spherical cell surrounded by a membrane (zona pellucida) and enclosed by a mantle of follicular cells (corona radiata). Processes from these follicular cells extend through the zona pellucida and are thought to supply the oocyte with nutrients.

Although as many as twenty primary follicles may begin the process of maturation at any one time, usually only one follicle reaches full development, and the others degenerate.

Fig. 20.8 As a follicle matures, the egg cell enlarges and becomes surrounded by a mantle of follicular cells and fluid. Eventually the mature follicle ruptures and the egg cell is released.

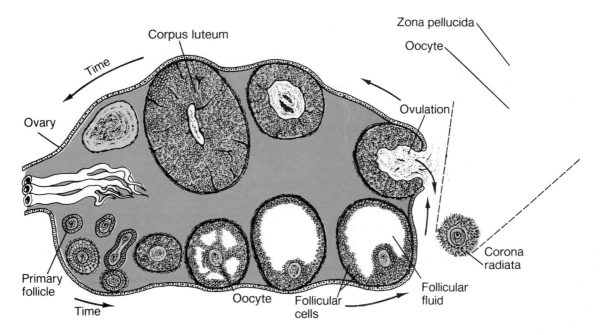

Ovulation. As a follicle matures, its primary oocyte undergoes oogenesis, giving rise to a secondary oocyte and a first polar body. These cells are released from the follicle by the process called ovulation.

Ovulation is stimulated by hormones from the anterior pituitary gland, which apparently cause the mature follicle to swell rapidly and the wall to weaken. Eventually the wall ruptures and the follicular fluid, accompanied by the oocyte, oozes outward from the surface of the ovary and enters the peritoneal cavity.

After it is expelled from the ovary, the oocyte and one or two layers of follicular cells surrounding it usually are propelled to the opening of the nearby *uterine tube.* If the oocyte is not fertilized by union with a sperm cell within a relatively short time, it will degenerate. The maturation of a follicle and the release of an oocyte are illustrated in figure 20.8.

---

1. What changes occur in a follicle and its oocyte during maturation?
2. What causes ovulation?
3. What happens to an egg cell following ovulation?

Female Internal Accessory Organs

The *internal accessory organs* of the female reproductive system include a pair of uterine tubes, a uterus, and a vagina.

The Uterine Tubes. The **uterine tubes** (fallopian tubes or oviducts), which convey egg cells toward the uterus, have openings near the ovaries. Each tube, which is about 10 centimeters (4 inches) long and 0.7 centimeter (0.3 inch) in diameter, passes medially to the uterus, penetrates its wall, and opens into the uterine cavity.

Near the ovary, the uterine tube expands to form a funnel-shaped *infundibulum,* which partially encircles the ovary, medially. On its margin the infundibulum bears a number of extensions called *fimbriae.* (See fig. 20.9.) Although the infundibulum generally does not touch the ovary, one of its larger extensions is connected directly to it.

Fig. 20.9 The funnel-shaped infundibulum of the uterine tube partially encircles the ovary. What factors aid the movement of an egg cell into the infundibulum following ovulation?

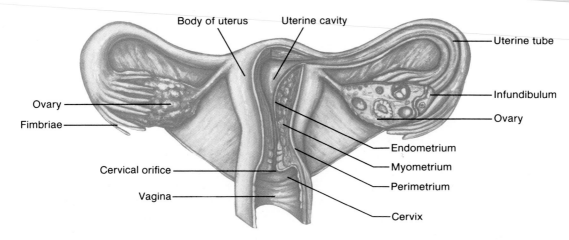

The wall of a uterine tube is lined with simple columnar epithelial cells, some of which are *ciliated*. The epithelium secretes mucus, and the cilia beat toward the uterus. These actions apparently help to draw the egg cell and expelled follicular fluid into the infundibulum following ovulation.

Ciliary action also aids the transport of the egg cell down the uterine tube, and peristaltic contractions in the tube's muscular layer help force the egg along.

The Uterus. The **uterus** functions to receive the embryo resulting from a fertilized egg cell and to sustain its life during development. It is a hollow, muscular organ shaped somewhat like an inverted pear.

Although the size of the uterus changes greatly during pregnancy, in its usual state it is about 7 centimeters (2.8 inches) long, 5 centimeters (2 inches) wide (at its broadest point), and 2.5 centimeters (1 inch) in diameter. The uterus is located medially within the anterior portion of the pelvic cavity, above the vagina, and usually is bent forward over the urinary bladder.

The upper two-thirds of the uterus, or *body,* has a dome-shaped top and is joined by the uterine tubes that enter its wall at its broadest part.

The lower one-third of the uterus is the tubular **cervix,** which extends downward into the upper portion of the vagina. (See fig. 20.9.)

Cancer developing within the tissues of the cervix usually can be detected by means of a relatively simple and painless procedure called the *Pap* (Papanicolaou) *smear test.* This technique involves removing a tiny sample of tissue by scraping, smearing the sample on a glass slide, staining it, and examining it for the presence of abnormal cells.

Since this test can reveal certain types of cervical cancer in the early stages of development when it may be cured completely, the American Cancer Society recommends that all women over 20 years of age have Pap tests at regular intervals.

The wall of the uterus is relatively thick and is composed of three layers—endometrium, myometrium, and perimetrium. The **endometrium,** the inner mucosal layer lining the uterine cavity, is covered with columnar epithelium and contains numerous tubular glands. The **myometrium,** a very thick, muscular layer, consists largely of bundles of smooth muscle fibers. During the monthly female reproductive cycles and during pregnancy, the endometrium and myometrium undergo extensive changes. The **perimetrium** is the outer serosal layer. It is composed of the peritoneal layer of the broad ligament that covers the body of the uterus and part of the cervix. (See fig. 20.9.)

The Vagina. The **vagina** is a fibromuscular tube, about 9 centimeters (3.6 inches) in length, extending from the uterus to the vestibule. It conveys uterine secretions to the outside, receives the erect penis during sexual intercourse, and transports the offspring during the birth process.

The vagina extends upward and back from the vestibule into the pelvic cavity. It is located posterior to the urinary bladder and urethra and anterior to the rectum, and is attached to these parts by connective tissues.

The *vaginal orifice* opens into the vestibule as a slit and is partially closed by a thin membrane of connective tissue and stratified squamous epithelium called the **hymen.** A central opening of varying size allows uterine and vaginal secretions to pass to the outside.

The vaginal wall consists of three layers—an inner mucosa, a middle muscularis, and an outer fibrous layer. The *mucosal layer* contains only a few glands. The mucus found in the lumen of the vagina comes primarily from the glands of the uterus.

The *muscular layer* consists mainly of smooth muscle fibers. At the lower end of the vagina there is a thin band of striated muscle. This band helps to close the vaginal opening, but a voluntary muscle (bulbospongiosus) is primarily responsible for closing this orifice.

The *fibrous layer* consists of dense fibrous connective tissue interlaced with elastic fibers, and it serves to attach the vagina to surrounding organs.

---

1. How is an egg cell moved along a uterine tube?
2. Describe the structure of the uterus.
3. Describe the structure of the vagina.

Female External Reproductive Organs

The *external organs* of the female reproductive system include the labia majora, labia minora, clitoris, and vestibular glands. As a group, these structures that surround the openings of the urethra and vagina compose the **vulva.** (See fig. 20.6.)

The Labia Majora. The **labia majora** enclose and protect the other external reproductive organs. They correspond to the scrotum of the male and are composed primarily of rounded folds of adipose tissue covered by skin.

The labia majora lie closely together and are separated longitudinally by a cleft that includes the urethral and vaginal openings. At their anterior ends, the labia merge to form a medial, rounded elevation of fatty tissue called the *mons pubis,* which overlies the symphysis pubis. (See fig. 20.6.)

The Labia Minora. The **labia minora** are flattened longitudinal folds located within the cleft between the labia majora. They are composed largely of connective tissue that is richly supplied with blood vessels, causing a pinkish appearance.

Posteriorly, the labia minora merge with the labia majora, while anteriorly they converge to form a hoodlike covering around the clitoris.

The Clitoris. The **clitoris** is a small projection at the anterior end of the vulva between the labia minora. Although most of it is embedded in surrounding tissues, it is usually about 2 centimeters (0.8 inch) long and 0.5 centimeter (0.2 inch) in diameter. The clitoris corresponds to the penis of the male and is somewhat similar in structure. More specifically, it is composed of two columns of erectile tissue called *corpora cavernosa,* and at its anterior end, a small mass of erectile tissue forms a **glans,** which is richly supplied with sensory nerve fibers. (See fig. 20.6.)

The Vestibule. The **vestibule** of the vulva is the space enclosed by the labia minora. The vagina opens into the posterior portion of the vestibule, while the urethra opens in the midline just anterior to the vagina and about 2.5 centimeters (1 inch) behind the glans of the clitoris.

A pair of **vestibular glands,** which correspond to the bulbourethral glands of the male, lie one on either side of the vaginal opening.

Beneath the mucosa of the vestibule on either side is a mass of vascular erectile tissue called *vestibular bulbs.*

---

1. What is the male counterpart of the labia majora? Of the clitoris?
2. What structures are located within the vestibule?

Erection, Lubrication, and Orgasm

As in the male, the erectile tissues of the female, located in the clitoris and around the entrance of the vagina, respond to sexual stimulation. Thus, during periods of sexual excitement, *parasympathetic nerve impulses* pass out from the sacral portion of the spinal cord, causing arteries associated with the erectile tissues to dilate and veins to constrict. As a result, the inflow of blood increases, the outflow decreases, and the tissues swell.

Chart 20.2 Functions of the female reproductive organs

Organ	Function
Ovary	Production of egg cells and female sex hormones
Uterine tube	Conveys egg cell toward uterus; site of fertilization; conveys developing embryo to uterus
Uterus	Protects and sustains life of embryo during pregnancy
Vagina	Conveys uterine secretions to outside of body; receives erect penis during sexual intercourse; transports fetus during birth process
Labia majora	Encloses and protects other external reproductive organs
Labia minora	Forms margins of vestibule; protects openings of vagina and urethra
Clitoris	Glans is richly supplied with sensory nerve endings associated with feeling of pleasure during sexual stimulation
Vestibule	Space between labia minora that includes vaginal and urethral openings
Vestibular glands	Secrete fluid that moistens and lubricates vestibule

From Hole, John W. Jr., *Human Anatomy and Physiology 3d ed.* © 1978, 1981, 1984 Wm. C. Brown Publishers, Dubuque, Iowa. All Rights Reserved. Reprinted by permission.

At the same time, the vagina expands and elongates. If the stimulation is sufficiently intense, parasympathetic impulses cause the vestibular glands to secrete mucus into the vestibule. This secretion moistens and lubricates the tissues surrounding the vestibule and the lower end of the vagina, facilitating the insertion of the penis into the vagina.

The clitoris is abundantly supplied with sensory nerve fibers that are especially sensitive to local stimulation, and the culmination of such stimulation is the pleasurable sense of physiological and psychological release called **orgasm.**

Just prior to orgasm, the tissues of the outer third of the vagina become engorged with blood and swell. This action serves to increase the friction on the penis during intercourse. As orgasm is triggered, a series of reflexes involving the sacral and lumbar portions of the spinal cord are initiated.

In response to these reflexes, muscles of the perineum contract rhythmically, and the muscular walls of the uterus and uterine tubes become active. These muscular contractions are thought to aid the transport of sperm cells through the female reproductive tract toward the upper ends of the uterine tubes where there may be an oocyte.

The various functions of the female reproductive organs are summarized in chart 20.2.

1. What events result from parasympathetic stimulation of the female reproductive organs?
2. What changes take place in the vagina just prior to and during female orgasm?

Hormonal Control of Female Reproductive Functions

Female reproductive functions are largely controlled by hormones secreted by the *hypothalamus, anterior pituitary gland,* and the *ovaries.* These hormones are responsible for the development and maintenance of female secondary sexual characteristics, the maturation of egg cells, and changes that occur during the monthly reproductive cycles.

Female Sex Hormones

A female child's body remains reproductively immature until about 8 years of age. About that time, the hypothalamus begins to secrete increasing amounts of gonadotropin-releasing hormone (GnRH), which, in turn, stimulates the anterior pituitary gland to release the gonadotropins FSH and LH. These hormones play primary roles in the control of sex cell maturation and the production of female sex hormones.

Several different female sex hormones are secreted by various tissues, including the ovaries, adrenal cortices, and the placenta (during pregnancy). These hormones belong to two major groups called **estrogen** and **progesterone.**

The primary source of *estrogen* (in a nonpregnant female) is the ovaries. At puberty, under the influence of the anterior pituitary gland, these organs secrete increasing amounts of the hormone. The estrogen stimulates enlargement of various accessory reproductive organs, including the vagina, uterus, uterine tubes, ovaries, and the external structures. Estrogen also is responsible for the development and maintenance of the female *secondary sexual characteristics,* which include:

1. Development of the breasts and the ductile system of the mammary glands within the breasts.
2. Increased deposition of adipose tissue in the subcutaneous layer generally, and particularly in the breasts, thighs, and buttocks.
3. Increased vascularization of the skin.

Certain other changes that occur in females at puberty seem to be related to *androgen* concentrations. For example, increased growth of hair in the pubic and axillary regions seems to be due to the presence of androgen secreted by the adrenal cortices. Conversely, the development of the female skeletal configuration, which includes narrow shoulders and broad hips, seems to be related to a lack of androgen.

The ovaries also are the primary source of *progesterone* (in a nonpregnant female). This hormone promotes changes that occur in the uterus during the rhythmic reproductive cycles. In addition, it affects the mammary glands and helps to regulate the secretion of gonadotropins from the anterior pituitary gland.

Excess sex hormones, as well as other steroids, are metabolized primarily in the liver and the products are excreted in the bile and urine. Thus, people with liver disorders sometimes experience the effects of increased sex hormone activity. For example, males or females with liver diseases may develop symptoms produced by the presence of excessive estrogen or androgen.

---

1. What factors initiate sexual maturity in a female?
2. What are the functions of estrogen?
3. What is the function of androgen in a female?

Female Reproductive Cycles

The female reproductive cycles, or **menstrual cycles,** are characterized by regular, recurring changes in the uterine lining that culminate in menstrual bleeding. Such cycles usually begin near the thirteenth year of life and continue into middle age.

A female's first menstrual cycle is initiated when the hypothalamus secretes gonadotropin-releasing hormone (GnRH), which in turn stimulates the anterior pituitary gland to secrete FSH (follicle-stimulating hormone) and LH (luteinizing hormone). As its name implies, FSH acts upon the ovary to stimulate the maturation of a *follicle,* and during this development, the follicular cells produce increasing amounts of estrogen. LH also plays a role in promoting this production of estrogen.

As mentioned, estrogen is responsible for the development of various secondary sexual characteristics in a young female. The estrogen secreted during

future menstrual cycles is responsible for continuing the development of these traits and for their maintenance.

The increasing concentration of estrogen during the first few days of a menstrual cycle causes changes in the uterine lining, including thickening of the glandular endometrium. This activity continues for about 14 days. Meanwhile, the developing follicle has completed its maturation, and by the fourteenth day of the cycle, it appears on the surface of the ovary as a blisterlike bulge.

For reasons that are not clear, about this time the anterior pituitary gland releases a relatively large quantity of LH and an increased amount of FSH. The resulting surge in concentrations of these hormones seems to cause the mature follicle to swell rapidly and rupture. Ovulation occurs as the follicular fluid, accompanied by the oocyte, leaves the follicle.

One form of birth control employs an *oral contraceptive,* commonly called the pill. The most common forms of such pills contain estrogen-like and progesterone-like substances. The mechanism by which these drugs produce their contraceptive effects is not well understood, but they are thought to inhibit the release of gonadotropins from the anterior pituitary gland and thus prevent ovulation.

Following ovulation, the remnants of the follicle remaining in the ovary undergo rapid changes. The space occupied by the follicular fluid fills with blood that soon clots, and the follicular cells enlarge greatly to form a new glandular structure within the ovary called **corpus luteum.**

Although the follicular cells secrete minute quantities of progesterone during the first part of the menstrual cycle, corpus luteum cells secrete very large quantities of progesterone and estrogen during the last half of the cycle. Consequently, as a corpus luteum becomes established, the blood concentration of progesterone increases sharply.

Progesterone acts on the endometrium of the uterus, causing it to become more vascular and glandular. It also stimulates the uterine glands to secrete increasing quantities of glycogen and lipids. As a result, the endometrial tissues of the uterus become filled with fluids containing nutrients and electrolytes that can provide a favorable environment for the development of an embryo.

Fig. 20.10 Major steps in the female reproductive cycle: (1) the hypothalamus releases GnRH; (2) the anterior lobe of the pituitary gland secretes FSH, which stimulates the maturation of an ovarian follicle; (3) follicular cells secrete estrogen, which causes the uterine wall to thicken; (4) the anterior pituitary gland secretes LH, which stimulates ovulation and causes follicular cells to become corpus luteum; (5) the corpus luteum secretes estrogen and progesterone, which stimulate further development of the uterine wall; (6) the corpus luteum degenerates, and as the concentrations of estrogen and progesterone decrease, tissues slough away from the uterine wall.

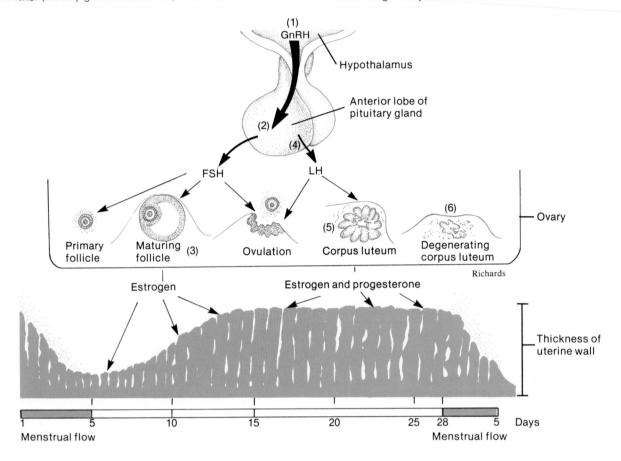

Since high blood concentrations of estrogen and progesterone inhibit the release of GnRH from the hypothalamus and gonadotropins from the pituitary gland, no other follicles are stimulated to develop during the time the corpus luteum is active. If the oocyte that was released at ovulation is not fertilized by a sperm cell, however, the corpus luteum begins to degenerate about the twenty-fourth day of the cycle.

When the corpus luteum ceases to function, the concentrations of estrogen and progesterone decline rapidly, and in response, blood vessels in the endometrium become constricted. This action reduces the supply of oxygen and nutrients to the thickened uterine lining, and these tissues soon disintegrate and slough away. At the same time, blood escapes from damaged capillaries, creating a flow of blood and cellular debris that passes through the vagina as the *menstrual flow* (menses). This flow usually begins about the twenty-eighth day of the cycle and continues for 3 to 5 days while the estrogen concentration is relatively low.

The beginning of the menstrual flow marks the end of a menstrual cycle and the beginning of a new cycle. This cycle is diagrammed in figure 20.10 and summarized in chart 20.3.

Since the blood concentrations of estrogen and progesterone are low at the beginning of the cycle, the hypothalamus and pituitary gland are no longer inhibited. Consequently, the concentrations of FSH and LH soon increase, and a new follicle is stimulated to

Fig. 20.11 Changes in relative concentrations of hormones during a menstrual cycle.

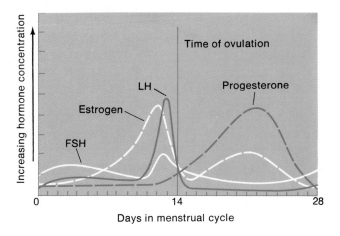

Days in menstrual cycle

mature. As this follicle secretes estrogen, the lining of the uterus undergoes repair and the endometrium begins to thicken again.

Figure 20.11 summarizes the changing hormone concentrations during the menstrual cycle.

After puberty, menstrual cycles usually continue to occur at more or less regular intervals into the late forties, at which time they usually become increasingly irregular and, after a few months or years, cease altogether. This period of life is called **menopause.**

The cause of menopause seems to be aging of the ovaries. After about 35 years of cycling, apparently few primary follicles remain to be stimulated by pituitary gonadotropins. Consequently, follicles no longer mature, ovulation does not occur, and the blood concentration of estrogen decreases greatly.

As a result of low estrogen concentration and lack of progesterone, the female secondary sexual characteristics usually undergo varying degrees of change. For example, the vagina, uterus, and uterine tubes may decrease in size, as may the external reproductive organs. The pubic and axillary hair may become thinner, and the breasts may regress.

1. *Trace the events of the female menstrual cycle.*
2. *What causes the menstrual flow?*

Mammary Glands

The **mammary glands** are accessory organs of the female reproductive system that are specialized to secrete milk following pregnancy. They are located in the subcutaneous tissue of the anterior thorax within elevations called *breasts.* The breasts overlie the *pectoralis major* muscles and extend from the second to the sixth ribs, and from the sternum to the axillae.

A *nipple* is located near the tip of each breast at about the level of the fourth intercostal space, and it is surrounded by a circular area of pigmented skin called the *areola.* (See fig. 20.12.)

A mammary gland is composed of fifteen to twenty lobes, each of which includes tubular glands, called *alveolar glands,* and a duct that leads to the nipple and opens to the outside. The lobes are separated from each other by dense connective and adipose tissues. These tissues also support the glands and attach them to the fascia of the underlying pectoral muscles. Other connective tissue, which forms dense strands called *suspensory ligaments,* extends inward from the dermis of the breast to the fascia, helping to support the weight of the breast.

Fig. 20.12 Sagittal section of a breast.

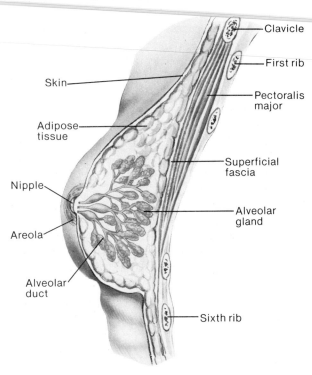

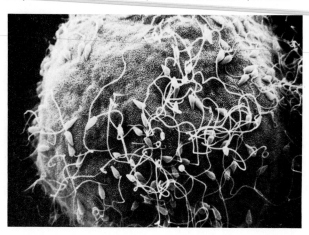

The mammary glands of male and female children are similar. As children reach *puberty,* however, the male glands fail to develop, and the female glands are stimulated to develop by ovarian hormones. As a result, the aveolar glands and ducts enlarge, and fat is deposited so that the glands become surrounded by adipose tissue.

---

1. *Describe the structure of a mammary gland.*
2. *What changes occur in these glands as a result of ovarian hormones?*

Pregnancy

Pregnancy is the condition characterized by the presence of a developing offspring within the uterus. It results from the union of an egg cell with a sperm cell (fertilization).

Transport of Sex Cells

Ordinarily, before fertilization can occur, an egg cell (secondary oocyte) must be released by ovulation and be carried into a uterine tube by the action of the ciliated epithelium lining the tube.

During sexual intercourse, seminal fluid containing sperm cells usually is deposited in the vagina near the cervix. To reach the egg cell, the sperm cells must then be transported upward through the uterus and uterine tube. This transport is aided by the lashing movements of the sperm's tail and by muscular contractions within the walls of the uterus and uterine tube, which are thought to be stimulated by prostaglandins in the seminal fluid. Also, under the influence of the high estrogen concentrations during the first part of the menstrual cycle, the uterus and cervix contain a thin, watery secretion that promotes sperm transport and survival. Conversely, during the latter part of the cycle, when progesterone is high in concentration, these parts secrete a viscous fluid that is unfavorable for sperm transport and survival.

Studies indicate that an egg cell may survive for only 12 to 24 hours following ovulation, while sperm cells may live up to 72 hours within the female reproductive tract. Consequently, sexual intercourse probably must occur no more than 72 hours before ovulation or 24 hours following ovulation if fertilization is to take place.

Sperm cells are thought to reach the upper portions of the uterine tube within an hour following sexual intercourse. Although many sperm cells may reach an egg cell, only one will participate in fertilization. (See fig. 20.13.)

Fig. 20.14 The process of fertilization: (a) after a sperm cell enters it, a secondary oocyte divides unequally to form a polar body and a zygote; (b) the nucleus of the sperm cell enlarges and approaches the nucleus of the egg cell; (c) the nuclear membranes disappear, and the chromosomes of the two nuclei combine; (d) the result is a cell with 46 chromosomes.

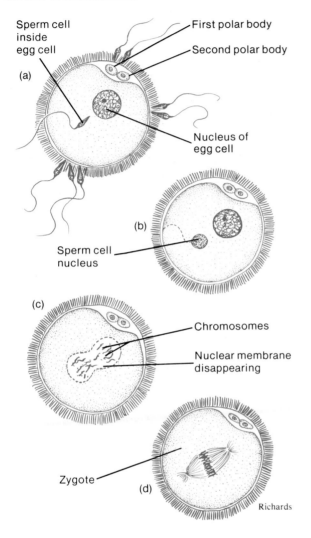

Sperm cell inside egg cell

(a)

First polar body

Second polar body

Nucleus of egg cell

(b)

Sperm cell nucleus

(c)

Chromosomes

Nuclear membrane disappearing

Zygote

(d)

Richards

Fertilization

When a sperm cell reaches an egg cell, it moves through the follicular cells that adhere to the egg's surface (corona radiata) and penetrates the *zona pellucida* that surrounds the egg cell membrane. This penetration seems to be aided by an enzyme (hyaluronidase), released by the acrosome of the sperm.

As the sperm cell penetrates the zona pellucida, this covering becomes impenetrable to any other sperm cells. The mechanism preventing the entrance of extra sperm cells seems to involve structural changes in the membrane that occur soon after the first sperm cell penetrates it and the release of certain enzymes that keep other sperm cells from attaching to the membrane.

Once a sperm cell reaches the egg cell membrane, it passes through the membrane and enters the cytoplasm. During this process, the sperm cell loses its tail, and its head swells to form a nucleus. As explained previously, the egg cell (secondary oocyte) then divides unequally to form a relatively large cell and a tiny second polar body, which is expelled. The nuclei of the egg cell and the sperm cell come together in the center of the larger cell. Their nuclear membranes disappear, and their chromosomes combine, thus completing the process of **fertilization.** This process is diagrammed in figure 20.14.

Since the sperm cell and the egg cell each provide 23 chromosomes, the result of fertilization is a cell with 46 chromosomes—the usual number of a human cell. This cell, called a **zygote,** is the first cell of the future offspring.

---

1. *What factors enhance motility of sperm cells following sexual intercourse?*
2. *Where does fertilization normally take place?*
3. *List the events that occur during fertilization.*

Early Embryonic Development

Shortly after it is formed, the zygote undergoes *mitosis,* giving rise to two daughter cells. These cells in turn divide into four cells, which divide into eight cells, and so forth. With each subsequent division, the resulting daughter cells are smaller and smaller. Consequently, this phase in development is termed **cleavage.**

Meanwhile, the tiny mass of cells is moved through the uterine tube to the cavity of the uterus. The trip to the uterus takes about 3 days, and at the end of this time the structure consists of a solid ball (morula) of about sixteen cells.

Fig. 20.15 Stages in early human development.

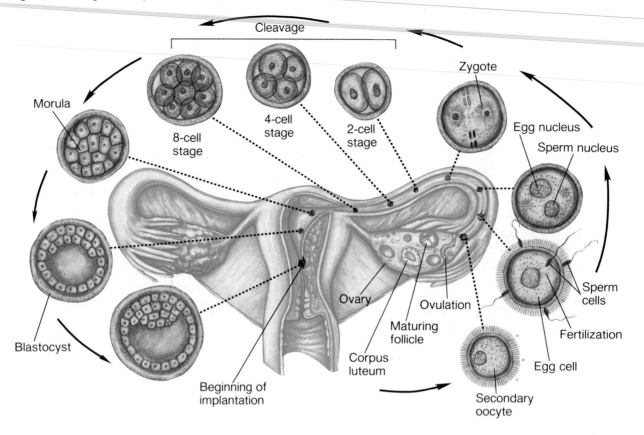

Cleavage

Morula

8-cell stage

4-cell stage

2-cell stage

Zygote

Egg nucleus

Sperm nucleus

Sperm cells

Blastocyst

Ovary

Ovulation

Fertilization

Maturing follicle

Egg cell

Corpus luteum

Secondary oocyte

Beginning of implantation

Once inside the uterus, it remains free within the uterine cavity for about 3 days, during which the zona pellucida of the original egg cell degenerates, and the structure, which now consists of a hollow ball of cells (blastocyst), begins to attach itself to the uterine lining. By the end of the first week of development, it is superficially *implanted* in the endometrium (fig. 20.15).

An *intrauterine device (IUD)* is a small, solid object that may be placed within the uterine cavity by a physician to prevent pregnancies. It is believed that the IUD somehow interferes with the implantation of embryos within the uterine wall, perhaps by causing inflammatory reactions in the uterine tissues.

Unfortunately, an IUD may be expelled from the uterus spontaneously, or it may produce side effects, such as pain or excessive bleeding. On occasion, it may injure the uterus or produce other serious health problems.

About the time of implantation, certain cells within the blastocyst organize themselves into a group (inner cell mass) that will give rise to the body of the offspring. This marks the beginning of the embryonic period of development. (See fig. 20.16.) The offspring is termed an **embryo** until the end of the eighth week, after which it is called a **fetus.**

Eventually, the outer cells of the embryo together with cells of the maternal endometrium form a complex vascular structure called the **placenta.** This organ serves to attach the embryo to the uterine wall, to exchange nutrients, gases, and wastes between the maternal blood and the embryonic blood, and to secrete hormones.

1. *What is meant by cleavage?*
2. *What is meant by implantation?*
3. *How does a developing offspring obtain nutrients and oxygen?*

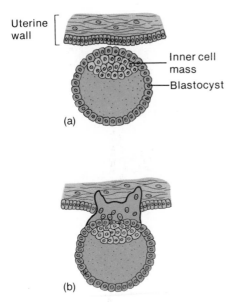

Fig. 20.16 (a) About the sixth day of development, the blastocyst contacts the uterine wall and (b) becomes implanted within the wall.

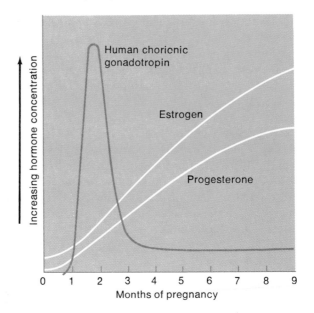

Fig. 20.17 Relative concentrations of three hormones in the blood during pregnancy.

Hormonal Changes during Pregnancy

During the usual menstrual cycle, the corpus luteum degenerates about two weeks after ovulation. Consequently, the estrogen and progesterone levels decline rapidly, the uterine lining is no longer maintained, and the endometrium sloughs away as menstrual flow. If this occurs following implantation, the embryo will be lost.

The mechanism that normally prevents such a termination of pregnancy involves a hormone, called HCG (human chorionic gonadotropin). This hormone is secreted by a layer of embryonic cells (trophoblast) that surrounds the developing embryo and later becomes involved with the formation of the placenta. HCG has properties similar to those of LH and causes the corpus luteum to be maintained and to continue secreting relatively large amounts of estrogen and progesterone. Thus, the uterine wall continues to grow and develop. At the same time, the estrogen and progesterone suppress the release of FSH and LH from the pituitary gland, so that normal menstrual cycles are inhibited.

The secretion of HCG continues at a high level for about two months, then declines to a relatively low level by the end of four months. Although the corpus luteum is maintained throughout the pregnancy, its function as a source of hormones becomes less important after the first three months of pregnancy (first trimester). This is due to the fact that the placenta usually is well developed by this time, and the placental tissues secrete high levels of estrogen and progesterone (fig. 20.17).

HCG secretion by the embryonic tissues begins shortly after fertilization and increases to a peak in about 50 to 60 days. Thereafter the concentration of HCG drops to a much lower level that remains relatively stable throughout the pregnancy.

Since HCG is excreted in the urine, its presence in urine can be used to detect pregnancy. Such a test may indicate positive results within about 28 days of fertilization.

For the remainder of the pregnancy, *placental estrogen* and *progesterone* maintain the uterine wall. The placenta also secretes a hormone called **placental lactogen.** This hormone is thought to stimulate breast development and preparation for milk secretion, a function that is aided by placental estrogen and progesterone. Placental progesterone and a polypeptide hormone from corpus luteum called *relaxin* inhibit the smooth muscles in the myometrium so that uterine contractions are suppressed until it is time for the birth process to begin.

The high concentration of placental estrogen during pregnancy causes enlargement of the vagina and external reproductive organs, as well as relaxation of the ligaments holding the symphysis pubis and sacroiliac joints together. This latter action allows for greater movement at these joints and thus aids the passage of the fetus through the birth canal. This relaxation of ligaments and the softening of the cervix near the time of delivery also may be aided by relaxin.

Other hormonal changes that occur during pregnancy include increased secretion of aldosterone from the adrenal cortex and parathormone from the parathyroid glands. The aldosterone promotes renal reabsorption of sodium, leading to fluid retention, while parathormone helps to maintain a high level of maternal blood calcium (see chapter 12).

Chart 20.4 summarizes the hormonal changes of pregnancy.

1. What mechanism is responsible for maintaining the uterine wall during pregnancy?
2. What is the source of hormones that sustain the uterine wall during pregnancy?
3. What other hormonal changes occur during pregnancy?

Embryonic Stage of Development

The **embryonic stage** extends from the second week through the eighth week of development and is characterized by the formation of the placenta, the development of the main internal organs, and the appearance of the major external body structures.

Early in this stage, the cells of the inner cell mass become organized into a flattened **embryonic disk** that consists of two distinct layers—an outer *ectoderm* and an inner *endoderm*. (See fig. 20.18.) A short time later, a third layer of cells, the *mesoderm,* forms between the ectoderm and endoderm. These three layers of cells are called the **primary germ layers,** and they are responsible for forming all of the body organs.

More specifically, *ectodermal cells* give rise to the nervous system, portions of special sensory organs, the epidermis, hair, nails, glands of the skin, and the linings of the mouth and anal canal. *Mesodermal*

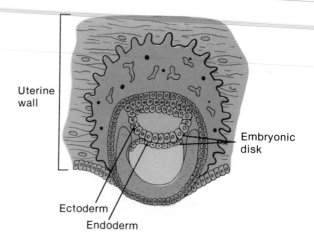

Fig. 20.18 The cells of the embryonic disk become organized into two layers.

Uterine wall

Embryonic disk

Ectoderm

Endoderm

Chart 20.4 Hormonal changes during pregnancy

1. Following implantation, embryonic cells begin to secrete HCG.
2. HCG causes the corpus luteum to be maintained and to continue secreting estrogen and progesterone.
3. As the placenta develops, it secretes large quantities of estrogen and progesterone.
4. Placental estrogen and progesterone
   a. Stimulate the uterine lining to continue development.
   b. Maintain the uterine lining.
   c. Inhibit the secretion of FSH and LH from the anterior pituitary gland.
   d. Stimulate development of the mammary glands.
   e. Progesterone inhibits uterine contractions.
   f. Estrogen causes enlargement of reproductive organs and relaxation of ligaments of pelvic joints.
5. Relaxin from corpus luteum also inhibits uterine contractions and causes pelvic ligaments to relax.
6. The placenta secretes placental lactogen that stimulates breast development.
7. Aldosterone from adrenal cortex promotes reabsorption of sodium.
8. Parathormone from parathyroid glands helps maintain a high concentration of maternal blood calcium.

cells form all types of muscle tissue, bone tissue, bone marrow, blood, blood vessels, lymphatic vessels, various connective tissues, internal reproductive organs, kidneys, and the epithelial linings of the body cavities. *Endodermal cells* produce the epithelial linings of the digestive tract, respiratory tract, urinary bladder, and urethra.

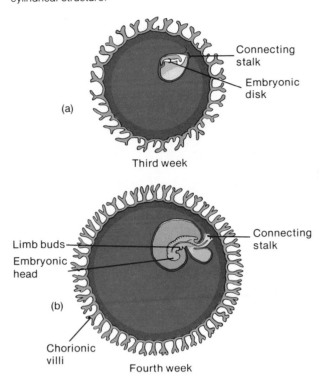

(a)

Connecting stalk

Embryonic disk

Third week

Limb buds

Embryonic head

Connecting stalk

(b)

Chorionic villi

Fourth week

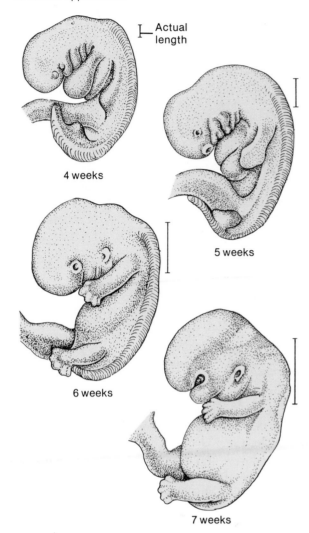

Actual length

4 weeks

5 weeks

6 weeks

7 weeks

During the fourth week of development (fig. 20.19), the flat embryonic disk is transformed into a cylindrical structure, which is attached to the developing placenta by a *connecting stalk*. By this time, the head and jaws are appearing, the heart is beating and forcing blood through blood vessels, and tiny buds that will give rise to the arms and legs are forming.

During the fifth through the seventh weeks, as shown in figure 20.20, the head grows rapidly and becomes rounded and erect. The face, which is developing eyes, nose, and mouth, becomes more humanlike. The arms and legs elongate, and fingers and toes appear.

By the end of the seventh week, all the main internal organs have become established, and as these structures enlarge, they affect the shape of the body. Consequently, the body takes on a humanlike appearance.

Meanwhile, the embryo continues to become implanted within the uterus. Early in this process slender projections grow out from the wall of the blastocyst into the surrounding endometrium of the uterine wall. These extensions, which are called **chorionic villi** (fig. 20.19), become branched, and by the end of the fourth week of development they are well formed.

While the chorionic villi are developing, embryonic blood vessels appear within them, and these vessels are continuous with those passing through the connecting stalk to the body of the embryo. At the same time, irregular spaces called **lacunae** are eroded around and between the villi. These spaces become filled with maternal blood that escapes from eroded endometrial blood vessels.

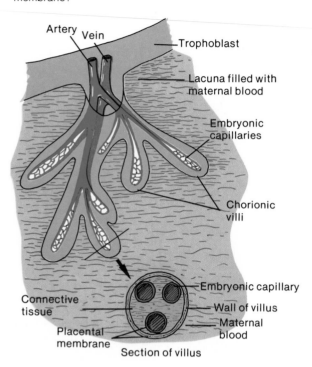

A thin membrane separates embryonic blood within the capillary of a chorionic villus from maternal blood in a lacuna. This membrane, called the **placental membrane,** is composed of the epithelium of the villus and the epithelium of the capillary. (See fig. 20.21.) Through this membrane, exchanges take place between maternal and embryonic blood. Oxygen and nutrients diffuse from the maternal blood into the embryonic blood, and carbon dioxide and other wastes diffuse from the embryonic blood into the maternal blood. Various substances also move through the placental membrane by active transport and pinocytosis.

Since most drugs are able to pass freely through the placental membrane, substances ingested by the mother may affect the fetus. Thus, fetal drug addiction may occur following the mother's use of drugs such as heroin.

Similarly, depressant drugs administered to the mother during labor can produce effects within the fetus and may, for example, depress the activity of its respiratory system.

1. What major events occur during the embryonic stage of development?
2. What tissues and structures develop from ectoderm? From mesoderm? From endoderm?
3. How are substances exchanged between the embryonic blood and the maternal blood?

Until about the end of the eighth week, the chorionic villi cover the entire surface of the former blastocyst. The membrane that contains these villi and surrounds the developing embryo is called the **chorion.** As the embryo and the chorion continue to enlarge, only those villi that remain in contact with the endometrium endure. The others degenerate, and the portions of the chorion to which they were attached become smooth. Thus the region of the chorion still in contact with the uterine wall is restricted to a disk-shaped area that becomes the **placenta.**

The embryonic portion of the placenta is composed of the chorion and its villi; the maternal portion is composed of the area of the uterine wall to which the villi are attached. When it is fully formed, the placenta appears as a reddish brown disk about 20 centimeters (8 inches) long and 2.5 centimeters (1 inch) thick. It usually weighs about 0.5 kilogram (1 pound). Figure 20.22 shows the structure of the placenta.

Fig. 20.23 (a, b, c) As the amnion develops, it surrounds the embryo and (d) the umbilical cord is formed. What structures comprise this cord?

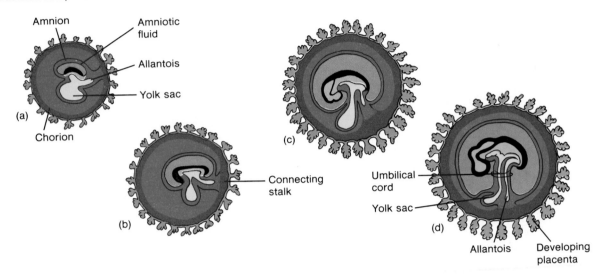

While the placenta is forming from the chorion, another membrane, called the **amnion,** develops around the embryo. This second membrane begins to appear during the second week. Its margin is attached around the edge of the embryonic disk, and fluid, called **amniotic fluid,** fills the space between them.

As the embryo is transformed into a cylindrical structure, the margins of the amnion are folded around it so that the embryo is enclosed by the amnion and surrounded by amniotic fluid. As this process continues, the amnion envelops the tissues on the underside of the embryo by which it is attached to the chorion and the developing placenta. In this manner, as figure 20.23 illustrates, the **umbilical cord** is formed.

The umbilical cord contains three blood vessels—two *umbilical arteries* and an *umbilical vein*—through which blood passes between the embryo and the placenta. (See fig. 20.22.)

The umbilical cord also functions to suspend the embryo in the amniotic cavity, while the amniotic fluid provides a watery environment in which the embryo can grow freely without being compressed by surrounding tissues. The fluid also protects the embryo against being jarred by movements of the mother's body.

In addition to the amnion and chorion, two other embryonic membranes appear during development. They are the yolk sac and the allantois. (See fig. 20.23.)

The **yolk sac** appears during the second week, and it is attached to the underside of the embryonic disc. It functions to form blood cells in the early stages of development and gives rise to the cells that later become sex cells.

The **allantois** forms during the third week as a tube extending from the early yolk sac into the connecting stalk of the embryo. It functions in the formation of blood cells and gives rise to the umbilical arteries and vein.

Agents that cause congenital malformations by affecting an embryo during its period of rapid development are called *teratogens*. Such factors include exposure to radiation, the presence of various drugs, and infections caused by certain microorganisms. Exposure to radiation, for example, tends to produce abnormalities within the central nervous system. The drug thalidomide interferes with the normal development of the limbs, and the presence of the virus causing rubella (German measles) may result in malformations of the heart as well as cataract and congenital deafness.

By the beginning of the eighth week, the embryo is usually 30 millimeters (1.2 inches) in length and weighs less than 5 grams (0.2 ounce). Although its body is quite unfinished, it is clearly recognizable as a human being.

1. Describe how the placenta forms.
2. What is the function of amniotic fluid?
3. What is the significance of the yolk sac?

Fetal Stage of Development

The **fetal stage** begins at the end of the eighth week of development and continues to the time of birth. During this period, the existing body structures continue to grow and mature, and only a few new parts appear. The rate of growth is great, however, and the body proportions change considerably. For example, at the beginning of the fetal stage, the head is disproportionately large and the legs are relatively short.

During the third lunar month, growth in body length is accelerated, while the growth of the head slows. (A lunar month equals 28 days and is commonly used in studies of human development.) The arms achieve the relative length they will maintain throughout development, and ossification centers appear in most of the bones. By the twelfth week, the external reproductive organs are distinguishable as male or female.

In the fourth lunar month, the body grows very rapidly and reaches a length of 13 to 17 centimeters (5 to 6.8 inches). The legs lengthen considerably, and the skeleton continues to ossify.

In the fifth lunar month, the rate of growth decreases somewhat. The legs achieve their final relative proportions, and the skeletal muscles become active, so that the mother may feel fetal movements. Some hair appears on the head, and the skin becomes covered with fine, downy hair. The skin also is coated with a cheesy mixture of sebum from the sebaceous glands and dead epidermal cells.

During the sixth lunar month, the body gains a substantial amount of weight. The eyebrows and eyelashes appear. The skin is quite wrinkled and translucent. It is also reddish, due to the presence of dermal blood vessels.

In the seventh lunar month, the skin becomes smoother as fat is deposited in the subcutaneous tissues. The eyelids, which fused together during the third month, reopen. At the end of this month a fetus is about 37 centimeters (14.8 inches) in length.

If a fetus is born prematurely, its chance of surviving increases directly with its age. One of the factors involved is the development of the lungs. Thus, fetuses have increased chances of surviving if their lungs are sufficiently developed so that they have the thin respiratory membranes necessary for rapid exchange of oxygen and carbon dioxide, and the lungs produce enough surfactant to reduce the alveolar surface tension (see chapter 14).

In the eighth lunar month, the fetal skin is still reddish and somewhat wrinkled. The testes of males descend from regions near the developing kidneys into the scrotum.

During the ninth lunar month, the fetus reaches a length of about 47 centimeters (18.8 inches). The skin is smooth, and the body appears chubby due to an accumulation of subcutaneous fat. The reddishness of the skin fades to pinkish or bluish pink, even in fetuses of dark-skinned parents, because melanin is not produced until the skin is exposed to light. Thus it may be difficult to determine a newborn's race by its skin color during the first few hours or days of life.

At the end of the tenth lunar month, the fetus is said to be *full term*. It is about 50 centimeters (19.5 inches) long and weighs 6 to 8 pounds. The skin has lost its downy hair but is still covered with sebum and dead epidermal cells. The scalp is usually covered with hair, the fingers and toes have well developed nails, and the bones of the head are ossified. As figure 20.24 shows, the fetus usually is positioned upside down with its head toward the cervix in preparation for the birth process.

1. What major changes characterize the fetal stage of development?
2. How is a fetus usually positioned within the uterus at the end of the tenth lunar month?

Fetal Circulation. Throughout the fetal stage of development, the maternal blood supplies the fetus with oxygen and nutrients and carries away its wastes. These substances diffuse between the maternal and fetal blood through the placental membrane, and they are carried to and from the fetal body by means of the umbilical blood vessels. Consequently, the fetal vascular system must be adapted to intrauterine life in special ways, and its pattern of blood flow must differ from that of an adult.

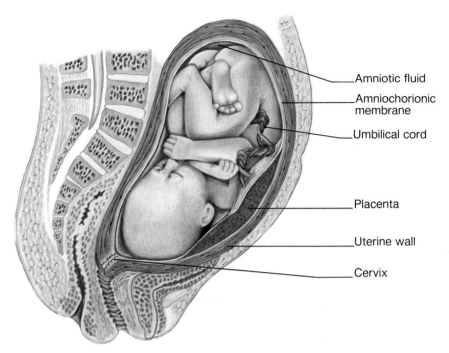

Amniotic fluid

Amniochorionic
membrane

Umbilical cord

Placenta

Uterine wall

Cervix

In fetal circulation, the *umbilical vein* transports blood rich in oxygen and nutrients from the placenta to the fetal body. This vein enters the body and travels along the anterior abdominal wall to the liver. About half the blood passes into the liver, and the rest enters a vessel called the **ductus venosus** that bypasses the liver. (See fig. 20.25.)

The ductus venosus travels a short distance and joins the inferior vena cava. There the oxygenated blood from the placenta is mixed with deoxygenated blood from the lower parts of the fetal body. This blood continues through the vena cava to the right atrium.

In an adult heart, blood from the right atrium enters the right ventricle and is pumped through the pulmonary artery to the lungs. (See chapter 16.) In the fetus however, the lungs are nonfunctional, and the blood largely bypasses them. More specifically, as blood relatively rich in oxygen enters the right atrium of the fetal heart, a large proportion of it is shunted directly into the left atrium through an opening in the atrial septum. This opening is called the **foramen ovale,** and blood passes through it because the blood pressure in the right atrium is somewhat greater than that in the left atrium. Furthermore, a small valve located on the left side of the atrial septum overlies the foramen ovale and helps to prevent blood from moving from left to right.

The rest of the blood entering the right atrium, as well as a large proportion of the deoxygenated blood entering from the superior vena cava, passes into the right ventricle and out through the pulmonary trunk.

The vessels of the pulmonary circuit have a high resistance to blood flow since the lungs are collapsed, and the vessels are somewhat compressed. Enough blood reaches the lung tissues, however, to sustain them.

Most of the blood in the pulmonary trunk bypasses the lungs by entering a fetal vessel called **ductus arteriosus,** which connects the pulmonary artery to the descending portion of the aortic arch. As a result, blood with a relatively low oxygen concentration returning to the heart in the superior vena cava bypasses the lungs. At the same time it is prevented from entering the portion of the aorta that provides branches leading to the heart and brain.

The more highly oxygenated blood that enters the left atrium through the foramen ovale is mixed with a small amount of deoxygenated blood returning from the pulmonary veins. This mixture moves into the left ventricle and is pumped into the aorta. Some of it reaches the myocardium by means of the coronary arteries, and some reaches the tissues of the brain through the carotid arteries.

Fig. 20.25 The general pattern of fetal circulation. How does this pattern of circulation differ from that of an adult?

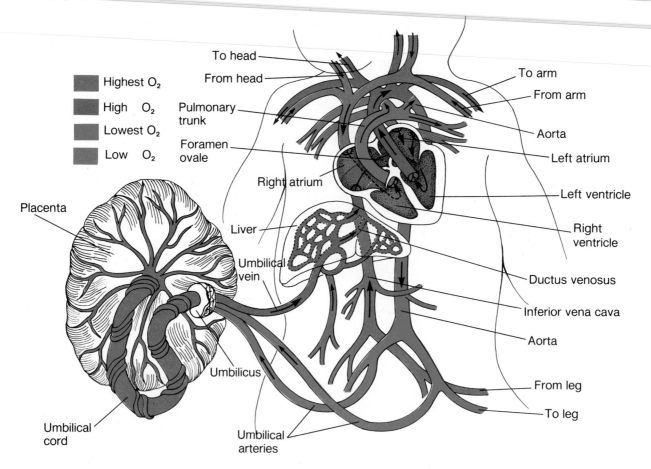

Highest O₂
High O₂
Lowest O₂
Low O₂

To head
From head
Pulmonary trunk
Foramen ovale
Right atrium
Placenta
Liver
Umbilical vein
Umbilicus
Umbilical cord
Umbilical arteries

To arm
From arm
Aorta
Left atrium
Left ventricle
Right ventricle
Ductus venosus
Inferior vena cava
Aorta
From leg
To leg

Chart 20.5	Fetal circulatory adaptations
Adaptation	**Function**
Umbilical vein	Carries oxygenated blood from placenta to fetus
Ductus venosus	Conducts about half the blood from the umbilical vein directly to the inferior vena cava, thus bypassing the liver
Foramen ovale	Conveys large proportion of blood entering the right atrium from the inferior vena cava, through the atrial septum, and into the left atrium, thus bypassing the lungs
Ductus arteriosus	Conducts some blood from the pulmonary artery to the aorta, thus bypassing the lungs
Umbilical arteries	Carry blood from the internal iliac arteries to the placenta for reoxygenation

From Hole, John W. Jr., *Human Anatomy and Physiology 3d ed.* © 1978, 1981, 1984 Wm. C. Brown Publishers, Dubuque, Iowa. All Rights Reserved. Reprinted by permission.

The blood carried by the descending aorta is partially oxygenated and partially deoxygenated. Some of it is carried into the branches of the aorta that lead to various parts in the lower regions of the body. The rest passes into the *umbilical arteries* which branch from the internal iliac arteries and lead to the placenta. There the blood is reoxygenated.

Chart 20.5 summarizes the major features of fetal circulation. At the time of birth, important adjustments must occur in this circulatory system when the placenta ceases to function.

1. *Which umbilical vessel carries oxygen-rich blood to the fetus?*
2. *What is the function of the ductus venosus?*
3. *How does fetal circulation allow blood to bypass the lungs?*

Fig. 20.26 Major steps in the birth process.

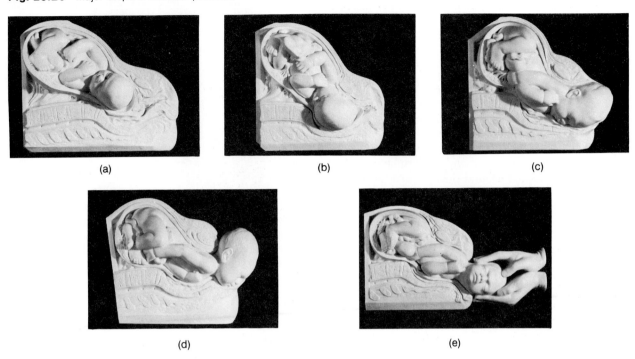

(a)

(b)

(c)

(d)

(e)

The Birth Process

Pregnancy usually continues for 40 weeks (280 days) or about 9 calendar months (10 lunar months) if it is measured from the beginning of the mother's last menstrual cycle. The pregnancy terminates with the *birth process.*

Although the mechanism causing birth is not well understood, a variety of factors seem to be involved. For example, progesterone suppresses uterine contractions during pregnancy. Estrogen, however, tends to excite such contractions. After the seventh month, the placental secretion of estrogen increases to a greater degree than the secretion of progesterone. As a result of the rising concentration of estrogen, the contractility of the uterine wall is enhanced.

The stretching of the uterine and vaginal tissues late in pregnancy is thought to initiate nerve impulses to the hypothalamus. The hypothalamus in turn signals the posterior pituitary gland, which responds by releasing the hormone **oxytocin.** (See chapter 12.)

Oxytocin is a powerful stimulator of uterine contractions, and its effect, combined with the greater excitability of the myometrium of the uterus, may be involved in initiating labor.

Labor is the term for the process whereby muscular contractions force the fetus through the birth canal. Once labor starts, rhythmic contractions that begin at the top of the uterus and travel down its length force the contents of the uterus toward the cervix.

Since the fetus is usually positioned with its head downward, labor contractions force the head against the cervix. This action causes stretching of the cervix, which is thought to elicit a reflex that stimulates still stronger labor contractions. Thus a *positive feedback system* (see chapter 12) operates in which uterine contractions result in more intense uterine contractions until a maximum effort is achieved. At the same time, dilation of the cervix reflexly stimulates an increased release of oxytocin from the pituitary gland.

As labor continues, abdominal wall muscles are stimulated to contract by a positive feedback mechanism, and they also aid in forcing the fetus through the cervix and vagina to the outside.

Following the birth of the fetus (usually within 10 to 15 minutes), the placenta, which remains inside the uterus, becomes separated from the uterine wall and is expelled by uterine contractions through the birth canal. This expulsion, which is termed the *afterbirth,* is accompanied by bleeding because vascular tissues are damaged in the process. This loss of blood is usually minimized by continued contraction of the uterus that constricts the bleeding vessels.

Figure 20.26 illustrates the steps of the birth process.

---

1. *Describe the events thought to initiate labor.*
2. *Explain how dilation of the cervix affects labor.*

Postnatal Changes

Following birth, a variety of physiological and structural changes occur in the mother's and the newborn's bodies.

Production and Secretion of Milk

During pregnancy, placental estrogen and progesterone stimulate further development of the mammary glands. Estrogen causes the ductile systems to grow and become branched and to have large quantities of fat deposited around them. Progesterone, on the other hand, stimulates the development of the alveolar glands. (See fig. 20.12.) These changes also are promoted by placental lactogen, which was mentioned previously.

As a consequence of hormonal activity, the breasts typically double in size during pregnancy, and the mammary glands become capable of secreting milk. No milk is produced, however, because the high concentrations of estrogen and progesterone that occur during pregnancy inhibit the hypothalamus. The hypothalamus in turn suppresses the secretion of the hormone **prolactin** by the anterior pituitary gland.

Following childbirth and the expulsion of the placenta, the maternal blood concentrations of estrogen and progesterone decline rapidly. Consequently, the hypothalamus is no longer inhibited, and it signals the pituitary gland to release prolactin. (See chapter 12.)

Prolactin stimulates the mammary glands to secrete large quantities of milk. This hormonal effect does not occur for 2 or 3 days following birth, and in the meantime, the glands secrete a few milliliters of a fluid called *colostrum* each day. Although colostrum contains some of the nutrients found in milk, it lacks fat.

The milk produced under the influence of prolactin does not flow readily through the ductile system of the mammary gland, but must be actively ejected by contraction of specialized *myoepithelial cells* surrounding the ducts. The contraction of these cells and the consequent ejection of milk through the nipple result from a reflex action.

This reflex is elicited when the breast is sucked or the nipple or areola is otherwise mechanically stimulated. Then sensory impulses travel to the hypothalamus, which signals the posterior pituitary gland to release oxytocin. The oxytocin reaches the breast by means of the blood, and it stimulates the myoepithelial cells of the ductile system to contract. Consequently, milk is ejected into a suckling infant's mouth in about 30 seconds.

As long as milk is removed from the breasts, prolactin and oxytocin continue to be released, and milk continues to be produced. If milk is not removed regularly, the hypothalamus causes the secretion of prolactin to be inhibited, and within about one week, the mammary glands lose their capacity to produce milk.

---

1. How does pregnancy affect the mammary glands?
2. What stimulates the mammary glands to produce milk?
3. How is milk stimulated to flow into the ductile system of a mammary gland?

Neonatal Period of Development

The **neonatal period,** which extends from birth to the end of the first four weeks, begins very abruptly at birth. At this time, physiological adjustments must be made quickly because the newborn must suddenly do for itself those things that the mother's body has been doing for it. The newborn must carry on respiration, obtain nutrients, digest nutrients, excrete wastes, regulate body temperature, and so forth. Its most immediate need, however, is to obtain oxygen and excrete carbon dioxide, so its first breath is critical.

The first breath must be particularly forceful, because the newborn's lungs are collapsed, and the airways are small and offer considerable resistance to air movement. Also, surface tension tends to hold the moist membranes of the lungs together.

Fortunately, the lungs of a full term fetus continuously secrete *surfactant* (see chapter 14), which reduces surface tension, and after the first powerful breath begins to expand the lungs, breathing becomes easier.

The newborn has a relatively high rate of metabolism, and its liver, which is not fully mature, may be unable to supply enough glucose to support the metabolic needs. Consequently, the newborn typically utilizes stored fat as an energy source.

---

1. Why must the first breath of an infant be particularly forceful?
2. What does a newborn use for an energy supply during the first few days after birth?

As a rule, the newborn's kidneys are unable to produce concentrated urine, so the newborn excretes a relatively dilute fluid. For this reason, the newborn may become dehydrated and develop a water and electrolyte imbalance. Also, some of the newborn's homeostatic control mechanisms may function imperfectly. The temperature regulating system, for example, may be unable to maintain a constant body temperature.

As was mentioned, when the circulation of blood through the placenta ceases and the lungs begin to function at birth, adjustments have to be made in the circulatory system because it is adapted to the needs of intrauterine life.

For example, following birth, the umbilical vessels constrict. The arteries close first, and if the umbilical cord is not clamped or severed for a minute or so, blood continues to flow from the placenta to the newborn through the umbilical vein, adding to the newborn's blood volume.

Similarly, the ductus venosus constricts shortly after birth and is represented in the adult as a fibrous cord (ligamentum venosum), which is superficially embedded in the wall of the liver.

The foramen ovale closes as a result of blood pressure changes occurring in the right and left atria as fetal vessels constrict. More precisely, as blood ceases to flow from the umbilical vein into the inferior vena cava, the blood pressure in the right atrium drops. Also, as the lungs expand with the first breathing movements, the resistance to blood flow through the pulmonary circuit decreases, more blood enters the left atrium through the pulmonary veins, and the blood pressure in the left atrium increases.

As pressure in the left atrium rises and that in the right atrium falls, the valve on the left side of the atrial septum closes the foramen ovale. In most individuals this valve gradually fuses with the tissues along the margin of the foramen. In an adult, the site of the previous opening is marked by a depression called the *fossa ovalis.*

In some newborns, the foramen ovale remains open, and since the blood pressure in the left atrium is greater than that in the right atrium, some blood may flow from the left chamber to the right chamber.

Although this condition usually produces no symptoms, it may cause a heart murmur.

The ductus arteriosus, like the other fetal vessels, constricts after birth. This constriction seems to be stimulated by a substance called *bradykinin,* which

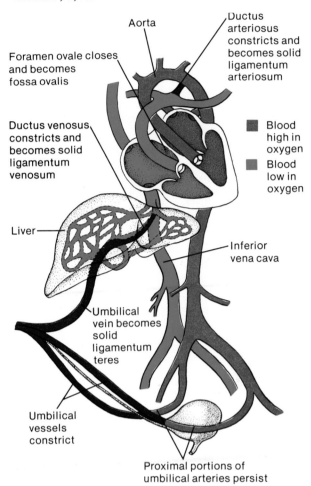

Fig. 20.27 Changes that occur in the newborn's circulatory system.

is released from the lungs during their initial expansions. Bradykinin is thought to act when the oxygen concentration of the aortic blood rises as a result of breathing. After the ductus arteriosus has closed it also becomes a solid cord (ligamentum arteriosum), and blood can no longer bypass the lungs by moving from the pulmonary artery directly into the aorta.

These changes in the newborn's circulatory system do not occur very rapidly. Although the constriction of the ductus arteriosus may be functionally complete within 15 minutes, the permanent closure of the foramen ovale may take up to a year. Figure 20.27 summarizes these circulatory changes.

---

1. How do the kidneys of a newborn differ from those of an adult?
2. What changes occur in the newborn's circulatory system?

Clinical Terms Related to Reproduction and Pregnancy

ablatio placentae (ab-la'she-o plah-cen'ta)—a premature separation of the placenta from the uterine wall.

abortion (ah-bor'shun)—the spontaneous or deliberate termination of pregnancy; a spontaneous abortion is commonly termed a miscarriage.

amniocentesis (am''ne-o-sen-te'sis)—a technique in which a sample of amniotic fluid is withdrawn from the amniotic cavity by inserting a hollow needle through the mother's abdominal wall.

conization (ko''nĭ-za'shun)—the surgical removal of a cone of tissue from the cervix for examination.

curettage (ku''rĕ-tahzh')—surgical procedure in which the cervix is dilated and the endometrium of the uterus is scraped (commonly called D and C).

dysmenorrhea (dis''men-ŏ-re'ah)—painful menstruation.

endometritis (en''do-me-tri'tis)—inflammation of the uterine lining.

epididymitis (ep''ĭ-did''ĭ-mi'tis)—inflammation of the epididymis.

hematometra (hem''ah-to-me'trah)—an accumulation of menstrual blood within the uterine cavity.

hysterectomy (his''tĕ-rek'to-me)—the surgical removal of the uterus.

mastitis (mas''ti'tis)—inflammation of a mammary gland.

oophorectomy (o''of-o-rek'to-me)—surgical removal of an ovary.

oophoritis (o''of-o-ri'tis)—inflammation of an ovary.

orchiectomy (or''ke-ek'to-me)—surgical removal of a testis.

orchitis (or-ki'tis)—inflammation of a testis.

prostatectomy (pros''tah-tek'to-me)—surgical removal of a portion or all of the prostate gland.

prostatitis (pros''tah-ti'tis)—inflammation of the prostate gland.

postpartum (pōst-par'tum)—occurring after birth.

teratology (ter''ah-tol'o-je)—the study of abnormal development and congenital malformations.

trimester (tri-mes'ter)—each third of the total period of pregnancy.

ultrasonography (ul''trah-son-og'rah-fe)—technique used to visualize the size and position of fetal structures by means of ultrasonic sound waves.

Chapter Summary

Introduction

Various reproductive organs produce sex cells and sex hormones, help sustain the lives of these cells, or transport them from place to place.

Organs of the Male Reproductive System

Primary organs are the testes, which produce sperm cells and male sex hormones.

The Testes

1. Structure of the testes
   a. The testes are composed of lobules separated by connective tissue and filled with seminiferous tubules.
   b. Seminiferous tubules are lined with germinal epithelium that produces sperm cells.
   c. Interstitial cells produce male sex hormones.

2. Germinal epithelium
   a. Germinal epithelium consists of supporting cells and spermatogenic cells.
      (1) Supporting cells support and nourish the spermatogenic cells.
      (2) Spermatogenic cells give rise to sperm cells.
   b. A sperm cell consists of a head, body, and tail.

3. Spermatogenesis
   a. Sperm cells are produced from spermatogonia.
   b. The number of chromosomes in sperm cells is reduced by one-half (46 to 23) by meiosis.
   c. The product of spermatogenesis is four sperm cells from each primary spermatocyte.

Male Internal Accessory Organs

1. The epididymis
   a. The epididymis is a tightly coiled tube that leads into the vas deferens.
   b. It stores immature sperm cells.

2. The vas deferens
   a. The vas deferens is a muscular tube that passes along the medial side of the testis.
   b. It passes through the inguinal canal.
   c. It fuses with the duct from the seminal vesicle to form the ejaculatory duct.

3. The seminal vesicle
   a. The seminal vesicle is a saclike structure attached to the vas deferens.
   b. It secretes alkaline fluid that contains nutrients, such as fructose.

4. The prostate gland
   a. This gland surrounds the urethra just below the urinary bladder.
   b. It secretes thin, milky fluid that neutralizes seminal fluid.

5. The bulbourethral glands
   a. These are two small structures beneath the prostate gland.
   b. They secrete fluid that serves as a lubricant for the penis.

6. Seminal fluid
   a. Seminal fluid is composed of sperm cells and secretions of seminal vesicles, the prostate gland, and bulbourethral glands.
   b. This fluid is slightly alkaline and contains nutrients and prostaglandins.

Male External Reproductive Organs

1. The scrotum
   The scrotum is a pouch of skin and subcutaneous tissue that encloses the testes.

2. The penis
   a. The penis is specialized to become erect for insertion into the vagina during sexual intercourse.
   b. The body is composed of three columns of erectile tissue.

Erection, Orgasm, and Ejaculation

1. During erection, vascular spaces within erectile tissue become engorged with blood.

2. Orgasm is the culmination of sexual stimulation and is accompanied by emission and ejaculation.

3. Movement of seminal fluid occurs as a result of sympathetic reflexes.

Hormonal Control of Male Reproductive Functions

1. Hypothalamic and pituitary hormones
   The male body remains reproductively immature until the hypothalamus releases GnRH, which stimulates the anterior pituitary gland to release gonadotropins.
   a. FSH stimulates spermatogenesis.
   b. LH (ICSH) stimulates the interstitial cells to produce male sex hormones.

2. Male sex hormones
   a. Male sex hormones are called androgens.
   b. Testosterone is the most important androgen.
   c. Androgen production increases rapidly at puberty.

3. Actions of testosterone
   a. Testosterone stimulates the development of the male reproductive organs.
   b. It is responsible for the development and maintenance of male secondary sexual characteristics.

4. Regulation of male sex hormones
   a. Testosterone concentration is regulated by a negative feedback mechanism.
      (1) As its concentration rises, the hypothalamus is inhibited and pituitary secretion of gonadotropins is reduced.
      (2) As the concentration falls, the hypothalamus signals the pituitary to secrete gonadotropins.
   b. The concentration of testosterone remains relatively stable from day to day.

Organs of the Female Reproductive System

The primary organs of the female reproductive system are the ovaries, which produce female sex cells and sex hormones. Accessory organs include internal and external reproductive organs.

The Ovaries

1. Structure of the ovaries
   a. The ovaries are divided into a medulla and a cortex.
   b. The medulla is composed of connective tissue, blood vessels, lymphatic vessels, and nerves.
   c. The cortex contains ovarian follicles and is covered by germinal epithelium.

2. Primordial follicles
   a. During development, groups of cells in the ovarian cortex form millions of primordial follicles.
   b. Each primordial follicle contains a primary oocyte and several follicular cells.
   c. The primary oocyte begins to undergo meiosis, but the process is soon halted and is not continued until puberty.
   d. The number of oocytes steadily decreases throughout the life of a female.

3. Oogenesis
   a. Beginning at puberty some oocytes are stimulated to continue meiosis.
   b. When a primary oocyte undergoes oogenesis it gives rise to a secondary oocyte in which the original chromosome number is reduced by one-half.
   c. A secondary oocyte represents an egg cell and can be fertilized to produce a zygote.

4. Maturation of a follicle
   a. At puberty, FSH stimulates primordial follicles to become primary follicles.
   b. During maturation, the oocyte enlarges, the follicular cells multiply, and a fluid-filled cavity appears.
   c. Usually only one follicle reaches full development.

5. Ovulation
   a. Oogenesis is completed as the follicle matures.
   b. The resulting oocyte is released when the follicle ruptures.
   c. After ovulation, the oocyte is drawn into the opening of the uterine tube.

Female Internal Accessory Organs

1. The uterine tubes
   a. The end of each tube is expanded and its margin bears irregular extensions.
   b. Movement of an egg cell into the uterine tube is aided by ciliated cells that line the tube and by peristaltic contractions in the wall of the tube.

2. The uterus
   a. The uterus receives the embryo and sustains its life during development.
   b. The uterine wall includes an endometrium, myometrium, and perimetrium.
3. The vagina
   a. The vagina serves to receive the erect penis, to convey uterine secretions to the outside, and to transport the offspring during birth.
   b. Its wall consists of a mucosa, muscularis, and outer fibrous coat.

Female External Reproductive Organs

1. The labia majora
   a. The labia majora are rounded folds of fatty tissue and skin.
   b. The upper ends form a rounded, fatty elevation over the symphysis pubis.
2. The labia minora
   a. The labia minora are flattened, longitudinal folds between the labia majora.
   b. They form the sides of the vestibule and anteriorly form the hoodlike covering of the clitoris.
3. The clitoris
   a. The clitoris is a small projection at the anterior end of the vulva that corresponds to the male penis.
   b. It is composed of two columns of erectile tissue.
4. The vestibule
   a. The vestibule is the space between the labia majora.
   b. Vestibular glands secrete mucus into the vestibule during sexual stimulation.

Erection, Lubrication, and Orgasm

1. During periods of sexual stimulation, erectile tissues of the clitoris and vestibular bulbs become engorged with blood and swell.
2. Vestibular glands secrete mucus into the vestibule and vagina, which lubricates these parts.
3. During orgasm, muscles of the perineum, uterine wall, and uterine tubes contract rhythmically.

Hormonal Control of Female Reproductive Functions

Hormones from the hypothalamus, anterior pituitary gland, and ovaries play important roles in the control of sex cell maturation and the development and maintenance of female secondary sexual characteristics.

1. Female sex hormones
   a. A female body remains reproductively immature until about 8 years of age when gonadotropin secretion increases.
   b. The most important female sex hormones are estrogen and progesterone.
      (1) Estrogen from the ovaries is responsible for development and maintenance of most female secondary sexual characteristics.
      (2) Progesterone functions to cause changes in the uterus.
2. Female reproductive cycles
   a. A menstrual cycle is initiated by FSH, which stimulates the maturation of a follicle.
   b. The maturing follicle secretes estrogen, which is responsible for maintaining the secondary sexual traits and causing the uterine lining to thicken.
   c. Ovulation is triggered when the anterior pituitary gland secretes a relatively large amount of LH and an increased amount of FSH.
   d. Following ovulation, follicular cells give rise to corpus luteum.
      (1) The corpus luteum secretes progesterone, which causes the uterine lining to become more vascular and glandular.
      (2) If an egg cell is not fertilized, the corpus luteum begins to degenerate.
      (3) As concentrations of estrogen and progesterone decline, the uterine lining disintegrates, causing menstrual flow.
   e. During this cycle, estrogen and progesterone inhibit the hypothalamus and the pituitary gland.

Mammary Glands

1. The mammary glands are located in subcutaneous tissue of the anterior thorax.
2. They are composed of lobes that contain tubular glands.
3. Lobes are separated by dense connective and adipose tissues.
4. Estrogen stimulates female breast development.
   a. Alveolar glands and ducts enlarge.
   b. Fat is deposited within the breasts.

Pregnancy

1. Transport of sex cells
   a. The movement of the egg cell to the uterine tube is aided by ciliary action.
   b. To move, a sperm cell lashes its tail; its movement within the female body is aided by muscular contractions in the uterus and uterine tube.
2. Fertilization
   a. A sperm cell penetrates an egg cell with the aid of an enzyme.
   b. When a sperm cell penetrates an egg cell, the entrance of any other sperm cells is prevented by structural changes that occur in the egg cell membrane and by enzyme actions.

c. When the nuclei of a sperm and an egg cell fuse, the process of fertilization is complete.

d. The product of fertilization is a zygote with 46 chromosomes.

3. Early embryonic development
   a. Cells undergo mitosis, giving rise to smaller and smaller cells.
   b. The developing offspring is moved down the uterine tube to the uterus, where it becomes implanted in the endometrium.
   c. The offspring is called an embryo from the second through the eighth week of development; thereafter it is a fetus.
   d. Eventually the embryonic and maternal cells together form a placenta.

4. Hormonal changes during pregnancy
   a. Embryonic cells produce HCG that causes the corpus luteum to be maintained.
   b. Placental tissue produces high concentrations of estrogen and progesterone.
      (1) Estrogen and progesterone maintain the uterine wall and inhibit the secretion of FSH and LH.
      (2) Progesterone and relaxin cause uterine contractions to be suppressed.
      (3) Estrogen causes enlargement of the vagina and relaxation of the ligaments that hold the pelvic joints together.
      (4) Relaxin also helps to soften the cervix and relax the pelvic ligaments.
   c. The placenta secretes placental lactogen that stimulates the development of the breasts.
   d. During pregnancy, increasing secretion of aldosterone promotes retention of sodium and body fluid, and increasing secretion of parathormone helps maintain a high concentration of maternal blood calcium.

5. Embryonic stage of development
   a. The embryonic stage extends from the second through the eighth week.
   b. It is characterized by development of the placenta and the main body structures.
   c. The cells of the inner cell mass become arranged into primary germ layers.
   d. The embryonic disk becomes cylindrical and is attached to the developing placenta.
   e. The placental membrane consists of the epithelium of chorionic villi and the epithelium of capillaries inside the villi.
      (1) Oxygen and nutrients diffuse from the maternal blood through the membrane and into the fetal blood.
      (2) Carbon dioxide and other wastes diffuse from the fetal blood through the membrane and into the maternal blood.
   f. A fluid-filled amnion develops around the embryo.

g. The umbilical cord is formed as the amnion envelops the tissues attached to the underside of the embryo.
h. The yolk sac forms on the underside of the embryonic disk.
i. The allantois extends from the yolk sac into the connecting stalk.
j. By the beginning of the eighth week, the embryo is recognizable as a human.

6. Fetal stage of development
   a. This stage extends from the end of the eighth week and continues until birth.
   b. Existing structures grow and mature; only a few new parts appear.
   c. A fetus is full term at the end of the tenth lunar month.

7. Fetal circulation
   a. Blood is carried between the placenta and fetus by umbilical vessels.
   b. Blood enters the fetus through the umbilical vein and partially bypasses the liver.
   c. Blood enters the right atrium and partially bypasses the lungs by means of the foramen ovale.
   d. Blood entering the pulmonary trunk partially bypasses the lungs by means of the ductus arteriosus.
   e. Blood enters the umbilical arteries by means of the internal iliac arteries.

8. The birth process
   a. During pregnancy, estrogen excites uterine contractions and progesterone inhibits uterine contractions.
   b. A variety of factors are involved with the birth process.
      (1) Secretion of progesterone decreases.
      (2) The posterior pituitary gland releases oxytocin.
      (3) Uterine muscles are stimulated to contract, and labor begins.
   c. Following the birth, the placental tissues are expelled.

Postnatal Changes

1. Production and secretion of milk
   a. Following birth, the maternal anterior pituitary secretes prolactin, and the mammary glands begin to secrete milk.
   b. Reflex response to mechanical stimulation of the nipple signals the release of oxytocin, which causes milk to be ejected from ducts.

2. Neonatal period
   a. This period extends from birth to the end of the fourth week.
   b. The newborn must begin to carry on respiration, obtain nutrients, excrete wastes, and regulate its body temperature.

c. The first breath must be powerful in order to expand the lungs.

d. The liver is immature and unable to supply sufficient glucose, so the newborn depends primarily on stored fat.

e. Immature kidneys cannot concentrate urine very well.

f. Homeostatic mechanisms may function imperfectly.

g. The circulatory system undergoes changes when placental circulation ceases.
   (1) Umbilical vessels constrict.
   (2) The ductus venosus constricts.
   (3) The foramen ovale is closed by a valve.
   (4) Bradykinin released from the lungs stimulates constriction of the ductus arteriosus.

Application of Knowledge

1. What changes, if any, might be expected to occur in the secondary sexual characteristics of an adult male following removal of one testis? Following removal of both testes? Following removal of the prostate gland?

2. How could you explain the observation that otherwise healthy females are sometimes infertile because their anterior pituitary glands fail to produce normal quantities of gonadotropic hormones?

Review Activities

1. List the general functions of the reproductive systems.

2. Distinguish between the primary and accessory male reproductive organs.

3. Describe the structure of a testis.

4. Describe the epididymis and explain its function.

5. Trace the path of the vas deferens from the epididymis to the ejaculatory duct.

6. On a diagram, locate the seminal vesicles, prostate gland, and bulbourethral glands, and describe the composition of their secretions.

7. Define seminal fluid.

8. Describe the structure of the penis.

9. Explain the mechanism that produces an erection of the penis.

10. Distinguish between emission and ejaculation.

11. Explain the role of GnRH in the control of male reproductive functions.

12. List several male secondary sexual characteristics.

13. Explain how the concentration of testosterone is regulated.

14. Describe the structure of an ovary.

15. Describe how a follicle matures.

16. On a diagram, locate the uterine tubes and explain their function.

17. Describe the structure of the uterus.

18. On a diagram, locate the clitoris and describe its structure.

19. Explain the role of GnRH in regulating female reproductive functions.

20. List several female secondary sexual characteristics.

21. Define menstrual cycle.

22. Explain the roles of estrogen and progesterone in the menstrual cycle.

23. Summarize the major events in a menstrual cycle.

24. Describe how male and female sex cells are transported within the female reproductive tract.

25. Describe the process of fertilization.

26. Define cleavage.

27. Define implantation.

28. Explain the major hormonal changes that occur in the maternal body during pregnancy.

29. Explain how the primary germ layers form.

30. List the major body parts derived from ectoderm, mesoderm, and endoderm.

31. Describe the formation of the placenta and explain its functions.

32. Define placental membrane.

33. Distinguish between the chorion and the amnion.

34. Explain the function of amniotic fluid.

35. Describe the formation of the umbilical cord.

36. Define fetus.

37. List the major changes that occur during the fetal stage of development.

38. Trace the pathway of blood from the placenta to the fetus and back to the placenta.

39. Discuss the events that occur during the birth process.

40. Explain the roles of prolactin and oxytocin in milk production and secretion.

41. Explain why the newborn's first breath must be particularly forceful.

42. Explain why newborns tend to develop water and electrolyte imbalances.

43. Describe the circulatory changes that occur in the newborn.

Appendix A
Units of Measurement and Their Equivalents

Apothecaries' Weights and Their Metric Equivalents

1 grain (gr) =
0.05 scruple (s)
0.017 dram (dr)
0.002 ounce (oz)
0.0002 pound (lb)
0.065 gram (g)
65. milligrams (mg)

1 scruple (s) =
20. grains (gr)
0.33 dram (dr)
0.042 ounce (oz)
0.004 pound (lb)
1.3 grams (g)
1,300. milligrams (mg)

1 dram (dr) =
60. grains (gr)
3. scruples (s)
0.13 ounce (oz)
0.010 pound (lb)
3.9 grams (g)
3,900. milligrams (mg)

1 ounce (oz) =
480. grains (gr)
24. scruples (s)
8. drams (dr)
0.08 pound (lb)
31.1 grams (g)
31,100. milligrams (mg)

1 pound (lb) =
5,760. grains (gr)
288. scruples (s)
96. drams (dr)
12. ounces (oz)
373. grams (g)
373,000. milligrams (mg)

Apothecaries' Volumes and Their Metric Equivalents

1 minim (min) =
0.017 fluid dram (fl dr)
0.002 fluid ounce (fl oz)
0.0001 pint (pt)
0.06 milliliter (ml)
0.06 cubic centimeter (cc)

1 fluid dram (fl dr) =
60. minims (min)
0.13 fluid ounce (fl oz)
0.008 pint (pt)
3.70 milliliters (ml)
3.70 cubic centimeters (cc)

1 fluid ounce (fl oz) =
480. minims (min)
8. fluid drams (fl dr)
0.06 pint (pt)
29.6 milliliters (ml)
29.6 cubic centimeters (cc)

1 pint (pt) =
7,680. minims (min)
128. fluid drams (fl dr)
16. fluid ounces (fl oz)
473. milliliters (ml)
473. cubic centimeters (cc)

Metric Weights and Their Apothecaries' Equivalents

1 gram (g) =
0.001 kilogram (kg)
1,000. milligrams (mg)
1,000,000. micrograms (μg)
15.4 grains (gr)
0.032 ounce (oz)

1 kilogram (kg) =
1,000. grams (g)
1,000,000. milligrams (mg)
1,000,000,000. micrograms (μg)
32. ounces (oz)
2.7 pounds (lb)

1 milligram (mg) =
0.000001 kilogram (kg)
0.001 gram (g)
1,000. micrograms (μg)
0.0154 grains (gr)
0.000032 ounce (oz)

Metric Volumes and Their Apothecaries' Equivalents

1 liter (l) =
1,000. milliliters (ml)
1,000. cubic centimeters (cc)
2.1 pints (pt)
270. fluid drams (fl dr)
34. fluid ounces (fl oz)

1 milliliter (ml) =
0.001 liter (l)
1. cubic centimeter (cc)
16.2 minims (min)
0.27 fluid dram (fl dr)
0.034 fluid ounce (fl oz)

Approximate Equivalents of Household Measures

1 teaspoon (tsp) =
4. milliliters (ml)
4. cubic centimeters (cc)
1. fluid dram (fl dr)

1 tablespoon (tbsp) =
15. milliliters (ml)
15. cubic centimeters (cc)
0.5 fluid ounce (fl oz)
3.7 teaspoons (tsp)

1 cup (c) =
240. milliliters (ml)
240. cubic centimeters (cc)
8. fluid ounces (fl oz)
0.5 pint (pt)
16. tablespoons (tbsp)

1 quart (qt) =
960. milliliters (ml)
960. cubic centimeters (cc)
2. pints (pt)
4. cups (c)
32. fluid ounces (fl oz)

Refer to the preceding equivalency lists when converting one unit to another equivalent unit.

To convert a unit shown in bold type to one of the equivalent units listed immediately below it, multiply the first number (bold type unit) by the appropriate equivalent unit listed below it.

Sample problems:

1. Convert 320 grains into scruples (1 gr = 0.05 s).

$$320 \text{ gr} \times \frac{0.05 \text{ s}}{1 \text{ gr}} = 16.0 \text{ s}$$

2. Convert 320 grains into drams (1 gr = 0.017 dr).

$$320 \text{ gr} \times \frac{0.017 \text{ dr}}{1 \text{ gr}} = 5.44 \text{ dr}$$

3. Convert 320 grains into grams (1 gr = 0.065 g).

$$320 \text{ gr} \times \frac{0.065 \text{ g}}{1 \text{ gr}} = 20.8 \text{ g}$$

°F	°C	°F	°C
95.0	35.0	100.0	37.8
95.2	35.1	100.2	37.9
95.4	35.2	100.4	38.0
95.6	35.3	100.6	38.1
95.8	35.4	100.8	38.2
96.0	35.5	101.0	38.3
96.2	35.7	101.2	38.4
96.4	35.8	101.4	38.6
96.6	35.9	101.6	38.7
96.8	36.0	101.8	38.8
97.0	36.1	102.0	38.9
97.2	36.2	102.2	39.0
97.4	36.3	102.4	39.1
97.6	36.4	102.6	39.2
97.8	36.6	102.8	39.3
98.0	36.7	103.0	39.4
98.2	36.8	103.2	39.6
98.4	36.9	103.4	39.7
98.6	37.0	103.6	39.8
98.8	37.1	103.8	39.9
99.0	37.2	104.0	40.0
99.2	37.3	104.2	40.1
99.4	37.4	104.4	40.2
99.6	37.6	104.6	40.3
99.8	37.7	104.8	40.4
		105.0	40.6

To convert °F to °C
Subtract 32 from °F and multiply by 5/9.

$$\underline{\hspace{2cm}} °F - 32 \times 5/9 = \underline{\hspace{2cm}} °C$$

To convert °C to °F
Multiply °C by 9/5 and add 32.

$$\underline{\hspace{2cm}} °C \times 9/5 + 32 = \underline{\hspace{2cm}} °F$$

Appendix B
Some Laboratory Tests of Clinical Importance

Common Tests Performed on Blood

Test	Normal Values* (adult)	Clinical Significance
Acetone and acetoacetate (serum)	0.3–2.0 mg/100 ml	Values increase in diabetic acidosis, toxemia of pregnancy, fasting, and high-fat diet.
Albumin-globulin ratio or A/G ratio (serum)	1.5:1 to 2.5:1	Ratio of albumin to globulin is lowered in kidney diseases and malnutrition.
Albumin (serum)	3.2–5.5 gm/100 ml	Values increase in multiple myeloma and decrease with proteinuria and as a result of severe burns.
Ammonia (plasma)	50–170 μg/100 ml	Values increase in severe liver disease, pneumonia, shock, and congestive heart failure.
Amylase (serum)	80–160 Somogyi units/100 ml	Values increase in acute pancreatitis, intestinal obstructions, and mumps. They decrease in chronic pancreatitis, cirrhosis of the liver, and toxemia of pregnancy.
Bilirubin, total (serum)	0.3–1.1 mg/100 ml	Values increase in conditions causing red blood cell destruction or biliary obstruction.
Blood urea nitrogen or BUN (plasma or serum)	10–20 mg/100 ml	Values increase in various kidney disorders and decrease in liver failure and during pregnancy.
Calcium (serum)	9.0–11.0 mg/100 ml	Values increase in hyperparathyroidism, hypervitaminosis D, and respiratory conditions that cause a rise in CO_2 concentration. They decrease in hypoparathyroidism, malnutrition, and severe diarrhea.
Carbon dioxide (serum)	24–30 mEq/l	Values increase in respiratory diseases, intestinal obstruction, and vomiting. They decrease in acidosis, nephritis, and diarrhea.
Chloride (serum)	96–106 mEq/l	Values increase in nephritis, Cushing's syndrome, and hyperventilation. They decrease in diabetic acidosis, Addison's disease, diarrhea, and following severe burns.

*These values may vary with hospital, physician, and type of equipment used to make measurements.

Test	Normal Values (adult)	Clinical Significance
Cholesterol, total (serum)	150–250 mg/100 ml	Values increase in diabetes mellitus and hypothyroidism. They decrease in pernicious anemia, hyperthyroidism, and acute infections.
Creatine phosphokinase or CPK (serum)	Men: 0–20 IU/l Women: 0–14 IU/l	Values increase in myocardial infarction and skeletal muscle diseases such as muscular dystrophy.
Creatine (serum)	0.2–0.8 mg/100 ml	Values increase in muscular dystrophy, nephritis, severe damage to muscle tissue, and during pregnancy.
Creatinine (serum)	0.7–1.5 mg/100 ml	Values increase in various kidney diseases.
Erythrocyte count or red cell count (whole blood)	Men: 4,600,000–6,200,000/cu mm Women: 4,200,000–5,400,000/cu mm Children: 4,500,000–5,100,000/cu mm (varies with age)	Values increase as a result of severe dehydration or diarrhea and decrease in anemia, leukemia, and following severe hemorrhage.
Fatty acids, total (serum)	190–420 mg/100 ml	Values increase in diabetes mellitus, anemia, kidney disease, and hypothyroidism. They decrease in hyperthyroidism.
Ferritin (serum)	Men: 10–270 μg/100 ml Women: 5–280 μg/100 ml	Values correlate with total body-iron store. They decrease with iron deficiency.
Globulin (serum)	2.5–3.5 gm/100 ml	Values increase as a result of chronic infections.
Glucose (plasma)	70–115 mg/100 ml	Values increase in diabetes mellitus, liver diseases, nephritis, hyperthyroidism, and pregnancy. They decrease in hyperinsulinism, hypothyroidism, and Addison's disease.
Hematocrit (whole blood)	Men: 40–54 ml/100 ml Women: 37–47 ml/100 ml Children: 35–49 ml/100 ml (varies with age)	Values increase in polycythemia due to dehydration or shock. They decrease in anemia and following severe hemorrhage.
Hemoglobin (whole blood)	Men: 14–18 gm/100 ml Women: 12–16 gm/100 ml Children: 11.2–16.5 gm/100 ml (varies with age)	Values increase in polycythemia, obstructive pulmonary diseases, congestive heart failure, and at high altitudes. They decrease in anemia, pregnancy, and as a result of severe hemorrhage or excessive fluid intake.
Iron (serum)	75–175 μg/100 ml	Values increase in various anemias and liver disease. They decrease in iron deficiency anemia.
Iron-binding capacity (serum)	250–410 μg/100 ml	Values increase in iron deficiency anemia and pregnancy. They decrease in pernicious anemia, liver disease, and chronic infections.
Lactic acid (whole blood)	6–16 mg/100 ml	Values increase with muscular activity and in congestive heart failure, severe hemorrhage, and shock.

Test	Normal Values (adult)	Clinical Significance
Lactic dehydrogenase or LDH (serum)	90–200 milliunits/ml	Values increase in pernicious anemia, myocardial infarction, liver diseases, acute leukemia, and widespread carcinoma.
Lipids, total (serum)	450–850 mg/100 ml	Values increase in hypothyroidism, diabetes mellitus, and nephritis. They decrease in hyperthyroidism.
Oxygen saturation (whole blood)	Arterial: 94–100% Venous: 60–85%	Values increase in polycythemia and decrease in anemia and obstructive pulmonary diseases.
pH (whole blood)	7.35–7.45	Values increase due to vomiting, Cushing's syndrome, and hyperventilation. They decrease as a result of hypoventilation, severe diarrhea, Addison's disease, and diabetic acidosis.
Phosphatase, acid (serum)	1.0–5.0 King-Armstrong units/ml	Values increase in cancer of the prostate gland, hyperparathyroidism, certain liver diseases, myocardial infarction, and pulmonary embolism.
Phosphatase, alkaline (serum)	5–13 King-Armstrong units/ml	Values increase in hyperparathyroidism (and in other conditions that promote resorption of bone), liver diseases, and pregnancy.
Phospholipids (serum)	6–12 mg/100 ml as lipid phosphorus	Values increase in diabetes mellitus and nephritis.
Phosphorus (serum)	3.0–4.5 mg/100 ml	Values increase in kidney diseases, hypoparathyroidism, acromegaly, and hypervitaminosis D. They decrease in hyperparathyroidism.
Platelet count (whole blood)	150,000–350,000/cu mm	Values increase in polycythemia and certain anemias. They decrease in acute leukemia and aplastic anemia.
Potassium (serum)	3.5–5.0 mEq/l	Values increase in Addison's disease, hypoventilation, and conditions that cause severe cellular destruction. They decrease in diarrhea, vomiting, diabetic acidosis, and chronic kidney disease.
Protein, total (serum)	6.0–8.0 gm/100 ml	Values increase in severe dehydration and shock. They decrease in severe malnutrition and hemorrhage.
Protein-bound iodine or PBI (serum)	3.5–8.0 μg/100 ml	Values increase in hyperthyroidism and liver disease. They decrease in hypothyroidism.
Prothrombin time (serum)	12–14 sec (one stage)	Values increase in certain hemorrhagic diseases, liver disease, vitamin K deficiency, and following the use of various drugs.
Sedimentation rate, Westergren (whole blood)	Men: 0–15 mm/hr Women: 0–20 mm/hr	Values increase in infectious diseases, menstruation, pregnancy, and as a result of severe tissue damage.

Test	Normal Values (adult)	Clinical Significance
Sodium (serum)	136–145 mEq/l	Values increase in nephritis and severe dehydration. They decrease in Addison's disease, myxedema, kidney disease, and diarrhea.
Thromboplastin time, partial (plasma)	35–45 sec	Values increase in deficiencies of blood factors VIII, IX, and X.
Thyroxine or T$_4$ (serum)	2.9–6.4 µg/100 ml	Values increase in hyperthyroidism and pregnancy. They decrease in hypothyroidism.
Transaminases or SGOT (serum)	5–40 units/ml	Values increase in myocardial infarction, liver disease, and diseases of skeletal muscles.
Uric acid (serum)	Men: 2.5–8.0 mg/100 ml Women: 1.5–6.0 mg/100 ml	Values increase in gout, leukemia, pneumonia, toxemia of pregnancy, and as a result of severe tissue damage.
White blood cell count, differential (whole blood)	Neutrophils 54–62% Eosinophils 1–3% Basophils 0–1% Lymphocytes 25–33% Monocytes 3–7%	Neutrophils increase in bacterial diseases; lymphocytes and monocytes increase in viral diseases; eosinophils increase in collagen diseases, allergies, and in the presence of intestinal parasites.
White blood cell count, total (whole blood)	5,000–10,000/cu mm	Values increase in acute infections, acute leukemia, and following menstruation. They decrease in aplastic anemia and as a result of drug toxicity.

Common Tests Performed on Urine

Test	Normal Values	Clinical Significance
Acetone and acetoacetate	0	Values increase in diabetic acidosis.
Albumin, qualitative	0 to trace	Values increase in kidney disease, hypertension, and heart failure.
Ammonia	20–70 mEq/l	Values increase in diabetes mellitus and liver diseases.
Bacterial count	Under 10,000/ml	Values increase in urinary tract infection.
Bile and bilirubin	0	Values increase in melanoma and biliary tract obstruction.
Calcium	Under 250 mg/24 hr	Values increase in hyperparathyroidism and decrease in hypoparathyroidism.
Creatinine clearance	100–140 ml/min	Values increase in renal diseases.
Creatinine	1–2 gm/24 hr	Values increase in infections and decrease in muscular atrophy, anemia, leukemia, and kidney diseases.
Glucose	0	Values increase in diabetes mellitus and various pituitary gland disorders.
17-hydroxycorticosteroids	2–10 mg/24 hr	Values increase in Cushing's syndrome and decrease in Addison's disease.
Phenylpyruvic acid	0	Values increase in phenylketonuria.
Urea clearance	Over 40 ml blood cleared of urea/min	Values increase in renal diseases.
Urea	25–35 gm/24 hr	Values increase as a result of excessive protein breakdown. They decrease as a result of impaired renal function.
Uric acid	0.6–1.0 gm/24 hr as urate	Values increase in gout and decrease in various kidney diseases.
Urobilinogen	0–4 mg/24 hr	Values increase in liver diseases and hemolytic anemia. They decrease in complete biliary obstruction and severe diarrhea.

Suggestions for Additional Reading

Chapter 1

Enger, E. D., et al. 1982. *Concepts in biology.* 3d ed. Dubuque, Ia.: Wm. C. Brown.

Langley, L. L. 1965. *Homeostasis.* New York: Reinhold Publishing.

Lockhart, R. D. 1974. *Living anatomy.* 5th ed. London: Faber and Faber.

Lyons, A. S., and Petrucelli, R. J. 1978. *Medicine: an illustrated history.* New York: Harry N. Abrams.

Majno, G. 1975. *The healing hand: man and wound in the ancient world.* Cambridge, Mass.: Harvard Univ. Press.

Singer, C. 1957. *A short history of anatomy and physiology.* New York: Dover Publications.

Singer, C. A., and Underwood, E. A. 1962. *A short history of medicine.* New York: Oxford Univ. Press.

Szent-Györgyi, A. 1972. *The living state.* New York: Academic Press.

Chapter 2

Baker, J. J. W., and Allen, G. E. 1981. *Matter, energy, and life.* 4th ed. Palo Alto: Addison-Wesley.

Borek, E. 1981. *The atoms within us.* 2d ed. New York: Columbia Univ. Press.

Frieden, E. July 1972. The chemical elements of life. *Scientific American.*

Holum, J. R. 1978. *Fundamentals of general, organic, and biological chemistry.* New York: John Wiley & Sons.

Koshland, D. E. October 1973. Protein shape and biological control. *Scientific American.*

Lehninger, A. 1975. *Biochemistry.* 2d ed. New York: Worth.

Maugh, T. H. 1973. Trace elements, a growing appreciation of effects on man. *Science* 181:253.

Orten, J. M., and Neuhaus, O. W. 1982. *Human biochemistry.* 10th ed. St. Louis: C. V. Mosby.

Sackheim, G. I. 1974. *Introduction to chemistry for biology students.* Chicago: Educational Methods.

Wood, W. B., et al. 1981. *Biochemistry.* 2d ed. Menlo Park, Calif.: W. A. Benjamin.

Chapter 3

Avers, C. J. 1982. *Cell biology.* 2d ed. New York: D. Van Nostrand Co.

Berns, M. W. 1977. *Cells.* New York: Holt, Rinehart and Winston.

DeWitt, W., and Brown, E. R. 1977. *Biology of the cell.* Philadelphia: W. B. Saunders.

Dustin, P. August 1980. Microtubules. *Scientific American.*

Lake, J. A. August 1981. The ribosome. *Scientific American.*

Mazia, D. January 1974. The cell cycle. *Scientific American.*

Novikoff, A. B., and Holtzman, E. 1976. *Cells and organelles.* 2d ed. New York: Holt, Rinehart and Winston.

Rothman, J. E. 1981. The golgi apparatus. *Science* 213.

Satir, B. October 1975. The final steps in secretion. *Scientific American.*

Sloboda, R. D. 1980. The role of microtubules in cell structure and cell division. *American Scientist* 68:290.

Swanson, C. P., and Webster, P. 1977. *The cell.* 4th ed. Englewood Cliffs, N.J.: Prentice-Hall.

Unwin, N., and Henderson, R. February 1984. The structure of proteins in biological membranes. *Scientific American.*

Wolfe, S. L. 1981. *Biology of the cell.* 2d ed. Belmont, Calif.: Wadsworth.

Chapter 4

Becker, W. M. 1977. *Energy and the living cell.* Philadelphia: Lippincott.

Darnell, J. E. October 1983. The processing of RNA. *Scientific American.*

Dickerson, R. E. December 1983. The DNA helix and how it is read. *Scientific American.*

Goldsby, R. A. 1977. *Cells and energy.* 3d ed. New York: Macmillan.

Hinkle, P. C. March 1978. How cells make ATP. *Scientific American.*

Kornberg, A. 1980. *DNA replication.* San Francisco: W. H. Freeman.

Lehninger, A. L. 1982. *Principles of biochemistry.* New York: Worth Publishers, Inc.

Mirsky, A. E. June 1968. The discovery of DNA. *Scientific American.*

Nomura, M. January 1984. The control of ribosome synthesis. *Scientific American.*

Pederson, T. January–February 1981. Messenger RNA biosynthesis and nuclear structure. *American Scientist.*

Smith, A. E. 1976. *Protein biosynthesis.* New York: Wiley.

Stryer, L. 1981. *Biochemistry.* 2d ed. San Francisco: W. H. Freeman.

Watson, J. D. 1968. *The double helix.* New York: Antheneum.

Chapter 5

Bevelander, G., and Ramaley, J. A. 1979. *Essentials of histology.* 8th ed. St. Louis: C. V. Mosby.

Bloom, W. B., and Fawcett, D. W. 1975. *A textbook of histology.* 10th ed. Philadelphia: W. B. Saunders.

Caplan, A. I. October 1984. Cartilage. *Scientific American.*

Copenhaver, W. M., et al. 1978. *Bailey's textbook of histology.* 17th ed. Baltimore, Md.: Williams and Wilkins.

DiFiore, M. S. M. 1981. *An atlas of human histology.* 5th ed. Philadelphia: Lea and Febiger.

Eyre, D. R. 1980. Collagen: molecular diversity in the body's protein scaffold. *Science* 207:1315.

Leeson, T. S., and Leeson, C. R. 1981. *Histology.* 4th ed. Philadelphia: W. B. Saunders.

Reith, E. J., and Ross, M. H. 1977. *Atlas of descriptive histology.* 3d ed. New York: Harper and Row.

Weiss, L., and Greep, R. O. 1983. *Histology.* 5th ed. New York: McGraw-Hill.

Chapter 6

Elden, H. R., ed. 1971. *Biophysical properties of the skin.* New York: John Wiley & Sons.

Hardy, J. D., et al. 1970. *Physiological and behavioral temperature regulation.* Springfield: Charles C. Thomas.

Heller, H. C., et al. August 1978. The thermostat of vertebrate animals. *Scientific American.*

Montagna, W. February 1965. The skin. *Scientific American.*

Nicoll, P. A., et al. 1972. The physiology of the skin. In *Annual Review of Physiology,* vol. 34, edited by J. H. Comroe, Jr., et al., p. 177. Palo Alto, Calif.: Annual Reviews.

Pawelek, J. M., and Korner, A. M. March–April 1982. The biosynthesis of mammalian melanin. *American Scientist.*

Rushmer, R. L., et al. 1966. The skin. *Science* 154:343.

Tregear, R. T. 1966. *Physical functions of skin.* New York: Oxford Univ. Press.

Chapter 7

Basmajian, J. V. 1980. *Grant's method of anatomy.* 10th ed. Baltimore: Williams and Wilkins.

Clemente, C. D. 1981. *Anatomy: a regional atlas of the human body.* 2d ed. Philadelphia: Lea and Febiger.

Goss, C. M., ed. 1973. *Gray's anatomy of the human body.* 29th American ed. Philadelphia: Lea and Febiger.

Hamilton, W. J., ed. 1976. *Textbook of human anatomy.* 2d ed. St. Louis: C. V. Mosby.

Hollinshead, W. H. 1974. *Textbook of anatomy.* 3d ed. New York: Harper and Row.

Jaffe, C. C. November–December 1982. Medical imaging. *American Scientist.*

Leeson, T. S., and Leeson, C. R. 1972. *Human structure.* Philadelphia: Saunders.

McMinn, R. M. H., and Hutchings, R. T. 1977. *Color atlas of human anatomy.* Chicago: Year Book Medical Publishers.

Pykett, I. L. May 1982. NMR imaging in medicine. *Scientific American.*

Snell, R. S. 1978. *Atlas of clinical anatomy.* Boston: Little, Brown and Co.

Woodburne, R. T. 1978. *Essentials of human anatomy.* 6th ed. New York: Oxford Univ. Press.

Chapter 8

Aufranc, O. E., and Turner, R. H. October 1971. Total replacement of the arthritic hip. *Hospital Practice.*

Bourne, G. W., ed. 1976. *The biochemistry and physiology of bone.* 2d ed. New York: Academic Press.

Evans, F. G., ed. 1966. *Studies in the anatomy and function of bones and joints.* New York: Springer-Verlag.

Hall, B. K. 1970. Cellular differentiation in skeletal tissue. *Biol. Rev.* 45:455.

Harris, W. H., and Heaney, R. P. 1970. *Skeletal renewal and metabolic bone disease.* Boston: Little, Brown and Co.

Raisz, L. G., and Kream, B. E. 1981. Hormonal control of skeletal growth. In *Annual Review of Physiology,* vol. 43, edited by I. S. Edelman and S. G. Schultz, p. 225. Palo Alto, Calif.: Annual Reviews.

Sonstegard, D. A., et al. January 1978. The surgical replacement of the human knee joint. *Scientific American.*

Vaughan, J. M. 1981. *The physiology of bone.* 3d ed. New York: Oxford Univ. Press.

Chapter 9

Carlson, F. D., and Wilkie, D. R. 1974. *Muscle physiology.* Englewood Cliffs, N.J.: Prentice-Hall.

Close, R. I. 1972. Dynamic properties of mammalian skeletal muscles. *Physiological Reviews* 52:129.

Cohen, C. November 1975. The protein switch of muscle contraction. *Scientific American.*

Endo, M. 1977. Calcium release from the sarcoplasmic reticulum. *Physiological Reviews* 57:71.

Huddart, H., and Hunt, S. 1975. *Visceral muscle.* New York: Halsted Press.

Lester, H. A. February 1977. The response to acetylcholine. *Scientific American.*

Margaria, R. March 1972. The sources of muscular energy. *Scientific American.*

Merton, P. A. May 1972. How we control the contraction of our muscles. *Scientific American.*

Murray, J. M., and Weber, A. February 1974. The cooperative action of muscle proteins. *Scientific American.*

Peachey, L. D. 1974. *Muscle and motility.* New York: McGraw-Hill.

Sandow, A. 1970. Skeletal muscle. In *Annual Review of Physiology,* vol. 32, p. 479. Palo Alto, Calif.: Annual Reviews.

Uehara, Y. 1976. *Muscle and its innervation.* Baltimore: Williams and Wilkins Co.

Weber, A., and Murray, J. M. 1973. Molecular control mechanisms in muscle contraction. *Physiological Reviews* 53:612.

Chapter 10

Bloom, F. E. October 1981. Neuropeptides. *Scientific American.*

Catterall, W. A. 1984. The molecular basis of neuronal excitability. *Science* 223:653.

Eccles, J. C. 1976. *The understanding of the brain.* 2d ed. New York: McGraw-Hill.

Geschwind, N. September 1979. Specialization of the human brain. *Scientific American.*

Guillemin, R. 1978. Peptides in the brain. *Science* 202:390.

Hubel, D. H. September 1979. The brain. *Scientific American.*

Iversen, L. L. September 1979. The chemistry of the brain. *Scientific American.*

Kandel, E. R. September 1979. Small systems of neurons. *Scientific American*.

Keynes, R. D. September 1979. Ion channels in the nerve cell membrane. *Scientific American*.

Krieger, D. T. 1983. Brain peptides: what, where, and why? *Science* 222:975.

Morell, P., and Norton, W. T. 1980. Myelin. *Scientific American*.

Nauta, W. J. H., and Feirtag, M. September 1979. The organization of the brain. *Scientific American*.

Noback, C. T., and Demarest, R. 1980. *The human nervous system*. 3d ed. New York: McGraw-Hill.

Schwartz, J. H. April 1980. The transport of substances in nerve cells. *Scientific American*.

Shepherd, G. M. 1983. *Neurobiology*. New York: Oxford Univ. Press.

Solomon, S. H. 1980. Brain peptides as neurotransmitters. *Science* 209:976.

Springer, S. P., and Deutsch, G. 1981. *Left brain, right brain*. San Francisco: W. H. Freeman.

Stevens, C. F. September 1979. The neuron. *Scientific American*.

Chapter 11

Cain, W. S. 1979. To know with the nose: keys to odor identification. *Science* 203:467.

Daw, N. W. 1973. Neurophysiology of color vision. *Physiological Reviews* 53:571.

Durrant, J. D., and Lovrinic, J. H. 1977. *Bases of hearing science*. Baltimore: Williams and Wilkins.

Freese, A. S. 1977. *The miracle of vision*. New York: Harper and Row.

Green, D. M. 1976. *An introduction to hearing*. New York: Halsted Press.

Hudspeth, A. J. January 1983. The hair cells of the inner ear. *Scientific American*.

O'Brien, D. F. 1982. The chemistry of vision. *Science* 218:513.

Regan, D., et al. July 1979. The visual perception of motion and depth. *Scientific American*.

Rushton, W. A. H. March 1975. Visual pigments and color blindness. *Scientific American*.

Toates, F. M. 1972. Accommodation function of the human eye. *Physiological Reviews* 52:828.

Werblin, F. S. January 1973. The control of sensitivity in the retina. *Scientific American*.

Wilson, R. W., and Elmassean, B. J. April 1981. Endorphins. *American Journal of Nursing*.

Wolfe, J. M. February 1983. Hidden visual processes. *Scientific American*.

Chapter 12

Brownstein, M. J., et al. 1980. Synthesis, transport, and release of posterior pituitary hormones. *Science* 207:373.

Frohman, L. A. 1975. Neurotransmitters as regulators of endocrine functions. *Hospital Practice* 10:54.

Handler, J. S., and Orloff, J. 1981. Antidiuretic hormone. In *Annual Review of Physiology*, vol. 43, edited by I. S. Edelman and S. G. Schultz, p. 611. Palo Alto, Calif.: Annual Reviews.

Katzenellenbogen, B. S. 1980. Dynamics of steroid hormone receptor action. In *Annual Review of Physiology*, vol. 42, edited by I. S. Edelman, et al., p. 17. Palo Alto, Calif.: Annual Reviews.

Krulich, L. 1979. Central neurotransmitters and the secretion of prolactin, GH, LH, and TSH. In *Annual Review of Physiology*, vol. 41, edited by I. S. Edelman, et al., p. 603. Palo Alto, Calif.: Annual Reviews.

Kuehl, F. A., and Egan, R. W. 1980. Prostaglandins, arachinodic acid, and inflammation. *Science* 210:978.

McEwen, B. S. July 1976. Interactions between hormones and nerve tissue. *Scientific American*.

O'Malley, B. W., and Schrader, W. T. February 1976. The receptors of steroid hormones. *Scientific American*.

Oppenheimer, J. H. 1979. Thyroid hormone action at the cellular level. *Science* 203:971.

Rasmussen, H., and Pechet, M. M. October 1970. Calcitonin. *Scientific American*.

Reichlin, S., et al. 1976. Hypothalamic hormones. In *Annual Review of Physiology*, vol. 39, edited by Ernst Knobil, et al., p. 389. Palo Alto, Calif.: Annual Reviews.

Schally, A. V. 1978. Aspects of hypothalamic regulation of the pituitary gland. *Science* 202:18.

Turner, C. D., and Bagnara, J. T. 1976. *General endocrinology*. 6th ed. Philadelphia: W. B. Saunders Co.

Chapter 13

Bortoff, A. 1972. Digestion. In *Annual Review of Physiology* 34:261.

Davenport, H. W. 1982. *Physiology of the digestive tract*. 5th ed. Chicago: Year Book Medical Publishers.

Deevey, E. S. September 1970. Mineral cycles. *Scientific American*.

Emery, T. November–December 1982. Iron metabolism in humans and plants. *American Scientist*.

Fleck, H. 1981. *Introduction to nutrition*. 4th ed. New York: Macmillan.

Freedland, R. A., and Briggs, S. 1977. *A biochemical approach to nutrition*. New York: Halsted Press.

Grossman, M. I. 1979. Neural and hormonal regulation of gastrointestinal function. *Annual Review of Physiology*, vol. 41, edited by I. S. Edelman, et al., p. 27. Palo Alto, Calif.: Annual Reviews.

Guthrie, H. A. 1979. *Introductory nutrition*. 4th ed. St. Louis: C. V. Mosby.

Marx, J. L. 1979. The HDL: the good cholesterol carrier. *Science* 205:677.

Maugh, T. H. 1973. Trace elements: a growing appreciation of the effects on man. *Science* 181:253.

Mayer, J. 1979. *Human nutrition*. Springfield, Ill.: Charles C. Thomas.

Moog, F. November 1981. The lining of the small intestine. *Scientific American*.

Robinson, C. H., and Weigley, E. S. 1978. *Fundamentals of normal nutrition*. 3d ed. New York: Macmillan.

Soll, A., and Walsh, J. H. 1979. Regulation of gastric acid secretion. In *Annual Review of Physiology*, vol. 41, edited by I. S. Edelman, et al., p. 35. Palo Alto, Calif.: Annual Reviews.

Chapter 14

Avery, M. E., et al. April 1973. The lung of the newborn infant. *Scientific American*.

Cohen, M. I. 1981. Central determinants of respiratory rhythm. In *Annual Review of Physiology*, vol. 43, edited by I. S. Edelman and S. G. Schultz, p. 91. Palo Alto, Calif.: Annual Reviews.

Comroe, J. H. 1974. *Physiology of respiration: an introduction.* 2d ed. Chicago: Year Book Medical Publishers.

Crandall, E. D. 1976. Pulmonary gas exchange. In *Annual Review of Physiology,* vol. 38, edited by Ernst Knobil, et al., p. 69. Palo Alto, Calif.: Annual Reviews.

Fishman, A. P. 1980. Vasomotor regulation of the pulmonary circulation. In *Annual Review of Physiology,* vol. 42, edited by I. S. Edelman, et al., p. 211. Palo Alto, Calif.: Annual Reviews.

Fraser, R. G., and Pare, J. A. 1977. *Structure and function of the lung.* 2d ed. Philadelphia: W. B. Saunders.

Guz, A. 1975. Regulation of respiration in man. In *Annual Review of Physiology,* vol. 37, edited by Julius H. Comroe, Jr., et al., p. 303. Palo Alto, Calif.: Annual Reviews.

Lehninger, A. L. September 1961. How cells transform energy. *Scientific American.*

Slonim, N. B. 1981. *Respiratory physiology.* 4th ed. St. Louis: C. V. Mosby.

West, J. B. 1972. Respiration. In *Annual Review of Physiology* 34:91.

Winter, P. M., and Lowenstein, E. November 1969. Acute respiratory failure. *Scientific American.*

Chapter 15

Adamson, J. W., and Finch, C. A. 1975. Hemoglobin function, oxygen affinity, and erythropoietin. In *Annual Review of Physiology,* vol. 37, edited by Julius H. Comroe, Jr., et al., p. 351. Palo Alto, Calif.: Annual Reviews.

Child, J., et al. 1972. Blood transfusions. *American Journal of Nursing* 72:1602.

Doolittle, R. F. December 1981. Fibrinogen and fibrin. *Scientific American.*

McDonagh, A. F., et al. 1980. Blue light and bilirubin excretion. *Science* 208:145.

Perutz, M. F. December 1978. Hemoglobin structure and respiratory transport. *Scientific American.*

Platt, W. R. 1979. *Color atlas and textbook of hematology.* 2d ed. Philadelphia: J. B. Lippincott Co.

Seegers, W. H. 1969. Blood clotting mechanism: three basic reactions. In *Annual Review of Physiology* 31:269.

Sergis, E., and Hilgartner, M. W. 1972. Hemophilia. *American Journal of Nursing* 72:11.

Till, J. E. September–October 1981. Cellular diversity in the blood forming system. *American Scientist.*

Wood, W. B. February 1971. White blood cells vs bacteria. *Scientific American.*

Zucker, M. B. June 1980. The functioning of blood platelets. *Scientific American.*

Chapter 16

Alpert, N. R., et al. 1979. Heart muscle mechanics. In *Annual Review of Physiology,* vol. 41, edited by I. S. Edelman, et al., p. 521. Palo Alto, Calif.: Annual Reviews.

Baez, S. 1977. Microcirculation. In *Annual Review of Physiology,* vol. 39, edited by Ernst Knobil, et al., p. 391. Palo Alto, Calif.: Annual Reviews.

Berne, R. M., and Levy, M. N. 1981. *Cardiovascular physiology.* 4th ed. St. Louis: C. V. Mosby.

DeBakey, M., and Gotton, A. 1977. *The living heart.* New York: David McKay Co.

Donald, D. E., and Shepherd, J. T. 1980. Autonomic regulation of the peripheral circulation. In *Annual Review of Physiology,* vol. 42, edited by I. S. Edelman, et al., p. 419. Palo Alto, Calif.: Annual Reviews.

Gore, R. W., and McDonagh, P. F. 1980. Fluid exchange across single capillaries. In *Annual Review of Physiology,* vol. 42, edited by I. S. Edelman, et al., p. 337. Palo Alto, Calif.: Annual Reviews.

Hurst, J. W., ed. 1979. *The heart.* New York: McGraw-Hill.

Jarvik, R. K. January 1981. The total artificial heart. *Scientific American.*

Johansen, K. July 1982. Aneurysms. *Scientific American.*

Lassen, N. A., et al. October 1978. Brain function and blood flow. *Scientific American.*

Olsson, R. A. 1981. Local factors regulating cardiac and skeletal muscle blood flow. In *Annual Review of Physiology,* vol. 43, edited by I. S. Edelman and S. G. Schultz, p. 385. Palo Alto, Calif.: Annual Reviews.

Rushmer, R. F. 1976. *Structure and function of the cardiovascular system.* 2d ed. Philadelphia: W. B. Saunders.

Scheuer, J., and Tipton, C. M. 1977. Cardiovascular adaptation to physical training. In *Annual Review of Physiology,* vol. 39, edited by Ernst Knobil, et al., p. 221. Palo Alto, Calif.: Annual Reviews.

Westfall, V. A. February 1976. Electrical and mechanical events in the cardiac cycle. *American Journal of Nursing.*

Chapter 17

Bellanti, J. A. 1978. *Immunology.* Philadelphia: W. B. Saunders.

Buisseret, P. D. August 1982. Allergy. *Scientific American.*

Capra, J. D., and Edmundson, A. B. January 1977. The antibody combining site. *Scientific American.*

Collier, R. J., and Kaplan, D. A. July 1984. Immunotoxins. *Scientific American.*

Cunningham, B. A. October 1977. The structure and functions of histocompatibility antigens. *Scientific American.*

Eisen, H. N. 1980. *Immunology.* New York: Harper and Row.

Golub, E. S. 1981. *The cellular basis of the immune response.* 2d ed. Sunderland, Mass.: Sinauer Associates.

Hood, L. E., et al. 1978. *Immunology.* Menlo Park, Calif.: Benjamin Cummings.

Milstein, C. October 1980. Monoclonal antibodies. *Scientific American.*

Playfair, J. H. L. 1982. *Immunology at a glance.* 2d ed. St. Louis: C. V. Mosby.

Roitt, I. M. 1981. *Essential immunology.* 4th ed. St. Louis: C. V. Mosby.

Sell, S. 1980. *Immunology, immunopathology, and immunity.* 3d ed. New York: Harper and Row.

Sibley, C. H. 1984. How do B-lymphocytes control antibody production? *BioScience* 34:30.

Talmage, D. W. 1979. Recognition and memory in the cells of the immune system. *American Scientist* 67:174.

Chapter 18

Anderson, B. 1977. Regulation of body fluids. In *Annual Review of Physiology,* vol. 39, edited by Ernst Knobil, et al., p. 185. Palo Alto, Calif.: Annual Reviews.

Baer, P. G., and McGiff, J. C. 1980. Hormonal systems and renal hemodynamics. In *Annual Review of Physiology,* vol. 42, edited by I. S. Edelman, et al., p. 589. Palo Alto, Calif.: Annual Reviews.

Bauman, J. W., and Chinard, F. P. 1975. *Renal function: physiological and medical aspects.* St. Louis: C. V. Mosby.

Beeuwkes, R. 1980. The vascular organization of the kidney. In *Annual Review of Physiology,* vol. 42, edited by I. S. Edelman, et al., p. 531. Palo Alto, Calif.: Annual Reviews.

Harvey, R. J. 1976. *The kidneys and the internal environment.* New York: Halsted Press.

Lassiter, W. E. 1975. Kidney. In *Annual Review of Physiology,* vol. 37, edited by Julius H. Comroe, et al., p. 371. Palo Alto, Calif.: Annual Reviews.

Pitts, R. F. 1974. *Physiology of the kidney and body fluids.* 3d ed. Chicago: Year Book Medical Publishers.

Sullivan, L. P. 1974. *Physiology of the kidney.* Philadelphia: Lea and Febiger.

Ullrich, K. J. 1979. Sugar, amino acid, and sodium transport in the proximal tubule. In *Annual Review of Physiology,* vol. 41, edited by I. S. Edelman, et al., p. 181. Palo Alto, Calif.: Annual Reviews.

Vander, A. J. 1980. *Renal physiology.* 2d ed. New York: McGraw-Hill.

Chapter 19

Anderson, B. 1977. Regulation of body fluids. In *Annual Review of Physiology,* vol. 39, edited by Ernst Knobil, et al., p. 185. Palo Alto, Calif.: Annual Reviews.

Burk, S. R. 1972. *The composition and function of body fluids.* St. Louis: C. V. Mosby.

Deetjen, P., et al. 1974. *Physiology of the kidney and water balance.* New York: Springer-Verlag.

Hills, A. G. 1973. *Acid-base balance.* Baltimore: Williams and Wilkins.

Pitts, R. F. 1974. *Physiology of the kidney and body fluids.* 3d ed. Chicago: Year Book Medical Publishers.

Share, L., et al. 1972. Regulation of body fluids. In *Annual Review of Physiology* 34:235.

Valtin, H. 1973. *Renal function: mechanism for preserving fluid and solute balance in health.* Boston: Little, Brown and Co.

Vander, A. J. 1980. *Renal physiology.* 2d ed. New York: McGraw-Hill.

Weldy, N. J. 1972. *Body fluids and electrolytes.* St. Louis: C. V. Mosby.

Chapter 20

Annis, L. F. 1978. *The child before birth.* New York: Cornell Univ. Press.

Balinsky, B. I. 1981. *An introduction to embryology.* 5th ed. Philadelphia: W. B. Saunders.

Beaconsfield, P., et al. August 1980. The placenta. *Scientific American.*

Epel, D. November 1977. The program of fertilization. *Scientific American.*

Heymann, M. A., et al. 1981. Factors affecting changes in the neonatal systemic circulation. In *Annual Review of Physiology,* vol. 43, edited by I. S. Edelman and S. G. Schultz, p. 371. Palo Alto, Calif.: Annual Reviews.

Lein, A. 1979. *The cycling female.* San Francisco: W. H. Freeman.

Means, A. R., et al. Regulation of the testis sertoli cell by follicle-stimulating hormone. In *Annual Review of Physiology,* vol. 42, edited by I. S. Edelman, et al., p. 59. Palo Alto, Calif.: Annual Reviews.

Moore, K. L. 1974. *Before we are born: basic embryology and birth defects.* Philadelphia: W. B. Saunders.

————. 1974. *The developing human: clinically oriented embryology.* Philadelphia: W. B. Saunders.

Nilsson, L. 1973. *Behold man.* Boston: Little, Brown and Co.

O'Malley, B. W., and Means, A. R. 1974. Female steroid hormones and target cell nuclei. *Science* 183:610.

Patten, B. M., and Carlson, B. M. 1974. *Foundations of embryology.* 3d ed. New York: McGraw-Hill.

Segal, S. J. September 1974. The physiology of human reproduction. *Scientific American.*

Shiu, P. C., and Friesen, H. G. 1980. Mechanism of action of prolactin in the control of mammary gland functions. In *Annual Review of Physiology,* vol. 42, edited by I. S. Edelman, et al., p. 83. Palo Alto, Calif.: Annual Reviews.

Simpson, E. R., and MacDonald, P. C. 1981. Endocrine physiology of the placenta. In *Annual Review of Physiology,* vol. 43, edited by I. S. Edelman and S. G. Schultz, p. 163. Palo Alto, Calif.: Annual Reviews.

Glossary

The words in this glossary are followed by a phonetic guide to pronunciation.

In this guide, any unmarked vowel that ends a syllable or stands alone as a syllable is long. Thus, the word *play* would be spelled *pla*.

Any unmarked vowel that is followed by a consonant has the short sound. The word *tough,* for instance, would be spelled *tuf*.

If a long vowel does appear in the middle of a syllable (followed by a consonant), then it is marked with the macron (-), the sign for a long vowel. For instance, the word *plate* would be phonetically spelled *plāt*.

Similarly, if a vowel stands alone or ends a syllable, but should have the short sound, it is marked with a breve (˘).

abdomen (ab-do′men) Portion of the body between the diaphragm and the pelvis.

abduction (ab-duk′shun) Movement of a body part away from the midline.

absorption (ab-sorp′shun) The taking in of substances by cells or membranes.

accessory organs (ak-ses′o-re or′ganz) Organs that supplement the functions of other organs.

accommodation (ah-kom″o-da′shun) Adjustment of the lens for close vision.

acetylcholine (as″ĕ-til-ko′lēn) Substance secreted at the axon ends of many neurons that transmits a nerve impulse across a synapse.

acetyl coenzyme A (as′ĕ-til ko-en′zim) An intermediate compound produced during the oxidation of carbohydrates and fats.

acid (as′id) A substance that ionizes in water to release hydrogen ions.

ACTH Adrenocorticotropic hormone.

actin (ak′tin) A protein in a muscle fiber that, together with myosin, is responsible for contraction and relaxation.

action potential (ak′shun po-ten′shal) The sequence of electrical changes occurring when a nerve cell membrane is exposed to a stimulus that exceeds its threshold.

active transport (ak′tiv trans′port) Process that requires an expenditure of energy to move a substance across a cell membrane.

adaptation (ad″ap-ta′shun) Adjustment to environmental conditions.

adduction (ah-duk′shun) Movement of a body part toward the midline.

adenosine diphosphate (ah-den′o-sēn di-fos′fāt) ADP; molecule created when the terminal phosphate is lost from a molecule of adenosine triphosphate.

adenosine triphosphate (ah-den′o-sēn tri-fos′fāt) An organic molecule that stores energy and releases energy for use in cellular processes.

adenylate cyclase (ah-den′ĭ-lāt si′klās) An enzyme that is activated when certain hormones combine with receptors on cell membranes, causing ATP to become cyclic AMP.

ADH Antidiuretic hormone.

adipose tissue (ad′ĭ-pōs tish′u) Fat-storing tissue.

ADP Adenosine diphosphate.

adrenal cortex (ah-dre′nal kor′teks) The outer portion of the adrenal gland.

adrenal glands (ah-dre′nal glandz) Endocrine glands located on the tops of the kidneys.

adrenal medulla (ah-dre′nal me-dul′ah) The inner portion of the adrenal gland.

adrenergic fiber (ad″ren-er′jik fi′ber) A nerve fiber that secretes norepinephrine at the terminal end of its axon.

adrenocorticotropic hormone (ah-dre″no-kor″te-ko-trōp′ik hor′mōn) ACTH; hormone secreted by the anterior lobe of the pituitary gland that stimulates activity in the adrenal cortex.

aerobic respiration (a″er-ōb′ik res″pĭ-ra′shun) Phase of cellular respiration that requires the presence of oxygen.

afferent arteriole (af′er-ent ar-te′re-ōl) Vessel that supplies blood to the glomerulus of a nephron.

agglutination (ah-gloo″ti-na′shun) Clumping together of blood cells in response to a reaction between an agglutinin and an agglutinogen.

agglutinin (ah-gloo′ti-nin) A substance that reacts with an agglutinogen; an antibody.

agglutinogen (ag″loo-tin′o-jen) A substance that stimulates the formation of agglutinins.

agranulocytes (a-gran′u-lo-sīt) A nongranular leukocyte.

albumin (al-bu′min) A plasma protein that helps to regulate the osmotic concentration of the blood.

aldosterone (al-dos′ter-ōn) A hormone, secreted by the adrenal cortex, that functions in regulating sodium and potassium concentrations.

alimentary canal (al″i-men′tar-e kah-nal′) The tubular portion of the digestive tract that leads from the mouth to the anus.

alkaline (al′kah-līn) Pertaining to or having the properties of a base.

allantois (ah-lan′to-is) A structure that appears during embryonic development and functions in the formation of umbilical blood vessels.

allergen (al'er-jen) A foreign substance capable of stimulating an allergic reaction.

all-or-none response (al'or-nun' re-spons') Phenomenon in which a muscle fiber contracts completely when it is exposed to a stimulus of threshold strength.

alveolar ducts (al-ve'o-lar dukts') Fine tubes that carry air to the air sacs of the lungs.

alveolus (al-ve'o-lus) An air sac of a lung; a saclike structure.

amino acid (ah-me'no as'id) An organic compound of relatively small molecular size that contains an amino group ($-NH_2$) and a carboxyl group ($-COOH$); the structural unit of a protein molecule.

amnion (am'ne-on) An embryonic membrane that encircles a developing fetus and contains amniotic fluid.

ampulla (am-pul'ah) An expansion at the end of each semicircular canal that contains a crista ampullaris.

amylase (am'i-lās) An enzyme that functions to hydrolyze starch.

anabolic metabolism (an''ah-bol'ik mē-tab'o-lizm) Metabolic process by which larger molecules are formed from smaller ones; anabolism.

anaerobic respiration (an-a''er-ōb'ik res''pi-ra'shun) Phase of cellular respiration that occurs in the absence of oxygen.

anaphase (an'ah-fāz) Stage in mitosis during which duplicate chromosomes move to opposite poles of the cell.

anatomy (ah-nat'o-me) Branch of science dealing with the form and structure of body parts.

androgen (an'dro-jen) A male sex hormone such as testosterone.

antagonist (an-tag'o-nist) A muscle that acts in opposition to a prime mover.

antebrachium (an''te-bra'ke-um) The forearm.

antecubital (an''te-ku'bi-tal) The region in front of the elbow joint.

anterior (an-te're-or) Pertaining to the front; the opposite of posterior.

anterior pituitary (an-te're-or pi-tu''i-tār''e) The front lobe of the pituitary gland.

antibody (an'ti-bod''e) A specific substance produced by cells in response to the presence of an antigen.

antibody-mediated immunity (an''ti-bod''e me'de-ātid i-mu'ni-te) Resistance to disease-causing agents resulting from the production of specific antibodies by B-lymphocytes; humoral immunity.

antidiuretic hormone (an''ti-di''u-ret'ik hor'mōn) Hormone released from the posterior lobe of the pituitary gland that enhances the conservation of water.

antigen (an'ti-jen) A substance that stimulates cells to produce antibodies.

aorta (a-or'tah) Major systemic artery that receives blood from the left ventricle.

aortic semilunar valve (a-or'tik sem''i-lu'nar valv) Flaplike structures in the wall of the aorta near its origin that prevent blood from returning to the left ventricle of the heart.

aortic sinus (a-or'tik si'nus) Swelling in the wall of the aorta that contains pressoreceptors.

apocrine gland (ap'o-krin gland) A type of sweat gland that responds during periods of emotional stress.

aponeurosis (ap''o-nu-ro'sis) A sheetlike tendon by which certain muscles are attached to other parts.

appendicular (ap''en-dik'u-lar) Pertaining to the arms or legs.

aqueous humor (a'kwe-us hu'mor) Watery fluid that fills the anterior and posterior chambers of the eye.

arachnoid (ah-rak'noid) Delicate, weblike middle layer of the meninges; arachnoid mater.

arrector pili muscle (ah-rek'tor pil'i mus'l) Smooth muscle in the skin associated with a hair follicle.

arrhythmia (ah-rith'me-ah) Abnormal heart action characterized by a loss of rhythm.

arteriole (ar-te're-ōl) A small branch of an artery that communicates with a capillary network.

artery (ar'ter-e) A vessel that transports blood away from the heart.

articular cartilage (ar-tik'u-lar kar'ti-lij) Hyaline cartilage that covers the ends of bones in synovial joints.

articulation (ar-tik''u-la'shun) The joining together of parts at a joint.

ascending tracts (ah-send'ing trakts) Groups of nerve fibers in the spinal cord that transmit sensory impulses upward to the brain.

ascorbic acid (as-kor'bik as'id) One of the water-soluble vitamins; vitamin C.

assimilation (ah-sim''i-la'shun) The action of changing absorbed substances into forms that differ chemically from those entering.

association area (ah-so''se-a'shun a're-ah) Region of the cerebral cortex related to memory, reasoning, judgment, and emotional feelings.

astrocyte (as'tro-sīt) A type of neuroglial cell that functions to connect neurons to blood vessels.

atmospheric pressure (at''mos-fer'ik presh'ur) Pressure exerted by the weight of the air.

atom (at'om) Smallest particle of an element that has the properties of that element.

atomic number (ah-tom'ik num'ber) Number equal to the number of protons in an atom of an element.

atomic weight (ah-tom'ik wāt) Number approximately equal to the number of protons plus the number of neutrons in an atom.

ATP Adenosine triphosphate.

atrioventricular bundle (a''tre-o-ven-trik'u-lar bun'dl) Group of specialized fibers that conduct impulses from the atrioventricular node to the ventricular muscle of the heart; A-V bundle.

atrioventricular node (a''tre-o-ven-trik'u-lar nōd) Specialized mass of muscle fibers located in the interatrial septum of the heart.

atrium (a'tre-um) A chamber of the heart that receives blood from veins.

atrophy (at'ro-fe) A wasting away or decrease in size of an organ or tissue.

auditory (aw'di-to''re) Pertaining to the ear or to the sense of hearing.

auditory ossicle (aw'di-to''re os'i-kl) A bone of the middle ear.

auricle (aw'ri-kl) An earlike structure; the portion of the heart that forms the wall of an atrium.

autonomic nervous system (aw''to-nom'ik ner'vus sis'tem) Portion of the nervous system that functions to control the actions of the visceral organs and skin.

A-V bundle (bun'dl) A group of fibers that conducts cardiac impulses from the A-V node to the Purkinje fibers; bundle of His.

A-V node (nōd) Atrioventricular node.

axial skeleton (ak'se-al skel'ĕ-ton) Portion of the skeleton that supports and protects the organs of the head, neck, and trunk.

axillary (ak'sĭ-ler''e) Pertaining to the armpit.

axon (ak'son) A nerve fiber that conducts a nerve impulse away from a neuron cell body.

basal ganglion (ba'sal gang'gle-on) Mass of gray matter located deep within a cerebral hemisphere of the brain.

base (bās) A substance that ionizes in water to release hydroxyl ions (OH⁻) or other ions that combine with hydrogen ions.

basement membrane (bās'ment mem'brān) A layer of nonliving material that anchors epithelial tissue to underlying connective tissue.

basophil (ba'so-fil) White blood cell characterized by the presence of cytoplasmic granules that become stained by basophilic dye.

beta oxidation (ba'tah ok''sĭ-da'shun) Chemical process by which fatty acids are converted to molecules of acetyl coenzyme A.

bicuspid valve (bi-kus'pid valv) Heart valve located between the left atrium and the left ventricle; mitral valve.

bile (bīl) Fluid secreted by the liver and stored in the gallbladder.

bilirubin (bil''ĭ-roo'bin) A bile pigment produced as a result of hemoglobin breakdown.

biliverdin (bil''ĭ-ver'din) A bile pigment produced as a result of hemoglobin breakdown.

biotin (bi'o-tin) A water-soluble vitamin; a member of the vitamin B complex.

blastocyst (blas'to-sist) An early stage of embryonic development that consists of a hollow ball of cells.

B-lymphocyte (lim'fo-sīt) Lymphocyte that reacts against foreign substances in the body by producing and secreting antibodies.

Bowman's capsule (bo'manz kap'sūl) Proximal portion of a renal tubule that encloses the glomerulus of a nephron.

brachial (bra'ke-al) Pertaining to the arm.

brain stem (brān stem) Portion of the brain that includes the midbrain, pons, and medulla oblongata.

Broca's area (bro'kahz a're-ah) Region of the frontal lobe that coordinates complex muscular actions of the mouth, tongue, and larynx, making speech possible.

bronchial tree (brong'ke-al tre) The bronchi and their branches that function to carry air from the trachea to the alveoli of the lungs.

bronchiole (brong'ke-ōl) A small branch of a bronchus within the lung.

bronchus (brong'kus) A branch of the trachea that leads to a lung.

buccal (buk'al) Pertaining to the mouth and the inner lining of the cheeks.

buffer (buf'er) A substance that can react with a strong acid or base to form a weaker acid or base and thus resist a change in pH.

bulbourethral glands (bul''bo-u-re'thral glandz) Glands that secrete a viscous fluid into the male urethra at times of sexual excitement.

bursa (bur'sah) A saclike, fluid-filled structure, lined with synovial membrane, that occurs near a joint.

calcitonin (kal''sĭ-to'nin) Hormone secreted by the thyroid gland that helps to regulate the level of blood calcium.

calorie (kal'o-re) A unit used in the measurement of heat energy and the energy values of foods.

canaliculus (kan''ah-lik'u-lus) Microscopic canals that interconnect the lacunae of bone tissue.

cancellous bone (kan'sĕ-lus bōn) Bone tissue with a lattice-work structure; spongy bone.

capillary (kap'ĭ-ler''e) A small blood vessel that connects an arteriole and a venule.

carbaminohemoglobin (kar''bah-me'no-he''mo-glo'bin) Compound formed by the union of carbon dioxide and hemoglobin.

carbohydrate (kar''bo-hi'drāt) An organic compound that contains carbon, hydrogen, and oxygen, with a 2:1 ratio of hydrogen to oxygen atoms.

carbonic anhydrase (kar-bon'ik an-hi'drās) Enzyme that promotes the reaction between carbon dioxide and water to form carbonic acid.

carboxypeptidase (kar-bok''se-pep'ti-dās) A protein-splitting enzyme found in pancreatic juice.

cardiac conduction system (kar'de-ak kon-duk'shun sis'tem) System of specialized muscle fibers that conducts cardiac impulses from the S-A node into the myocardium.

cardiac cycle (kar'de-ak si'kl) A series of myocardial contractions that constitutes a complete heartbeat.

cardiac muscle (kar'de-ak mus'l) Specialized type of muscle tissue found only in the heart.

cardiac output (kar'de-ak owt'poot) A quantity calculated by multiplying the stroke volume by the heart rate in beats per minute.

carpals (kar'pals) Bones of the wrist.

carpus (kar'pus) The wrist; the wrist bones as a group.

cartilage (kar'tĭ-lij) Type of connective tissue in which cells are located within lacunae and are separated by a semi-solid matrix.

cartilaginous bone (kar''tĭ-laj'ĭ-nus bōn) Bone that appears in the form of a cartilaginous model, which is largely replaced by bony tissue during development.

catabolic metabolism (kat''ah-bol'ik mĕ-tab'o-lism) Metabolic process by which large molecules are broken down into smaller ones; catabolism.

celiac (se'le-ak) Pertaining to the abdomen.

cell (sel) The structural and functional unit of an organism.

cell body (sel bod'e) Portion of a nerve cell that includes a cytoplasmic mass and a nucleus and from which the nerve fibers extend.

cell-mediated immunity (sel me'de-ātid ĭ-mu'nĭ-te) Resistance to invasion by foreign cells characterized by direct attack of T-lymphocytes.

cellular respiration (sel'u-lar res''pi-ra'shun) Process by which energy is released from organic compounds within cells.

cellulose (sel'u-lōs) A polysaccharide that is very abundant in plant tissues, but cannot be digested by human enzymes.

cementum (se-men'tum) Bonelike material that surrounds the root of a tooth.

central canal (sen'tral kah-nal') Tube within the spinal cord that is continuous with the ventricles of the brain and contains cerebrospinal fluid.

central nervous system (sen'tral ner'vus sis'tem) Portion of the nervous system that consists of the brain and spinal cord; CNS.

centriole (sen'tre-ōl) A cellular organelle that functions in the organization of the spindle during mitosis.

centromere (sen'tro-mēr) Portion of a chromosome to which the spindle fiber attaches during mitosis.

centrosome (sen'tro-sōm) Cellular organelle consisting of two centrioles.

cephalic (se-fal'ik) Pertaining to the head.

cerebellar cortex (ser''e-bel'ar kor'teks) The outer layer of the cerebellum.

cerebellum (ser''e-bel'um) Portion of the brain that coordinates skeletal muscle movement.

cerebral cortex (ser'e-bral kor'teks) Outer layer of the cerebrum.

cerebral hemisphere (ser'e-bral hem'i-sfēr) One of the large, paired structures that together constitute the cerebrum of the brain.

cerebrospinal fluid (ser''e-bro-spi'nal floo'id) Fluid that occupies the ventricles of the brain, the subarachnoid space of the meninges, and the central canal of the spinal cord.

cerebrum (ser'e-brum) Portion of the brain that occupies the upper part of the cranial cavity.

cervical (ser'vi-kal) Pertaining to the neck or to the cervix of the uterus.

cervix (ser'viks) Narrow, inferior end of the uterus that leads into the vagina.

chemoreceptor (ke''mo-re-sep'tor) A receptor that is stimulated by the presence of certain chemical substances.

chief cell (chēf sel) Cell of gastric gland that secretes various digestive enzymes, including pepsinogen.

cholecystokinin (ko''le-sis''to-ki'nin) Hormone secreted by the small intestine that stimulates the release of pancreatic juice from the pancreas and bile from the gallbladder.

cholesterol (ko-les'ter-ol) A lipid produced by body cells that is used in the synthesis of steroid hormones.

cholinergic fiber (ko''lin-er'jik fi'ber) A nerve fiber that secretes acetylcholine at the terminal end of its axon.

cholinesterase (ko''lin-es'ter-ās) An enzyme that causes the decomposition of acetylcholine.

chorion (ko're-on) Embryonic membrane that forms the outermost covering around a developing fetus.

chorionic villi (ko''re-on'ik vil'i) Projections that extend from the outer surface of the chorion.

choroid coat (ko'roid kōt) The vascular, pigmented middle layer of the wall of the eye.

choroid plexus (ko'roid plek'sus) Mass of specialized capillaries from which cerebrospinal fluid is secreted.

chromatid (kro'mah-tid) A member of a duplicate pair of chromosomes.

chromatin (kro'mah-tin) Nuclear material that gives rise to chromosomes during mitosis.

chromosome (kro'mo-sōm) Rodlike structure that appears in the nucleus of a cell during mitosis.

chylomicron (ki''lo-mi'kron) A microscopic droplet of fat, found in the blood following the digestion of fats.

chyme (kīm) Semifluid mass of food material that passes from the stomach to the small intestine.

chymotrypsin (ki''mo-trip'sin) A protein-splitting enzyme found in pancreatic juice.

cilia (sil'e-ah) Microscopic, hairlike processes on the exposed surfaces of certain epithelial cells.

ciliary body (sil'e-er''e bod'e) Structure associated with the choroid layer of the eye that secretes aqueous humor and contains the ciliary muscle.

circle of Willis (sir'kl uv wil'is) An arterial ring located on the ventral surface of the brain.

circular muscles (ser'ku-lar mus'lz) Muscles whose fibers are arranged in circular patterns, usually around an opening or in the wall of a tube.

circumduction (ser''kum-duk'shun) Movement of a body part, such as a limb, so that the end follows a circular path.

cisternae (sis-ter'ne) Enlarged portions of the sarcoplasmic reticulum near the actin and myosin filaments of a muscle fiber.

citric acid cycle (sit'rik as'id si'kl) A series of chemical reactions by which various molecules are oxidized and energy is released from them; Kreb's cycle.

cleavage (klēv'ij) The early successive divisions of embryonic cells into smaller and smaller cells.

clitoris (kli'to-ris) Small erectile organ located in the anterior portion of the female vulva; corresponds to the penis of the male.

clone (klōn) A group of identical cells that originated from a single early cell.

CNS Central nervous system.

coagulation (ko-ag''u-la'shun) The clotting of blood.

cochlea (kok'le-ah) Portion of the inner ear that contains the receptors of hearing.

coenzyme (ko-en'zim) A nonprotein substance that is necessary to complete the structure of an enzyme molecule.

cofactor (ko'fak-tor) A nonprotein substance that must be combined with the protein portion of an enzyme before the enzyme can act.

collagen (kol'ah-jen) Protein that occurs in the white fibers of connective tissues.

common bile duct (kom'mon bil dukt) Tube that transports bile from the cystic duct to the duodenum.

complete protein (kom-plēt' pro'te-in) A protein that contains adequate amounts of the essential amino acids.

compound (kom'pownd) A substance composed of two or more elements joined by chemical bonds.

condyle (kon'dil) A rounded process of a bone, usually at the articular end.

cones (kōns) Color receptors located in the retina of the eye.

conjunctiva (kon''junk-ti'vah) Membranous covering on the anterior surface of the eye.

connective tissue (kŏ-nek'tiv tish'u) One of the basic types of tissue that includes bone, cartilage, and various fibrous tissues.

convergence (kon-ver'jens) The coming together of nerve impulses from different parts of the nervous system so that they reach the same neuron.

convolution (kon''vo-lu'shun) An elevation on the surface of a structure.

cornea (kor'ne-ah) Transparent anterior portion of the outer layer of the eye wall.

coronary artery (kor'o-na''re ar'ter-e) An artery that supplies blood to the wall of the heart.

coronary sinus (kor'o-na''re si'nus) A large vessel on the posterior surface of the heart into which the cardiac veins drain.

corpus callosum (kor'pus kah-lo'sum) A mass of white matter within the brain, composed of nerve fibers connecting the right and left cerebral hemispheres.

corpus luteum (kor'pus lut'e-um) Structure that forms from the tissues of a ruptured ovarian follicle and functions to secrete female hormones.

cortex (kor'teks) Outer layer of an organ.

cortisol (kor'ti-sol) A glucocorticoid secreted by the adrenal cortex.

costal (kos'tal) Pertaining to the ribs.

covalent bond (ko'va-lent bond) Chemical bond created by the sharing of electrons between atoms.

cranial (kra'ne-al) Pertaining to the cranium.

cranial nerve (kra'ne-al nerv) Nerve that arises from the brain.

crenation (kre-na'shun) Shrinkage of a cell caused by contact with a hypertonic solution.

crest (krest) A ridgelike projection of a bone.

cricoid cartilage (kri'koid kar'ti-lij) A ringlike cartilage that forms the lower end of the larynx.

crista ampullaris (kris'tah am-pul'ar-is) Sensory organ located within a semicircular canal.

cubital (ku'bi-tal) Pertaining to the forearm.

cutaneous (ku-ta'ne-us) Pertaining to the skin.

cystic duct (sis'tik dukt) Tube that connects the gallbladder to the common bile duct.

cytoplasm (si'to-plazm) The contents of a cell surrounding its nucleus.

deamination (de-am''i-na'shun) Chemical process by which amino groups ($-NH_2$) are removed from amino acid molecules.

deciduous teeth (de-sid'u-us tēth) Teeth that are shed and replaced by permanent teeth.

decomposition (de-kom''po-zish'un) The breakdown of molecules into simpler compounds.

defecation (def''ĕ-ka'shun) The discharge of feces from the rectum through the anus.

dehydration synthesis (de''hi-dra'shun sin'thĕ-sis) Anabolic process by which molecules are joined together to form larger molecules.

dendrite (den'drīt) Nerve fiber that transmits impulses toward a neuron cell body.

dentine (den'tēn) Bonelike substance that forms the bulk of a tooth.

deoxyhemoglobin (de-ok''si-he''mo-glo'bin) Hemoglobin that lacks oxygen.

depolarization (de-po''lar-i-za'shun) The loss of an electrical charge on the surface of a membrane.

dermis (der'mis) The thick layer of the skin beneath the epidermis.

descending tracts (de-send'ing trakts) Groups of nerve fibers that carry nerve impulses downward from the brain through the spinal cord.

detrusor muscle (de-trūz'or mus'l) Muscular wall of the urinary bladder.

diaphragm (di'ah-fram) A sheetlike structure that separates the thoracic and abdominal cavities.

diaphysis (di-af'i-sis) The shaft of a long bone.

diastole (di-as'tol-le) Phase of the cardiac cycle during which a heart chamber wall is relaxed.

diastolic pressure (di-a-stol'ik presh'ur) Arterial blood pressure during the diastolic phase of the cardiac cycle.

diencephalon (di''en-sef'ah-lon) Portion of the brain in the region of the third ventricle that includes the thalamus and hypothalamus.

differentiation (dif''er-en''she-a'shun) Process by which cells become structurally and functionally specialized.

diffusion (di-fu'zhun) Random movement of molecules from a region of higher concentration toward one of lower concentration.

digestion (di-jes'chun) The process by which larger molecules of food substances are broken down into smaller molecules that can be absorbed; hydrolysis.

dipeptide (di-pep'tid) A molecule composed of two amino acids joined together.

disaccharide (di-sak'ah-rīd) Sugar produced by the union of two monosaccharide molecules.

distal (dis'tal) Further from the midline or origin; opposite of proximal.

divergence (di-ver'jens) Arrangement of nerve fibers by which a nerve impulse can spread to increasing numbers of neurons within a neuronal pool.

DNA Deoxyribonucleic acid.

dorsal root (dor'sal root) The sensory branch of a spinal nerve by which it joins the spinal cord.

dorsal root ganglion (dor'sal root gang'gle-on) Mass of sensory neuron cell bodies located in the dorsal root of a spinal nerve.

dorsum (dors'um) Pertaining to the back surface of a body part.

ductus arteriosus (duk'tus ar-te''re-o'sus) Blood vessel that connects the pulmonary artery and the aorta in a fetus.

ductus venosus (duk'tus ven-o'sus) Blood vessel that connects the umbilical vein and the inferior vena cava in a fetus.

dura mater (du'rah ma'ter) Tough outer layer of the meninges.

dynamic equilibrium (di-nam'ik e''kwi-lib're-um) The maintenance of balance when the head and body are suddenly moved or rotated.

eccrine gland (ek'rin gland) Sweat gland that functions in the maintenance of body temperature.

ECG Electrocardiogram; EKG.

ectoderm (ek'to-derm) The outermost layer of the primary germ layers, responsible for forming certain embryonic body parts.

edema (ĕ-de'mah) An excessive accumulation of fluid within the tissue spaces.

effector (ĕ-fek'tor) Organ, such as a muscle or gland, that responds to stimulation.

efferent arteriole (ef'er-ent ar-te're-ol) Arteriole that conducts blood away from the glomerulus of a nephron.

ejaculation (e-jak''u-la'shun) Discharge of sperm-containing seminal fluid from the male urethra.

elastin (e-las'tin) Protein that comprises the yellow, elastic fibers of connective tissue.

electrocardiogram (e-lek″tro-kar′de-o-gram″) A recording of the electrical activity associated with the heartbeat; ECG or EKG.

electrolyte (e-lek′tro-līt) A substance that ionizes in water solution.

electrolyte balance (e-lek′tro-līt bal′ans) Condition that exists when the quantities of electrolytes entering the body equal those leaving it.

electron (e-lek′tron) A small, negatively charged particle that revolves around the nucleus of an atom.

electrovalent bond (e-lek″tro-va′lent bond) Chemical bond formed between two ions as a result of the transfer of electrons.

element (el′ĕ-ment) A basic chemical substance.

embolus (em′bo-lus) A substance, such as a blood clot or bubble of gas, that is carried by the blood and obstructs a blood vessel.

embryo (em′bre-o) An organism in its earliest stages of development.

emission (e-mish′un) The movement of sperm cells from the vas deferens into the ejaculatory duct and urethra.

emulsification (e-mul″sĭ-fĭ-ka′shun) Process by which fat globules are caused to break up into smaller droplets.

enamel (e-nam′el) Hard covering on the exposed surface of a tooth.

endocardium (en″do-kar′de-um) Inner lining of the heart chambers.

endocrine gland (en′do-krin gland) A gland that secretes hormones directly into the blood or body fluids.

endocytosis (en″do-si-to′sis) Process by which substances move through a cell membrane that involves the formation of tiny vacuoles.

endoderm (en′do-derm) The innermost layer of the primary germ layers responsible for forming certain embryonic body parts.

endometrium (en″do-me′tre-um) The inner lining of the uterus.

endoplasmic reticulum (en-do-plaz′mic rĕ-tik′-u-lum) Cytoplasmic organelle composed of a system of interconnected membranous tubules and vesicles.

endothelium (en″do-the′le-um) The layer of epithelial cells that forms the inner lining of blood vessels and heart chambers.

energy (en′er-je) An ability to cause something to move and thus to do work.

enterogastrone (en″ter-o-gas′trōn) A hormone secreted from the intestinal wall that inhibits gastric secretion and motility.

enzyme (en′zīm) A protein that is synthesized by a cell and acts as a catalyst in a specific cellular reaction.

eosinophil (e″o-sin′o-fil) White blood cell characterized by the presence of cytoplasmic granules that become stained by acidic dye.

ependyma (ĕ-pen′dĭ-mah) Neuroglial cells that line the ventricles of the brain.

epicardium (ep″ĭ-kar′de-um) The visceral portion of the pericardium.

epicondyle (ep″ĭ-kon′dīl) A projection of a bone located above a condyle.

epidermis (ep″ĭ-der′mis) Outer epithelial layer of the skin.

epididymis (ep″ĭ-did′ĭ-mis) Highly coiled tubule that leads from the seminiferous tubules of the testis to the vas deferens.

epidural space (ep″ĭ-du′ral spās) The space between the dural sheath of the spinal cord and the bone of the vertebral canal.

epigastric region (ep″ĭ-gas′trik re′jun) The upper middle portion of the abdomen.

epiglottis (ep″ĭ-glot′is) Flaplike cartilaginous structure located at the back of the tongue near the entrance to the trachea.

epinephrine (ep″ĭ-nef′rin) A hormone secreted by the adrenal medulla during times of stress.

epiphyseal disk (ep″ĭ-fiz′e-al disk) Cartilaginous layer within the epiphysis of a long bone that functions as a growing region.

epiphysis (ĕ-pif′ĭ-sis) The end of a long bone.

epithelium (ep″ĭ-the′le-um) The type of tissue that covers all free body surfaces.

equilibrium (e″kwĭ-lib′re-um) A state of balance between two opposing forces.

erythroblast (ĕ-rith′ro-blast) An immature red blood cell.

erythrocyte (ĕ-rith′ro-sīt) A red blood cell.

erythropoiesis (ĕ-rith″ro-poi-e′sis) Red blood cell formation.

erythropoietin (ĕ-rith″ro-poi′ĕ-tin) Substance released by the kidneys and liver that promotes red blood cell formation.

esophagus (ĕ-sof′ah-gus) Tubular portion of the digestive tract that leads from the pharynx to the stomach.

essential amino acid (ĕ-sen′shal ah-me′no as′id) Amino acid required for health that cannot be synthesized in adequate amounts by body cells.

essential fatty acid (ĕ-sen′shal fat′e as′id) Fatty acid required for health that cannot be synthesized in adequate amounts by body cells.

estrogen (es′tro-jen) Hormone that stimulates the development of female secondary sexual characteristics.

eustachian tube (u-sta′ke-an tūb) Tube that connects the middle ear to the pharynx.

evaporation (e″vap′o-ra-shun) Process by which a liquid changes into a gas.

eversion (e-ver′zhun) Movement in which the sole of the foot is turned outward.

excretion (ek-skre′shun) Process by which metabolic wastes are eliminated.

exocrine gland (ek′so-krin gland) A gland that secretes its products into a duct or onto a body surface.

expiration (ek″spī-ra′shun) Process of expelling air from the lungs.

extension (ek-sten′shun) Movement by which the angle between parts at a joint is increased.

extracellular (ek″strah-sel′u-lar) Outside of cells.

extrapyramidal tract (ek″strah-pi-ram′i-dal trakt) Nerve tracts, other than the corticospinal tracts, that transmit impulses from the cerebral cortex into the spinal cord.

extremity (ek-strem′ĭ-te) A limb; an arm or leg.

facet (fas′et) A small, flattened surface of a bone.

facilitated diffusion (fah-sil″ĭ-tat′ed dĭ-fu′zhun) Movement of molecules from a region of higher concentration toward one of lower concentration that involves special carrier molecules within a cell membrane.

facilitation (fah-sil″ĭ-ta′shun) Process by which a neuron becomes more excitable as a result of incoming stimulation.

fallopian tube (fah-lo′pe-an tūb) Tube that transports an egg cell from the region of the ovary to the uterus; oviduct or uterine tube.

fascia (fash′e-ah) A sheet of fibrous connective tissue that encloses a muscle.

fat (fat) Adipose tissue; or an organic substance whose molecules contain glycerol and fatty acids.

fatty acid (fat'e as'id) An organic substance that serves as a building block for a fat molecule.

feces (fe'sēz) Material expelled from the digestive tract during defecation.

fertilization (fer''ti-li-za'shun) The union of an egg cell and a sperm cell.

fetus (fe'tus) A human embryo after eight weeks of development.

fibril (fi'bril) A tiny fiber or filament.

fibrin (fi'brin) Insoluble, fibrous protein formed from fibrinogen during blood coagulation.

fibrinogen (fi-brin'o-jen) Plasma protein that is converted into fibrin during blood coagulation.

fibroblast (fi'bro-blast) Cell that functions to produce fibers and other intercellular materials in connective tissues.

filtration (fil-tra'shun) Movement of a material through a membrane as a result of hydrostatic pressure.

fissure (fish'ur) A narrow cleft separating parts, such as the lobes of the cerebrum.

flexion (flek'shun) Bending at a joint so that the angle between bones is decreased.

follicle (fol'ĭ-kl) A pouchlike depression or cavity.

follicle-stimulating hormone (fol'ĭ-kl stim'u-la''ting hor'mōn) A substance secreted by the anterior pituitary gland that stimulates the development of an ovarian follicle in a female or the production of sperm cells in a male; FSH.

follicular cells (fō-lik'u-lar selz) Ovarian cells that surround a developing egg cell and secrete female sex hormones.

fontanel (fon''tah-nel') Membranous region located between certain cranial bones in the skull of a fetus.

foramen (fo-ra'men) An opening, usually in a bone or membrane (plural, *foramina*).

foramen magnum (fo-ra'men mag'num) Opening in the occipital bone of the skull through which the spinal cord passes.

foramen ovale (fo-ra'men o-val'e) Opening in the interatrial septum of the fetal heart.

formula (fōr'mu-lah) A group of symbols and numbers used to express the composition of a compound.

fossa (fos'ah) A depression in a bone or other part.

fovea (fo've-ah) A tiny pit or depression.

fovea centralis (fo've-ah sen-tral'is) Region of the retina, consisting of densely packed cones.

frontal (frun'tal) Pertaining to the region of the forehead.

FSH Follicle-stimulating hormone.

gallbladder (gawl'blad-er) Saclike organ associated with the liver that stores and concentrates bile.

ganglion (gang'gle-on) A mass of neuron cell bodies, usually outside the central nervous system.

gastric gland (gas'trik gland) Gland within the stomach wall that secretes gastric juice.

gastric juice (gas'trik joos) Secretion of the gastric glands within the stomach.

gastrin (gas'trin) Hormone secreted by the stomach lining that stimulates the secretion of gastric juice.

gene (jēn) Portion of a DNA molecule that contains the information needed to synthesize an enzyme.

genetic code (jě-net'ik kōd) System by which information for synthesizing proteins is built into the structure of DNA molecules.

germinal epithelium (jer'mi-nal ep''ĭ-the'le-um) Tissue within an ovary or testis that gives rise to sex cells.

germ layers (jerm la'ers) Layers of cells within an embryo that form the body organs during development.

globin (glo'bin) The protein portion of a hemoglobin molecule.

globulin (glob'u-lin) A type of protein that occurs in blood plasma.

glomerulus (glo-mer'u-lus) A capillary tuft located within the Bowman's capsule of a nephron.

glottis (glot'is) Slitlike opening between the true vocal folds.

glucagon (gloo'kah-gon) Hormone secreted by the pancreatic islets of Langerhans that causes the release of glucose from glycogen.

glucocorticoid (gloo''ko-kor'tĭ-koid) Any one of a group of hormones secreted by the adrenal cortex that influences carbohydrate, fat, and protein metabolism.

glucose (gloo'kōs) A monosaccharide found in the blood that serves as the primary source of cellular energy.

gluteal (gloo'te-al) Pertaining to the buttocks.

glycerol (glis'er-ol) An organic compound that serves as a building block for fat molecules.

glycogen (gli'ko-jen) A polysaccharide that functions to store glucose in the liver and muscles.

glycoprotein (gli''ko-pro'te-in) A substance composed of a carbohydrate combined with a protein.

goblet cell (gob'let sel) An epithelial cell that is specialized to secrete mucus.

Golgi apparatus (gol'je ap''ah-ra'tus) A cytoplasmic organelle that functions in preparing cellular products for secretion.

gonadotropin (go-nad''o-trōp'in) A hormone that stimulates activity in the gonads.

granulocyte (gran'u-lo-sīt) A leukocyte that contains granules in its cytoplasm.

gray matter (gra mat'er) Region of the central nervous system that generally lacks myelin.

groin (groin) Region of the body between the abdomen and thighs.

growth (grōth) Process by which a structure enlarges.

growth hormone (grōth hor'mōn) A hormone released by the anterior lobe of the pituitary gland that promotes growth.

hair follicle (hār fol'ĭ-kl) Tubelike depression in the skin in which a hair develops.

haversian canal (ha-ver'shan kah-nal') Tiny channel in bone tissue that contains a blood vessel.

haversian system (ha-ver'shan sis'tem) A group of bone cells and the haversian canal that they surround; the basic unit of structure in osseous tissue.

head (hed) An enlargement on the end of a bone.

hematocrit (he-mat'o-krit) The percentage by volume of red blood cells in a blood sample.

hematoma (he''mah-to'mah) A mass of coagulated blood within tissues or a body cavity.

hematopoiesis (hem''ah-to-poi-e'sis) The production of blood and blood cells; hemopoiesis.

heme (hem) The iron-containing portion of a hemoglobin molecule.

hemoglobin (he''mo-glo'bin) Pigment of red blood cells responsible for the transport of oxygen.

hemorrhage (hem'o-rij) Loss of blood from the circulatory system; bleeding.

hemostasis (he''mo-sta'sis) The stoppage of bleeding.

hepatic (hĕ-pat′ik) Pertaining to the liver.

hepatic lobule (hĕ-pat′ik lob′ul) A functional unit of the liver.

homeostasis (ho″me-o-sta′sis) A state of equilibrium in which the internal environment of the body remains relatively constant.

hormone (hor′mōn) A substance secreted by an endocrine gland that is transmitted in the blood or body fluids.

humoral immunity (hu′mor-al ĭ-mu′nĭ-te) Resistance to the effects of specific disease-causing agents due to the presence of circulating antibodies.

hydrolysis (hi-drol′ĭ-sis) The splitting of a molecule into smaller portions by the addition of a water molecule.

hydrostatic pressure (hi″dro-stat′ik presh′ur) Pressure exerted by fluids.

hydroxyl ion (hi-drok′sil i′on) OH^-.

hymen (hi′men) A membranous fold of tissue that partially covers the vaginal opening.

hypertonic (hi″per-ton′ik) Condition in which a solution contains a greater concentration of dissolved particles than the solution with which it is compared.

hypertrophy (hi-per′tro-fe) Enlargement of an organ or tissue.

hyperventilation (hi″per-ven″tĭ-la′shun) Breathing that is abnormally deep and prolonged.

hypochondriac region (hi″po-kon′dre-ak re′jun) The portion of the abdomen on either side of the middle or epigastric region.

hypogastric region (hi″po-gas′trik re′jun) The lower middle portion of the abdomen.

hypothalamus (hi″po-thal′ah-mus) A portion of the brain located below the thalamus.

hypotonic (hi″po-ton′ik) Condition in which a solution contains a lesser concentration of dissolved particles than the solution to which it is compared.

iliac region (il′e-ak re′jun) Portion of the abdomen on either side of the lower middle or hypogastric region.

ilium (il′e-um) One of the bones of a coxal bone or hipbone.

immunity (ĭ-mu′nĭ-te) Resistance to the effects of specific disease-causing agents.

immunoglobulin (im″u-no-glob′u-lin) Globular plasma proteins that function as antibodies of immunity.

implantation (im″plan-ta′shun) The embedding of an embryo in the lining of the uterus.

impulse (im′puls) A wave of depolarization conducted along a nerve fiber or muscle fiber.

incomplete protein (in″kom-plet′ pro′te-in) A protein that lacks essential amino acids.

inferior (in-fer′e-or) Situated below something else; pertaining to the lower surface of a part.

inflammation (in″flah-ma′shun) A tissue response to stress that is characterized by dilation of blood vessels and an accumulation of fluid in the affected region.

inguinal (ing′gwi-nal) Pertaining to the groin region.

inorganic (in″or-gan′ik) Pertaining to chemical substances that lack carbon.

insertion (in-ser′shun) The end of a muscle that is attached to a movable part.

inspiration (in″spĭ-ra′shun) Act of breathing in; inhalation.

insula (in′su-lah) A cerebral lobe located deep within the lateral sulcus.

insulin (in′su-lin) A hormone secreted by the pancreatic islets of Langerhans that functions in the control of carbohydrate metabolism.

integumentary (in-teg-u-men′tar-e) Pertaining to the skin and its accessory organs.

intercalated disk (in-ter″kah-lat′ed disk) Membranous boundary between adjacent cardiac muscle cells.

intercellular (in″ter-sel′u-lar) Between cells.

intercellular fluid (in″ter-sel′u-lar floo′id) Tissue fluid located between cells other than blood cells.

interneuron (in″ter-nu′ron) A neuron located between a sensory neuron and a motor neuron.

interphase (in′ter-faz) Period between two cell divisions.

interstitial cell (in″ter-stish′al sel) A hormone-secreting cell located between the seminiferous tubules of the testis.

interstitial fluid (in″ter-stish′al floo′id) Same as intercellular fluid.

intervertebral disk (in″ter-ver′tĕ-bral disk) A layer of fibrocartilage located between the bodies of adjacent vertebrae.

intestinal gland (in-tes′tĭ-nal gland) Tubular gland located at the base of a villus within the intestinal wall.

intestinal juice (in-tes′tĭ-nal joos) The secretion of the intestinal glands.

intracellular (in″trah-sel′u-lar) Within cells.

intracellular fluid (in″trah-sel′u-lar floo′id) Fluid within cells.

intrinsic factor (in-trin′sik fak′tor) A substance produced by the gastric glands that promotes the absorption of vitamin B_{12}.

inversion (in-ver′zhun) Movement in which the sole of the foot is turned inward.

involuntary (in-vol′un-tar″e) Not consciously controlled; functions automatically.

ion (i′on) An atom or a group of atoms with an electrical charge.

ionization (i″on-ĭ-za′shun) Chemical process by which substances dissociate into ions.

iris (i′ris) Colored muscular portion of the eye that surrounds the pupil.

irritability (ir″ĭ-tah-bil′ĭ-te) The ability of an organism to react to changes taking place in its environment.

ischemia (is-ke′me-ah) A deficiency of blood in a body part.

isotonic solution (i″so-ton′ik so-lu′shun) A solution that has the same concentration of dissolved particles as the solution with which it is compared.

joint (joint) The union of two or more bones; an articulation.

juxtaglomerular apparatus (juks″tah-glo-mer′u-lar ap′ah-ra′tus) Structure located in the walls of arterioles near the glomerulus that plays an important role in regulating renal blood flow.

keratin (ker′ah-tin) Protein present in the epidermis, hair, and nails.

keratinization (ker″ah-tin″ĭ-za′shun) The process by which cells form fibrils of keratin.

ketone body (ke′tōn bod′e) Type of compound produced during fat catabolism.

Kupffer cell (koop′fer sel) Large, fixed phagocyte in the liver that removes bacterial cells from the blood.

labor (la'bor)   The process of childbirth.

labyrinth (lab'ĭ-rinth)   The system of interconnecting tubes within the inner ear.

lacrimal gland (lak'rĭ-mal gland)   Tear-secreting gland.

lactation (lak-ta'shun)   The production of milk by the mammary glands.

lacteal (lak'te-al)   A lymphatic vessel associated with a villus of the small intestine.

lactic acid (lak'tik as'id)   An organic substance formed from pyruvic acid during anaerobic respiration.

lacuna (lah-ku'nah)   A hollow cavity.

laryngopharynx (lah-ring''go-far'ingks)   The lower portion of the pharynx near the opening to the larynx.

larynx (lar'ingks)   Structure located between the pharynx and trachea that houses the vocal cords.

lateral (lat'er-al)   Pertaining to the side.

leukocyte (lu'ko-sīt)   A white blood cell.

lever (lev'er)   A simple mechanical device consisting of a rod, fulcrum, weight, and a source of energy.

ligament (lig'ah-ment)   A cord or sheet of connective tissue by which two or more bones are bound together at a joint.

limbic system (lim'bik sis'tem)   A group of interconnected structures within the brain that functions to produce various emotional feelings.

lingual (ling'gwal)   Pertaining to the tongue.

lipase (li'pās)   A fat-digesting enzyme.

lipid (lip'id)   A fat, oil, or fatlike compound.

lipoprotein (lip''o-pro'te-in)   A complex of lipid and protein.

lumbar (lum'bar)   Pertaining to the region of the loins.

lumen (lu'men)   Space within a tubular structure.

luteinizing hormone (lu'te-in-iz''ing hor'mōn)   A hormone secreted by the anterior pituitary gland that controls the formation of corpus luteum in females and the secretion of testosterone in males; LH.

lymph (limf)   Fluid transported by the lymphatic vessels.

lymph node (limf nōd)   A mass of lymphoid tissue located along the course of a lymphatic vessel.

lymphocyte (lim'fo-sīt)   A type of white blood cell produced in lymphatic tissue.

lysosome (li'so-sōm)   Cytoplasmic organelle that contains digestive enzymes.

macrophage (mak'ro-fāj)   A large phagocytic cell.

macroscopic (mak''ro-skop'ik)   Large enough to be seen with the unaided eye.

macula lutea (mak'u-lah lu'te-ah)   A yellowish depression in the retina of the eye.

mammary (mam'ar-e)   Pertaining to the breast.

marrow (mar'o)   Connective tissue that occupies the spaces within bones.

mast cell (mast sel)   A cell to which antibodies, formed in response to allergens, become attached.

mastication (mas''tĭ-ka'shun)   Chewing movements.

matrix (ma'triks)   The intercellular substance of connective tissue.

matter (mat'er)   Anything that has weight and occupies space.

meatus (me-a'tus)   A passageway or channel, or the external opening of such a passageway.

mechanoreceptor (mek''ah-no-re-sep'tor)   A sensory receptor that is sensitive to mechanical stimulation.

medial (me'de-al)   Toward or near the midline.

mediastinum (me''de-ah-sti'num)   Tissues and organs of the thoracic cavity that form a septum between the lungs.

medulla (mĕ-dul'ah)   The inner portion of an organ.

medulla oblongata (mĕ-dul'ah ob''long-gah'tah)   Portion of the brain stem located between the pons and the spinal cord.

medullary cavity (med'u-lār''e kav'ĭ-te)   Cavity within the diaphysis of a long bone occupied by marrow.

meiosis (mi-o'sis)   Process of cell division by which egg and sperm cells are formed.

melanin (mel'ah-nin)   Dark pigment normally found in skin and hair.

melanocyte (mel'ah-no-sīt'')   Melanin-producing cell.

membranous bone (mem'brah-nus bōn)   Bone that develops from layers of membranous connective tissue.

memory cell (mem'o-re sel)   B- or T- lymphocyte produced in response to a primary immune response that remains dormant and can respond rapidly if the same antigen is encountered in the future.

meninges (mĕ-nin'jēz)   A group of three membranes that covers the brain and spinal cord (singular, *meninx*).

menisci (men-is'si)   Pieces of fibrocartilage that separate the articulating surfaces of bones in the knee.

menopause (men'o-pawz)   Termination of menstrual cycles.

menstrual cycle (men'stroo-al si'kl)   The female reproductive cycle that is characterized by regularly reoccurring changes in the uterine lining.

menstruation (men''stroo-a'shun)   Loss of blood and tissue from the uterus at the end of a female reproductive cycle.

mesentery (mes'en-ter''e)   A fold of peritoneal membrane that attaches an abdominal organ to the abdominal wall.

mesoderm (mez'o-derm)   The middle layer of the primary germ layers, responsible for forming certain embryonic body parts.

messenger RNA (mes'in-jer)   Molecule of RNA that transmits information for protein synthesis from the nucleus of a cell to the cytoplasm.

metabolic rate (met''ah-bol'ic rāt)   The rate at which chemical changes occur within the body.

metabolism (mĕ-tab'o-lizm)   All of the chemical changes that occur within cells considered together.

metacarpals (met''ah-kar'pals)   Bones of the hand between the wrist and finger bones.

metaphase (met'ah-fāz)   Stage in mitosis when chromosomes become aligned in the middle of the spindle.

metatarsals (met''ah-tar'sals)   Bones of the foot between the ankle and toe bones.

microfilament (mi''kro-fil'ah-ment)   Tiny rod of protein that occurs in cytoplasm and functions in causing various cellular movements.

microglia (mi-krog'le-ah)   A type of neuroglial cell that helps support neurons and acts to carry on phagocytosis.

microscopic (mi″kro-skop′ik) Too small to be seen with the unaided eye.

microtubule (mi″kro-tu′būl) A minute, hollow rod found in the cytoplasm of cells.

microvilli (mi″kro-vil′i) Tiny, cylindrical processes that extend outward from some epithelial cell membranes and increase the membrane surface area.

micturition (mik″tu-rish′un) Urination.

midbrain (mid′brān) A small region of the brain stem located between the diencephalon and pons.

mineralocorticoid (min″er-al-o-kor′tĭ-koid) Any one of a group of hormones secreted by the adrenal cortex that influences the concentrations of electrolytes in body fluids.

mitochondrion (mi″to-kon′dre-on) Cytoplasmic organelle that contains enzymes responsible for aerobic respiration (plural, *mitochondria*).

mitosis (mi-to′sis) Process by which body cells divide to form two identical daughter cells.

mixed nerve (mikst nerv) Nerve that includes both sensory and motor nerve fibers.

molecule (mol′ĕ-kūl) A particle composed of two or more atoms bonded together.

monocyte (mon′o-sīt) A type of white blood cell that functions as a phagocyte.

monosaccharide (mon″o-sak′ah-rid) A simple sugar, such as glucose or fructose, that represents the structural unit of a carbohydrate.

motor area (mo′tor a′re-ah) A region of the brain from which impulses to muscles or glands originate.

motor end plate (mo′tor end plāt) Specialized region of a muscle fiber where it is joined by the end of a motor nerve fiber.

motor nerve (mo′tor nerv) A nerve that consists of motor nerve fibers.

motor neuron (mo′tor nu′ron) A neuron that transmits impulses from the central nervous system to an effector.

motor unit (mo′tor unit) A motor neuron and the muscle fibers associated with it.

mucosa (mu-ko′sah) The membrane that lines tubes and body cavities that open to the outside of the body; mucous membrane.

mucous cell (mu′kus sel) Glandular cell that secretes mucus.

mucous membrane (mu′kus mem′brān) Mucosa.

mucus (mu′kus) Fluid secretion of the mucous cells.

mutagenic (mu″tah-jen′ik) Pertaining to a factor that can cause mutations.

mutation (mu-ta′shun) A change in the genetic information of a chromosome.

myelin (mi′ĕ-lin) Fatty material that forms a sheathlike covering around some nerve fibers.

myocardium (mi″o-kar′de-um) Muscle tissue of the heart.

myofibril (mi″o-fi′bril) Contractile fibers found within muscle cells.

myoglobin (mi″o-glo′bin) A pigmented compound found in muscle tissue that acts to store oxygen.

myogram (mi′o-gram) A recording of a muscular contraction.

myometrium (mi″o-me′tre-um) The layer of smooth muscle tissue within the uterine wall.

myosin (mi′o-sin) A protein that, together with actin, is responsible for muscular contraction and relaxation.

nasal cavity (na′zal kav′ĭ-te) Space within the nose.

nasal concha (na′zal kong′kah) Shelllike bone extending outward from the wall of the nasal cavity.

nasal septum (na′zal sep′tum) A wall of bone and cartilage that separates the nasal cavity into two portions.

nasopharynx (na″zo-far′ingks) Portion of the pharynx associated with the nasal cavity.

negative feedback (neg′ah-tiv fēd′bak) A mechanism that is activated by an imbalance and acts to correct it.

neonatal (ne″o-na′tal) Pertaining to the period of life from birth to the end of 4 weeks.

nephron (nef′ron) The functional unit of a kidney, consisting of a renal corpuscle and a renal tubule.

nerve (nerv) A bundle of nerve fibers.

neuroglia (nu-rog′le-ah) The supporting tissue within the brain and spinal cord, composed of neuroglial cells.

neurolemma (nu″ro-lem′ah) Sheath on the outside of some nerve fibers due to the presence of Schwann cells.

neuromuscular junction (nu″ro-mus′ku-lar jungk′shun) Union between a nerve fiber and a muscle fiber.

neuron (nu′ron) A nerve cell that consists of a cell body and its processes.

neuronal pool (nu′ro-nal pool) A group of neurons within the central nervous system that receives impulses from input nerve fibers, processes the impulses, and conducts impulses away on output fibers.

neurotransmitter (nu″ro-trans-mit′er) Chemical substance secreted by the terminal end of an axon that stimulates a muscle fiber contraction or an impulse in another neuron.

neutral (nu′tral) Neither acid nor alkaline.

neutron (nu′tron) An electrically neutral particle found in an atomic nucleus.

neutrophil (nu′tro-fil) A type of phagocytic leukocyte.

niacin (ni′ah-sin) A vitamin of the B-complex group; nicotinic acid.

nonelectrolyte (non″e-lek′tro-līt) A substance that does not dissociate into ions when it is dissolved.

nonprotein nitrogenous substance (non-pro′te-in ni-troj′ĕ-nus sub′stans) A substance, such as urea or uric acid, that contains nitrogen but is not a protein.

norepinephrine (nor″ep-i-nef′rin) A neurotransmitter substance released from the axon ends of some nerve fibers.

nuclease (nu′kle-ās) An enzyme that causes nucleic acids to decompose.

nucleic acid (nu-kle′ik as′id) A substance composed of nucleotides bonded together; RNA or DNA.

nucleolus (nu-kle′o-lus) Small structure that occurs within the nucleus of a cell and contains RNA.

nucleotide (nu′kle-o-tīd″) A component of a nucleic acid molecule, consisting of a sugar, a nitrogenous base, and a phosphate group.

nucleus (nu′kle-us) A body that occurs within a cell and contains relatively large quantities of DNA; the dense core of an atom that is composed of protons and neutrons.

nutrient (nu′tre-ent) A chemical substance that must be supplied to the body from its environment.

occipital (ok-sip′ĭ-tal) Pertaining to the lower, back portion of the head.

olfactory (ol-fak′to-re) Pertaining to the sense of smell.

olfactory nerves (ol-fak'to-re nervz) The first pair of cranial nerves that conduct impulses associated with the sense of smell.

oligodendrocyte (ol''ĭ-go-den'dro-sīt) A type of neuroglial cell that functions to connect neurons to blood vessels and to form myelin.

oocyte (o'o-sīt) An immature egg cell.

oogenesis (o''o-jen'ĕ-sis) The process by which an egg cell forms from an oocyte.

ophthalmic (of-thal'mik) Pertaining to the eye.

optic (op'tik) Pertaining to the eye.

optic chiasma (op'tik ki-az'mah) X-shaped structure on the underside of the brain created by a partial crossing over of fibers in the optic nerves.

optic disk (op'tik disk) Region in the retina of the eye where nerve fibers leave to become part of the optic nerve.

oral (o'ral) Pertaining to the mouth.

organ (or'gan) A structure consisting of a group of tissues that performs a specialized function.

organelle (or''gah-nel') A living part of a cell that performs a specialized function.

organic (or-gan'ik) Pertaining to carbon-containing substances.

organism (or'gah-nizm) An individual living thing.

orifice (or'ĭ-fis) An opening.

origin (or'ĭ-jin) End of a muscle that is attached to a relatively immovable part.

oropharynx (o''ro-far'ingks) Portion of the pharynx in the posterior part of the oral cavity.

osmoreceptor (oz''mo-re-sep'tor) Receptor that is sensitive to changes in the osmotic pressure of body fluids.

osmosis (oz-mo'sis) Diffusion of water through a selectively permeable membrane.

osmotic (oz-mot'ik) Pertaining to osmosis.

osmotic pressure (oz-mot'ik presh'ur) The amount of pressure needed to stop osmosis; the potential pressure of a solution due to the presence of nondiffusible solute particles in the solution.

ossification (os''ĭ-fĭ-ka'shun) The formation of bone tissue.

osteoblast (os'te-o-blast'') A bone-forming cell.

osteoclast (os'te-o-klast'') A cell that causes the erosion of bone.

osteocyte (os'te-o-sīt) A bone cell.

otolith (o'to-lith) A small particle of calcium carbonate associated with the receptors of equilibrium.

oval window (o'val win'do) Opening between the stapes and the inner ear.

ovarian (o-va're-an) Pertaining to the ovary.

ovary (o'var-e) The primary reproductive organ of a female; an egg-cell-producing organ.

oviduct (o'vĭ-dukt) A tube that leads from the ovary to the uterus; uterine tube or fallopian tube.

ovulation (o''vu-la'shun) The release of an egg cell from a mature ovarian follicle.

oxidation (ok''sĭ-da'shun) Process by which oxygen is combined with a chemical substance.

oxygen debt (ok'sĭ-jen det) The amount of oxygen that must be supplied following physical exercise to convert accumulated lactic acid to glucose.

oxyhemoglobin (ok''sĭ-he''mo-glo'bin) Compound formed when oxygen combines with hemoglobin.

oxytocin (ok''sĭ-to'sin) Hormone released by the posterior lobe of the pituitary gland that causes contraction of smooth muscles in the uterus and mammary glands.

pacemaker (pās'māk-er) Mass of specialized muscle tissue that controls the rhythm of the heartbeat; the sinoatrial node.

pain receptor (pān re''sep'tor) Sensory nerve ending associated with the feeling of pain.

palate (pal'at) The roof of the mouth.

palatine (pal'ah-tīn) Pertaining to the palate.

palmar (pahl'mar) Pertaining to the palm of the hand.

pancreas (pan'kre-as) Glandular organ in the abdominal cavity that secretes hormones and digestive enzymes.

pancreatic (pan''kre-at'ik) Pertaining to the pancreas.

pantothenic acid (pan''to-then'ik as'id) A vitamin of the B-complex group.

papilla (pah-pil'ah) Tiny nipplelike projection.

papillary muscle (pap'ĭ-ler''e mus'l) Muscle that extends inward from the ventricular walls of the heart.

parasympathetic division (par''ah-sim''pah-thet'ik dĭ-vizh'un) Portion of the autonomic nervous system that arises from the brain and sacral region of the spinal cord.

parathormone (par''ah-thor'mōn) Hormone secreted by the parathyroid glands that helps to regulate the level of blood calcium and phosphate.

parathyroid glands (par''ah-thi'roid glandz) Small endocrine glands that are embedded in the posterior portion of the thyroid gland.

parietal (pah-ri'ĕ-tal) Pertaining to the wall of an organ or cavity.

parietal cell (pah-ri'ĕ-tal sel) Cell of a gastric gland that secretes hydrochloric acid and intrinsic factor.

parietal pleura (pah-ri'ĕ-tal ploo'rah) Membrane that lines the inner wall of the thoracic cavity.

parotid glands (pah-rot'id glandz) Large salivary glands located on the sides of the face just in front and below the ears.

partial pressure (par'shal presh'ur) The pressure produced by one gas in a mixture of gases.

pectoral (pek'tor-al) Pertaining to the chest.

pectoral girdle (pek'tor-al ger'dl) Portion of the skeleton that provides support and attachment for the arms.

pelvic (pel'vik) Pertaining to the pelvis.

pelvic girdle (pel'vik ger'dl) Portion of the skeleton to which the legs are attached.

pelvis (pel'vis) Bony ring formed by the sacrum and coxal bones.

penis (pe'nis) External reproductive organ of the male through which the urethra passes.

pepsin (pep'sin) Protein-splitting enzyme secreted by the gastric glands of the stomach.

pepsinogen (pep-sin'o-jen) Inactive form of pepsin.

peptide (pep'tid) Compound composed of two or more amino acid molecules joined together.

peptide bond (pep'tid bond) Bond that forms between the carboxyl group of one amino acid and the amino group of another.

pericardial (per''ĭ-kar'de-al) Pertaining to the pericardium.

pericardium (per''ĭ-kar'de-um) Serous membrane that surrounds the heart.

perichondrium (per″ĭ-kon′dre-um) Layer of fibrous connective tissue that encloses cartilaginous structures.

perilymph (per′ĭ-limf) Fluid contained in the space between the membranous and osseous labyrinths of the inner ear.

perineal (per″ĭ-ne′al) Pertaining to the perineum.

perineum (per″ĭ-ne′um) Body region between the scrotum or urethral opening and the anus.

periodontal ligament (per″e-o-don′tal lig′ah-ment) Fibrous membrane that surrounds a tooth and attaches it to the bone of the jaw.

periosteum (per″e-os′te-um) Covering of fibrous connective tissue on the surface of a bone.

peripheral (pĕ-rif′er-al) Pertaining to parts located near the surface or toward the outside.

peripheral nervous system (pĕ-rif′er-al ner′vus sis′tem) The portions of the nervous system outside the central nervous system.

peripheral resistance (pĕ-rif′er-al re-zis′tans) Resistance to blood flow due to friction between the blood and the walls of the blood vessels.

peristalsis (per″ĭ-stal′sis) Rhythmic waves of muscular contraction that occur in the walls of various tubular organs.

peritoneal (per″ĭ-to-ne′al) Pertaining to the peritoneum.

peritoneal cavity (per″ĭ-to-ne′al kav′ĭ-te) The potential space between the parietal and visceral peritoneal membranes.

peritoneum (per″ĭ-to-ne′um) A serous membrane that lines the abdominal cavity and encloses the abdominal viscera.

peritubular capillary (per″ĭ-tu′bu-lar kap′ĭ-ler″e) Capillary that surrounds a renal tubule of a nephron.

permeable (per′me-ah-bl) Open to passage or penetration.

pH The measurement unit of the hydrogen ion concentration used to indicate the acid or alkaline condition of a solution.

phagocytosis (fag″o-si-to′sis) Process by which a cell engulfs and digests solid substances.

phalanx (fa′langks) A bone of a finger or toe.

pharynx (far′ingks) Portion of the digestive tube between the mouth and esophagus.

phospholipid (fos″fo-lip′id) A lipid that contains phosphorus.

photoreceptor (fo″to-re-sep′tor) A nerve ending that is sensitive to light energy.

physiology (fiz″e-ol′o-je) The branch of science dealing with the study of body functions.

pia mater (pi′ah ma′ter) Inner layer of meninges that encloses the brain and spinal cord.

pineal gland (pin′e-al gland) A small structure located in the central part of the brain.

pinocytosis (pin″o-si-to′sis) Process by which a cell engulfs droplets of fluid from its surroundings.

pituitary gland (pi-tu′ĭ-tār″e gland) Endocrine gland that is attached to the base of the brain and consists of anterior and posterior lobes.

placenta (plah-sen′tah) Structure by which an unborn child is attached to its mother's uterine wall and through which it is nourished.

plantar (plan′tar) Pertaining to the sole of the foot.

plasma (plaz′mah) Fluid portion of circulating blood.

plasma cell (plaz′mah sel) Antibody-producing cell that is formed as a result of the proliferation of sensitized B-lymphocytes.

plasma protein (plaz′mah pro′te-in) Any of several proteins normally found dissolved in blood plasma.

platelet (plāt′let) Cytoplasmic fragment formed in the bone marrow that functions in blood coagulation.

pleural (ploo′ral) Pertaining to the pleura or membranes investing the lungs.

pleural cavity (ploo′ral kav′ĭ-te) Potential space between the pleural membranes.

pleural membranes (ploo′ral mem′brānz) Serous membranes that enclose the lungs.

plexus (plek′sus) A network of interlaced nerves or blood vessels.

polar body (po′lar bod′e) Small, nonfunctional cell produced as a result of meiosis during egg cell formation.

polarization (po″lar-ĭ-za′shun) The development of an electrical charge on the surface of a cell membrane.

polypeptide (pol″e-pep′tid) A compound formed by the union of many amino acid molecules.

polysaccharide (pol″e-sak′ah-rid) A carbohydrate composed of many monosaccharide molecules joined together.

pons (ponz) A portion of the brain stem above the medulla oblongata and below the midbrain.

popliteal (pop″li-te′al) Pertaining to the region behind the knee.

positive feedback (poz′ĭ-tiv fēd′bak) Process by which changes cause more changes of a similar type, producing unstable conditions.

posterior (pos-tēr′e-or) Toward the back; opposite of anterior.

postganglionic fiber (pōst″gang-gle-on′ik fi′ber) Autonomic nerve fiber located on the distal side of a ganglion.

postnatal (pōst-na′tal) After birth.

preganglionic fiber (pre″gang-gle-on′ik fi′ber) Autonomic nerve fiber located on the proximal side of a ganglion.

pregnancy (preg′nan-se) The condition in which a female has a developing offspring in her uterus.

prenatal (pre-na′tal) Before birth.

pressoreceptor (pres″o-re-sep′tor) A receptor that is sensitive to changes in pressure.

primary reproductive organs (pri′ma-re re″pro-duk′tiv or′ganz) Sex-cell-producing parts; testes in males and ovaries in females.

prime mover (prim moov′er) Muscle that is mainly responsible for a particular body movement.

process (pros′es) A prominent projection on a bone.

progesterone (pro-jes′tĕ-rōn) A female hormone secreted by the corpus luteum of the ovary and by the placenta.

projection (pro-jek′shun) Process by which the brain causes a sensation to seem to come from the region of the body being stimulated.

prolactin (pro-lak′tin) Hormone secreted by the anterior pituitary gland that stimulates the production of milk from the mammary glands.

pronation (pro-na′shun) Movement in which the palm of the hand is moved downward or backward.

prophase (pro′fāz) Stage of mitosis during which chromosomes become visible.

proprioceptor (pro″pre-o-sep′tor) A sensory nerve ending that is sensitive to changes in tension of a muscle or tendon.

prostaglandins (pros″tah-glan′dins) A group of compounds that have powerful, hormonelike effects.

prostate gland (pros′tāt gland) Gland located around the male urethra below the urinary bladder.

protein (pro′te-in) Nitrogen-containing organic compound composed of amino acid molecules joined together.

prothrombin (pro-throm′bin) Plasma protein that functions in the formation of blood clots.

proton (pro′ton) A positively charged particle found in an atomic nucleus.

protraction (pro-trak′shun) A forward movement of a body part.

proximal (prok′sĭ-mal) Closer to the midline or origin; opposite of distal.

puberty (pu′ber-te) Stage of development in which the reproductive organs become functional.

pulmonary (pul′mo-ner″e) Pertaining to the lungs.

pulmonary circuit (pul′mo-ner″e ser′kit) System of blood vessels that carries blood between the heart and lungs.

pulse (puls) The surge of blood felt through the walls of arteries due to the contraction of the ventricles of the heart.

pupil (pu′pil) Opening in the iris through which light enters the eye.

Purkinje fibers (pur-kin′je fi′berz) Specialized muscle fibers that conduct the cardiac impulse from the A-V bundle into the ventricular walls.

pyramidal cell (pĭ-ram′ĭ-dal sel) A large, pyramid-shaped neuron found within the cerebral cortex.

pyruvic acid (pi-roo′vik as′id) An intermediate product of carbohydrate oxidation.

receptor (re″sep′tor) Part located at the distal end of a sensory dendrite that is sensitive to stimulation.

red marrow (red mar′o) Blood-cell-forming tissue located in spaces within bones.

referred pain (re-ferd′ pān) Pain that feels as if it is originating from a part other than the site being stimulated.

reflex (re′fleks) A rapid, automatic response to a stimulus.

reflex arc (re′fleks ark) A nerve pathway, consisting of a sensory neuron, interneuron, and motor neuron, that forms the structural and functional bases for a reflex.

refraction (re-frak′shun) A bending of light as it passes from one medium into another medium with a different density.

relaxin (re-lak′sin) Hormone from corpus luteum that inhibits uterine contractions during pregnancy.

renal (re′nal) Pertaining to the kidney.

renal corpuscle (re′nal kor′pusl) Part of a nephron that consists of a glomerulus and a Bowman's capsule.

renal cortex (re′nal kor′teks) The outer portion of a kidney.

renal medulla (re′nal mĕ-dul′ah) The inner portion of a kidney.

renal pelvis (re′nal pel′vis) The hollow cavity within a kidney.

renal tubule (re′nal tu′būl) Portion of a nephron that extends from the renal corpuscle to the collecting duct.

renin (re′nin) Enzyme released from the kidneys that triggers a mechanism leading to a rise in blood pressure.

reproduction (re″pro-duk′shun) The process by which an offspring is formed.

resorption (re-sorp′shun) The process by which something that has been secreted is absorbed.

respiration (res″pĭ-ra′shun) Cellular process by which energy is released from nutrients.

respiratory center (re-spi′rah-to″re sen′ter) Portion of the brain stem that controls the depth and rate of breathing.

respiratory membrane (re-spi′rah-to″re mem′brān) Membrane composed of a capillary and an alveolar wall through which gases are exchanged between the blood and air.

response (re-spons′) The action resulting from a stimulus.

resting potential (res′ting po-ten′shal) The difference in electrical charge between the inside and outside of an undisturbed nerve cell membrane.

reticular formation (rĕ-tik′u-lar for-ma′shun) A complex network of nerve fibers within the brain stem that functions in arousing the cerebrum.

reticuloendothelial tissue (rĕ-tik″u-lo-en″do-the′le-al tish′u) Tissue composed of widely scattered phagocytic macrophages.

retina (ret′ĭ-nah) Inner layer of the eye wall that contains the visual receptors.

retinene (ret′ĭ-nēn) Substance used in the production of rhodopsin by the rods of the eye.

retraction (rĕ-trak′shun) Movement of a part toward the back.

rhodopsin (ro-dop′sin) Light-sensitive substance that occurs in the rods of the retina.

riboflavin (ri″bo-fla′vin) A vitamin of the B-complex group; vitamin B_2.

ribonucleic acid (ri″bo-nu-kle′ik as′id) A nucleic acid that contains ribose sugar; RNA.

ribose (ri′bōs) A five-carbon sugar found in RNA molecules.

ribosome (ri′bo-sōm) Cytoplasmic organelle that functions in the synthesis of proteins.

RNA Ribonucleic acid.

rod (rod) A type of light receptor that is responsible for colorless vision.

rotation (ro-ta′shun) Movement by which a body part is turned on its longitudinal axis.

round window (rownd win′do) A membrane-covered opening between the inner ear and the middle ear.

sagittal (saj′i-tal) A plane or section that divides a structure into right and left portions.

S-A node (nōd) Sinoatrial node.

sarcomere (sar′ko-mēr) The structural and functional unit of a myofibril.

sarcoplasmic reticulum (sar″ko-plaz′mik rĕ-tik′u-lum) Membranous network of channels and tubules within a muscle fiber.

saturated fatty acid (sat′u-rāt″ed fat′e as′id) Fatty acid molecule that lacks double bonds between the atoms of its carbon chain.

Schwann cell (shwahn sel) Cell that surrounds a fiber of a peripheral nerve.

sclera (skle′rah) White fibrous outer layer of the eyeball.

scrotum (skro′tum) A pouch of skin that encloses the testes.

sebaceous gland (se-ba′shus gland) Gland of the skin that secretes sebum.

sebum (se′bum) Oily secretion of the sebaceous glands.

secretin (se-kre′tin) Hormone secreted from the small intestine that stimulates the release of pancreatic juice.

semicircular canal (sem″ĭ-ser′ku-lar kah-nal′) Tubular structure within the inner ear that contains the receptors responsible for the sense of dynamic equilibrium.

seminal fluid (sem′ĭ-nal floo′id) Fluid discharged from the male reproductive tract at ejaculation.

seminiferous tubule (sem″ĭ-nif′er-us tu′būl) Tubule within the testes in which sperm cells are formed.

semipermeable (sem″ĭ-per′me-ah-bl) Condition in which a membrane is permeable to some molecules and not to others; selectively permeable.

sensation (sen-sa′shun) A feeling resulting from the interpretation of sensory nerve impulses by the brain.

sensory area (sen′so-re a′re-ah) A portion of the cerebral cortex that receives and interprets sensory nerve impulses.

sensory nerve (sen′so-re nerv) A nerve composed of sensory nerve fibers.

sensory neuron (sen′so-re nu′ron) A neuron that transmits an impulse from a receptor to the central nervous system.

serotonin (se″ro-to′nin) A vasoconstricting substance that is released by blood platelets when blood vessels are broken.

serous cell (se′rus sel) A glandular cell that secretes a watery fluid with a high enzyme content.

serous fluid (ser′us floo′id) The secretion of a serous membrane.

serous membrane (ser′us mem′brān) Membrane that lines a cavity without an opening to the outside of the body.

serum (se′rum) The fluid portion of coagulated blood.

sesamoid bone (ses′ah-moid bōn) A round bone that may occur in tendons adjacent to joints.

simple sugar (sim′pl shoog′ar) A monosaccharide.

sinoatrial node (si″no-a′tre-al nōd) Group of specialized tissue in the wall of the right atrium that initiates cardiac cycles; the pacemaker; S-A node.

sinus (si′nus) A cavity or hollow space in a bone or other body part.

skeletal muscle (skel′ĕ-tal mus′l) Type of muscle tissue found in muscles attached to skeletal parts.

smooth muscle (smooth mus′l) Type of muscle tissue found in the walls of hollow visceral organs; visceral muscle.

solute (sol′ūt) The substance that is dissolved in a solution.

solvent (sol′vent) The liquid portion of a solution in which a solute is dissolved.

somatic cell (so-mat′ik sel) Any cell of the body other than the sex cells.

special sense (spesh′al sens) Sense that involves receptors associated with specialized sensory organs such as the eyes and ears.

spermatid (sper′mah-tid) An intermediate stage in the formation of sperm cells.

spermatocyte (sper-mat′o-sīt) An early stage in the formation of sperm cells.

spermatogenesis (sper″mah-to-jen′ĕ-sis) The production of sperm cells.

spermatogonium (sper″mah-to-go′ne-um) Undifferentiated spermatogenic cell.

sphincter (sfingk′ter) A circular muscle that functions to close an opening or the lumen of a tubular structure.

spinal (spi′nal) Pertaining to the spinal cord or to the vertebral canal.

spinal cord (spi′nal kord) Portion of the central nervous system extending downward from the brain stem through the vertebral canal.

spinal nerve (spi′nal nerv) Nerve that arises from the spinal cord.

spleen (splēn) A large, glandular organ located in the upper left region of the abdomen.

spongy bone (spunj′e bōn) Bone that consists of bars and plates separated by irregular spaces; cancellous bone.

squamous (skwa′mus) Flat or platelike.

starch (starch) A polysaccharide that is common in foods of plant origin.

static equilibrium (stat′ik e″kwĭ-lib′re-um) The maintenance of balance when the head and body are motionless.

stimulus (stim′u-lus) A change in the environmental conditions that is followed by a response by an organism or cell.

stomach (stum′ak) Digestive organ located between the esophagus and the small intestine.

stratified (strat′ĭ-fid) Arranged in layers.

stratum corneum (stra′tum kor′ne-um) Outer horny layer of the epidermis.

stratum germinativum (stra′tum jer′mĭ-na″tiv-um) The deepest layer of the epidermis in which the cells undergo mitosis.

stroke volume (strōk vol′um) The amount of blood discharged from the ventricle with each heartbeat.

structural formula (struk′cher-al for′mu-lah) A representation of the way atoms are bound together within a molecule, using the symbols for each element present and lines to indicate chemical bonds.

subarachnoid space (sub″ah-rak′noid spās) The space within the meninges between the arachnoid mater and the pia mater.

subcutaneous (sub″ku-ta′ne-us) Beneath the skin.

sublingual (sub-ling′gwal) Beneath the tongue.

submucosa (sub″mu-ko′sah) Layer of connective tissue that underlies a mucous membrane.

substrate (sub′strāt) The substance upon which an enzyme acts.

sucrose (soo′krōs) A disaccharide; table sugar.

sulcus (sul′kus) A shallow groove, such as that between adjacent convolutions on the surface of the brain.

superficial (soo″per-fish′al) Near the surface.

superior (su-pe′re-or) Pertaining to a structure that is higher than another structure.

supination (soo″pĭ-na′shun) Rotation of the forearm so that the palm faces upward when the arm is outstretched.

surface tension (ser′fas ten′shun) Force that tends to hold moist membranes together.

surfactant (ser-fak'tant) Substance produced by the lungs that reduces the surface tension within the alveoli.

suture (soo'cher) An immovable joint, such as that between adjacent flat bones of the skull.

sympathetic nervous system (sim''pah-thet'ik ner'vus sis'tem) Portion of the autonomic nervous system that arises from the thoracic and lumbar regions of the spinal cord.

symphysis (sim'fi-sis) A slightly movable joint between bones separated by a pad of fibrocartilage.

synapse (sin'aps) The junction between the axon end of one neuron and the dendrite or cell body of another neuron.

synaptic knob (si-nap'tik nob) Tiny enlargement at the end of an axon that secretes a neurotransmitter substance.

syncytium (sin-sish'e-um) A mass of merging cells.

synovial fluid (si-no've-al floo'id) Fluid secreted by the synovial membrane.

synovial joint (si-no've-al joint) A freely movable joint.

synovial membrane (si-no've-al mem'bran) Membrane that forms the inner lining of the capsule of a freely movable joint.

synthesis (sin'the-sis) The process by which substances are united to form more complex substances.

system (sis'tem) A group of organs that act together to carry on a specialized function.

systemic circuit (sis-tem'ik ser'kit) The vessels that conduct blood between the heart and all body tissues except the lungs.

systole (sis'to-le) Phase of cardiac cycle during which a heart chamber wall is contracted.

systolic pressure (sis-tol'ik presh'ur) Arterial blood pressure during the systolic phase of the cardiac cycle.

target tissue (tar'get tish'u) Specific tissue on which a hormone acts.

tarsus (tar'sus) The bones that form the ankle.

taste bud (tast bud) Organ containing the receptors associated with the sense of taste.

telophase (tel'o-faz) Stage in mitosis during which daughter cells become separate structures.

tendon (ten'don) A cordlike or bandlike mass of white fibrous connective tissue that connects a muscle to a bone.

testis (tes'tis) Primary reproductive organ of a male; a sperm-cell-producing organ.

testosterone (tes-tos'te-ron) Male sex hormone.

tetany (tet'ah-ne) A continuous, forceful muscular contraction.

thalamus (thal'ah-mus) A mass of gray matter located at the base of the cerebrum in the wall of the third ventricle.

thermoreceptor (ther''mo-re-sep'tor) A sensory receptor that is sensitive to changes in temperature; a heat receptor.

thiamine (thi'ah-min) Vitamin B_1.

thoracic (tho-ras'ik) Pertaining to the chest.

threshold stimulus (thresh'old stim'u-lus) The level of stimulation that must be exceeded to elicit a nerve impulse or a muscle contraction.

thrombus (throm'bus) A blood clot in a blood vessel that remains at its site of formation.

thymus (thi'mus) A two-lobed glandular organ located in the mediastinum behind the sternum and between the lungs.

thyroid gland (thi'roid gland) Endocrine gland located just below the larynx and in front of the trachea.

thyroxine (thi-rok'sin) A hormone secreted by the thyroid gland.

tissue (tish'u) A group of similar cells that performs a specialized function.

T-lymphocyte (lim'fo-sit) Lymphocytes that interact directly with antigen-bearing particles and are responsible for cell-mediated immunity.

trachea (tra'ke-ah) Tubular organ that leads from the larynx to the bronchi.

transcellular fluid (trans''sel'u-lar floo'id) A portion of the extracellular fluid, including the fluid within special body cavities.

transfer RNA (trans'fer) Molecule of RNA that carries an amino acid to a ribosome in the process of protein synthesis.

transverse tubule (trans-vers' tu'bul) Membranous channel that extends inward from a muscle fiber membrane and passes through the fiber.

tricuspid valve (tri-kus'pid valv) Heart valve located between the right atrium and the right ventricle.

triglyceride (tri-glis'er-id) A lipid composed of three fatty acids combined with a glycerol molecule.

triiodothyronine (tri''i-o''do-thi'ro-nen) One of the thyroid hormones.

trochanter (tro-kan'ter) A broad process on a bone.

trochlea (trok'le-ah) A pulley-shaped structure.

tropic hormone (trop ik hor'mon) A hormone that has an endocrine gland as its target tissue.

trypsin (trip'sin) An enzyme in pancreatic juice that acts to break down protein molecules.

tubercle (tu'ber-kl) A small, rounded process on a bone.

tuberosity (tu''be-ros'i-te) An elevation or protuberance on a bone.

twitch (twich) A brief muscular contraction followed by relaxation.

tympanic membrane (tim-pan'ik mem'bran) A thin membrane that covers the auditory canal and separates the external ear from the middle ear; the eardrum.

umbilical cord (um-bil'i-kal kord) Cordlike structure that connects the fetus to the placenta.

umbilical region (um-bil'i-kal re'jun) The central portion of the abdomen.

unsaturated fatty acid (un-sat'u-rat''ed fat'e as'id) Fatty acid molecule that has one or more double bonds between the atoms of its carbon chain.

urea (u-re'ah) A nonprotein nitrogenous substance produced as a result of protein metabolism.

ureter (u-re'ter) A muscular tube that carries urine from the kidney to the urinary bladder.

urethra (u-re'thrah) Tube leading from the urinary bladder to the outside of the body.

uterine (u'ter-in) Pertaining to the uterus.

uterine tube (u'ter-in tub) Tube that extends from the uterus on each side toward an ovary and functions to transport sex cells; fallopian tube or oviduct.

uterus (u'ter-us) Hollow muscular organ located within the female pelvis in which a fetus develops.

utricle (u'tri-kl)   An enlarged portion of the membranous labyrinth of the inner ear.

uvula (u'vu-lah)   A fleshy portion of the soft palate that hangs down above the root of the tongue.

vaccine (vak'sēn)   A substance that contains antigens and is used to stimulate the production of antibodies.

vacuole (vak'u-ōl)   A space or cavity within the cytoplasm of a cell.

vagina (vah-ji'nah)   Tubular organ that leads from the uterus to the vestibule of the female reproductive tract.

vascular (vas'ku-lar)   Pertaining to blood vessels.

vas deferens (vas def'er-ens)   Tube that leads from the epididymis to the urethra of the male reproductive tract (plural, *vasa deferentia*).

vasoconstriction (vas''o-kon-strik'shun)   A decrease in the diameter of a blood vessel.

vasodilation (vas''o-di-la'shun)   An increase in the diameter of a blood vessel.

vein (vān)   A vessel that carries blood toward the heart.

vena cava (ven'ah kav'ah)   One of two large veins that convey deoxygenated blood to the right atrium of the heart.

ventral root (ven'tral root)   Motor branch of a spinal nerve by which it is attached to the spinal cord.

ventricle (ven'tri-kl)   A cavity, such as those of the brain that are filled with cerebrospinal fluid or those of the heart that contain blood.

venule (ven'ūl)   A vessel that carries blood from capillaries to a vein.

villus (vil'us)   Tiny, fingerlike projection.

visceral (vis'er-al)   Pertaining to the contents of a body cavity.

visceral peritoneum (vis'er-al per''i-to-ne'um)   Membrane that covers the surfaces of organs within the abdominal cavity.

visceral pleura (vis'er-al ploo'rah)   Membrane that covers the surfaces of the lungs.

viscosity (vis-kos'i-te)   The tendency for a fluid to resist flowing due to the friction of its molecules on one another.

vitamin (vi'tah-min)   An organic substance other than a carbohydrate, lipid, or protein that is needed for normal metabolism but cannot be synthesized in adequate amounts by the body.

vitreous humor (vit're-us hu'mor)   The substance that occupies the space between the lens and retina of the eye.

vocal cords (vo'kal kordz)   Folds of tissue within the larynx that create vocal sounds when they vibrate.

voluntary (vol'un-tār''e)   Capable of being consciously controlled.

vulva (vul'vah)   The external reproductive parts of the female that surround the opening of the vagina.

water balance (wot'er bal'ans)   Condition in which the quantity of water entering the body is equal to the quantity leaving it.

water of metabolism (wot'er uv mē-tab'o-lizm)   Water produced as a by-product of metabolic processes.

yellow marrow (yel'o mar'o)   Fat storage tissue found in the cavities within bones.

zygote (zi'gōt)   Cell produced by the fusion of an egg and sperm; a fertilized egg cell.

zymogen granule (zi-mo'jen gran'ūl)   A cellular structure that stores inactive forms of protein-splitting enzymes in a pancreatic cell.

Credits

Photos

Chapter 3

Figure 3.14: © Edwin A. Reschke.

Chapter 5

Figure 5.1(B): © Edwin A. Reschke; figure 5.2(B): © Warren Rosenberg, Iona College/BPS; figure 5.3(B): © Manfred Kage/Peter Arnold, Inc.; figures 5.4(B), 5.5(B), 5.6(A), 5.7(B), 5.8(B), 5.9(B), 5.10(B), 5.11(B), 5.12(B), 5.13(B), 5.14(B), 5.15(B), 5.16(B): © Edwin A. Reschke; figures 5.17(B), 5.18(B): © Manfred Kage/Peter Arnold, Inc.

Chapter 6

Figure 6.4: from *Tissues and Organs: A Text-Atlas of Scanning Electron Microscopy* by Richard G. Kessel and Randy H. Kardon. W. H. Freeman and Company. Copyright © 1979.

Chapter 7

Figure 7.14(both): © courtesy of Ohio Nuclear.

Chapter 8

Figure 8.3: © Ted Conde, St. Joseph's Hospital; figure 8.13: © Martin M. Rotker/Taurus Photos; figure 8.20: © Victor B. Eichler/BIO-ART; figure 8.22: © courtesy of Kodak; figures 8.27, 8.30(both): © Martin M. Rotker/Taurus Photos.

Chapter 9

Figure 9.3: © courtesy of H. E. Huxley, Cambridge University; figure 9.C(both): © John Carl Mese.

Chapter 10

Figure 10.3: © courtesy of J. D. Robertson.

Chapter 11

Figure 11.10 © Per H. Kjeldsen; University of Michigan; figure 11.20: © reproduced with permission from Asbury T. Vaughn, GENERAL OPTHALMOLOGY, 8th Ed., Lange, 1977.

Chapter 13

Figure 13.15: © Carrol H. Weiss, RBP, 1973; figure 13.20: © from *Tissues and Organs: A Text-Atlas of Scanning Electron Microscopy* by Richard G. Kessel and Randy H. Kardon. W. H. Freeman and Company. Copyright © 1979; figure 13.23: © James L. Shaffer.

Chapter 14

Figure 14.2: © American Lung Association; figure 14.16: © from *Tissues and Organs; A Text-Atlas of Scanning Electron Microscopy* by Richard G. Kessel and Randy H. Kardon. W. H. Freeman and Company. Copyright © 1979.

Chapter 15

Figure 15.5(both): Armed Forces Institute of Pathology; figure 15.6: © Warren Rosenberg, Iona College/BPS; figures 15.7–15.10: © Edwin A. Reschke; figure 15.13: "Scanning Electron Micrograph of an Erythrocyte Enmeshed in Fibrin", Bernstein, E. and Kairnen, E., *Science* Vol. 173, Cover, 27 August 1971. Figure 15.14: © Camera M. D. Studios, Inc. 1974.

Chapter 16

Figure 16.19: from *Tissues and Organs: A Text-Atlas of Scanning Electron Microscopy* by Richard G. Kessel and Randy H. Kardon. W. H. Freeman and Company. Copyright © 1979; figure 16.21(A): © Edwin A. Reschke; figure 16.21(B): © Victor B. Eichler, BIO-ART.

Chapter 17

Figure 17.3: © courtesy of Carolina Biological Supply; figure 17.12: © courtesy of Memorial Sloan-Kettering Cancer Center.

Chapter 18

Figure 18.4: from *Tissues and Organs: A Text-Atlas of Scanning Electron Microscopy*, by Richard G. Kessel and Randy H. Kardon. W. H. Freeman and Company: Copyright © 1979.

Chapter 20

Figure 20.3(B): from *Tissues and Organs: A Text-Atlas of Scanning Electron Microscopy* by Richard G. Kessel and Randy H. Kardon. W. H. Freeman and Company. Copyright © 1979; figure 20.13: © "Sea Urchin Sperm-Egg Interactions Studied with the Scanning Electron Microscope," Tegner, M. J. and Epel, D. *Science* Vol 179 pages 685–688, fig. 2, 16 February 1973; figure 20.26: © Dickinson-Belskie Birth Models, The Cleveland Health Education Museum.

Illustrations

Chapter 1

Figures 1.2, 1.3, and 1.4 From Hole, John W., Jr., *Human Anatomy and Physiology 3d ed.* © 1978, 1981, 1984 Wm. C. Brown Publishers, Dubuque, Iowa. All Rights Reserved. Reprinted by permission.

Chapter 2

Figures 2.1–2.15 From Hole, John W., Jr., *Human Anatomy and Physiology 3d ed.* © 1978, 1981, 1984 Wm. C. Brown Publishers, Dubuque, Iowa. All Rights Reserved. Reprinted by permission.

Chapter 3

Figures 3.1, 3.6–3.13, and 3.15 From Hole, John W., Jr., *Human Anatomy and Physiology 3d ed.* © 1978, 1981, 1984 Wm. C. Brown Publishers, Dubuque, Iowa. All Rights Reserved. Reprinted by permission.
Figures 3.2–3.5 From Mader, Sylvia S., *Inquiry Into Life 4th ed.* © 1976, 1979, 1982, 1985 Wm. C. Brown Publishers, Dubuque, Iowa. All Rights Reserved. Reprinted by permission.

Chapter 4

Figures 4.1–4.3, 4.5–4.11, and 4.13–4.18 From Hole, John W., Jr., *Human Anatomy and Physiology 3d ed.* © 1978, 1981, 1984 Wm. C. Brown Publishers, Dubuque, Iowa. All Rights Reserved. Reprinted by permission.
Figure 4.4 From Volpe, E. Peter, *Biology and Human Concerns 3d ed.* © 1975, 1979, 1983 Wm. C. Brown Publishers, Dubuque, Iowa. All Rights Reserved. Reprinted by permission.

Chapter 5

Figures 5.1a–5.5a, 5.6b&c, and 5.7a–5.18a From Hole, John W., Jr., *Human Anatomy and Physiology 3d ed.* © 1978, 1981, 1984 Wm. C. Brown Publishers, Dubuque, Iowa. All Rights Reserved. Reprinted by permission.

Chapter 6

Figures 6.1–6.3 and 6.5 From Hole, John W., Jr., *Human Anatomy and Physiology 3d ed.* © 1978, 1981, 1984 Wm. C. Brown Publishers, Dubuque, Iowa. All Rights Reserved. Reprinted by permission.

Chapter 7

Figures 7.1–7.10 and 7.13 From Hole, John W., Jr., *Human Anatomy and Physiology 3d ed.* © 1978, 1981, 1984 Wm. C. Brown Publishers, Dubuque, Iowa. All Rights Reserved. Reprinted by permission.

Chapter 8

Figures 8.1–8.5, 8.14–8.18, 8.21, 8.23–8.26, 8.28, 8.29, and 8.31–8.42 From Hole, John W., Jr., *Human Anatomy and Physiology 3d ed.* © 1978, 1981, 1984 Wm. C. Brown Publishers, Dubuque, Iowa. All Rights Reserved. Reprinted by permission.
Figures 8.6, 8.7, and 8.9–8.12 From Van De Graaff, Kent M., *Human Anatomy.* © 1984 Wm. C. Brown Publishers, Dubuque, Iowa. All Rights Reserved. Reprinted by permission.

Index

Index

Index